NINTH EDITION

Successful Writing at Work

Philip C. Kolin
University of Southern Mississippi

WADSWORTH
CENGAGE Learning™

Australia • Brazil • Japan • Korea • Mexico • Singapore • Spain • United Kingdom • United States

WADSWORTH
CENGAGE Learning™

Successful Writing at Work, Ninth Edition
Philip C. Kolin

To Kristin, Eric, and Theresa
Evan Philip and Megan Elise
Julie and Loretta
Diane
and
MARY

Publisher: Michael Rosenberg

Development Editor: Bruce Cantley

Assistant Editor: Megan Garvey

Editorial Assistant: Rebekah Matthews

Media Editor: Amy Gibbons

Marketing Manager: Christina Shea

Marketing Assistant: Ryan Ahern

Marketing Communications Manager:
    Elizabeth Rodio

Senior Content Project Manager, Editorial
Production: Margaret Park Bridges

Art and Design Manager: Jill Haber

Manufacturing Buyer: Arethea L. Thomas

Senior Rights Acquisition Account Manager:
    Katie Huha

Text Researcher: Mary Dalton Hoffman

Production Service: Alison Fields

Senior Photo Editor: Jennifer Meyer Dare

Cover Design Manager: Anne S. Katzeff

Cover Images: Businesspeople Working
    Together © Corbis; Co-Workers Writing on
    Paper © Simon Jarratt/Corbis; Business-
    people Working in Conference Room ©
    Tim Pannell/Corbis

Compositor: Pre-PressPMG

For product information and technology assistance, contact us at
Cengage Learning Customer & Sales Support, 1-800-423-0563

For permission to use material from this text or product,
submit all requests online at **www.cengage.com/permissions**
Further permissions questions can be emailed to
permissionrequest@cengage.com

Library of Congress Control Number: 2008929476

ISBN-13: 978-0-547-14791-8

ISBN-10: 0-547-14791-0

Wadsworth
20 Channel Center Street
Boston, MA 02210-1202
USA

Cengage Learning is a leading provider of customized learning solutions with office locations around the globe, including Singapore, the United Kingdom, Australia, Mexico, Brazil, and Japan. Locate your local office at **international.cengage.com/region**

Cengage Learning products are represented in Canada by Nelson Education, Ltd.

For your course and learning solutions, visit **www.cengage.com.**

Purchase any of our products at your local college store or at our preferred online store **www.ichapters.com.**

Microsoft® Clip Art reprinted by permission from Microsoft Corporation.

Printed in the United States of America
1 2 3 4 5 6 7 13 12 11 10 09

CONTENTS

Preface xv

PART I: Backgrounds 1

Chapter 1: Getting Started: Writing and Your Career 3

Writing—An Essential Job Skill 3

Writing for the Global Marketplace 6
Competing for International Business 6
Communicating with Global Audiences: Seeing the World
  Through Their Eyes 6
Cultural Diversity at Home 8
Using International English 8

Four Keys to Effective Writing 8
Identifying Your Audience 10
Establishing Your Purpose 15
Formulating Your Message 16
Selecting Your Style and Tone 18

Characteristics of Job-Related Writing 20
1. Providing Practical Information 20
2. Giving Facts, Not Impressions 21
3. Providing Visuals to Clarify and Condense Information 21
4. Giving Accurate Measurements 23
5. Stating Responsibilities Precisely 24
6. Persuading and Offering Recommendations 24

Ethical Writing in the Workplace 27
Ethical Requirements on the Job 29
International Readers and Ethics 29
Employers Insist On and Monitor Ethical Behavior 30
Some Guidelines to Help You Reach Ethical Decisions 31
Ethical Dilemmas 33
Writing Ethically on the Job 34

Successful Employees Are Successful Writers 36

Revision Checklist 37
Exercises 38

Chapter 2: **The Writing Process at Work**   43

What Writing Is and Is Not   43
    What Writing Is   43
    What Writing Is Not   44

The Writing Process   44
    Researching   44
    Planning   45
    Drafting   50
    Revising   55
    Editing   59

Revision Checklist   70
Exercises   70

Chapter 3: **Collaborative Writing and Meetings at Work**   76

Teamwork Is Crucial to Business Success   76

Advantages of Collaborative Writing   78

Collaborative Writing and the Writing Process   80
    Case Study: Collaborative Editing   81

Guidelines for Setting Up a Successful Writing Group   83

Ten Proven Ways to Be a Valuable Team Player   84

Sources of Conflict in Collaborative Groups and How to Solve Them   85
    Common Problems, Practical Solutions   85

Models for Collaboration   87
    Cooperative Model   88
    Sequential Model   88
    Functional Model   88
    Integrated Model   91

Case Study: Evolution of a Collaboratively Written Document   91
    First Draft   93
    Subsequent Drafts   95
    Final Copy   97

Computer-Supported Collaboration   97
    Advantages of Computer-Supported Collaboration   99
    Types of Groupware   100
    Models for Computer-Supported Collaboration   105
    Avoiding the Pitfalls of Computer-Supported Collaboration   106

Meetings   107
    Planning a Meeting   107
    Creating an Agenda   108
    Observing Courtesy at a Group Meeting   109
    Writing the Minutes   110
    Taking Notes   113

Revision Checklist   114
Exercises   115

PART II: Correspondence   119

Chapter 4: Writing Routine Business Correspondence: Memos, Faxes, E-Mails, IMs, and Blogs   121

What Memos, Faxes, E-Mails, IMs, and Blog Posts Have in Common   121

Memos   122
  Memo Protocol   122
  E-Mail or Hard Copy?   124
  Memo Audience, Style, and Tone   124
  Memo Format   127
  Strategies for Organizing a Memo   129
  Routing Memos   129

Faxes   130
  Guidelines for Sending Faxes   130

E-Mail   131
  Business E-Mail Versus Personal E-Mail   131
  E-Mails Are Legal Records   132
  Guidelines for Using E-Mail   134
  When Not to Use E-Mail   140

Instant Messages (IMs) for Business Use   140
  When to Use IMs Versus E-Mails   140
  Guidelines on Using IMs in the Workplace   142

Blogs   142
  Guidelines for Writing Internal Blogs   143
  Guidelines for Writing External Blogs   145

Revision Checklist   149
Exercises   150

Chapter 5: Writing Letters: Some Basics for Audiences Worldwide   153

Letters in the Age of the Internet   153

Letter Formats   154
  Full Block Format   154
  Modified Block Format   154
  Continuing Pages   157

Parts of a Letter   157
  Date Line   157
  Inside Address   157
  Salutation   159
  Body of the Letter   159
  Complimentary Close   159
  Signature   160
  Enclosure(s) Line   160
  Copy Notation   160

The Appearance of Your Letter 161

Organizing a Standard Business Letter 162

Making a Good Impression on Your Reader 164
  Achieving the "You Attitude": Four Guidelines 165

International Business Correspondence 170
  Guidelines for Communicating with International Readers 173
  Respecting Your Reader's Nationality and Ethnic/Racial Heritage 181
  Case Study: Writing to a Client from a Different Culture 183
  Two Versions of a Sales Letter 183

Revision Checklist 187
Exercises 189

Chapter 6: Types of Business Letters 195

Formulating Your Message 195

The Four Most Common Types of Business Letters 196

Inquiry Letters 197

Special Request Letters 197

Sales Letters 201
  Preliminary Guidelines 201
  The Four A's of Sales Letters 203

Customer Relations Letters 207
  Diplomacy and Reader Psychology 207
  The Customers Always Write 208
  Planning Your Customer Relations Letters—Preliminary Guidelines 208
  Follow-Up Letters 212
  Complaint Letters 212
  Adjustment Letters 218
  Refusal-of-Credit Letters 226
  Collection Letters 228

Sending Letter-Quality Messages: Final Advice to Seal Your Success 233

Revision Checklist 233
Exercises 234

Chapter 7: How to Get a Job: Searches, Dossiers, Portfolios, Résumés, Letters, and Interviews 241

Steps the Employer Takes to Hire 241

Steps to Follow to Get Hired 242

Analyzing Your Strengths and Restricting Your Job Search 243

Looking in the Right Places for a Job 243

Dossiers and Career Portfolios/Webfolios 246
  Dossiers 247
  Career Portfolios/Webfolios 248

Preparing a Résumé   251
   What Employers Like to See in a Résumé   251
   The Process of Writing Your Résumé   253
   Parts of a Résumé   254
   Organizing Your Résumé   261
   The Online Résumé   266

Letters of Application   273
   How Application Letters and Résumés Differ   273
   Writing the Letter of Application   273

Going to an Interview   281
   Preparing for the Interview   281
   Questions to Expect at Your Interview   282
   What Do I Say About Salary?   283
   Questions You May Ask the Interviewer(s)   283
   What Interviewer(s) Can't Ask You   283
   Ten Interview Dos and Don'ts   284
   The Follow-Up Letter   286

Accepting or Declining a Job Offer   286

Searching for the Right Job Pays   286

Revision Checklist   289
Exercises   289

PART III: Gathering and Summarizing Information   295

Chapter 8: Doing Research and Documentation on the Job   297

Some Research Scenarios   298

The Differences Between School and Workplace Research   299

Characteristics of Effective Workplace Research   300

The Research Process   301

Two Types of Research: Primary and Secondary   303
   Case Study: Primary and Secondary Research in the Real World   305

Primary Research   310
   Direct Observation, Site Visits, and Tests   310
   Interviews and Focus Groups   311
   Surveys   318

Secondary Research   325
   Libraries   325
   Periodical Databases   330
   Reference Materials   334
   Internet Searches   340
   Evaluating Websites   346

Note Taking   348
  The Importance of Note Taking   348
  How to Take Effective Notes   348
  What to Record   349
  To Quote or Not to Quote   350

Documenting Sources   353
  The Ethics of Documentation: Determining What to Cite   353
  Documentation Styles   354
  What MLA and APA Have in Common   355
  Following MLA Style   355
  Following APA Style   365

A Business Research Report   374

Conclusion   389
  Revision Checklist   389
  Exercises   390

Chapter 9: Summarizing Information at Work   398

The Importance of Summaries in Business   398

Contents of a Summary   399
  What to Include in a Summary   400
  What to Omit from a Summary   400

Preparing a Summary   401

Case Study: Summarizing an Original Article   403

Executive Summaries   408
  What Managers Want to See in an Executive Summary   408
  Organization of an Executive Summary   412

Evaluative Summaries   412
  Guidelines for Writing a Successful Evaluative Summary   413
  Evaluating the Content   414
  Evaluating the Style   414

Abstracts   417
  Differences Between a Summary and an Abstract   417
  Writing the Informative Abstract   419
  Writing the Descriptive Abstract   420

Writing Successful News Releases   420
  Subjects Appropriate for News Releases   420
  News Releases About Bad News   422
  Topics That Do Not Warrant a News Release   423
  Organization of a News Release   423
  Style and Tone of a News Release   427

Revision Checklist   427
Exercises   428

PART IV: Preparing Documents and Visuals   435

Chapter 10: **Designing Clear Visuals**   437

Visual Thinking in the Workplace   437

The Purpose of Visuals   438

Types of Visuals and Their Functions   439

Choosing Effective Visuals   439
    Ineffective Visuals: What *Not* to Do   443

Inserting and Writing About Visuals: Some Guidelines   444
    Identify Your Visuals   444
    Cite the Source for Visuals   445
    Insert Your Visuals Appropriately   445
    Introduce Your Visuals   447
    Interpret Your Visuals   447

Two Categories of Visuals: Tables and Figures   448

Tables   448
    Parts of a Table   448
    Guidelines for Using Tables   449

Figures   451
    Graphs   451
    Charts   455
    Maps   463
    Photographs   464
    Drawings   469
    Clip Art   470

Using Visuals Ethically   472
    Guidelines for Using Visuals Ethically   474

Using Appropriate Visuals for International Audiences   479
    Visuals Do Not Always Translate from One Culture
      to Another   479
    Guidelines for Using Visuals for International Audiences   480

Conclusion   482

Revision Checklist   483

Exercises   483

Chapter 11: **Designing Successful Documents and Websites**   491

Organizing Information Visually   491

Characteristics of Effective Design   492

Desktop Publishing   493
    Type   493
    Templates   493
    Graphics   494

Before Choosing a Design   495

The ABCs of Print Document Design   496
    Page Layout   496
    Typography   501
    Graphics   505
    Using Color   506

Poor Document Design: What *Not* to Do   507

Writing For and Designing Websites   508
    The Organization of a Website—The Basics   508
    Web Versus Paper Pages   509
    Web Versus Print Readers   510
    Case Study: Converting a Print Document
       into a Website   510
    Guidelines for Designing and Writing a Successful
       Homepage   512
    Creating Storyboards for Websites and Other Documents   518

Four Rules of Effective Page Design: A Wrap-Up   520

Revision Checklist   521
Exercises   522

Chapter 12: Writing Instructions and Procedures   526

Instructions and Your Job   526

Why Instructions Are Important   526
    Safety   527
    Efficiency   528
    Convenience   528

The Variety of Instructions: A Brief Overview   528

Assessing and Meeting Your Audience's Needs   528
    Key Questions to Ask About Your Audience   529
    Writing Instructions for International Audiences   530
    Two Case Studies on Meeting Your Audience's Needs   532

The Process of Writing Instructions   533
    Plan Your Steps   533
    Do a Trial Run   536
    Write and Test Your Draft   536
    Revise and Edit   536

Using the Right Style   537

Using Visuals Effectively   538
    Guidelines for Using Visuals in Instructions   538

The Five Parts of Instructions   539
    Introduction   539
    List of Equipment and Materials   541
    Steps for Your Instructions   542

Warnings, Cautions, and Notes   544
Conclusion   546

Model of Full Set of Instructions   546

Writing Procedures for Policies and Regulations   554
Some Examples of Procedures   554
Meeting the Needs of Your Marketplace   554

Some Final Advice   555

Revision Checklist   558
Exercises   558

Chapter 13: **Writing Winning Proposals**   561

Writing Successful Proposals   561

Characteristics of Proposals   562
Proposals Vary in Size and Scope   562
Proposals Are Persuasive Plans   563
Proposals Frequently Are Collaborative Efforts   563

Types of Proposals   564
Solicited Proposals and Requests for Proposals   564
Unsolicited Proposals   565
Internal and External Proposals   569

Guidelines for Writing a Successful Proposal   570

Internal Proposals   573
Typical Topics for Internal Proposals   573
Following Corporate Policy   578
Ethically Resolving and Anticipating Readers' Problems   578
Organization of an Internal Proposal   579

Sales Proposals   581
The Audience and Its Needs   581
Organizing a Sales Proposal   581

Proposals for Research Reports   587
Organization of a Proposal for a Research Report   587

A Final Reminder   592

Revision Checklist   592
Exercises   593

Chapter 14: **Writing Effective Short Reports**   599

Types of Short Reports   599

Guidelines for Writing Short Reports   601
Do the Necessary Research   601
Anticipate How an Audience Will Use Your Report   602
Be Objective and Ethical   603
Organize Carefully   604

Write Clearly and Concisely   604
Use an Appropriate Format and Visuals   605

Case Study: A Poor and an Effective Short Report   606

Periodic Reports   610

Sales Reports   610

Progress Reports   610
Audience and Length for a Progress Report   611
Frequency of Progress Reports   616
Parts of a Progress Report   616

Trip/Travel Reports   617
Questions Trip/Travel Reports Answer   617
Common Types of Trip/Travel Reports   618

Test Reports   625
Questions Your Report Needs to Answer   625
Case Study: Two Sample Test Reports   625

Incident Reports   629
When to Submit an Incident Report   630
Parts of an Incident Report   630

Protecting Yourself Legally   632

Short Reports: Some Final Thoughts   633

Revision Checklist   634
Exercises   634

Chapter 15: Writing Careful Long Reports   638

How a Long Report Differs from a Short Report   638
Scope   639
Research   639
Format   641
Timetable   641
Audience   641
Collaborative Effort   641

The Process of Writing a Long Report   642

Parts of a Long Report   643
Front Matter   643
Text of the Report   647
Back Matter   651

A Model Long Report   651

Final Words of Advice About Long Reports   668

Revision Checklist   668
Exercises   669

Chapter 16: Making Successful Presentations at Work   670

Types of Presentations   670

Informal Briefings   670

Guidelines for Preparing Informal Briefings   671
Using Telephones and Cell Phones Effectively   671
Telephone Etiquette   671

Formal Presentations   673

Analyzing Your Audience   673
The Parts of Formal Presentations   675
Presentation Software   680
Noncomputerized Presentations   683
Rehearsing Your Presentation   684
Delivering Your Presentation   685
Evaluating Presentations   688

Revision Checklist   691
Exercises   691

A Writer's Brief Guide to Paragraphs, Sentences, and Words   693

Paragraphs   693

Writing a Well-Developed Paragraph   693
Supply a Topic Sentence   693
Three Characteristics of an Effective Paragraph   695

Sentences   697

Constructing and Punctuating Sentences   697
What Makes a Sentence   698
Avoiding Sentence Fragments   699
Avoiding Comma Splices   700
Avoiding Run-On Sentences   701
Making Subjects and Verbs Agree in Your Sentences   702
Writing Sentences That Say What You Mean   703
Correct Use of Pronoun References in Sentences   704

Words   705

Spelling Words Correctly   705
Using Apostrophes Correctly   705
Using Hyphens Properly   706
Using Ellipses   706
Using Numerals Versus Words   707
Matching the Right Word with the Right Meaning   707

Proofreading Marks   711
Index   712

Successful Writing at Work is a comprehensive introductory text for business, technical, professional, and occupational writing courses. As readers of earlier editions of this text have learned, *Successful Writing at Work* can help students develop key communication skills essential for a successful career. Writing is a vital part of virtually every job today—and this real-world, practical textbook will guide students to become better writers while they also learn to develop and design effective workplace documents, visuals, and presentations.

Successful Writing at Work continues to take students step by step through each type of workplace writing by giving them detailed guidelines for preparing clear, well organized, and readable documents, websites, and presentations for a wide variety of readers. Moreover, because effective models are critical to learning new skills, students will—as in past editions—find a wide and up-to-date range of realistic and rhetorically diverse examples (all of them annotated and visually varied) demonstrating the function, scope, format, and organization of various types of documents. All of these models are focused directly on practical issues in the world of work and portray employees as successful writers.

This new ninth edition is as versatile as it is comprehensive. Full enough for a 16-week semester, it can also be easily adapted to a shorter 6-, 8-, or 10-week course. Furthermore, *Successful Writing at Work* is designed to go beyond classroom applications: It is a ready reference that students can easily carry with them as they begin or advance in the workplace. As students will quickly find, this edition—like earlier ones—is rich in practical applications. It can be as useful to readers with little or no job experience as to those with years of experience in one or several fields. This edition also takes into consideration the needs of students reentering the job market or changing careers.

Distinctive Features of *Successful Writing at Work*

The distinctive features that in the past have made *Successful Writing at Work* a user-friendly text in the contemporary workplace continue to be emphasized in the ninth edition. These features include those skills that are central to success in the world of work—analyzing audiences, approaching writing as a problem-solving activity, using the latest workplace technologies, being an ethical employee, and writing for the global marketplace.

- **Analyzing audiences.** The ninth edition once again stresses the importance of audience analysis and the writer's obligation to achieve the "you attitude" in every workplace document. Moreover, the concept of audience is extended to include readers worldwide, as well as non-native speakers of English, whether as co-workers, employers, clients, or representatives of various agencies and organizations. Memos, e-mails, letters, résumés, application letters, summaries, visuals, websites, instructions, proposals, short reports, long reports, and presentations are introduced and illustrated with the intended audience(s) in mind.

- **Approaching writing as a problem-solving activity.** The ninth edition continues to emphasize workplace writing as a problem-solving activity in which employees meet the needs of their employers, co-workers, customers, clients, and vendors worldwide by getting to the bottom line. *Successful Writing at Work* presents multiple situations and problems that students will have to address, and it highlights the rhetorical and technical options available for solving these problems. It teaches students how to develop the critical skills necessary for planning, researching, drafting, revising, and editing a variety of documents from memos to reports. In addition, in-depth case studies throughout the book demonstrate how writers find resources and develop rhetorical strategies to solve problems in the business world.

- **Using the latest workplace technologies.** This ninth edition offers the most current and extensive coverage of communication technologies for writing successfully in the rapidly changing world of work—from the Internet, e-mail, presentation software, and instant messaging to the newest workplace technologies, including blogs, wikis, document tracking systems, intranets, whiteboards, web and video conferencing tools, and smartphones. You'll find coverage of these technologies not only integrated into each chapter, but also highlighted in the Tech Notes boxes in every chapter.

- **Being an ethical employee.** Companies expect their employees to behave and write ethically. Often, though, writers are faced with ethical dilemmas in the workplace. As in earlier editions, the ninth edition of *Successful Writing at Work* reinforces the importance of ethical workplace writing. Not only are ethics stressed in Chapter 1, but all chapters offer guidelines and examples on the need to write and act ethically in the world of work. Special attention to ethics can be found in the sections on editing to avoid sexism and biased language; writing e-mails, instant messages, and blogs; drafting diplomatic and respectful letters; preparing honest and realistic résumés; conducting accurate, objective, and documented research; using and constructing unaltered and unbiased visuals; preparing safe and effective instructions; and writing truthful proposals and reports.

- **Writing for the global marketplace.** In today's global workplace, effective employees must be consistently aware of and know how to write for a wide variety of readers, both in the United States and across the globe. Consequently, almost every chapter in this new edition includes increased coverage of writing for international readers and non-native speakers of English. The needs and expectations of these international audiences receive special attention starting with Chapter 1 and continuing throughout the chapters focusing on correspondence, visuals, instructions, short and long reports, and presentations.

New and Updated Material in the Ninth Edition

To meet the needs of employees in today's workplace, the ninth edition of *Successful Writing at Work* is one of the most extensively revised editions yet. It has been carefully streamlined and updated to make it the most contemporary tool possible for instructors and students alike. Throughout this edition you will find strengthened coverage, new topics, updated guidelines, and a wealth of new annotated examples of workplace documents, plus new Tech Notes, case studies, and exercises to make the presentation of workplace writing more relevant and current. Here is an overview of what has been added or updated to make *Successful Writing at Work*, Ninth Edition, a comprehensive and up-to-date text:

- **New chapter on research and documentation.** The ninth edition includes a new Chapter 8, "Doing Research and Documentation on the Job," focusing on the tools and strategies students need for workplace success. Beginning with a discussion of how research done in school differs from that in the workplace, the chapter gives students a thorough and realistic introduction to the types of problems they will need to research and solve in the world of work. Special attention is given to the methods and types of primary research most often used in the workplace (with new annotated examples), including observations, interviews, focus groups, and surveys. Secondary research is covered in even greater detail than in previous editions—again from a workplace perspective—with an emphasis on navigating different types of libraries (including Internet and corporate libraries); using general and business-specific periodical databases; consulting reference materials with demographic and government statistics; conducting business-oriented Internet searches; taking effective notes; and documenting sources properly according to the latest MLA and APA guidelines.

- **New real-world extended examples in the research chapter.** The most innovative features in Chapter 8 are its two new extended examples, both adapted from real-world sources: "The Ways Research Is Conducted at the B&L Stores," written by a retail executive with years of practical experience, is a behind-the-scenes look at how and why research is necessary for both day-to-day and long-range decision making. "A Marketing Plan for Sawmill Ridge," written by a collaborative real estate marketing team, is an actual business research report—fully annotated—that illustrates the range of research methods and materials covered earlier in the chapter and demonstrates how writers organize the results of their research for decision makers.

- **New and expanded coverage of workplace technologies**. In addition to new and revised Tech Notes in every chapter, business communication technologies are highlighted in many of the chapters: Chapter 3, "Collaborative Writing and Meetings at Work," now includes a new section on collaborative editing using e-mail, document tracking systems, and wikis. Chapter 4, "Writing Routine Business Correspondence," provides updated coverage on writing and organizing e-mails, along with new sections on (and examples of) instant messages and internal and external business blogs. Chapter 7, "How to Get a Job," features updated coverage of online résumés, along with new sections on developing and designing career portfolios. Chapter 8, "Doing Research and Documentation on the Job," details the latest strategies for doing research with electronic and print sources and documenting a wide range of sources—print, electronic, personal, statistical—in the business world. Chapter 11, "Designing Successful Documents and Websites," contains a new streamlined discussion of writing and designing texts for an online environment. Finally, Chapter 16, "Making Successful Presentations at Work," helps students understand how to be better, more persuasive speakers using PowerPoint technology and includes a revised sample slide show.

- **New and Updated Tech Notes.** Thirteen new Tech Notes have been added to this edition, covering important technologies ranging from coordinating virtual meetings to finding gray literature and using RSS feeds, PDF files, and whiteboards. In addition, all other Tech Notes have been updated, making this edition a valuable introduction to the workplaces of the new millennium. See the inside back cover for a complete list of the Tech Notes, both new and revised, in this edition.

- **New and updated material on collaborative writing and meetings.** In addition to streamlined and updated guidelines for setting up, conducting, and avoiding conflicts in group settings, Chapter 3, "Collaborative Writing and Meetings at Work," now includes a section on ten ways to be a team player, a discussion of collaborating electronically with new figures showing how documents are collaboratively edited using e-mail and document tracking systems, plus further guidelines on planning virtual and face-to-face meetings.

- **New and enhanced discussion of workplace correspondence.** Each of the three chapters on workplace correspondence contains new and updated material to help students become diplomatic and proficient writers. Chapter 4, "Writing Routine Business Correspondence," features updated coverage of memos, e-mails, and faxes, along with new sections (with annotated figures) on writing instant messages and internal and external blogs. Chapter 5, "Writing Letters," supplies updated guidelines for writing standard business letters, as well as enhanced coverage of international business correspondence. And Chapter 6, "Types of Business Letters," gives students practical, updated guidelines on writing various types of letters, with enhanced attention to how a corporate culture promotes its identity.

- **New and expanded coverage of employment correspondence.** Already praised for its helpful coverage of the job search, Chapter 7, "How to Get a Job," offers the most current advice on searching for and applying for a job. It includes a new section on helping students prepare for their careers while they are still in college; updated advice on searching for a job; streamlined discussions and numerous annotated examples of résumés and letters of application; cutting-edge coverage of online résumés; and new sections on dossiers and career portfolios, with an annotated sample webfolio. Reflecting changes in how companies interview and hire job candidates, the chapter closes with new, highly practical advice on interview strategies.

- **New model internal proposal.** The new unsolicited internal proposal in Chapter 13, "Writing Winning Proposals," underscores a theme in this new edition that technology and effective writing are inseparable in the world of work. It is written to persuade a company to purchase GPS tracking systems for its fleet of vehicles, a topic not only highly relevant in this era of escalating energy costs, but one that effectively shows how proposal writers use research and audience analysis.

- **New and stronger emphasis on greening the workplace.** This edition gives greater attention to the importance of protecting and preserving the environment, both in the workplace and at off-site locations. Chapter 1 includes a major example of how a power company and its employees safeguard the natural resources their customers need and then describes the ethical responsibilities companies and their employees have to respect the environment. Subsequent chapters include examples of how they can go about this. For instance, Chapter 3 offers drafts of a report on the importance of recycling, Chapter 8 includes a screen shot of the EPA website; and Chapter 14 shows a progress report from a contractor remodeling an office space to save energy as well as a test report on soil conservation.

Organization of *Successful Writing at Work,* Ninth Edition

The following overview of the organization of this new edition briefly outlines how this text can help both students and instructors.

Overview of Part I: Backgrounds

Part I (Chapters 1–3) explains the foundational concepts of occupational writing, the basics of the writing process, and the importance of collaborative writing in the workplace.

Chapter 1, "Getting Started," sets the stage for all occupational writing by introducing writing as an essential job skill vitally important in the global marketplace, then defines and illustrates the basic concepts of audience analysis, purpose, message, persuasion, style, and tone. The chapter then turns to the characteristics of job-related writing and concludes with a strong emphasis on being an ethical writer and employee.

Chapter 2, "The Writing Process at Work," introduces students to the process of researching, planning, drafting, revising, and editing their written work and then shows them how to troubleshoot some of the most common writing errors found in job-related writing. The information on effective writing in this chapter is threaded throughout the ninth edition.

Chapter 3, "Collaborative Writing and Meetings at Work," emphasizes the importance of teamwork, giving students valuable and easy-to-apply guidelines for being productive, cooperative collaborative writers. This chapter also explores some of the major problems writers face when working together and gives helpful, realistic advice on how they can use software to become effective editors during the collaboration process. The final section of the chapter focuses on planning, attending, and recording meetings.

Overview of Part II: Correspondence

Part II (Chapters 4–7) concentrates on how to write various types of business correspondence, from informal documents and electronic correspondence to formal letters and job application materials.

Chapter 4, "Writing Routine Business Correspondence," offers abundant examples of and guidelines for writing workplace e-correspondence, including e-mails, IMs, and internal and external blogs, as well as memos and faxes.

Chapter 5, "Writing Letters," introduces students to the basics of letter writing, concentrating on why letters remain important in the Internet age, how to select appropriate letter formats, how to organize a letter, what constitutes a clear and concise style, and why the audience's needs must always be at the forefront. The chapter also includes an in-depth section on writing for international readers.

Chapter 6, "Types of Business Letters," examines the rhetorical strategies for producing a variety of business letters, including inquiry, special request, sales, and several types of customer relations letters. The chapter also gives students organizational strategies for preparing good news or bad news letters, along with annotated examples of both appropriate and inappropriate letters.

Chapter 7, "How to Get a Job," takes students step by step through the process of finding a job, from preparing for a career while they are still in college through looking in the right places for a job, creating both print and online résumés, writing application letters, preparing a dossier and webfolio, and interviewing. The chapter provides a wide variety of sample documents—particularly résumés and letters of application—from applicants with varying degrees of experience.

Overview of Part III: Gathering and Summarizing Information

Part III (Chapters 8 and 9) occupies a key position in the ninth edition because it helps students identify and employ the strategies and tools to do careful research and write clear summaries and abstracts essential in the busy world of work.

Chapter 8, "Doing Research and Documentation on the Job," focuses exclusively on the hows and whys of research in the workplace, from understanding the research process as a whole to preparing a fully documented business research report. The process of doing research is simplified by dividing it into three components: doing primary research in a business context, doing business-focused secondary research, and avoiding plagiarism by properly documenting a wide range of sources using MLA and APA styles. An annotated real-world business research report illustrates how writers find, organize, and analyze information to help decision makers solve workplace problems.

Chapter 9, "Summarizing Information at Work," shows students how to write clear and succinct summaries, including executive and evaluative summaries, abstracts, and news releases. Particularly helpful in this chapter is an extended, annotated example showing students how to summarize—from an original document with important points underscored to a working draft summary to a final, effective summary.

Overview of Part IV: Preparing Documents and Visuals

In Part IV (Chapters 10–16), students are given continued opportunities to apply the skills they learned in Parts II and III to ever more complex writing assignments. The section focuses first on visuals and document design, then on major business and technical writing documents (instructions, proposals, short reports, and long reports), and finally on presentations.

Chapter 10, "Designing Clear Visuals," offers practical advice on designing, inserting, and writing about the numerous types of visuals (tables, graphs, charts, maps, photographs, clip art) students will use in the world of work. An especially helpful feature of this chapter is its emphasis on using visuals ethically and appropriately, especially for international audiences.

Chapter 11, "Designing Successful Documents and Websites," stresses the significance of document design and gives students practical advice and pertinent examples for making their work more reader-friendly and visually appealing. The chapter also contains detailed guidelines on writing for websites, emphasizing the differences between print and Web messages.

Chapter 12, "Writing Instructions and Procedures," underscores the importance of writing efficient, safe, convenient, and audience-appropriate instructions. Incorporated into the discussion are guidelines on planning, writing, and testing instructions, illustrated with a complete set of annotated instructions. Procedures, a frequent business genre distinct from standard instructions, are discussed and illustrated in depth at the end of the chapter.

Chapter 13, "Writing Winning Proposals," describes the persuasive nature of proposals and explores how to write three common types of proposals: internal, sales, and research proposals. Annotated examples of each type of proposal give students positive, effective models to help strengthen their skills.

Chapter 14, "Writing Effective Short Reports," outlines the principles common to all short reports and then discusses specific types of reports, including periodic, sales, progress, trip/travel, test, and incident reports. This chapter again cautions students about the ethical implications of what they write and shows them how to avoid typical pitfalls.

Chapter 15, "Writing Careful Long Reports," encourages students to see the long report as the culmination of all of their work in the course. The chapter looks at individual parts of the long report illustrated in detail in the fully annotated model report that closes the chapter.

Chapter 16, "Making Successful Presentations at Work," offers common sense advice on how to use the phone, conduct informal briefings, and prepare for, organize, deliver, and evaluate a formal presentation with appropriate software.

Supplements

The *Successful Writing at Work*, Ninth Edition, **Online Study Center (www .cengage.com/english/kolin)** includes the following resources for students:

- **Improve Your Grade:** *Online exercises* for each chapter are designed to help students simultaneously practice chapter skills and become effective writers using the latest technologies—from word processing features such as report templates and document tracking, to presentation software, to Internet technologies like mind-mapping software and résumé-, survey-, and blog builders. In addition, annotated *Web links* accompanying every chapter enable students to explore chapter topics even further.
- **ACE the Test:** Two gradable 10-question *ACE self-tests* per chapter are provided to help students test their full understanding of chapter topics.

The **Online Teaching Center (www.cengage.com/english/kolin)** provides plentiful material for instructors looking for ideas and aids to teach the course:

- **Correlation Guide.** For those instructors transitioning from either *Successful Writing at Work*, Eighth Edition, or the Concise Second Edition, this guide provides side-by-side content comparisons for easy updating of course syllabi.
- **Sample Syllabi.** Two syllabi are provided, one for a 15-week course incorporating research and long reports, and one for a shorter 10-week course. Both syllabi provide course goals and week-by-week strategies, but they can also be downloaded and adapted to the particular needs of various courses.
- **Some Suggestions on How to Teach Job-Related Writing.** This helpful guide provides ideas for simulating real-world experience in the classroom, enhancing classes by bringing in outside speakers and examples, and highlighting the crucial topics of ethics, global audience, technology, and collaboration.
- **PowerPoint Slides.** Slide shows for each chapter thoroughly cover all chapter topics and allow for enhanced classroom presentation.
- **Suggested Approaches to Exercises.** Since most of the exercises in *Successful Writing at Work* are designed to elicit a variety of responses from students, suggested approaches to evaluating and grading exercises are provided, rather than "right" or "wrong" answers.

The **InSite** online writing and research tool includes electronic peer review, an originality checker, an assignment library, help with common grammar and writing errors, and access to InfoTrac® College Edition. Portfolio management gives you the ability to grade papers, run originality reports, and offer feedback in an easy-to-use online course management system. Using InSite's peer review feature, students can easily review and respond to their classmates' work. Other features include fully integrated discussion boards, streamlined assignment creation, and more. Visit **www.academic.cengage.com/insite** to view a demonstration.

Acknowledgments

In a very real sense, the ninth edition has profited from a collaboration of various reviewers with the author. I am, therefore, honored to thank the following reviewers who have helped me improve this edition significantly:

- Lynette L. Emanuel, *Wisconsin Indianhead Technical College*
- Darrell Fike, *Valdosta State University*
- Julie Freeman, *Indiana University-Purdue University Indianapolis*
- Patrick L. Green, *Aiken Technical College*
- Elizabeth Heath, *Florida Gulf Coast University*
- Michael Piotrowski, *The University of Toledo*
- Sheila Squillante, *The Pennsylvania State University*
- Kelly Wilkinson, *Indiana State University*

I am also deeply grateful to the following individuals at the University of Southern Mississippi for their help as I prepared the ninth edition. From the Department of English, I thank my Chair, Michael Mays, as well as Lori Brister, Ben Geddes, Anna Gibson, Josh Grey, Erin Smith, Sherry Smith, and Danielle Sypher-Haley. I am especially grateful to Denise von Herrmann, Dean of the College of Arts and Letters, for her continued appreciation of my work. My thanks also go to Cliff Burgess (Department of Computer Science), Mary Lux (Department of Medical Technology), Andreas Skalko (Office of Professional, Developmental and Educational Outreach), and Naofumi Tatsumi (Department of Foreign Languages). I am grateful to Dean Carole Kiehl for her assistance. To the following librarians at Cook Library goes my gratitude for their help—Mary Beth Applin, Sherry Laughlin, Edward McCormack, Steven Turner, and especially Maria Englert for her cooperation.

I am also grateful to Terri Smith Ruckel at Louisiana State University for assistance with Chapter 15, and Jianqing Zheng at Mississippi Valley State University and Fatih Uzuner for their help with the Chinese and Turkish translations used in Chapter 12.

Several individuals from the business world also gave me wise counsel, for which I am thankful. They include Sally Eddy at Georgia Pacific; Cathy and Hilary J. Englert at Rice's Potato Chips; Richard E. Kramer, New York City; John Krumpos at Gulf Paper Company; Kirk Woodward at Visiting Nurse Services of New York; and Joycelyn Woolfolk at the Federal Reserve Bank in Atlanta.

I am also especially grateful to Father Michael Tracey for his counsel and his contributions to Chapter 11 on document design, particularly on websites.

My thanks go to my editors at Cengage Learning for their assistance, encouragement, and friendship—Margaret Bridges, Judith Fifer, Megan Garvey, Philip Lanza, Michael Rosenberg, Sarah Truax, and Helen Wood, and to freelance development editor and consultant Bruce Cantley, photo researcher Bruce Carson, and Alison Fields, who handled the book's production.

I deeply thank my extended family—Margie and Al Parish, Sister Carmelita Stinn, and Mary and Ralph Torrelli—for their prayers and love.

Finally, I am deeply grateful to my son Eric and my daughter-in-law Theresa for their enthusiastic and invaluable assistance as I prepared Chapter 8; to my grandson Evan Philip and granddaughter Megan Elise for their love and encouragement. My daughter Kristin also merits loving praise for her help throughout this new edition by doing various searches and revisions and by offering practical advice on successful writing at work. And to Diane Dobson I say thank you for bringing so much peace, music, and love into my life.

P.C.K.

PART I

Backgrounds

1 **Getting Started**
Writing and Your Career

2 **The Writing Process at Work**

3 **Collaborative Writing and Meetings at Work**

Getting Started

Writing and Your Career

What skills have you learned in school or on the job? Perhaps you have learned health care techniques to become a nurse, respiratory therapist, or dental hygienist. Maybe you have received training in law enforcement to work in crime detection or traffic control. Possibly you have studied or worked in software technology, agriculture, information science, hotel and restaurant management, or forestry. Or maybe you have improved skills that will make you a better marketing specialist, salesperson, office manager, programmer, paralegal, or accountant. Whatever your area of accomplishment, the practical know-how you have acquired is crucial for your career.

Visit www
.cengage.com/
english/kolin for
this chapter's
online exercises,
ACE quizzes,
and Web links.

Writing—An Essential Job Skill

Writing is also a part of every job, from your initial letter of application conveying first impressions to the memos, e-mails, blogs, business letters, proposals, and reports that will earn you raises and promotions.

The Associated Press reported in a recent survey that "most American businesses say workers need to improve their writing . . . skills." The same report cited a survey of 402 companies that identified writing as "the most valued skill of employees." Still, the employers polled in that survey indicated that 80 percent of their employees need to work on their writing skills. Clearly, then, writing is an essential skill for employers and employees alike. Figure 1.1 is an e-mail from a human resources director offering an incentive to employees to improve their writing skills by taking a college writing course.

As Rowe Pinkerton, the author of that e-mail, realizes, among the most cost-effective skills you can offer a prospective employer is your writing ability. Businesses pay a premium price for good writing. According to Don Bagin, a communications consultant, most people need an hour or more to write a typical business letter. If an employer is paying someone $30,000 a year, one letter costs $14 of that employee's time; for someone who earns $50,000 a year, the cost of the average letter jumps to $24. Mistakes in letters are also costly. As David F. Noble cautions in his book

Figure 1.1   An employer's view of the importance of writing.

| Reply | Save | Forward | Print | Delete |

To: All Employees
From: <rpinkerton@greer.com>
Date: 5/10/2009
Subject: New company benefit to improve writing

Stresses the importance of writing

I am pleased to announce a new company benefit approved by the Board at its meeting last week. In its continuing effort to improve writing in the workplace, Greer, Inc., will offer tuition reimbursement to any employee who takes a course in business, technical, or occupational writing, starting this fall.

Three Requirements:
To qualify, employees must do the following:

(1) Submit a two-page proposal on how such a course will improve the employee's job performance here at Greer.

Clearly explains how to take advantage of company offer

(2) Take the class at one of the approved colleges or universities in the Cleveland area listed on the attachment to this e-mail.

(3) Provide proof (through a transcript or final grade report) that he or she has successfully completed the course.

To apply, please submit your proposal to Dawn Wagner-Lawlor in Human Resources (<u>dwlawlor@greer.com</u>) at least one month before you intend to enroll in the course.

Closes on an upbeat note

Here's to productive writing!

Rowe Pinkerton
Human Resources Director
781-555-3692
<rpinkerton@greer.com>

Gallery of Best Cover Letters, "the cost of a cover letter (in applying for a job) might be as much as a third of a million dollars—even more—if you figure the amount of income and benefits you don't receive over, say, a 10-year period for a job you don't get because of an error that got you screened out."

One obvious indication of how essential writing is to any job is the technology found in the workplace—PCs and laptops process and display writing; hard drives and disks store writing; printers produce hardcopies of writing; and scanners, e-mails, faxes, IMs, and text messages transmit writing. Writing keeps business

Tech Note

www

Visit www .cengage.com/ english/kolin for an online exer- cise, "Getting to Know Your Computer."

Know Your Computer at Work

A major part of any job is knowing your computer thoroughly. You need to know not just how to use the applications installed on your computer, but also what to do if there is a computer emergency.

Given the kinds of security risks businesses face today, employees have to be especially careful. As Kim Becker cautions in a recent issue of *Nevada Business*, "With malware, spyware, adware, viruses, Trojans, worms, phishing, and server problems, it's time for every business to review its IT strategy and security before a loss occurs."*

Here are some guidelines on how to use your computer effectively on the job:

- **Understand how the software on your computer works.** Your office will most likely require employees to use Microsoft Word or WordPerfect. But make sure you know how to use the entire software package, not just the word-processing feature, but also the filing, formatting, spreadsheet, presentation, and tables/graphics software.
- **Get training on how to use company-specific applications.** You will be expected to know how to use company-created databases, templates, and other customized applications on the job. If your company offers classes on how to use these programs, take them. Otherwise, ask for the advice of a co-worker or someone in i-Tech who knows these procedures.
- **Learn how to back up your files.** If your computer crashes, you will save yourself, your boss, your co-workers, and your clients time and stress by backing up your essential files regularly, so that you have them easily on hand rather than risk losing them.
- **Make sure that your computer is virus protected.** Ask a co-worker or someone from your company's i-Tech department to help you ensure that your computer is virus free and that current anti-virus software has been in- stalled. Also, find out how to avoid computer viruses and what to do if your computer becomes infected.
- **Set up an alternate e-mail address.** If you cannot access your e-mail on the job because of a server that is down, sign up for a free e-mail account with a provider such as Hotmail, Yahoo! Mail, or G-mail that you can use until the server is up.

*Kim Becker, "Security in the Workplace: Technology Issues Threaten Business Prosperity," *Nevada Business* (July 2008).

moving. It allows employees to communicate with one another, with management, and with the customers and clients that the company must serve to stay in business. In your job, you can expect to write e-mails, letters, memos, summaries, instructions, procedures, questionnaires, proposals, reports, and much more. This book will show you, step by step, how to write those and other job-related communications easily and competently.

Chapter 1 gives you some basic information about writing in the global marketplace and offers major questions you can ask yourself to make the writing process easier and the results more effective. It also describes the basic functions of on-the-job writing and introduces you to one of the most important requirements in the business world—writing ethically.

Writing for the Global Marketplace

Visit www
.cengage.com/
english/kolin for
an online exer-
cise, "Exploring
the Online
Global Market-
place."

The Internet, e-mail, express delivery, teleconferencing, and e-commerce have shrunk the world into a global village. Accordingly, it is no longer feasible to think of business in exclusively regional or even national terms. Many companies are multinational corporations with offices throughout the world. In fact, many U.S. businesses are branches of international firms. A large, multinational corporation may have its equipment designed in Japan; built in Bangladesh; and sold in Detroit, Atlanta, and Los Angeles. Its stockholders may be in Mexico City as well as Saudi Arabia—in fact, anywhere. In this global economy, every country is affected by every other one, and all of them are connected by the Internet.

Competing for International Business

Companies must compete for international sales to stay in business. Every business, whether large or small, has to appeal to diverse international markets to be competitive. Each year a larger share of the U.S. gross national product (GNP) depends on global markets. Some U.S. firms estimate that 40 to 50 percent of their business is conducted outside of the United States. Wal-Mart, for example, has opened hundreds of stores in mainland China, and General Electric has plants in over fifty countries. In fact, Jupiter Research estimates that 75 percent of the global Internet population lives outside the United States. If your company, however small, has a website, then it is an international business.

Communicating with Global Audiences:
Seeing the World Through Their Eyes

To be a successful employee in this highly competitive, global market, you have to communicate clearly and diplomatically with a host of readers from different cultural backgrounds. Adopting a global perspective on business will help you communicate and build goodwill with the customers you write to, no matter where they live—across town, in another state, or on other continents, miles and time zones away. As a result, don't presume that you will be writing only to native speakers of American English. As a part of your job, you may communicate with readers in

Native and non-native speakers of English across the world.   Figure 1.2

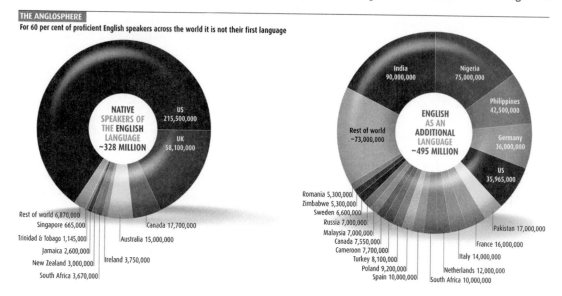

Source: *New Scientist* magazine.

Singapore, Jamaica, and South Africa, for example, who speak varieties of English quite different from American English, as illustrated in the first pie chart in Figure 1.2. You will also very likely be writing to readers for whom English is not their first (or native) language, as shown in the second pie chart in Figure 1.2. These international readers will have varying degrees of proficiency in English, from a fairly good command (as with many readers in India and the Philippines, where English is widely spoken), to little comprehension without the use of a foreign language dictionary and a grammar book to decode your message (as in countries where English is widely taught in schools and recommended for success in the business world but not spoken on a regular basis). These individuals, who may reside either in the United States or in a foreign country, will constitute a large and important audience for your work.

Writing to these international readers with proper business etiquette means first learning about their cultural values and assumptions—what they value and also what they regard as communication taboos. They may not conduct business exactly the way it is done in the United States, and to think they should is wrong. Your international audience is likely to have different expectations of how they want a letter addressed or written to them, how they prefer a proposal to be submitted, how they wish a business meeting to be conducted, or how they think questions should be asked and agreements reached. Their concepts of time, family, money, the world, the environment, managers, and communication itself may be nothing like those in the United States. Visuals, including icons, which are easily understood in the United States, may be baffling elsewhere in the world. If you misunderstand your audience's culture by inadvertently writing, creating, or saying something inappropriate, it can cost your company a contract and you your job.

Cultural Diversity at Home

Cultural diversity exists inside as well as outside the company you work for. Don't conclude that your boss or co-workers are all native speakers of English, either, or that they come from the same cultural background that you do. In the next decade, as much as 40 to 50 percent of the U.S. skilled work force may be composed of recent immigrants who bring their own traditions and languages with them. These are highly educated, multicultural, and multinational individuals who have acquired English as a second language.

For the common good of your company, you need to be respectful of these international colleagues. In fact, multinational employees can be tremendously important for your company in making contacts in their native country and in helping your firm understand and appreciate ethical/cultural differences among customers. The long report in Chapter 15 (pages 652–666) describes some ways in which a company can both acknowledge and respect the different cultural traditions of its international employees. Businesses want to emphasize their international commitments. A large corporation such as Citibank, for instance, is eager to promote its image of helping customers worldwide, as Figure 1.3 shows.

Using International English

Whether international readers are your customers or co-workers, you will have to adapt your writing to respect their language needs and communication protocols. To communicate with non-native speakers, you should use "international English," a way of writing that is easily understood, culturally tactful, and diplomatic. To write international English means you reexamine your own writing. The words, idioms, phrases, and sentences you choose instinctively for U.S. readers may not be appropriate for an audience for whom English is a second, or even third, language. If you find a set of directions accompanying your computer or software package confusing, imagine how much more intimidating such a document would be for non-native speakers of English. But with international English your message is clear, straightforward, and appropriate for readers who are not native speakers of English. International English is user-friendly in terms of the words, formats, and visuals you choose. It is free from complex, hard-to-process sentences as well as from cultural bias.

International English is discussed in detail on pages 170–187. Later chapters of this book also give you practical guidelines on writing or creating e-mails, correspondence, instructions, reports, websites, and other work-related documents suitable for a global audience.

▌ Four Keys to Effective Writing

Effective writing on the job is carefully planned, thoroughly researched, and clearly presented. Its purpose is always to accomplish a specific goal and to be as persuasive as possible. Whether you send a routine e-mail to a co-worker in Cincinnati or

A company's dedication to globalization.   Figure 1.3

How Citigroup Meets Banking Needs Around the World

WITH A BANKING EMPIRE that spans more than 100 countries, Citigroup is experienced at meeting the diverse financial services needs of businesses, individuals, customers, and governments. The bank is headquartered in New York City but has offices in Africa, Asia, Central and South America, Europe, the Middle East, as well as throughout North America. Live or work in Japan? You can open a checking account at Citigroup's Citibank branch in downtown Tokyo. How about Mexico? Visit a Grupo Financiero Banamex-Accival branch, owned by Citigroup. Citigroup owns European American Bank and has even bought a stake in a Shanghai-based bank with an eye toward attracting more of China's $1 trillion in bank deposits. Between acquisitions and long-established branches, Citigroup covers the globe from the Atlantic to the Pacific and the Indian Oceans.

Individuals can use Citigroup for all the usual banking services . . . . Personalized service is the hallmark of . . . the bank, which can help prepare customized financial plans for . . . customers, manage their securities trading activities, provide trust services, and much more. What's more, Citigroup is active in communities around the world through philanthropic contributions and grants, financial literacy seminars, volunteerism, and supplier diversity programs. This financial services giant strives for the best of both worlds, wielding its global presence and resources to meet banking needs locally, one customer at a time.

Source: From Pride, Hughes, and Kapoor, *Business,* 8th ed., p. 587. Copyright © 2005 by Houghton Mifflin Company. Used by permission.

in Shanghai or a special report to the president of the company, your writing will be more effective if you ask yourself four questions:

1. *Who* will read what I write? (Identify your *audience*.)
2. *Why* should they read what I write? (Establish your *purpose*.)
3. *What* do I have to say to them? (Formulate your *message*.)
4. *How* can I best communicate? (Select your *style* and *tone*.)

The questions *who? why? what?* and *how?* do not function independently; they are all related. You write (1) for a specific audience (2) with a clearly defined purpose in mind (3) about a topic your readers need to understand (4) in language appropriate for the occasion. Once you answer the first question, you are off to a good start toward answering the other three. Now let us examine each of the four questions in detail.

Identifying Your Audience

Knowing *who* makes up your audience is one of your most important responsibilities as a writer. Expect to analyze your audience throughout the composing process.

Look at the posters in Figures 1.4, 1.5, and 1.6. The main purpose of all three posters is the same: to discourage people from smoking. The essential message in

Figure 1.4   No-smoking poster aimed at fathers who smoke.

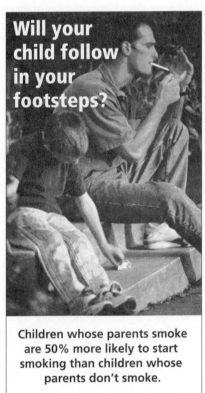

Will your child follow in your footsteps?

Children whose parents smoke are 50% more likely to start smoking than children whose parents don't smoke.

Photo by Peter Poulides/Getty Images.

No-smoking poster directed at pregnant women.   **Figure 1.5**

**Smoking
Puts
Both
Mother**

and **Child
at Risk**

Photo by Bill Crump/Brand X Pictures/Fotosearch/Royalty-Free Image.

each poster—smoking is dangerous to your health—is also the same. But note how the different details—words, photographs, situations—have been selected to appeal to three different audiences.

The poster in Figure 1.4 is aimed at fathers who smoke. As you can see, it is an image of a father smoking next to his son, who is reaching for his pack of cigarettes. Note how the caption "Will your child follow in your footsteps?" plays on the fact that the father and son are literally sitting on steps, but at the same time the word *footsteps* implies "following another's lead." The statistic at the bottom of the poster reinforces both the photo caption and the image, hitting home the point that parental behavior strongly influences children's behavior. Already the child in the photograph is following his father by showing a clear interest in smoking.

The poster in Figure 1.5 is aimed at an audience of pregnant women and appropriately shows a pregnant woman with a lit cigarette. The words on the poster appeal to a mother's sense of responsibility as the reason to stop smoking, a reason to which this audience would be most likely to respond.

Figure 1.6 is directed toward young athletes. The word *smoke* is aimed directly at their game and their goal. In fact, the poster writers appropriately made the goal the same for the game as for the players' lives. Note, too, how this image is suitable for an international audience.

Figure 1.6   No-smoking poster appealing to young athletes.

The copywriters who created these posters have chosen appropriate details—words, pictures, captions, and so on—to convince each audience not to smoke. With their careful choices, they successfully answered the question "How can we best communicate with each audience?" Note that details relevant for one audience (athletes, for example) could not be used as effectively for another audience (such as fathers).

The three messages illustrate some fundamental points to keep in mind when identifying your audience.

- Members of each audience differ in backgrounds, experiences, needs, and opinions.
- How you picture your audience will determine what you say to them.
- Viewing something from the audience's perspective will help you to select the most relevant details for that audience.

Some Questions to Ask About Your Audience

You can form a fairly accurate picture of your audience by asking yourself some questions *before* you write. For each audience you need to reach, consider the following questions.

1. **Who is my audience?** What individual(s) will most likely be reading my work?

   If you are writing for individuals at work:

   - What is my reader's job title? Co-worker? Immediate supervisor? Vice president?
   - What kind of job experience, education, and interests does my reader have?

   If you are writing for clients or consumers (a very large, sometimes fragmented audience):

   - How can I find out about their interest in my product or service?
   - How much will this audience know about my company? About me?

2. **How many people will make up my audience?**

   - Will just one individual read what I write (the nurse on the next shift, the production manager) or will many people read it (all the consumers of a product or service)?
   - Will my boss want to see my work (say, a letter to a consumer in response to a complaint) to approve it?
   - Will I be sending my message to a large group of people sharing a similar interest in my topic, such as a listserv?

3. **How well does my audience understand English?**

   - Are all my readers native speakers of English?
   - Will I be communicating with people all around the globe?
   - Will some of my readers speak English as a second or even third language, and so require extra sensitivity on my part to their needs as non-native speakers of English?
   - Are my English-speaking readers all located within the United States, or are some from other English-speaking countries, such as England, Australia, Singapore, Jamaica, Ireland, or New Zealand?
   - Will some of my readers speak no English, but instead use an English grammar book and foreign language dictionary to understand what I've written?

4. **How much does my audience already know about my writing topic?**

   - Will my audience know as much as I do about the particular problem or issue, or will they need to be briefed or updated?
   - Are my readers familiar with, and do they expect me to use, technical terms and descriptions, or will I have to provide easy-to-understand and nontechnical wording?
   - Will I need to include detailed tables and visuals, or will a map, photograph, or simple drawing be enough?

5. **What is my audience's reason for reading my work?**

   - Is my communication part of their routine duties, or are they looking for information to solve a problem or make a decision?
   - Am I writing to describe benefits that another writer or company cannot offer?

- Will my readers expect complete details, or will a short summary be enough?
- Are they reading my work to make an important decision affecting a co-worker, a client, or a community?
- Are they reading something I write because they must (a legal notification, for instance)?

6. **What are my audience's expectations about my written work?**

- Do they want an e-mail or will they expect a formal letter?
- Will they expect me to follow a company format and style?
- Are they looking for a one-page memo or for a comprehensive report?
- Should I use a formal tone or a more relaxed and conversational style?

7. **What is my audience's attitude toward me and my work?**

- Will I be writing to a group of disgruntled and angry customers or vendors about a sensitive issue (a product recall, a refusal of credit, or a shipment delay)?
- Will I have to be sympathetic while at the same time give firm reasons for my company's (or my) decision?
- Will my readers be skeptical, indifferent, or friendly about what I write?
- Will my readers feel guilty that they have not answered an earlier message of mine, not paid a bill now overdue, or not kept a promise or commitment?

8. **What do I want my audience to do after reading my work?**

- Do I want my readers to purchase something from me, approve my plan, or send me additional documentation?
- Do I simply want my readers to get my message and not respond at all?
- Do I expect my readers to get my message, acknowledge it, save it for future reference, or review it and e-mail it to another individual or office?
- Do my readers have to take immediate action, or do they have several days or weeks to respond?

As your answers to these questions will show, you may have to communicate with many different audiences on your job. Each group of readers will have different expectations and requirements; you need to understand those audience differences if you want to supply relevant information.

Let's say you work for a manufacturer of heavy-duty equipment and that you have to write for many individuals in that organization. Here are the priorities of five different audiences with appropriate information to give each one.

Audience	Information to Communicate
Owner or principal executive	The writer stresses financial benefits, indicating that the machine is a "money-maker" and is compatible with other existing equipment.
Production engineer	The writer emphasizes "state-of-the-art" transmissions, productivity, availability of parts.

Operator	The writer focuses on information about how easy and safe it is to run.
Maintenance worker	The writer provides key details about routine maintenance as well as troubleshooting advice on problems.
Production supervisor	The writer emphasizes the speed and efficiency the machine offers.

As these examples show, to succeed in the world of work, give each reader the details he or she needs to accomplish a given job.

Establishing Your Purpose

By knowing *why* you are writing, you will communicate better and find writing itself to be an easier process. The reader's needs and your goal in communicating will help you to formulate your purpose. It will guide you in determining exactly what you can and must say. With your purpose clearly identified, you are on the right track.

Make sure you follow the most important rule in occupational writing: **Get to the point right away.** At the start of your message, state your goal clearly. Don't feel as if you have to entertain or impress your reader.

I want new employees to know how to log on to the computer.

Think over what you have written. Rewrite your purpose statement until it states precisely why you are writing and what you want your readers to do or to know.

I want to teach new employees the security code for logging on to the company computer.

Since your purpose controls the amount and order of information you include, state it clearly at the beginning of every e-mail, memo, letter, and report.

This memo will acquaint new employees with the security measures they must take when logging on to the company computer.

In the opening purpose statement that follows, note how the author clearly informs the reader what the report will and will not cover.

As you requested at last month's organizational meeting, I have conducted a survey of how well our websites advertise our products. This survey describes users' responses but does not prioritize them.

The following preface to a publication on architectural casework details contains a model statement of purpose suited to a particular audience.

This publication has been prepared by the Architectural Woodwork Institute to provide a source book of conventional details and uniform detail terminology. For this purpose a series of casework detail drawings, . . . representative of the best industry-wide practice, has been prepared and is presented here. By supplying both architect and woodwork manufacturer with a common authoritative reference, this work will enable architects and woodworkers to communicate in a common technical language. . . . Besides serving as a basic reference for architects and architectural drafters, this guide will be an effective educational tool for the beginning drafter-architect-in-training. It should also be a valuable aid to the project manager in coordinating the work of many drafters on large projects.[1]

After seeing that preface, readers have a clear sense of why they should use this resource and what to do with the material they find in it.

Formulating Your Message

Your message is the sum of *what* facts, responses, and recommendations you put into writing. A message includes the details and scope of your communication.

- The *details* are those key points you think readers need to know.
- *Scope* refers to how much information you give readers about those key details.

Some messages will consist of one or two phrases or sentences: "Do not touch; wet paint." "Order #756 was sent this afternoon by FedEx. It should arrive at your office on March 22." At the other extreme, messages may extend over many pages. Messages may carry good news or bad news. They may deal with routine matters, or they may handle changes in policy, special situations, or problems.

Keep in mind that you will need to adapt your message to fit your audience. For some audiences, such as engineers or technicians, you may have to supply a complete report with every detail noted or contained in an appendix. For other readers—busy executives, for example—include only a short discussion or summary of financial or managerial significance.

Consider the message of the excerpt in Figure 1.7 from a section called "Technology in the Grocery Store" included in a consumer handbook. The message provides factual information and a brief explanation of how a clerk scans an item, informing consumers about how and why they may have to wait longer in line. It also tells readers that the process is not as simple as it looks.

This bar code message is appropriate for consumers who do not need or desire more information. Individuals responsible for entering data into the computer or doing inventory control, however, would need more detailed instructions on how to program the supermarket's computer so it automatically tells the point-of-sale (POS) terminal what price and product match each bar code.

But technicians responsible for affixing the bar codes at the manufacturer's plant would require much more detailed information than would consumers or store cashiers. These technicians must be familiar with the Universal Product Code (UPC), which specifies bar codes worldwide. They would also have to know about the UPC binary code formulas and how they work—that is, the number of lines,

---

[1]Reprinted by permission of Architectural Woodwork Institute.

Bar Code Readers

Every time you check out at the grocery store, many of your purchases are scanned to record the price. The scanner uses a laser beam to read the bar codes, those zebra-striped lines imprinted on packages or canned goods. These codes are fed into the store's computer, which provides the price that matches the product code. The product and its price are then recorded on your receipt.

Scanning an item requires more skill than you might think. To make sure that the scanner accurately reads the bar codes, the clerk has to take into account the following four conditions:

Speed: The clerk has to pass the item across the scanner at a particular speed. If the item moves too slowly, the bars will look too long or too wide and the computer will reject the item. If the item is moved too fast, the scanner cannot identify the code.

Angle: The clerk needs to gauge the exact angle at which to pass the item across the scanner. If the angle is wrong there will be insufficient reflection of the laser beam back to the scanner and so it will not be able to read the code. Since it is best for the item to reflect as much of the laser as possible, the clerk should try to hold the code at right angles to the laser.

Distance: Moving the item too close to the scanner is as unproductive as holding it too far away. Either way the code can be out of focus for the scanner reader. Holding the item about 3–4 inches away is best; holding it out more than 8–9 inches ensures that the scanner will not read the code.

Rotation: The code needs to be facing the scanner so that the lines can be read properly.

Direction in which clerk moves product

If your clerk makes a mistake in any one of these calibrations, your wait in line is sure to be longer.

width of spacing, and the framework to indicate to the scanner when to start reading the code and when to stop. Such formulas, technical details, and functions of photoelectric scanners are appropriate for this audience.

Selecting Your Style and Tone

Style

Style refers to *how* something is written rather than what is written. Style helps to determine how well you communicate with an audience and how well your readers understand and receive your message. It involves the choices you make about

- the construction of your paragraphs
- the length and patterns of your sentences
- your choice of words

You will have to adapt your style to take into account different messages, different purposes, and different audiences. Your words, for example, will certainly vary with your audience. If all your readers are specialists in your field, you may safely use the technical language and symbols of your profession. Nonspecialists, however, will be confused and annoyed if you write to them in the same way. The average consumer, for example, will not know what a *potentiometer* is, but by writing "volume control on a radio" instead, you will be using words that the general public can understand. And, as we saw, when you write for an international audience you have to take into account their proficiency in English and choose your words and sentences with their needs in mind (see pp. 170–187).

Tone

Tone in writing, like tone of voice, expresses your attitude toward a topic and toward your audience. Your tone can range from formal and impersonal (a scientific report) to informal and personal (e-mail to a friend or a how-to article for consumers). It can be unprofessionally sarcastic or diplomatically agreeable.

Tone, like style, is indicated in part by the words you choose. For example, saying that someone is "interested in details" conveys a more positive tone than saying the person is a "nitpicker." The word *economical* is more positive than *stingy* or *cheap*.

The tone of your writing is especially important in occupational writing, because it reflects the image you project to your readers and thus determines how they will respond to you, your work, and your company. Depending on your tone, you can appear sincere and intelligent or angry and uninformed. Of course, in all your written work, you need to sound professional and knowledgeable. The wrong tone in a letter or a proposal might cost you a customer. Sarcastic or hostile language will at once alienate you from your readers, as the letters in Figures 5.5 (p. 166) and 5.6 (p. 167) demonstrate.

Case Study: A Description of Heparin for Two Different Audiences

To better understand the effects of style and tone on writing, read the following two excerpts. In both the message is basically the same, but because the audiences differ, so do the style and the tone. The two pieces both describe *heparin*, a drug used to prevent blood clots.

The first description of heparin appears in a reference work for physicians and other health care providers and is written in a highly technical style with an impersonal tone.

HEPARIN SODIUM INJECTION, USP
STERILE SOLUTION

Description: Heparin Sodium Injection, USP is a sterile solution of heparin sodium derived from bovine lung tissue, standardized for anticoagulant activity.

Each ml of the 1,000 and 5,000 USP units per ml preparations contains: heparin sodium 1,000 or 5,000 USP units; 9 mg sodium chloride; 9.45 mg benzyl alcohol added as preservative. Each ml of the 10,000 USP units per ml preparations contains: heparin sodium 10,000 units; 9.45 mg benzyl alcohol added as preservative.

When necessary, the pH of Heparin Sodium Injection, USP was adjusted with hydrochloric acid and/or sodium hydroxide. The pH range is 5.0–7.5.

Clinical pharmacology: Heparin inhibits reactions that lead to the clotting of blood and the formation of fibrin clots both *in vitro* and *in vivo*. Heparin acts at multiple sites in the normal coagulation system. Small amounts of heparin in combination with antithrombin III (heparin cofactor) can inhibit thrombosis by inactivating activated Factor X and inhibiting the conversion of prothrombin to thrombin.

Dosage and administration: Heparin sodium is not effective by oral administration and should be given by intermittent intravenous injection, intravenous infusion, or deep subcutaneous (intrafrat, i.e., above the iliac crest or abdominal fat layer) injection. **The intramuscular route of administration should be avoided because of the frequent occurrence of hematoma at the injection site.**[2]

The writer has made the appropriate stylistic choices for the audience, the purpose, and the message. Physicians and other health care providers understand and expect the technical vocabulary and the scientific and lengthy explanations to prescribe and/or administer heparin correctly. The author's authoritative, impersonal tone is coldly clinical, which, of course, is also correct because the purpose is to convey the accurate, complete scientific facts about this drug, not the writer's or reader's opinions or beliefs. The author sounds both knowledgeable and appropriately objective.

The description of heparin on the next page, however, is written in a nontechnical style and with an informal, caring tone. This description is similar to those found on information cards given to patients about the drugs they are receiving in a hospital.

---

[2]Copyright © *Physicians' Desk Reference*® 45th edition, 1991, published by Medical Economics, Montvale, New Jersey 07645. Reprinted by permission. All rights reserved.

> Your doctor has prescribed a drug called *heparin* for you. This drug will prevent any new blood clots from forming in your body. Since heparin cannot be absorbed from your stomach or intestines, you will not receive it in a capsule or tablet. Instead, it will be given into a vein or the fatty tissue of your abdomen. After several days, when the danger of clotting is past, your dosage of heparin will be gradually reduced. Then another medication you can take by mouth will be started.

The writer of this description has also made appropriate choices for nonspecialists such as patients who do not need elaborate descriptions of the origin and composition of the drug. Using familiar words and adopting a personal tone help to win the patients' confidence and enable them to understand why they should take the drug.

■ Characteristics of Job-Related Writing

Job-related writing characteristically serves six basic functions: (1) to provide practical information, (2) to give facts rather than impressions, (3) to provide visuals to clarify and condense information, (4) to give accurate measurements, (5) to state responsibilities precisely, and (6) to persuade and offer recommendations. These six functions tell you what kind of writing you will produce after you successfully answer the *who? why? what?* and *how?*

1. Providing Practical Information

On-the-job writing requires a practical "here's what you need to do or to know" approach. One such practical approach is *action oriented*. You instruct the reader to do something—assemble a ceiling fan, test for bacteria, perform an audit, or create a website. Another practical approach of job-related writing is *knowledge oriented*. You explain what you want the reader to understand—why a procedure was changed, what caused a problem or solved it, how much progress was made on a job site, or why a new piece of equipment should be purchased.

The following description of Energy Efficiency Ratio combines both the action-oriented and knowledge-oriented approaches of practical writing.

> Whether you are buying window air-conditioning units or a central air-conditioning system, consider the performance factors and efficiency of the various units on the market. Before you buy, determine the Energy Efficiency Ratio (EER) of the units under consideration. The EER is found by dividing the BTUs (units of heat) that the unit removes from the area to be cooled by the watts (amount of electricity) the unit consumes. The result is usually a number between 5 and 12. The higher the number, the more efficiently the unit will use electricity.
>
> You'll note that EER will vary considerably from unit to unit of a given manufacturer, and from brand to brand. As efficiency is increased, you may find the purchase price is higher; however, your operating costs will be lower. Remember, a good rule to follow is to choose the equipment with the highest EER. That way you'll get efficient equipment and enjoy greater operating economy.[3]

---

[3]Reprinted by permission of New Orleans Public Services, Inc.

2. Giving Facts, Not Impressions

Occupational writing is concerned with what can be seen, heard, felt, tasted, or smelled. The writer uses *concrete language* and specific details. The emphasis is on facts rather than on the writer's feelings or guesses.

The discussion below addressed to a group of scientists about the sources of oil spills and their impact on the environment is an example of writing with objectivity. It describes events and causes without anger or tears. Imagine how much emotion could have been packed into this paragraph by the residents of the coastal states who have watched such spills come ashore.

> The most critical impact results from the escapement of oil into the ecosystem, both crude oil and refined fuel oils, the latter coming from sources such as marine traffic. Major oil spills occur as a result of accidents such as blowout, pipeline breakage, etc. Technological advances coupled with stringent regulations have helped to reduce the chances of such major spills; however, there is a chronic low-level discharge of oil associated with normal drilling and production operations. Waste oils discharged through the river systems and practices associated with tanker transports dump more significant quantities of oils into the ocean, compared to what is introduced by the offshore oil industry. All of this contributes to the chronic low-level discharge of oil into world oceans. The long-range cumulative effect of these discharges is possibly the most significant threat to the ecosystem.[4]

3. Providing Visuals to Clarify and Condense Information

Visuals are indispensable partners of words in conveying information to your readers. On-the-job writing makes frequent use of visuals such as tables, charts, photographs, flow charts, diagrams, and drawings to clarify and condense information. Thanks to various software packages, you can easily create and insert visuals into your writing. The use of visuals is discussed in detail in Chapters 10 and 11, as well as in PowerPoint presentations in Chapter 16.

Visuals play an important role in the workplace. Note how the photograph in Figure 1.8 can help computer users better understand and follow the accompanying written ergonomics guidelines. A visual like this, reproduced in an employee handbook or displayed as a poster, can significantly reduce stress and increase productivity.

Visuals are extremely useful in making detailed relationships clear to readers. The information in Table 1.1 (p. 23) on the world's ten most populous countries in 2008 and those projected for 2050 would be very difficult to discuss and follow if it were not in twin tables. When that information is presented in two tables, the writer makes it easy for the reader to see and understand relationships. If such information were just written in prose, it would be much harder to compare, contrast, and summarize.

In addition to the visuals already mentioned, the following graphic devices in your letters, reports, and websites can make your writing easier to read and follow:

- headings, such as **Four Keys to Effective Writing** or **Characteristics of Job-Related Writing**

---

[4]*Source:* The Offshore Ecology Investigation. Reprinted by permission of Gulf Universities Research Consortium.

Figure 1.8   Use of a visual to convey information.

An Ergonomic Workstation

Ergonomics is the science of designing workplace equipment to conform to human needs. To maintain comfort while you work and avoid workplace injuries, follow the principles shown in the photograph at right and discussed in the bulleted items below.

- To reduce the possibility of eye damage, maintain a distance of 18 to 24 inches between your eyes and the computer screen and keep your work area well lit.

- To reduce neck strain, position your computer screen so that the top of the screen is at or just below eye level.

- To avoid back and shoulder stress, sit up straight in your chair with your shoulders relaxed and your lower back firmly supported (with a cushion, if necessary).

- To prevent carpal tunnel syndrome and arm strain, rest your elbows at your side

and keep your forearms and wrists level with the keyboard.

- To avoid bumping your thighs and knees, allow 1–2 inches of space between your thighs and the keyboard shelf.

- To prevent leg and back strain, adjust your chair height so that your upper and lower legs form a 90-degree angle and your feet are either flat on the floor or on a footrest.

Photo courtesy Ergo Concepts, LLC.

- subheadings to divide major sections into parts, such as "Providing Practical Information" or "Giving Facts, Not Impressions"
- numbers within a paragraph, or even a line, such as (1) this, (2) this, and (3) also this
- different types of s p a c i n g
- CAPITALIZATION
- *italics* (easily made by a word processing command or indicated in typed copy by <u>underscoring</u>)
- **boldface** (darker print for emphasis)
- icons (visual markers such as →)
- HYPERTEXT (the use of color or boldface to indicate links to the Internet)
- asterisks to * separate * items * or to note key items *
- lists with "bullets" (like those before each entry in this list)

TABLE 1.1   Ten Most Populous Countries in 2008 vs. Those Projected in 2050

2008	(Millions)	2050	(Millions)
China	1,322	India	1,808
India	1,130	China	1,424
United States	301	United States	420
Indonesia	235	Nigeria	356
Brazil	190	Indonesia	313
Pakistan	165	Bangladesh	280
Bangladesh	150	Pakistan	278
Russia	141	Brazil	228
Nigeria	135	Kinshasa (Congo)	203
Japan	127	Mexico	148

Source: U.S. Census Bureau, Jan. 2008.

Keep in mind that graphic devices should be used carefully and with moderation, not just for decoration or to dress up a letter or report. Used properly, they can help you to

- organize, arrange, and emphasize your ideas
- make your work easier to read and to recall
- preview and summarize your ideas, for example, headings
- list related items to help readers distinguish, follow, compare, and recall them—as this bulleted list does

4. Giving Accurate Measurements

Much of your work will depend on measurements—acres, bytes, calories, centimeters, degrees, dollars and cents, grams, percentages, pounds, square feet, units, etc. Numbers are clear and convincing. However, you must be sensitive to which units of measurement you use when writing to international readers. Not every culture computes in dollars or records temperatures in degrees Fahrenheit. See pages 176–177.

The following discussion of mixing colored cement for a basement floor would be useless to readers if it did not supply accurate quantities.

> Including permanent color in a basement floor is a good selling point. One way of doing this is by incorporating commercially pure mineral pigments in a topping mixture placed to a 1-inch depth over a normal base slab. The topping mix should range in volume between 1 part portland cement, 1¼ parts sand, and 1¼ parts gravel or crushed stone and 1 part portland cement, 2 parts sand, and 2 parts gravel or crushed stone. Maximum size gravel or crushed stone should be ⅜ inch.
>
> Mix cement and pigment before aggregate and water are added and be very thorough to secure uniform dispersion and the full color value of the pigment. The proportion varies from 5 to 10 percent of pigment by weight of cement, depending on the shade desired. If carbon black is used as a pigment to obtain grays or black, a proportion of

---

[5] Reprinted by permission from *Concrete Construction Magazine*, World of Concrete Center, 426 South Westgate, Addison, Illinois 60101.

from ½ to 1 percent will be adequate. Manufacturers' instructions should be followed closely; care in cleanliness, placing, and finishing are also essential. Colored topping mixes are available from some suppliers of ready mixed concrete.[5]

5. Stating Responsibilities Precisely

Job-related writing, because it is directed to a specific audience, must make absolutely clear what it expects of, or can do for, that audience. Misunderstandings waste time and cost money. Directions on order forms, for example, should indicate how and where information is to be listed and how it is to be routed and acted on. The following three directions show readers how to perform different tasks and/or explain why.

- Enter agency code numbers in the message box.

- Items 1 through 16 of this form should be completed by the injured employee or by someone acting on his or her behalf, whenever an injury is sustained on the job. The term *injury* includes occupational disease caused by the employment. The form should be given to the employee's official superior within 24 hours following the injury. The official superior is that individual having responsible supervision over the employee.

- What is a **credit report**? A credit report is a record of how you've paid bills with credit grantors such as stores and banks. Credit grantors use credit reports to determine whether or not you will be extended credit. The report identifies you by information such as your name and address, credit accounts, and payment history. Your credit report also includes public record data, such as bankruptcies, court judgments, and tax liens. A list of those who have recently requested a copy of your credit report is also included. A credit report does not contain information on arrest records, specific purchases, or medical records.[6]

Other kinds of job-related writing deal with the writer's responsibilities rather than the reader's, for example, "Tomorrow I will meet with the district sales manager to discuss (1) July's sales, (2) the necessity of expanding our market, and (3) next fall's production schedule. I will e-mail a report of our discussion by August 5."

6. Persuading and Offering Recommendations

Persuasion is a vital part of writing on the job. In fact, persuasion is one of the most crucial skills you can learn. It determines how successful you and your company or agency will be. Persuasion is at the heart of the world of work, whether you are writing to someone outside or inside your company.

Writing Persuasively to Clients and Customers
Much of your writing in the business world will promote your company's image by persuading customers and clients (a) to buy a product or service or (b) to adopt

---

[6]Reprinted by permission of Associated Credit Bureaus, Inc.

An advertisement using arguments based on cost, time, efficiency, safety, and convenience to persuade a potential customer to use a service.

Figure 1.9

**GENERAL MEDICAL
WILL STOP THE
UNNECESSARY TRANSPORTING
OF YOUR INMATES.**

- We'll bring our X-ray services to your facility,
  7 days a week, 24 hours a day.

    We can reduce your X-ray costs by a minimum of 28%.

    X-ray cost includes radiologist's interpretation and written report.

    Same day service with immediate results telephoned to your facility.
- Save correctional officers' time, thereby saving your facility money.
- Avoid chance of prisoner's escape and possible danger to the public.
- Avoid long waits in overcrowded hospitals.
- Reduce your insurance liabilities.
- Other Services Available: Ultrasound, Two Dimensional Echocardiogram, C.T. Scan, EKG, Blood Lab and Holter Monitor.

*General Medical Is Your On-Site
Medical Problem Solver*

General Medical Services Corp.
A subsidiary of

Federal Medical Industries, Inc. O.T.C.
950 S.W. 12th Avenue, 2nd Floor Suite, Pompano, Florida 33069
(305) 942-1111   FL WATS: 1-800-654-8282

a plan of action endorsed by your employer. You will have to convince readers that you (and your company) can save them time and money, increase efficiency, reduce risks, or improve their image, and that you can do this better than your competitors can.

Expect also to be called on to write convincingly about your company's image, as in the case of product recalls, customer complaints, or damage control after a corporate mistake affecting the environment. You may also have to convince customers or organizations around the globe that your company respects cultural diversity and that your products and services can appeal to specific ethnic values. As we saw, the global marketplace is diversity driven.

A large part of persuasion is supporting your claims with evidence. You will have to conduct research, provide logical arguments, supply examples of appropriate data, and identify the most relevant information for your particular audience. Notice how the advertisement in Figure 1.9 offers a bulleted list of persuasive reasons—based on cost, time, safety, efficiency, and convenience—to convince correctional officials that they should use General Medical's services rather than those of a hospital or clinic.

Writing Persuasively In-House

As much as 60 percent of your writing may be to individuals you work with and for. In-house writing also requires you to develop your persuasive skills. In fact, your very first job-related writing will likely be a persuasive letter of application to obtain a job interview with a potential employer.

On the job, you will have to evaluate various products or options by studying, analyzing, and deciding on the most relevant one(s) for your boss. Your reader will expect you to offer clear-cut, logical, and convincing reasons for your choice.

The following summary concludes why it is better for a company to lease a truck than to purchase one. Note its persuasive tone and logical presentation.

After studying the pros and cons of buying or leasing a company truck, I recommend that we lease it for the following five reasons.

1. We will not have to expend any of our funds for a down payment, which is being waived.
2. Our monthly payments for leasing the vehicle will be at least $150 less than the payments we would have to make if we purchased the truck on a 3-year contract.
3. All major and minor maintenance (up to 36,000 miles) is included as part of our monthly leasing payment.
4. Insurance (theft and damage) is also part of our monthly leasing payment.
5. We have the option of trading in the truck every 16 months for a newer model or trading up for a more expensive model in the line every 12 months.

As part of your job, too, you will be asked to write persuasive memos, e-mails, and letters to boost the morale of employees, encourage them to be more productive, and compliment them on jobs well done. You can also expect to write persuasively to explain and solve problems your company faces. Finally, you may even have to write persuasively against the merits of a change proposed by a manager or another department in your company's organization.

Figure 1.10 is a persuasive e-mail from an employee to a manager reporting a payroll mistake and convincing the reader to correct it. The e-mail also contains many of the other characteristics of job-related writing we have discussed. Note how the writer provides factual, not subjective, information, attaches a time sheet (a type of visual), gives accurate details, and identifies his own and his immediate supervisor's responsibilities.

A persuasive e-mail from an employee to a business manager. **Figure 1.10**

| Reply | Save | Forward | Print | Delete |

To:	<lgriffin@starinstruments.com> (Lee Griffin)
From:	<rburke@starinstruments.com> (R. Burke)
Date:	19 October 2009 10:05 a.m. EST
Subject:	Shortage in October Paycheck

My paycheck for the two-week period ending October 15 was $60.00 short. For this period I should have been paid $525.00. Instead, my check was for only $480.00. I think I know why there may have been a discrepancy. The $45.00 additional pay was the result of my having put in five hours of overtime on October 8 and October 12 (2½ hours each day @ $9.00 per hour). This overtime was not reflected on my current pay stub.

Clearly explains and documents the problem

I have double-checked with my supervisor, Gloria Arrelo, who assured me that she recorded my overtime on the timesheets she sent to your office. She has kindly given me a copy that I have scanned and attached to this e-mail to verify my hours.

Offers further evidence

Thank you for correcting your records and for crediting me with the additional $45.00 for my overtime.

Closes politely with request

Ethical Writing in the Workplace

One of your most important job obligations is to ensure that your writing and behavior are ethical. Being ethical means doing what is right and fair and being honest and just with your employer, co-workers, and customers. Your reputation and character plus your employer's corporate image will depend on your following an ethical course of action. Note how in Figure 1.11 the Southern Company and its employees are proud of their ethical commitments to the environment, the community, and the country. The Southern Company projects an image of itself as being concerned about and dedicated to following the highest corporate ethical goals.

Many of the most significant phrases in the world of business reflect an ethical commitment to honesty and fairness: *accountability, public trust, equal opportunity employer, core values, global citizenship, good faith effort, truth in lending, fair play, honest advertising, full disclosure, high professional standards, industry standards, fair trade, community involvement,* and *social responsibility.*

Figure 1.11   A company's commitment to ethical responsibility.

Our Environmental Responsibility

Southern Company is not only a leader in the energy market, but also a leader in protecting the environment. We believe our environmental initiatives and our strong compliance record will give us a competitive advantage.

Emphasizes corporate commitment to ethical conduct

The Southern Company's environmental policy spells out each company's commitment to protecting the environment. The first and foremost goal is to meet or exceed all regulatory requirements for domestic and international operations. To do that, we're using a combination of the best technologies and voluntary pollution-prevention programs. We also set aggressive environmental goals and make sure employees are aware of their individual environmental responsibilities. We are good citizens wherever we serve.

Links good business practices to good ethical behavior

As an affiliate of Southern Company, Mississippi Power's environmental issues are business issues. In addition to regulatory obligations, our employees carry out a most active grassroots environmental program. It's this employee involvement and strong environmental commitment that gives our commitment life and promises future generations a healthy environment.

Praises employees for their contributions to both the community and the company

For example, one employee's concern that motor oil is properly discarded led to the founding of a countrywide annual household hazardous waste collection program. Thousands of tons of waste have been collected, including jars of DDT, mercury, paint, batteries, pesticides, and other poisons.

Scores of employees participate in island, beach, and river cleanups throughout Mississippi Power's 23-county service area. More than 30 employees compiled "The Wolf River Environmental Monitoring Program."

Assures readers that corporate ethical behavior extends to the entire community

This report is the first-ever historical, biological assessment completed on the Wolf River by scientists and engineers. Employees volunteered countless hours to compile the statistical data. Today, Mississippi Power employees continue to support the Wolf River Project by producing photographs and slides as an educational and community awareness project.

Our commitment to the environment goes beyond our business. By sponsoring a variety of programs, we're helping to teach the public, students, and teachers about environmental responsibility.

Reprinted by permission of Mississippi Power Company.

Unethical business dealings, on the other hand, are stigmatized in *cover-ups, shady deals, spin doctors, foul play, bid rigging, employee raiding, misrepresentations, kickbacks, hostile takeovers, planned obsolescence, price gouging, bias,* and *unfair advantage.* Those are the activities that make customers angry and that Better Business Bureaus investigate.

Ethical Requirements on the Job

In the workplace, you will be expected to meet the highest ethical standards by fulfilling the following requirements:

- Supplying honest and up-to-date information about yourself in your résumé and job applications. The résumé (see pages 251–273) is one key place where you must make ethical decisions about candor and honesty.
- Respecting co-workers, customers, and suppliers in conduct that avoids bullying, discrimination, or any other unfair and unprofessional action.
- Avoiding language that excludes others on the basis of gender, race, national origin, religion, age, physical ability, or sexual orientation (see pages 181–182).
- Maintaining accurate and current records at work. Remember: "If it isn't written, it didn't happen."
- Complying with all local, state, and federal regulations, especially those ensuring a safe, healthy work environment, products, and/or services, for example, by the Occupational Safety and Health Administration.
- Adhering to your profession's code or standard of ethics, internal audits, licenses, and certificate requirements.
- Following your company's policies and procedures.
- Honoring guarantees and warranties and meeting customer needs impartially.
- Cooperating fairly and on a timely basis with your collaborative team.
- Treating international colleagues with respect, as you would any peers.
- Respecting all copyright obligations and privileges.

Following these guidelines is not only an ethical requirement; it could also be a legal one. For example, doing personal (or outside consulting) work on company time is unethical and illegal. It would also be neglectful and unethical to allow an unsafe product to stay on the market just to spare your company the expense and embarrassment of a product recall.

Computer Ethics

Computer ethics are essential in the e-commerce world of work. A good rule to follow is never to do anything online that you wouldn't do offline. Never use a company computer for any activity not directly related to your job. It would be grossly unethical to erase a computer program intentionally, violate a software licensing agreement, or misrepresent (by fabrication or exaggeration) the scope of a database. Follow the Ten Commandments of Computer Ethics prepared by the Computer Ethics Institute listed in Figure 1.12.

International Readers and Ethics

Writing for the world of multinational corporations places additional ethical demands on you as a writer. You have to make sure that you respect the ethics of all of the countries where your firm does business. Some behaviors regarded as normal or routine in the United States might be seen as highly unethical elsewhere, or vice

Figure 1.12   The Ten Commandments of Computer Ethics

1. Thou shalt not use a computer to harm other people.

2. Thou shalt not interfere with other people's computer work.

3. Thou shalt not snoop around in other people's computer files.

4. Thou shalt not use a computer to steal.

5. Thou shalt not use a computer to bear false witness.

6. Thou shalt not copy or use proprietary software for which you have not paid.

7. Thou shalt not use other people's computer resources without authorization or proper compensation.

8. Thou shalt not appropriate other people's intellectual output.

9. Thou shalt think about the social consequences of the program you are writing or the system you are designing.

10. Thou shalt always use a computer in ways that ensure consideration and respect for your fellow humans.

Source: Computer Ethics Institute, London.

versa. In many countries, accepting a gift to initiate or conclude a business agreement is considered not only proper but also honorable. This is not the case in the United States, where a "bribe" is seen as bad business or even illegal. Moreover, you should be on your ethical guard not to take advantage of a host country, such as allowing or encouraging poor environmental control because regulatory and inspection procedures are not as strict as those of the United States, or using pesticides or conducting experiments outlawed in the United States.

Employers Insist On and Monitor Ethical Behavior

Ethical behavior is crucial to your success in the workplace. Your employer will insist that you are honest, show integrity, and exhibit loyalty in your professional relationships with clients, co-workers, supervisors, and vendors. These ethical values are stressed in orientation/training sessions and through every level of management in the business world. You will be expected to know and comply with your company policies and procedures as outlined in the employee or agency handbook and you will also have to follow the professional codes, regulations, and methods that affect your job. Operating with ethical behavior and sound judgment is necessary so that companies can:

- create strategic advantages for healthy competition
- champion change and innovation
- create customer loyalty
- promote global perspective of their "brand"

On the job, employers can legally monitor their employees' work—electronically, through cameras, or by personal visits. Some of these visits are not announced (such as the secret shopper who reports on the customer service he/she received). How many times have you made a call to an organization and heard, "This call may be monitored for quality assurance"? According to a 2008 survey conducted by the American Management Association, monitoring employees has risen 45 percent in the past few years and extends to voice mail, e-mail, IM, and Internet use.

Employers monitor the behavior of their employees for several reasons:

- to determine if a worker is doing his/her job correctly
- to find out how well a worker is doing a job
- to identify wrongdoing on the employee's part
- to improve service, production, communication, transportation
- to ensure legal compliance with federal and state codes
- to limit liability
- to heighten security measures

Monitoring gives management solid facts about employee training, performance reviews, and promotions. But working with integrity means doing the right thing—even when no one is watching.

Some Guidelines to Help You Reach Ethical Decisions

The workplace presents conflicts over who is right and who is wrong, what is best for the company and what is not, and whether a service or product should be changed and why. You will be asked to make a decision and justify it. Here are some guidelines to help you comply with the ethical requirements of your job.

1. **Follow your conscience and "to thine own self be true."** Do not authorize something that you believe is wrong, dangerous, unfair, contradictory, or incomplete. But don't be hasty. Leave plenty of room for diplomacy and for careful questioning. Recall the story of Chicken Little, who always cried that the sky was falling. Don't blow a small matter out of proportion.

2. **Be suspicious of convenient (and false) appeals that go against your beliefs.** Watch out for these red flags that anyone places in the way of your conscience: "No one will ever know." "It's OK to cut corners every once in a while." "We got away with it last time." "Don't rock the boat." "No one's looking." "As long as the company makes money, who cares?" These excuses are traps you must avoid.

3. **Meet your obligations to your employer, your co-workers, your customers, and the global community.** Examine their reasons for reading your communication and consider how and in what contexts they will use it. Keeping information from a co-worker who needs it, omitting a fact, justifying unnecessary expenses, concealing something risky about a product from an international customer that you otherwise would disclose to a U.S. customer—all of these are unethical acts.

4. **Take responsibility for your actions.** Saying "I do not know" when you do know can constitute a serious ethical violation. Keep your records up-to-date and accurate, sign and date your work, and never backdate a document to delete

information or to fix an error that you committed. Always do what is expected in terms of documentation and notification. Failing to test a set of instructions thoroughly, for example, might endanger readers around the globe.

5. Honor confidentiality at work. Never share sensitive/confidential information with individuals who are *not* entitled to see or hear it. One of a company's most valuable assets is its intellectual property. Companies guard trade secrets and oppose unfair competition to protect their rights and to ensure that they can patent their work successfully. You violate corporate trust by telling others about your company's research, marketing strategies, sales records, personnel decisions, or customer/client interactions. It is equally unethical to divulge personal information that a co-worker or supervisor has asked you to keep confidential. In fact, your employer may insist that you sign a confidentiality agreement when you are hired. Your employer rightfully demands that you act honorably from your first day at work until your last. Respecting a former employer's confidentially extends to even after you leave for another job.

6. Document your work completely, carefully, and honestly. Rely on hard evidence in your written work and document where you found that information (see Chapter 8, pages 352–388). Always give credit to your sources—those in print, on the Web, or from discussions with individuals who contributed to your work. Be sure, too, that your documentation complies with appropriate city, state, federal, and international codes and regulations. FDA and OSHA inspectors, for instance, look closely at a company's documentation to ensure public safety. Not documenting sources is unethical, and so is inventing information or falsifying a document, whether a financial statement or a report. Such actions constitute fraud.

7. Keep others in the loop. Confer regularly with your collaborative writing team (see Chapter 3, pages 76–107) and any other co-workers affected by your job. Report to your boss as often as you are instructed to give progress reports, to alert him or her about problems, and to help you coordinate your duties with co-workers. If you experience a problem at work, don't wait to tell your supervisor and/or co-workers until it gets worse. Prompt and honest notifications are essential to the safety, security, progress, and ultimate success of a company. Also, never keep a customer/vendor waiting; call in advance if you are going to be delayed.

8. Treat company property respectfully. Use company supplies, networks/computers, equipment, and vehicles responsibly and only for work-related business. Taking supplies home, making personal long-distance calls on a company cell phone, charging non–work related expenses (meals, clothes, travel) on a company credit card, surfing the Internet when you are at work—these are just a few instances of unethical behavior. Being wasteful (of paper, supplies, ink) also disrespects your company's resources.

9. Think green in the workplace. Closely related to #8 above is being respectful of the environment—whether at the office, at a work site, or in the community your company serves. Many companies have adopted a green philosophy, as Figure 1.11 demonstrates. At your office, conserve energy by turning off all computers, copiers, and other office machines when you leave work; replace old light bulbs with longer-life ones; recycle paper; copy and print on both sides of paper; view documents on your computer screen instead of printing them; adjust thermostats when you are gone for the day or weekend, and car pool or van pool. You can also reduce toxic chemicals in the atmosphere by using soy-based ink and inspecting vehicles regularly and reporting any pollution.

10. Weigh all sides before you commit to a conclusion. You may think a particular course of action is right at the time, but don't overlook the possibility that your decision may create a bigger problem in the future. For example, you hear that a co-worker is involved in wrongdoing; you report it to your boss, and a reprimand is placed in that worker's file. Later you learn that what was reported to you was malicious gossip or only a small part of a much larger but very ethical picture. Give people the benefit of the doubt until you have sufficient facts to the contrary. Giving incomplete information on an incident report may temporarily protect you but may falsely incriminate someone else or unfairly increase your company's liability insurance rates.

Ethical Dilemmas

Sometimes in the workplace you will face situations where there is no clear-cut right or wrong choice. Here are a few scenarios, similar to ones in which you may find yourself, that are gray areas, ethically speaking, along with some possible solutions.

- You work with someone who is the office bully, who often intimidates co-workers, including you, by talking down to them, interrupting them when they are speaking, or even insulting them for their suggestions. At times, this bully has even sent sarcastic e-mails and IMs. You are upset that this behavior has not yet been reported to management. Yet you are concerned that if the bully finds out that you have reported the situation to management the entire office may suffer. How should you handle the problem?

  You cannot allow such rude, insulting behavior to go unreported. But first you need to provide documentation about where, when, and how often the bullying has occurred. You may want to speak directly to the bully, but if you feel uncomfortable doing this, go directly to your boss and report how the bully's actions have negatively affected the workplace, and to ask for assistance. You may also get help from your company's employee assistance program or from someone in human resources. In accordance with state and federal laws, companies must provide a safe work enviornment, free from intimidation, harrassment, or threats of dismissal for reporting bullying.

- You work very closely with an individual who takes frequent extended lunch breaks, often comes in late and leaves early, and even misses deadlines. As a result, you are put in an awkward position. Sometimes you cover for him when he is not at the office to answer questions, and often you take on additional work he should be doing to keep your department running smoothly and efficiently. Your department is under minimal supervision from an off-site manager, so there is no boss looking over your colleague's shoulder. You like your co-worker and do not want him to be reprimanded or, worse yet, fired, but he is taking advantage of your friendship and unfairly expecting you to cover for him. What should you do?

The best route is to take your co-worker aside and speak with him before informing management. Let him know you value working with him but firmly explain that you no longer will cover for him or take on his workload. If he does not agree with you, let him know that you will be forced to discuss the problem with your manager. If the problem persists, and you go to your boss, bring documentation—dates, duties not performed, etc., with you.

- You see an opening for a job in your area but the employer wants someone with a minimum of two years of field experience. You have just completed an internship and had one summer's (12 weeks) experience, which together total almost 7 months. Should you apply for the job, describing yourself as "experienced"?

Yes, but honestly state the type and the extent of your field experience and the conditions under which you obtained it.

- You work for a company that usually assigns commissions to the salesperson for whom a customer asks. One afternoon a customer asks for a salesperson who happens to have the day off. You assist the customer all afternoon and even arrange to have an item shipped overnight so that the customer can have it in the morning. When you ring up the sale, should you list your employee number for the commission or the off-duty employee's?

You probably should defer crediting the sale to either number until you speak to the absent employee and suggest a compromise—splitting the commission.

- A piece of computer equipment, scheduled for delivery to your customer the next day, arrives with a damaged part. You decide to replace it at your store before the customer receives it. Should you inform the customer?

Yes, but assure the customer that the equipment is still under the same warranty and that the replacement part is new and also under the same warranty. If the customer protests, agree to let him or her use the computer until a new unit arrives.

As these brief scenarios suggest, sometimes you have to make concessions and compromises to be ethical in the world of work.

Writing Ethically on the Job

Your writing as well as your behavior must be ethical. Words, like actions, have implications and consequences. If you slant your words to conceal the truth or gain an unfair advantage, you are not being ethical. False advertising is false writing. Bias and omission of facts are wrong. Strive to be fair, reliable, and accurate in reporting events, statistics, and trends.

Unethical writing is usually guilty of one or more of the following faults, which can conveniently be listed as the three *M*'s: misquotation, misrepresentation, and manipulation. Here are eight examples:

1. Plagiarism is stealing someone else's words and claiming them as your own without documenting the source. Do not think that by changing a few words of someone else's writing here and there you are not plagiarizing. Give proper credit to your source, whether in print, in person (through an interview), or online. The penalties for plagiarism are severe—a reprimand or even the loss of your job. See pages 353–354 for further advice on how to avoid plagiarism.

2. Selective misquoting deliberately omits damaging or unflattering comments to paint a better (but untruthful) picture of you or your company. By picking and choosing only a few words from a quotation, you unethically misrepresent what the speaker or writer originally intended.

> **Full Quotation:** I've enjoyed at times our firm's association with Technology, Inc., although I was troubled by the uneven quality of their service. At times, it was excellent while at others it was far less so.
>
> **Selective Misquotation:** I've enjoyed . . . our firm's association with Technology, Inc. The quality of their service was . . . excellent.

The dots, called *ellipses*, unethically suggest that only extraneous or unimportant details were omitted.

3. Arbitrary embellishment of numbers unethically misrepresents, by increasing or decreasing percentages or other numbers, statistical or other information. It is unethical to stretch the differences between competing plans or proposals to gain an unfair advantage or to express accurate figures in an inaccurate way.

> **Embellishment:** An overwhelming majority of residents voted for the new plan.
> **Ethical:** The new plan was passed by a vote of 53 to 49.

> **Embellishment:** Our competitor's sales volume increased by only 10 percent in the preceding year while ours doubled.
> **Ethical:** Our competitor controls 90 percent of the market, yet we increased our share of that market from 5 percent to 10 percent last year.

4. Manipulation of information or context, closely related to #3, is the misrepresentation of events, usually to "put a good face" on a bad situation. The writer here unethically uses slanted language and intentionally misleading euphemisms to misinterpret events for readers.

> **Manipulation:** Looking ahead to 2010, the United Funds Group is exceptionally optimistic about its long-term prospects in an expanding global market. We are happy to report steady to moderate activity in an expanding sales environment last year. The United Funds Group seeks to build on sustaining investment opportunities beneficial to all subscribers.
> **Ethical:** Looking ahead to 2010, the United Funds Group is optimistic about its long-term prospects in an expanding global market. Though the market suffered from inflation this year, the United Funds Group hopes to recoup its losses in the year ahead.

The writer minimizes the negative effects of inflation by calling it "an expanding sales environment."

5. Using fictitious benefits to promote a product or service seemingly promises customers advantages but delivers none.

> False Benefit: Our bottled water is naturally hydrogenated from clear underground springs.
> Truth: All water is hydrogenated because it contains hydrogen.

6. Unfairly characterizing (by exaggerating or minimizing) hiring or firing conditions is unethical.

> Unethical: One of the benefits of working for Spelco is the double pay you earn for overtime.
> Truth: Overtime is assigned on the basis of seniority.
>
> Unethical: Our corporate restructuring will create a more efficient and streamlined company, benefiting management and workers alike.
> Truth: Downsizing has led to 150 layoffs this quarter.

Companies faced with laying off employees want to protect their corporate image and maintain their stockholders' good faith, so they often "put the best face" on such an action.

7. Manipulating international readers by adopting a condescending view of their culture and economy is unethical.

> Unethical: Since our product has appealed to U.S. customers for the last sixteen months, there's no doubt that it will be popular in your country as well.
> Fair: Please let us know if any changes in product design or construction may be necessary for customers in your country.

8. Misrepresenting through distortion or slanted visuals is one of the most common types of unethical writing. Making a visual appear bigger, smaller, or more or less favorable is all too easy with graphics software packages. Making warning or caution statements the same size and type font as ingredients or directions or enlarging advertising hype (Double Your Money Back) is unethical if major points are then reduced to small print. See pages 472–478 of Chapter 10 for guidelines on how to prepare ethical visuals.

Ethical writing is clear, accurate, fair, and honest. These are among the most important goals of any workplace communication. Because ethics is such an important topic in writing for the business world, it will be emphasized throughout this book.

Successful Employees Are Successful Writers

As this chapter has stressed, being a successful employee means being a successful writer at work. The following guidelines, which summarize the key points of this chapter, will help you to be both.

1. Know your job—assignments, roles, responsibilities, goals, what you need to write and what you **shouldn't**.
2. Be prepared to give and to receive feedback.
3. Work toward and meet all deadlines.
4. Analyze your audience's needs, and what they will expect to find in your written work.

5. Respect the cultural diversity and contributions of your customers, vendors, and co-workers.
6. Write clearly, concisely, and appropriately for readers in the global marketplace. Adapt your message for your audience's background.
7. Document, document, document. Submit everything with clear-cut evidence based on factual details and persuasive, logical interpretations.
8. Include clear, relevant visuals to help readers understand your message.
9. Follow your company policy and promote your company's image.
10. Be ethical in what you say, write, and do.

 ## Revision Checklist

At the end of each chapter is a checklist you should review before you submit the final copy of your work, either to your instructor or to your boss. The checklists specify the types of research, planning, drafting, editing, and revising you should do to ensure the success of your work. Regard each checklist as a summary of the main ideas in the chapter as well as a handy guide to quality control. You may find it helpful to check each box as you verify that you have performed the necessary revision/review. Effective writers are also careful editors.

☐ Showed respect for and appropriately shaped my message for a global marketplace.
☐ Identified my audience—their background, knowledge of English, reason for reading my work, and likely response to my work and me.
☐ Tailored my message to my audience's needs and background, giving them neither too little nor too much information.
☐ Pushed to the main point right away; did not waste my reader's time.
☐ Selected the most appropriate language, technical level, tone, and level of formality.
☐ Did not waste my audience's time with unsupported generalizations or opinions; instead gave them accurate measurements, facts, and carefully researched material.
☐ Used appropriate visuals to make my work easier for my audience to follow.
☐ Used persuasive reasons and data to convince my reader to accept my plan or work.
☐ Ensured that my writing and visuals are ethical—accurate, fair, honest, a true reflection of the situation or condition I am explaining or describing, for U.S. as well as global audiences.
☐ Followed the Ten Commandments of Computer Ethics.
☐ Adhered to the ethical codes of my profession as well as those policies and regulations set down by my employer.
☐ Gave full and complete credit to any sources I used, including resource people.
☐ Avoided plagiarism and unfair or dishonest use of copyrighted materials, both written and visual, including all electronic media.

Exercises

1. What is your chosen career? Make a list of the types of writing you think you will do, or have already been assigned, on the job.

2. Make a list of the kinds of writing you have done in a history or English class or for a laboratory or shop/studio course.

3. Compare your lists for Exercises 1 and 2. How do the two types of writing differ?

4. Write a memo (see pp. 122–130 for format) addressed to a prospective supervisor to introduce yourself. Your memo should have four headings: **education**—including goals and accomplishments; **job information**—where you have worked and your responsibilities; **community service**—volunteer work, church work, youth groups; and **writing experience**—your strengths and what you would like to see improved.

5. Write a memo in response to Rowe Pinkerton's e-mail in Figure 1.1. Explain how you will use the skills you learn in the tuition-reimbursed writing course on your job.

6. Bring to class a set of printed instructions, a memo, a sales letter, or a brochure. Comment on how well the printed material answers the following questions.
   a. Who is the audience?
   b. Why was the material written?
   c. What is the message?
   d. Are the style and tone appropriate for the audience, the purpose, and the message? Why?
   e. Discuss the use of any visuals and color in the document. How does color (or the lack of it) affect an audience's response to the message?

7. Cut out a newspaper ad that contains a drawing or photograph. Bring it to class together with a paragraph of your own (75–100 words) describing how the message of the ad is directed to a particular audience and commenting on why the illustration was selected for that audience.

8. Pick one of the following topics and write two descriptions of it. In the first description, use technical vocabulary. In the second, use language suitable for the general public.
   a. spark plug
   b. blood pressure cuff
   c. carburetor
   d. computer chip
   e. smartphone
   f. legal contract
   g. electric sander
   h. cyberspace
   i. muscle
   j. protein
   k. BlackBerry
   l. e-mail
   m. bread
   n. money
   o. iPod
   p. soap
   q. blogging
   r. computer virus
   s. AIDS
   t. thermostat
   u. trees
   v. food processor
   w. earthquake
   x. recycling

9. Redo Exercise 8 as a collaborative writing project.

10. Select one article from a newspaper and one article either from a professional journal in your major field or from one of the following journals: *Advertising Age, American Journal of Nursing, Business Marketing, Business Week, Computer, Computer Design, Construction Equipment, Criminal Justice Review, E-Commerce, Food Service Marketing, Journal of Forestry, Journal of Soil and Water Conservation, National Safety News, Nutrition Action, Office Machines, Park Maintenance, Scientific American.* State how the two articles you selected differ in terms of audience, purpose, message, style, and tone.

11. Assume that you work for Appliance Rentals, Inc., a company that rents TVs, microwave ovens, stereo components, and the like. Write a persuasive letter to the members of a campus organization or civic club urging them to rent an appropriate appliance or appliances. Include details in your letter that might have special relevance to members of this specific organization.

12. How do the visuals and the text of the Sodexho advertisement on page 40 stress to current (and potential) employees, customers, and stockholders that the company is committed to human diversity in the workplace? Also explain how the ad illustrates the functions of on-the-job writing defined on pages 20–26.

13. Read the article, "Microwaves," on pages 41–42 and identify its audience (technical or general), purpose, message, style, and tone.

14. Write a letter to an Internet service provider that has mistakenly billed you for caller ID equipment that you never ordered, received, or needed.

15. The following statements contain embellishments, selected misquotations, false benefits, and other types of unethical tactics. Revise each statement to eliminate the unethical aspects.
    a. Hurricane damage done to water filtration plant #3 was minimal. While we had to shut down temporarily, service resumed to meet residents' needs.
    b. All customers qualify for the maximum discount available.
    c. The service contract . . . on the whole . . . applied to upgrades.
    d. We followed the protocols precisely with test results yielding further opportunities for experimentation.
    e. All our costs were within fair-use guidelines.
    f. Customers' complaints have been held to a minimum.
    g. All the lots we are selling offer relatively easy access to the lake.

16. You work for a large international firm and a co-worker tells you that he has no plans to return to his job after he takes his annual two-week vacation. You know that your department cannot meet its deadline short-handed and that your company will need at least two or three weeks to recruit and hire a qualified replacement. You also know that it is your company's policy not to give paid vacations to employees who do not agree to work for at least three months following a vacation. What should you do? What points would you make in a confidential, ethical memo to your boss (see pp. 26–36)? What points would you make to your co-worker?

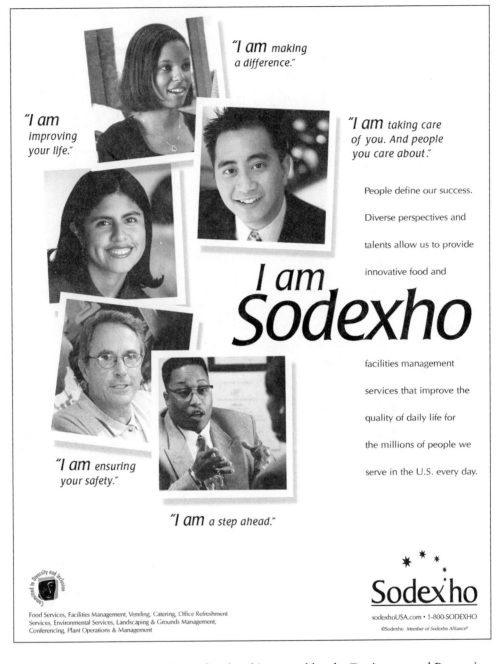
17. Your company is regulated and inspected by the Environmental Protection Agency. In ninety days, the EPA will relax a particular regulation about dumping industrial waste. Your company's management is considering cutting costs by relaxing the standard now, before the new, easier regulation is in place. You know that the EPA inspector probably will not return before the ninety-day period elapses. What do you recommend to management?

18. Write a memo to your boss about one of the following unethical activities you
have either witnessed or been a victim of in your workplace. Your memo must
be carefully documented, fair, and persuasive.
   a. bullying
   b. surfing pornography websites
   c. falsifying compensatory or travel time
   d. telling sexist, off-color jokes
   e. concealing the use of company funds for personal gifts (birthdays,
      anniversaries) for fellow employees
   f. misdating or backdating company records
   g. sharing privileged information with individuals outside your department or
      company

Microwaves

Much of the world around us is in motion. A wave-like motion. Some waves are big
like tidal waves and some are small like the almost unseen footprints of a waterspi-
der on a quiet pond. Other waves can't be seen at all, such as an idling truck sending
out vibrations our bodies can feel. Among these are electromagnetic waves. They
range from very low frequency sound waves to very high frequency X-rays, gamma
rays, and even cosmic rays.

Energy behaves differently as its frequency changes. The start of audible
sound—somewhere around 20 cycles per second—covers a segment at the low end
of the electromagnetic spectrum. Household electricity operates at 60 hertz (cycles
per second). At a somewhat higher frequency we have radio, ranging from short-
wave and marine beacons, through the familiar AM broadcast band that lies between 500 and 1600 kilohertz, then to citizen's band, FM, television, and up to the
higher frequency police and aviation bands.

Even higher up the scale lies visible light with its array of colors best seen when
light is scattered by raindrops to create a rainbow.

Lying between radio waves and visible light is the microwave region—from
roughly one gigahertz (a billion cycles per second) up to 3000 gigahertz. In this region the electromagnetic energy behaves in special ways.

Microwaves travel in straight lines, so they can be aimed in a given direction.
They can be *reflected* by dense objects so that they send back echoes—this is the
basis for radar. They can be *absorbed*, with their energy being converted into heat—
the principle behind microwave ovens. Or they can pass *through* some substances
that are transparent to the energy—this enables food to be cooked on a paper plate
in a microwave oven.

Microwaves for Radar

World War II provided the impetus to harness microwave energy as a means of
detecting enemy planes. Early radars were mounted on the Cliffs of Dover to
bounce their microwave signals off Nazi bombers that threatened England. The
word *radar* itself is an acronym for *RA*dio *D*etection *A*nd *R*anging.

Radars grew more sophisticated. Special-purpose systems were developed to
detect airplanes, to scan the horizon for enemy ships, to paint finely detailed electronic pictures of harbors to guide ships, and to measure the speeds of targets.
These were installed on land and aboard warships. Radar—especially shipboard

radar—was surely one of the most significant technological achievements to tip the scales toward an Allied victory in World War II.

Today, few mariners can recall what it was like before radar. It is such an important aid that it was embraced universally as soon as hostilities ended. Now, virtually every commercial vessel in the world has one, and most larger vessels have two radars: one for use on the open sea and one, operating at a higher frequency, to "paint" a more finely detailed picture, for use near shore.

Microwaves are also beamed across the skies to fix the positions of aircraft in flight, obviously an essential aid to controlling the movement of aircraft from city to city across the nation. These radars have also been linked to computers to tell air traffic controllers the altitude of planes in the area and to label them on their screens.

A new kind of radar, phased array, is now being used to search the skies thousands of miles out over the Atlantic and Pacific oceans. Although these advanced radars use microwave energy just as ordinary radars do, they do not depend upon a rotating antenna. Instead, a fixed antenna array, comprising thousands of elements like those of a fly's eye, looks everywhere. It has been said that these radars roll their eyes instead of turning their heads.

High-Speed Cooking

During World War II Raytheon had been selected to work with M.I.T. and British scientists to accelerate the production of magnetrons, the electron tubes that generate microwave energy, in order to speed up the production of radars. While testing some new, higher-powered tubes in a laboratory at Raytheon's Waltham, Massachusetts, plant, Percy L. Spencer and several of his staff engineers observed an interesting phenomenon. If you placed your hand in a beam of microwave energy, your hand would grow pleasantly warm. It was not like putting your hand in a heated oven that might sear the skin. The warmth was deep-heating and uniform.

Spencer and his engineers sent out for some popcorn and some food, then piped the energy into a metal wastebasket. The microwave oven was born.

From these discoveries, some 35 years ago, a new industry was born. In millions of homes around the world, meals are prepared in minutes using microwave ovens. In many processing industries, microwaves are being used to perform difficult heating or drying jobs. Even printing presses use microwaves to speed the drying of ink on paper.

In hospitals, doctors' offices, and athletic training rooms, that deep heat that Percy Spencer noticed is now used in diathermy equipment to ease the discomfort of muscle aches and pains.

Telephones Without Cable

The third characteristic of microwaves—that they pass undistorted through the air—makes them good messengers to carry telephone conversations as well as live television signals—without telephone poles or cables—across town or across the country. The microwave signals are beamed via satellite or by dish reflectors mounted atop buildings and mountaintop towers.

Microwaves take their name from the Greek *mikro* meaning very small. While the waves themselves may be very small, they play an important role in our world today: in defense; in communications; in air, sea, and highway safety; in industrial processing; and in cooking. At Raytheon the applications expand every day.

Reprinted by permission of *Raytheon Magazine.*

CHAPTER 2

The Writing Process at Work

In Chapter 1 you learned about the different functions of writing for the world of work and also explored some basic concepts all writers must master. To be a successful writer, you need to

Visit www .cengage.com/ english/kolin for this chapter's online exercises, ACE quizzes, and Web links.

- identify your audience's needs
- determine your purpose in writing to that audience
- make sure your message meets your audience's needs
- use the most appropriate style and tone for your message
- format your work so that it clearly reflects your message to your audience

Just as significant to your success is knowing how effective writers actually create their work for their audiences. This chapter gives you some practical information about the strategies and techniques careful writers use when they work. These procedures are a vital part of what is known as the *writing process.* This process involves such matters as how writers gather information, how they transform their ideas into written form, and how they organize and revise what they have written to make it suitable for their audiences.

What Writing Is and Is Not

As you begin your study of writing for the world of work, it might be helpful to identify some notions about what writing is and what it is not.

What Writing Is

- **Writing is a dynamic process; it is not static.** It enables you to discover and evaluate your thoughts as you draft and revise.
- **A piece of writing changes as your thoughts and information change** and as your view of the material changes.
- **Writing takes time.** Some people think that revising and polishing are too time consuming. But poor writing actually takes more time and costs more money in the end. It can lead to misunderstandings, lost sales, product recalls, and even damage to your reputation and that of your company.
- Writing means making a number of judgment calls.
- Writing grows sometimes in bits and pieces and sometimes in great spurts. It needs many revisions; an early draft is never a final copy.

What Writing Is Not

- **Writing is not a mysterious process, known only to a few.** Even if you have not done much writing before, you can learn to do it effectively.
- **Writing is not simply following a magical formula.** Successful writing requires hard work and thoughtful effort, not simply following a formula, as if you were painting by numbers. Writing does not proceed in some predictable way, in which introductions are always written first and conclusions last.
- **Writing is not completed in a first attempt.** Just because you put something down on paper or on a computer screen does not mean it is permanent and unchangeable. Writing means *re*writing, *re*vising, and *re*thinking. The better a piece of writing is, the more the writer has reworked it.

■ The Writing Process

The writing process we have just discussed is something fluid, not static. Think of it as a back and forth process rather than following a formula—do this, then do that. To move from a blank sheet of paper or computer screen to a successful piece of writing, you need to follow a process. The parts of that process include researching, planning, drafting, revising, and editing. See how each part of the process is illustrated in Figures 2.2 through 2.5. As these figures show, office manager Melissa Hill asked employee Marcus Weekley to write a short report recommending ways to improve office efficiency. In doing so, Weekley followed the steps in the writing process from research and planning to drafting, revising, and editing.

Researching

Before you start to compose any e-mail, memo, letter, report, or proposal, you'll need to do research. Research is crucial so that you obtain the right information for your audience. Information must be factually correct and intellectually significant. The world of work is based on conveying information—the logical presentation and sensible interpretation of facts. Chapter 8 will introduce you to the variety of research strategies and tools you can expect to use in the world of work.

Don't ever think you are wasting time by not starting to write your e-mail, memo, letter, or short report immediately. Actually, you will waste time and risk doing a poor job if you do not find out as much as possible about your topic (and your audience's interest in it). Find out about your readers' needs and how to meet them.

Then you can determine the kind of research you must do to gather and interpret the information your audience needs. Depending on the length of your written work and on your audience's needs, your research may include:

- interviewing people inside and outside your company
- consulting notes from conferences and/or meetings
- collaborating in person, by e-mail, or by instant messaging (IM)
- doing Internet searches
- locating and evaluating websites

- searching abstracts, indexes, and other references on the Web or in print
- reading current periodicals, trade journals, reports, and other documents
- evaluating reports, products, and services
- getting briefings from sales or technical staff
- conferring with co-workers, customers, or vendors
- participating in a focus group
- surveying customers' views

Keep in mind that research is not confined to the beginning of the writing process; it is an ongoing process.

As Figures 2.2, 2.3, and 2.4 demonstrate, Marcus Weekley knew he would have to do research to make his recommendations to manager Melissa Hill about how the office could run more efficiently. To locate information about different communication technologies that might improve office efficiency, he searched for and read relevant materials in print and online sources, spoke with co-workers as well as with individuals in other departments, and contacted office supply representatives. As he researched various options, he discovered that switching to a multi-function printer would save the office time and money. He determined that purchasing this piece of equipment would be the most sensible and economically feasible way to improve efficiency and so based his report on the research he found about this new technology. The results of his research are reflected throughout the drafts of his work.

Planning

At the planning stage in the writing process, your goal is to get something—anything—down on paper or on your screen. For most writers, getting started is the hardest part of the job. But you will feel more comfortable and confident once you begin to see your ideas before your eyes. It is always easier to clarify and criticize something you can see.

Getting started is also easier if you have researched your topic, because then you have something to say and to build on. Each part of the process relates to and supports the next. Careful research prepares you to begin writing.

Still, getting started is not easy. Take advantage of a number of widely used strategies to develop, organize, and tailor the right information for your audience. Use the following techniques, alone or in combination.

Visit www
.cengage.com/
english/kolin
for two online
exercises,
"Brainstorming
and Outlining
on Your
Computer" and
"Using Mind
Mapping
Software."

　1. **Clustering.** In the middle of a sheet of paper, write the word or phrase that best describes your topic, then start writing other words or phrases that come to mind. As you write, circle each word or phrase and connect it to the word from which it sprang. Note the clustered grouping in Figure 2.1 for a report encouraging a manager to switch to flextime, a system in which employees work on a flexible time schedule within certain limits. The resulting diagram gives the writer a rough sense of some of the major divisions of the topic and where they may belong in the report.

　2. **Brainstorming.** At the top of a sheet of paper or your computer screen, describe your topic in a word or phrase and then list any information you know or

found out about that topic—in any order and as quickly as you can. Brainstorming is like thinking aloud except that you are recording your thoughts.

- Don't stop to delete, rearrange, or rewrite anything, and don't dwell on any one item.
- Don't worry about spelling, punctuation, grammar, or whether you are using words and phrases instead of complete sentences.
- Keep the ideas flowing. The result may well be an odd assortment of details, comments, and opinions.
- After stepping away from the list for a few minutes (or hours) and returning with fresh eyes, expect to add and delete ideas or combine and rearrange others as you start to develop them in more detail.

Figure 2.2 shows Marcus Weekley's initial brainstormed list of how and why purchasing a multifunction printer would help his office. After he began to revise his list, he realized that some items were not relevant for his audience (6, 8, and 13) and that another was pertinent but needed to be adapted for his reader (5). He also recognized that some items were repetitious (1, 2, and 11). Further investigation revealed that his company could purchase the new equipment for far less than his initial high guess (17).

Figure 2.1   Clustering on the topic of flextime.

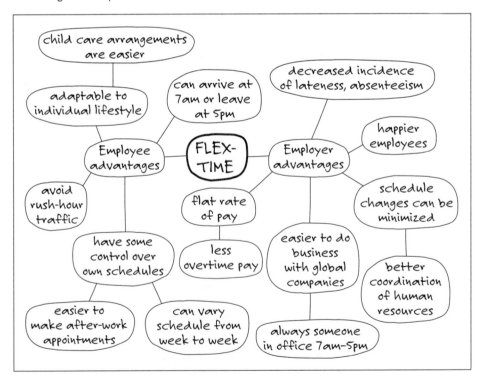

Marcus Weekley's initial, unrevised brainstormed list.   Figure 2.2

1. combines four separate pieces of equip—printer, copier, fax, and scanner

2. more comprehensive than our current configuration of four pieces of equip

3. would coordinate with office furniture

4. energy efficiency increased due to fewer machines being used

5. more scalable fonts

6. one machine interfaces with all others in same-case housing

7. scanner makes photographic-quality pictures

8. print capabilities are a real contribution to technology

9. increase communication abilities through fax machine

10. new scanner picture quality better than current scanner

11. only have to buy one machine as opposed to four

12. speed of fax allows quick response time

13. stock is doing better on Wall Street compared to other stocks

14. reducing our advertising costs through use of color printer

15. increase work area available

16. would help us do our work better

17. top of line models can be bought for $4,500

List is not organized but simply reflects writer's ideas about possible topics

1, 2, and 11 are repetitious

6, 8, and 13 are not relevant for audience or purpose

4, 9, 12, and 14 are of special interest to decision makers concerned about costs

Research will show price is too high

3. Outlining. For most writers, outlining may be the easiest and most comfortable way to begin or to continue planning their report or letter. Outlines can go through stages, so don't worry if your first attempt is brief and messy. It does not have to be formal (with roman and arabic numerals), complete, or pretty. It is intended for no one's eyes but yours. Use your preliminary outline as a quick way to sketch in some ideas, a convenient container into which you can put information. You might simply jot down a few major points and identify a few subpoints. Note that Marcus Weekley organized his revised brainstormed list into an outline (Figure 2.3).

Tech Note

Planning on Your Computer

You can brainstorm, outline, and create clusters on paper or you can use your computer to help you plan your document. Here are some tips to use these text-generating strategies on your computer:

- To use your word processor to brainstorm, simply key in your ideas as quickly as you can. Your word processor makes it easy for you to go back and add, delete, combine, and rearrange your ideas when you revisit the document later.
- You can make easy-to-read and easy-to-change outlines on your computer as well. A word processing program can help you create an outline with main points and subpoints that help you to organize your thoughts when you start to write a draft. This is a good way to get your ideas down fast and efficiently. These programs provide outline views allowing you to select the number of levels to display in an outline and to change levels in the hierarchy (for example, from a I to an A or from a to i). You can also add notes to yourself reminding you where you may have to fill in gaps or where you may need to transition from one idea to another. Keep in mind that there is no one right way to outline. You might simply write down large concepts as in the list in Figure 2.2 or use complete sentences for your major headings. But always leave plenty of space between major points so you can fill them in later with relevant information, as in Figure 2.3.
- A "mind-mapping" software program such as MindGenius, MindMapper, or NovaMind will help you to create clusters. This mind-mapping software automatically produces cluster diagrams as you keyboard information, giving you a graphic way to present groupings and connections, as in Figure 2.1. You can then convert your cluster diagrams into outlines by exporting them into your word processor.

Marcus Weekley's early outline after revising his brainstormed list.   Figure 2.3

I. Convenience/Capabilities

    A. Would reduce number of machines required to service

    B. Can be configured easily for our network system

    C. Easy to install and operate

    D. 35.6 Kbps fax machine would increase our communication

    E. 2,400 $\times$ 1,200 dpi copier means better copy quality

    F. 4,800 $\times$ 4,800 dpi scanner means higher quality pictures than current scanner

Outline form helps writer group ideas/topics and go to next step in revision

II. Time/Efficiency

    A. 50 ppm printer is nearly twice as fast as current printer

    B. 35.6 Kbps speed fax allows quick response time

    C. Greater graphics capability—90 scalable fonts

    D. Scanner compatible with our current PhotoEdit imaging/graphics software

    E. 100,000-page monthly duty cycle means less maintenance

Headings correspond to major sections of report

III. Money

    A. Costs less for multitasking printer than combined four machines

    B. Reduced monthly power bill by using one machine rather than four

    C. Included ScanText and WordPort software means not buying new software to network office computers

    D. Save on service costs

    E. Reduced advertising costs through printer 50–1,000% enlargement/reduction options, which allow for more in-house advertising

Outline reflects scope and details of writer's research

Drafting

If you have done your planning carefully, you will find it easier to start your first draft. When you draft, you convert the words and phrases from your outlines, brainstormed lists, or clustered groups into paragraphs. During drafting, as elsewhere in the writing process, you will see some overlap as you look back over your lists or outlines to shape your text.

Don't expect to wind up with a polished, complete version of your letter or report after working on only one draft. In most cases, you will have to work through many drafts, but each draft should be less rough and more acceptable than the preceding one.

Key Questions to Ask as You Draft

As you work on your drafts, ask yourself the following questions about your content and organization:

- Am I giving my readers too much or too little information?
- Does this point belong where I have it, or would it more logically follow or precede something else?
- Is this point necessary and relevant?
- Am I repeating or contradicting myself?
- Have I ended appropriately for my audience?

Tech Note

Drafting on Your Computer

Here are a few tips to make the process go more smoothly when you draft your document using your word processor:

- Get your thoughts down on the screen as quickly as possible, without stopping to worry about spelling, punctuation, or spacing. If you stop to correct these small errors, you may forget to write down larger points.
- Save your document regularly (every few minutes) so that you do not lose valuable work and time in the event of a power outage or computer malfunction. Back up your document at the end of each day.
- If you run into trouble completing a point or writing a transition, or if you know you will need to add documentation later, do not lose momentum by stopping. Instead, add notes to yourself (in parentheses, in italics, or in a different color) reminding you to "add transition," "back up this point," "clarify," "check with boss or co-worker," "add documentation."
- Use the Save As option on your computer to save each draft of your document. That way, you will be able to salvage versions of sentences or paragraphs from one draft and incorporate them into another. Give each draft a clear title, such as "Office Equipment, 3rd draft," so you can tell at a glance which version is which.

To answer those questions successfully, you may have to continue researching your topic and reexamining your audience's needs. But in the process new and even better ideas will come to you, and ideas that you once thought were essential may in time appear unworkable or unnecessary.

Guidelines for Successful Drafting

Following are some suggestions to help your drafting go more smoothly and efficiently:

- In an early draft, write the easiest part first. Some writers feel more comfortable drafting the body (or middle) of their work first.
- As you work on a later draft, write straight through. Do not worry about spelling, punctuation, or the way a word or sentence sounds. Save those concerns for later stages.
- Allow enough time between drafts so you can evaluate your work with fresh eyes and a clear mind.
- Get frequent outside opinions. Show, e-mail, or fax a draft to a co-worker or a supervisor for comment. A new pair of eyes will see things you missed. Collaboration is essential in the workplace, as you will see in Chapter 3.
- Start considering if visuals would enhance the quality of your work and, if so, decide on what types of visuals to use and where.

Figure 2.4 shows one of the several drafts that Marcus Weekley prepared. Because he wisely recognized that his outline was not final, he continued to work on it during the drafting stage. Note that he has added an introduction and a conclusion, which were not part of his original outline (Figure 2.3), to convince Melissa Hill to purchase a new multifunction printer. Even so, Weekley recognized that his draft was still not ready for his boss to see, so he showed it to a co-worker for suggestions.

Intermediate draft of Marcus Weekley's report.   Figure 2.4

To: Melissa Hill, Office Manager
From: Marcus Weekley
Date: May 25, 2009
Subject: Improving Efficiency

As you requested, I have been researching what to do about improving efficiency and customer relations. One of the most beneficial and immediate solutions I have found is to replace our current laser printer with a new multifunction printer. More and more businesses today are incorporating multifunction printers into their information technology systems because of ease and efficiency. With the advances in printer technology in the past three years, it makes good business sense to replace our Van Eisen 4200 laser printer with a Lightech 520 multifunction printer. This multifunction printer would reduce business costs and increase business efficiency at a total cost significantly lower than the combined price of a laser printer, fax machine, copier, and scanner.

Vague subject line

Wordy opening paragraph takes too long to get to the point

Continued

Figure 2.4   (Continued)

Page 2

No headings or bullets make information hard to find

printer, fax machine, copier, and scanner, all of which are compatible with our operating systems. Addition of a 35.6 Kbps fax machine to our office, as opposed to our current 20.5 Kbps, would increase operating efficiency by allowing quicker response times. With the new multifunction color printer, our office can print up to 40 high-resolution color pages per minute, compared with our current printer's 20 color ppm. And the new printer's 3,000 MB of memory helps to manage multiple jobs easily. The Lightech 520 printer also holds 3,100 sheets of paper, as opposed to our current printer's 500-page capacity.

Does not supply source of research

The multifunction printer would provide higher quality printing through a 4,800 dpi printer, whereas our current Image 4200 laser printer only has 900 dpi. The new printer also offers over 90 scalable type fonts, whereas our current printer only offers 60. Also, our current scanners scan images in at 1,200 × 1,200 dpi, whereas the new scanner operates at 4,800 × 4,800 dpi, so scanned images will be of an even better quality.

Includes most important point for reader— costs—last

Paragraph is too long and hard to follow. Needs to be broken into 2 or 3 paragraphs

The greatest benefit a multifunction printer would provide our company is monetary. The price of a new multifunctional business printer ranges from $3,000 to $9,000 depending on the model. Lightech's 520 multifunction color laser printer (including 2,400 × 1,200 dpi copier, 4,800 × 4,800 dpi scanner, and 35.6 Kbps fax) costs only $3,175 not including shipping and handling purchased from Computerbuyers.com. This cost nearly equals the price of our own Van Eisen 4200 printer and West 400 scanners, but combines the equipment into one more efficient machine. Purchasing the multifunction printer would not only save our business money on the initial purchase, but use of a multitasking unit that combines four machines into one would also save on subsequent servicing and maintenance, as well as decreasing our monthly electric bill by $50–$150 per month. Use of the multifunction printer's 50–1,000% enlargement/reduction options should also save us an additional $300–$500 each month by using less outside advertising. Computerbuyers.com also offers a two-year warranty on all products sold through its website. What better way to begin improving business efficiency and customer relations than through the purchase of a new multifunction color printer?

Ends with question rather than plan for how to make change

From discussions with his co-worker and after further work on his draft, Weekley realized that he had placed one of the most important considerations for his audience (savings) last. In his final version, shown in Figure 2.5 (pp. 53–54), he moved that section to the beginning of his report because he realized that

Melissa Hill would be most concerned about costs. Weekley thus paid attention to his audience's priorities and needs.

He also added headings and bulleted lists to help his reader find information. The design of his earlier draft in Figure 2.4 did not assist his readers in finding information quickly or reflect a convincing organizational plan.

Final version of Marcus Weekley's report.   Figure 2.5

To: Melissa Hill, Office Manager
From: Marcus Weekley
Date: May 25, 2009
Subject: Purchasing a New Multifunction Printer

As you requested, I have investigated some ways to improve office efficiency and customer relations. The best solution I have found is to replace our current Van Eisen 4200 laser printer, two Readrite 400 scanners, and XL290 fax machine with a new generation multifunction printer.

Introduction gets to the point quickly

With advances in printer technology over the previous three years (enclosed is a copy of a review article "New Technology Means Office Efficiency" from *Computer World* [Mar. 2009]: 96–98), it makes good business sense to replace our less-efficient laser printer with a Lightech 520 multifunction printer. Our laser printer does only one task, while the Lightech will give us higher-resolution color printing, a high-speed fax machine, a copier, and a scanner all in one. This new unit is economical, more efficient, and will significantly improve the transmission and design of our documents.

Documentation shows research on subject

Cost
The greatest benefit of the multifunction Lightech 520 is cost. We can purchase this printer for only $3,175, plus shipping and handling, totaling $3,298 when ordered through **Computerbuyers.com**. Purchasing a Lightech 520 would allow us to recoup that cost easily in just a few months because we would

Excellent use of headings and bulleted lists

- realize a savings in the purchase price—one Lightech costs less than the four current machines combined
- receive Smartext and WordPort software free with the printer
- decrease our monthly electricity bill by $50–$150 by reducing four pieces of office equipment to one

Continued

Figure 2.5   (Continued)

Justification for new purchase clearly laid out

- have less maintenance, saving $150 per month in service calls
- receive a two-year warranty with guaranteed overnight service, decreasing any downtime
- be able to trade in our current printer for $375 and our two scanners for $200 each
- save an additional $300–$500 each month by not having to use outside advertising thanks to Lightech's 50–1,000% enlargement/reduction capabilities

Efficiency

This multifunction printer accomplishes all four tasks—printing, faxing, scanning, copying—simultaneously, and each task is compatible with our current networking system. Other key features include

Gives reader right amount of detail

- allows quicker faxing (3 seconds per page) with 35.6 Kbps and greater storage (600 pages)
- prints twice as fast—40 high-resolution color pages per minute as opposed to our current printer's 20
- enhanced memory of 3,000 MB manages multiple jobs easily
- expanded paper capacity (3,100 sheets) to handle our heavy quarterly mailings easily
- 100,000-page monthly duty cycle means less maintenance
- all four tasks compatible with our PhotoEdit software

Contrasts current equipment with new model

Quality

The Lightech 520 prints and processes higher quality and quantity of work because it

Selects only most relevant points for reader

- has a 4,800 dpi printer whereas our current printer offers only 900 dpi
- exhibits same color density on the 1,000th copy as on the first; solid ink sticks ensure no toner spills
- offers a 300 ¥ 600 dpi high-resolution copier
- offers a 2,400 × 1,200 dpi high-resolution copier
- would increase the quality of manipulated scanned images with a 4,800 × 4,800 dpi whereas our current scanner is only 1,200 × 1,200
- offers 90 scalable fonts as opposed to the 60 we now have, making our publications more professional looking

Recommenda-tion ends report persuasively

I recommend we purchase a new Lightech 520 multifunction printer from **Computerbuyers.com** to save us money, improve office efficiency, and help us integrate our office systems to better project our corporate image.

Revising

Revision is an essential stage in the writing process. It requires more than giving your work one more quick glance. Do not be tempted to skip the revision stage just because you have written the required number of words or sections or because you think you have put in too much time already. Revision is done *after* you produce a draft that you think conveys the appropriate message for your audience. The quality of your letter or report depends on the revisions you make. **Revision** gives you a second (or third or fourth) chance to get things right for your audience and to clarify your purpose in writing to them, as Marcus Weekley did.

Like planning or drafting, revision is not done well in one big push. It evolves over a period of time. Allow yourself enough time to do it carefully.

- Avoid drafting and revising in one sitting. If possible, wait at least a day before you start to revise. (In the busy work world, waiting a couple of hours may have to suffice.)
- Ask a co-worker or friend familiar with your topic to comment on your work, as Marcus Weekley did to make sure his report was convincing.
- Plan to read your revised work more than once.

Tech Note

Revising on Your Computer

Here are some suggestions on how you can effectively revise a document in your word processor:

- During this part of the writing process, respond to the notes you left yourself when you were drafting the document. Use the Copy, Paste, and Move commands to move sentences and paragraphs to more appropriate places/sections, add clarifications and transitions, insert parenthetical documentation, create headers, or add visuals. Again, at this stage do not worry about spelling and punctuation, which are fine-tuning matters you can resolve during the editing stage.
- As you did when you drafted, save your document regularly (every few minutes) so that you do not lose your work if there's a power outage or your computer malfunctions. Always back up your document at the end of each day.
- Continue to use the Save As option for each revision of your document, and title each revision, such as "Office Equipment, Revision #1."
- Take advantage of your document tracking options (see pages 103–104) such as Track Changes and Edit that allow you to see your edits (additions, cuts, and moves) in a different color, rather than deleting existing text. They also allow you to insert comments to yourself that appear at the bottom of your document or boxed within it, rather than interrupting the flow of your writing.

Continued

Continued

■   Use the Print Preview option to see a version of your document, enabling you to see quickly if your document is too long or too short or if some sections of your document appear to be too dense with text. If they are, you may need to reduce text, add headings and headers, or include visuals to make your document attractive and easy to read and follow (see pages 496–506 on document design).

Key Questions to Ask as You Revise

By asking and successfully answering the following questions as you revise, you can discover gaps or omissions, points to change, and errors to correct in your draft.

Content

1. Is it accurate? Are my facts (figures, names, addresses, dates, costs, references, warranty terms, statistics) correct? Have I documented them?
2. Is it relevant for my audience and purpose? Have I included information that is unnecessary, too technical, or not appropriate?
3. Have I given enough evidence to explain things adequately and to persuade my readers? (Too little information will make readers skeptical about what you are describing or proposing.) Have I left anything out?

Organization

1. Have I clearly identified my main points and shown my readers why those points are important?
2. Is everything in the correct, most effective order? Should anything be switched or moved—closer to the beginning or toward the end of my document?
3. Have I spent too much (or too little) effort on one section? Do I repeat myself? What can be cut?
4. Have I grouped related items in the same part of my report or letter, or have I scattered details that really need to appear together in one paragraph or section?

Tone

1. How do I sound to my readers—professional and sincere, or arrogant and unreliable? What attitude do my words or expressions convey?
2. How will my readers think I perceive them—honest and intelligent or unprofessional and uncooperative?

Case Study: A "Before" and "After" Revision

Mary Fonseca, a staff member at Seacoast Labs, was asked by her supervisor to prepare a short report for the general public on the lab's most recent experiments. Figures 2.6 and 2.7 show her "before" and "after" revisions.

Opening, unorganized paragraphs of Mary Fonseca's "before" draft.   **Figure 2.6**

Drag is an important concept in the world of science and technology. It has many implications. Drag occurs when a ship moves through the water and eddies build up. Ships on the high seas have to fight the eddies, which results in drag. In the same way, an airplane has to fight the winds at various altitudes at which it flies; these winds are very forceful, moving at many knots per hour. All these forces of nature are around us. Sometimes we can feel them, too. We get tired walking against a strong wind. The eddies around a ship are the same thing. These eddies form various barriers around the ship's hull. They come from a combination of different molecules around the ship's hull and exert quite a force. Both types of molecules pull against the ship. This is where the eddies come in.

Information hard to follow and not relevant for audience

Scientists at Seacoast Labs are concerned about drag. Dr. Karen Runnels, who joined Seacoast about three years ago, is the chief investigator. She and her team of highly qualified experts have constructed some fascinating multilevel water tunnels. These tunnels should be useful to ship owners. Drag wastes a ship's fuel.

Important point not developed

When starting to revise Figure 2.6, Fonseca realized that it lacked focus. It jumped back and forth between drag on ships and drag on airplanes. Since Seacoast Labs did not work on planes, she wisely decided to drop that idea. She also realized that the information on the effects of drag, something Seacoast was working on, was so important it deserved a separate paragraph. In light of this key idea, she realized that her explanation of molecules, eddies, and drag needed to be made more reader-friendly. Researching further, she decided to add a new paragraph on the causes and effects of drag, which became paragraph 2 in Figure 2.7.

Yet by pulling ideas about drag and its effects from her longish first paragraph in Figure 2.6, Fonseca was left with the job of finding an opening for her report. Buried in her original opening paragraph was the idea that we cannot always see the forces of nature but "we can feel them." She thought this analogy of walking against the wind and drag would work better for her audience of nonspecialists than the original wooden remarks she had started with.

Although her organization and ideas were far better now than those in her "before" draft in Figure 2.6, she realized she had said very little about her employer, Seacoast Labs. Doing more research, she found information about Seacoast's experiments and why they were so important in saving money. This information was far more significant and relevant than saying Dr. Runnels had been at Seacoast for three years.

Through revision and further research, then, Mary Fonseca transformed two poorly organized and incomplete paragraphs into three separate yet logically connected ones that highlighted her employer's work. Thanks to her revision (Figure 2.7), she came up with two very helpful headings—"What is drag?" and "How drag works"—for her nonspecialist readers.

Figure 2.7   A revision of Mary Fonseca's "after" draft in Figure 2.6.

Effective use of definition and headings

What is drag?

We cannot see or hear many of the forces around us, but we can certainly detect their presence. Walking or running into a strong wind, for example, requires a great deal of effort and often quickly leaves us feeling tired. When a ship sails through the water, it also experiences these opposing forces known as **drag**. Overcoming drag causes a ship to reduce its energy efficiency, which leads to higher fuel costs.

Describes cause and effect of drag

How drag works

It is not easy for a ship to fight drag. As the ship moves through the water, it drags the water molecules around its hull at the same rate the ship is moving. Because of the cohesive force of those molecules, other water molecules immediately outside the ship's path get pulled into its way. All the molecules become tangled rather than simply sliding past each other. The result is an eddy, or small circling burst of water around the ship's hull, which intensifies the drag. Dr. Jorge Fröes, a noted structural engineer, explains the process using an analogy: "When you put a spoon in honey and pull it out, half the honey comes out with the spoon. That's what is happening to ships. The ship is moving and at the same time dragging the ocean with it."

Uses an easy-to-follow analogy

Clearly explains the research and its importance for readers

At Seacoast Labs, scientists are working to find ways to reduce drag on ships. Dr. Karen Runnels, the principal investigator, and a team of researchers have constructed water tunnels to simulate the movement of ships at sea. The drag a ship encounters is measured from the tiny air bubbles emitted in the water tunnel. Dr. Runnels's team has also developed the use of polymers, or long carbon chain molecules, to reduce drag. The polymers act like a slimy coating for the ship's hull to help it glide through the water more easily. When asbestos fibers were added to the polymer solutions, the investigators measured a 90 percent reduction in drag. The team has also experimented with an external pump attached to the hull of a ship, which pushes the water away from a ship's path.

Editing

Editing is quality control for your reader. This last stage in the writing process might be compared to detailing an automobile—the preparation a dealer goes through to ready a new car for prospective buyers. Editing is done only after you are completely satisfied that you have made all of the big decisions about content and organization—that you have said what you wanted to, where and how you intended, for your audience.

When you edit, you will check your work for

- sentences
- word choices
- punctuation
- spelling
- grammar and usage
- tone

As with revising, don't skip or rush through the editing process, thinking that once your ideas are down, your work is done. If your work is hard to read or contains mistakes in spelling or punctuation, readers will think that your ideas and your research are also faulty.

The following sections will give you basic guidelines about what to look for when you edit your sentences and words. The appendix, "A Writer's Brief Guide to Paragraphs, Sentences, and Words" (pp. 693–710), also contains helpful suggestions on using correct spelling and punctuation.

Tech Note

Editing on Your Computer

Software programs make editing online easy and efficient. These programs flag errors in spelling, punctuation, word choice, subject-verb agreement, wordiness, and sentence construction. But keep the following guidelines in mind when you use these programs. Be aware of what your computer can and cannot do for you.

- Customize your spell checker. You can set your spell checker to flag words that you frequently misspell and to ignore words that aren't in its dictionary (such as proper names, brand names, technical terms, concepts, etc.).
- Don't accept everything your spell check or grammar check features tell you. While spell checkers are helpful, do not rely on them exclusively. For instance, a grammar check may fail to recognize homonyms (their/there or its/it's), and a spell check may highlight a proper name or industry jargon that is spelled correctly but is not included in your spell checker's dictionary. When in doubt, use a dictionary or consult a grammar handbook to verify the

Continued

Continued

accuracy of your spelling and grammar rather than relying on your word processor. (See also "A Writer's Brief Guide to Paragraphs, Sentences, and Words" on pages 693–710.)

- Don't use global Search and Replace. If your spell checker spots a word that is in fact misspelled or used incorrectly in one place, don't assume that all occurrences of the word are misspelled. For instance, if your spell checker tells you to capitalize the word "South" (as in "South Carolina") in one instance, the word may not need to be capitalized elsewhere (e.g., "south of the highway"). Check each correction individually.

- Use autoformatting to make sure you correctly and consistently format such things as quotations and paragraph indents throughout your document.

- Pay attention to the verbs and adjectives you use. Do not keep repeating the same ones throughout your document. This makes for very boring reading for your audience.

- When you edit online, be careful not to focus only on the lines you can see on the screen and neglect the larger organization of your document. Scroll down or print out your document to see how a change in one place may affect (by contradicting, duplicating, or weakening) something earlier or later.

Editing Guidelines for Writing Lean and Clear Sentences

Here are four of the most frequent complaints readers voice about poorly edited writing in the world of work:

- The sentences are too long. I could not follow the writer's meaning.
- The sentences are too complex. I could not understand what the writer meant the first time I read the work; I had to reread it several times.
- The sentences are unclear. Even after I reread them, I am not sure I understood the writer's message.
- The sentences are too short and simplistic. The writing felt "dumbed down."

Writing clear, readable sentences is not always easy. It takes effort, but the time you spend editing will pay off in rich dividends for you and your readers. The seven guidelines that follow should help with the editing phase of your work.

1. **Avoid needlessly complex or lengthy sentences.** Do not pile words on top of words. Instead, edit one overly long sentence into two or even three more manageable ones.

> Too long: The planning committee decided that the awards banquet should be held on March 15 at 6:30, since the other two dates (March 7 and March 22) suggested by the hospitality committee conflict with local sports events, even though one of those events could be changed to fit our needs.

> Edited for easier reading: The planning committee has decided to hold the awards banquet on March 15 at 6:30. The other dates suggested by the hospitality committee—March 7 and March 22—conflict with two local sports events. Although the date of one of those sports events could be changed, the planning committee still believes that March 15 is our best choice.

2. **Combine short, choppy sentences.** Don't shorten long, complex sentences, only to turn them into choppy, simplistic ones. A memo, e-mail, or letter written exclusively in short, staccato sentences sounds immature.

   When you find yourself looking at a series of short, blunt sentences, as in the following example, combine them where possible and use connective words similar to those italicized in the edited version.

> Choppy: Medical transcriptionists have many responsibilities. Their responsibilities are important. They must be familiar with medical terminology. They must listen to dictation. Sometimes physicians talk very fast. Then the transcriptionist must be quick to transcribe what is heard. Words could be missed. Transcriptionists must forward reports. These reports have to be approved. This will take a great deal of time and concentration. These final reports are copied and stored properly for reference.
>
> Edited: Medical transcriptionists have many important responsibilities. *These* include transcribing physicians' orders using correct medical terminology. *When* physicians talk rapidly, transcriptionists have to keyboard accurately *so* that no words are omitted. *Among their most demanding* duties are keyboarding and forwarding transcriptions *and then,* after approval, storing copies properly for future reference.

3. **Edit sentences to tell who does what to whom or what.** The clearest sentence pattern in English is the subject-verb-object (s-v-o) pattern.

   s   v   o
   Sue mowed the grass.

   s   v   o
   Our website contains a link to key training software programs.

   Readers find this pattern easiest to understand because it provides direct and specific information about the action. Hard-to-read sentences obscure or scramble information about the subject, the verb, or the object. In the following unedited sentence, subjects are hidden in the middle rather than being placed in the most crucial subject position.

> Unclear: The control of the ceiling limits of glycidyl ethers on the part of the employers for the optimal safety of workers in the workplace is necessary. (Who is responsible for taking action? What action must they take? For whom is such action taken?)
>
> Edited: Employers must control the ceiling limits of glycidyl ethers for their workers' safety.

4. **Use strong, active verbs rather than verb phrases.** In trying to sound important, many bureaucratic writers avoid using simple, graphic verbs. Instead,

these writers use a weak verb phrase (for example, *provide maintenance of* instead of *maintain*, *work in cooperation with* instead of *cooperate*). Such verb phrases imprison the active verb inside a noun format and slow a reader down. Note how the edited version here rewrites the weak verb phrase.

> Weak: The city provided the employment of two work crews to assist the strengthening of the dam.
> Strong: The city employed two work crews to strengthen the dam.

5. **Avoid piling modifiers in front of nouns.** Putting too many modifiers (words used as adjectives) in the reader's path to the noun is confusing for readers, who will have trouble deciphering how one modifier relates to another modifier or to the noun. To avoid that problem, edit the sentence to place some of the modifiers after or before the nouns they modify.

> Crowded: The vibration noise control heat pump condenser quieter can make your customer happier.
> Readable: The quieter on the condenser for the heat pump will make your customer happier by controlling noise and vibrations.

6. **Replace wordy phrases or clauses with one- or two-word synonyms.**

> Wordy: The college has parking zones for different areas for people living on campus as well as for those who do not live on campus and who commute to school.
> Edited: The college has different parking zones for resident and commuter students. (Twenty words of the original sentence—everything after "areas for"—have been reduced to four words: "resident and commuter students.")

7. **Combine sentences beginning with the same subject or ending with an object that becomes the subject of the next sentence.**

> Wordy: I asked the inspector if she were going to visit the plant this afternoon. I also asked her if she would come alone.
> Edited: I asked the inspector if she were going to visit the plant alone this afternoon.

> Wordy: Homeowners want to buy low-maintenance bushes. These low-maintenance bushes include the ever-popular holly and boxwood varieties. These bushes are also inexpensive.
> Edited: Homeowners want to buy low-maintenance and inexpensive bushes such as holly and boxwood. (This revision combines three sentences into one, condenses twenty-four words into fourteen, and joins three related thoughts.)

Editing Guidelines for Cutting Out Unnecessary Words

Too many people in business think the more words, the better. Nothing could be more self-defeating. Your readers are busy; unnecessary words slow them down. Make every word work. Cut out any words you can from your sentences. If the sentence still makes sense and reads correctly, you have eliminated wordiness.

1. **Replace wordy phrases with precise ones.** For example, on the next page, the phrases on the left should be replaced with the precise words on the right.

Wordy	Concisely Edited
at a slow rate	slowly
at this point in time	now
be in agreement with	agree
bring to a conclusion	conclude, end
come to terms with	agree, accept
due to the fact that	because
express an opinion that	affirm
for the period of	for
in such a manner that	so that
in the area /case / field of	in
in the neighborhood of	approximately
look something like	resemble
show a tendency to	tend
with reference to	regarding, about
with the result that	so

2. Use concise, not redundant, phrases. Another kind of wordiness comes from using redundant expressions—saying the same thing a second time, only in different words. "Fellow colleague," "component parts," "corrosive acid," and "free gift" are phrases that contain this kind of double speech; a fellow *is* a colleague, a component *is* a part, acid *is* corrosive, and a gift *is* free. In the examples below, the suggested changes on the right are preferable to the redundant phrases on the left.

Redundant	Concise
absolutely essential	essential
advance reservations	reservations
basic necessities	necessities, needs
close proximity	proximity, nearness
end result	result
final conclusions/final outcome	conclusions/outcome
first and foremost	first
full and complete	full, complete
personal opinion	opinion
tried and true	tried, proven

3. Watch for repetitious words, phrases, or clauses within a sentence. Sometimes one sentence or one part of a sentence needlessly duplicates another.

Redundant: To provide more room for employees' cars, the security department is studying ways to expand the employees' parking lot.

Edited: The security department is studying ways to expand the employees' parking lot. (Since the first phrase says nothing that the reader does not know from the independent clause, cut it.)

4. Avoid unnecessary prepositional phrases. Adding a prepositional phrase can sometimes contribute to redundancy. The italicized words below are unnecessary. Be on the lookout for the italicized phrases on the next page and delete them.

audible *to the ear*	second *in sequence*	soft *in texture*
bitter *in taste*	short *in duration*	tall *in height*
fly *through the air*	hard *to the touch*	twenty *in number*
orange *in color*	honest *in character*	visible *to the eye*
rectangular *in shape*	light *in weight*	loud *in volume*

Figure 2.8 shows an e-mail that Trudy Wallace wants to send to her boss, Lee Chadwick, about issuing smartphones to the entire sales force. Her unedited work is bloated with unnecessary words, expendable phrases, and repetitious ideas.

Figure 2.8   Wordy, unedited e-mail.

Wordy and unfocused subject

One long, unbroken paragraph is hard to follow

Repeats same idea in two or three sentences

Uses awkward and wordy sentences

Does not say what writer will do about problem

Reply   Save   Forward   Print   Delete

To:	<lchadwick@transtech.org>
From:	<twallace@transtech.org>
Date:	11/7/2009 2:45 PM
Subject:	Issuing Smartphones and Today's Technology

Due to the inescapable reliance on technology, specifically on e-mail and Internet communication, within our company, I believe it would be beneficial to look into the possibility of issuing smartphones such as BlackBerries, iPhones, or Palm Treos to our employees. Issuing smartphones would have a variety of positive implications for efficiency of our company. Unlike normal cell phones, these smartphones have many new features that will help our employees in their daily work, since they combine cellular phone technology with e-mail and document transfer capabilities, as well as many other important features. The employees could increase their efficiency due to the fact that they could constantly keep track of appointments on their schedule for each day. The employees would also benefit from smartphones by having access to their contacts even when they are out on the road traveling, whether they are at local office meetings or on cross-country business trips. By means of smartphones I feel quite certain that our company's correspondence would be dealt with much more speedily, since these devices will allow our employees to access their e-mail at all times. I think it would be absolutely essential for the satisfaction of our customers and to the ongoing operation of our company's business today to respond fully and completely to the possibility such a proposal affords us. It would therefore appear safe to conclude that with reference to the issue of smartphones that every means at our disposal would be brought to bear on issuing such smartphones to our employees.

After careful editing, Wallace streamlined her e-mail to Chadwick. Note how, in Figure 2.9, she pruned wordy expressions and combined sentences to cut out duplication. The revised version is only 86 words, as opposed to the 253 words in the draft. Not only has Wallace shortened her message, she has made it easier to read.

Editing Guidelines to Eliminate Sexist Language

Editing involves far more than just making sure that your sentences are readable. It also reflects your professional style—how you see and characterize the world of work and the individuals in it, not to mention how you want your readers to see you. Your words should reflect a high degree of ethics and honesty, free from bias and offense. They need to be sensitive to your international audience's needs as well.

Sexist language offers a distorted view of our society and discriminates in favor of one sex at the expense of another, usually women. Using sexist language offends and demeans female readers by depriving them of their equal rights. It may cost your company business. Avoid gender bias by using inclusive language for women and men alike.

Sexist language is often based on sexist stereotypes that depict men as superior to women. For example, calling politicians *city fathers* or *favorite sons* follows the stereotypical picture of seeing politicians as male. Such phrases discriminate against women who do or could hold public office at all levels of government. Similarly, avoid such sexist designations as *male nurse, male models, or male secretary.* Never assume or imply a person's gender based upon his or her profession.

The e-mail in Figure 2.8 edited for conciseness.   Figure 2.9

Reply   Save   Forward   Print   Delete

To:	<lchadwick@transtech.org>
From:	<twallace@transtech.org>
Date:	11/7/2009 3:15 PM
Subject:	Issuing Smartphones to Staff

Focused subject

Because the sales force spends so much time out of the office, I think we should issue smartphones such as BlackBerries and Apple iPhones to increase productivity and efficiency. Combining laptop and cell phone technology, smartphones will allow our employees to e-mail clients about appointments, to access accounts, and to transfer documents anywhere, thus increasing customer satisfaction.

Gets to point quickly

Provides convincing reasons

I think switching to smartphones is both a necessary and cost-effective investment. With your approval, I will obtain more information from vendors to prepare a proposal to request bids.

Closes with plan and request

Such language also prejudiciously labels some professions as masculine and others as feminine. For example, sexist phrases assume engineers, physicians, and pilots are male (*he, his,* and *him* are often linked with these professions in descriptions) while social workers, nurses, and secretaries are female (*she, her*), although members of both sexes work in all those professions. Sexist language also wrongly points out gender identities when such roles do not seem to follow biased expectations—*lady lawyer, male secretary, female surgeon,* or *female astronaut.* Such offensive distinctions reflect prejudiced attitudes that you should eliminate from your writing.

Always prune the following sexist phrases: *every man for himself, gal Friday, little woman, lady of the house, old maid, the best man for the job, to man a desk* or *post, the weaker sex, woman's work, working wives, a manly thing to do,* and *young man on the way up.* These and other sexist terms will not only offend but also exclude many of the members of the audience you want to reach.

Finally, don't assume all employees are male by writing "All staff members and their wives are invited to attend." Simply say, "All staff members and their guests are invited to attend."

Ways to Avoid Sexist Language

1. Replace sexist words with neutral ones. Neutral words do *not* refer to a specific sex; they are genderless. The sexist words on the left in the following list can be replaced by the neutral nonsexist substitutes on the right.

Sexist	Neutral
alderman; assemblyman	representative
businessman	businessperson
cameraman	photographer
chairman	chair, chairperson
congressman	representative
craftsman	skilled worker
divorcée	divorced person
fireman	firefighter
foreman	supervisor
housewife	homemaker
janitress	cleaning person
landlord, landlady	owner
maiden name	family name
mailman/postman	mail carrier
man-hours	work-hours
mankind	humanity, human beings
manmade	synthetic, artificial
manpower	strength, power
man to man	candidly
men	human beings, people
policeman	police officer

Sexist	Neutral
repairman	repair person
salesman	salesperson, clerk
spokesman	spokesperson
waitress	server
weatherman	meteorologist
woman's intuition	intuition
workman	worker

2. Watch masculine pronouns. Avoid using the masculine pronouns (*he, his, him*) when referring to a group that includes both men and women.

Every worker must submit his travel expenses by Monday.

Workers may include women as well as men, and to assume that all workers are men is misleading and unfair to women. You can edit such sexist language in several ways.

a. Make the subject of your sentence plural and thus neutral.

Workers must submit their travel expenses by Monday.

b. Replace the pronoun *his* with *the* or *a* or drop it altogether.

Every employee is to submit a travel expense report by Monday.
Every worker must submit travel expenses by Monday.

c. Use *his or her* instead of *his*.

Every worker must submit his or her travel expenses by Monday.

d. Reword the sentence using the passive voice.

All travel expenses must be submitted by Monday.

Moreover, in some contexts exclusive use of the masculine pronoun might invite a lawsuit. For example, you would be violating federal employment laws prohibiting discrimination on the basis of sex if you wrote the following in a help-wanted notice for your company.

Each applicant must submit his transcript with his application. He must also supply three letters of recommendation from individuals familiar with his work.

The language of such a notice implies that only men can apply for the position.

Keep in mind that international readers may find these guidelines on using *him/her/his* confusing since many languages (e.g., French, Spanish) follow grammatical instead of natural gender. In French, the word for *doctor* is masculine.

3. Avoid using sexist words that end in -ess or -ette: *stewardess* (use *flight attendant*); *poetess* (for *poet*); *usherette* (for *usher*); or *drum majorette* (for *drummer*).

4. Eliminate sexist salutations. Never use the following salutations when you are unsure of who your readers are:

- Dear Sir
- Gentlemen
- Dear Madam

Any woman in the audience will surely be offended by the first two greetings and may also be unhappy with the pompous and obsolete *madam.* It is usually best to write to a specific individual, but if you cannot do that, direct your letter to a particular department or office: *Dear Warranty Department* or *Dear Selection Committee.*

Be careful, too, about using the titles *Miss, Mr.,* and *Mrs.* Sexist distinctions are unjust and insulting. It would be preferable to write *Dear Ms. McCarty* rather than *Dear Miss or Mrs. McCarty.* A woman's marital status should not be an issue. Try to find out if the person prefers *Ms.* to another courtesy title (e.g., Editor Hawkins, Supervisor Jones). If you are in doubt, write *Dear Indira Kumar.* Chapter 5 shows you acceptable salutations to use in your letters (see pp. 159 and 178).

5. Never single out a person's physical appearance. Such sexist physical references negatively draw attention to a woman's gender. Sexist writers would not describe a male manager in the sentence below.

The manager is a tall blonde who received training at Mason Technical Institute.

Avoiding Other Types of Stereotypical Language

In addition to sexist language, avoid any references that stereotype an individual because of race, national origin, age, or disability. Not only are such references almost always irrelevant in the workplace (except for Equal Employment Opportunity Commission reports or health care), they are discriminatory, culturally insensitive, and ethically wrong.

To eliminate biased language in your workplace writing, follow the guidelines below.

1. Do not single out an individual because of race or national origin or stereotype him or her because of it. Be especially sensitive when referring to someone's ethnic identity.

Wrong: Bill, who is African American, is one of the company's top sales reps.
Right: Bill is one of the company's top sales reps.

Wrong: The Chinese computer whiz was able to find the problem.
Right: The programmer was able to find the problem.

2. Identify members of an international community accurately. Not every native Spanish-speaker is Latin American or Hispanic. There are significant cultural differences you need to be sensitive to, e.g., Cuban American, Mexican American.

3. Avoid words or phrases that discriminate against an individual because of age. For example, do not use *elderly, up in years, geezer, old timer, over the hill, senior moment,* or the adjectives *spry* or *frail* when they are applied to someone's age: "a spry 67." Similarly, don't refer to employees as *kids, youngsters, juveniles, wet behind the ears,* or a *middle-aged supervisor.* Making someone's age an issue is unfair, whatever it may be.

Wrong: Jerry Fox, who will be 57 next month, comes up with obsolete plans from time to time.
Right: Some of Jerry Fox's plans have not been adopted.

Wrong: Our company keeps hiring youngsters who lack experience.
Right: Our firm recruits individuals with little or no experience.

4. Respect individuals who may have a disability. Avoid derogatory words such as *amputee, crippled, handicapped, impaired,* or *lame* (physical disabilities) or *retarded* or *slow* (mental disabilities). Stay away from terms such as these because they identify the entire individual rather than just the aspects that the disability affects. Emphasize the individual instead of the physical or mental condition as if it solely determined the person's abilities.

Wrong: Tom suffers from MS.
Right: Tom is a person living with MS.

Wrong: Sarah, who is crippled, still does an excellent job of keyboarding.
Right: Sarah's disability does not prevent her from keyboarding.

Keep in mind that the Americans with Disabilities Act (1990) prohibits employers from asking if a job applicant has a disability.

Also avoid using discriminatory expressions such as a *crippled economy, lame excuse, mentally challenged, mental midget,* or *crazy scheme.*

5. Don't stereotype based on sexual orientation. Avoid unnecessarily labeling a person by sexual orientation as if that is the only significant aspect of that person's life.

Wrong: Paula Smith, a strong supporter of gays and lesbians, hosts a successful daytime talk show.
Right: Paula Smith hosts a successful daytime talk show.

In addition, avoid derogatory innuendos, comments, or jokes about gay men, lesbians, or bisexuals (e.g., "That's so gay"), and don't assume that all of your readers are heterosexual (e.g., asking about marital status).

Revision Checklist

☐ Investigated the research, drafting, revising, and editing benefits available through computer software.

☐ Researched my topic carefully to obtain enough information to answer all my readers' questions—online searches, interviews, questionnaires, personal observations.

☐ Before writing, determined how much and what kind of information are needed to complete writing task.

☐ Spent important time planning—brainstorming, outlining, clustering, or a combination of those techniques. Produced substantial material from which to shape a draft. Documented sources.

☐ Prepared enough drafts to decide on major points in message to readers. Made major changes and deletions where necessary in drafts to strengthen document.

☐ Revised drafts carefully to successfully answer reader questions about content, organization, and tone. Formatted text to make it easy to follow.

☐ Made time to edit work so that style is clear and concise and sentences are readable and varied. Checked punctuation, sentences, and words to make sure they are spelled correctly and appropriate for audience.

☐ Eliminated sexist and other biased language that unfairly stereotypes individuals because of race, ethnicity, sexual orientation, or disability.

Exercises

1. Below is a writer's initial brainstormed list on stress in the workplace. Revise the brainstormed list, eliminating repetition and combining related items.

> leads to absenteeism
> high costs for compensation for stress-related illnesses
> proper nutrition
> numerous stress reduction techniques
> good idea to conduct interviews to find out levels, causes, and extent of stress in the workplace
> low morale caused by stress
> higher insurance claims for employees' physical ailments
> myth to see stress leading to greater productivity
> various tapes used to teach relaxation
> environmental factors—too hot? too cold?
> teamwork intensifies stress
> counseling

      work overload
      setting priorities
      wellness campaign
      savings per employee add up to $6,150 per year
      skills to relax
      learning to get along with co-workers
      need for privacy
      interpersonal communication
      employee's need for clear policies on transfers, promotion
      stress management workshops very successful in California
      physical activity to relieve stress
      affects management
      breathing exercises

2. Prepare a suitable outline from your revised list in Exercise 1 for a report to a decision maker on the problems of stress in the workplace and the necessity of creating a stress management program.

3. From the revised brainstormed list in Exercise 1, write a one-page memo to a decision maker about how the problems of stress negatively affect workplace production.

4. Write a short report (2–3 pages) to the manager of the small company you work for on a topic of your choice. Prepare an outline similar to that on flextime in Figure 2.1 (p. 46). Add, delete, or rearrange anything in this clustered grouping to complete your outline. Submit your final outline along with your report to your instructor.

5. Compare the draft of Marcus Weekley's report in Figure 2.4 (pp. 51–52) with the final copy of his report in Figure 2.5 (pp. 53–54). What kinds of changes did he make? Were they appropriate and effective for his audience and purpose? Why or why not?

6. Assume you have been asked to write a short report (similar to Marcus Weekley's in Figure 2.5) to a decision maker (the manager of a business you work for or have worked for; the director of your campus union, library, or security force; a city official) about one of the following topics.
   a. recruitment of more specialists in your field.
   b. Internet resources
   c. security lighting
   d. food service
   e. health care plans
   f. public transportation
   g. sporting events/activities
   h. team building
   i. morale
   j. hiring more part-time student workers

Do relevant research and planning about one of those topics and the audience for whom it is intended by answering the following questions:

- What is my precise purpose in writing to my audience?
- What do I know about the topic?
- What information will my audience expect me to know?
- Where can I obtain relevant information about my topic to meet my audience's needs?

7. Using one or more of the planning strategies discussed in this chapter (clustering, brainstorming, outlining), generate a group of ideas for the topic you chose in Exercise 6. Work on your planning activities for about 15–20 minutes or until you have about 10–15 items. At this stage do not worry about how appropriate the ideas are or even if some of them overlap. Just get some thoughts down on paper.

8. Go through the list you prepared in Exercise 7 and eliminate any entries that are inappropriate for your topic or audience or that overlap. Try to see how many of them you might expand or rearrange into categories or subcategories. Then create an outline similar to the one in Figure 2.3 (p. 49).

9. Using your outline in Exercise 8, prepare some drafts of your memo report. Submit at least two drafts to your instructor.

10. Revise your drafts as much as necessary to create the final copy of your report.

11. In a few paragraphs, explain to your instructor the changes you made between your early drafts and your revised drafts. Explain why you made them. Concentrate on major changes—adding and moving paragraphs—as well as matters of style, tone, and even format.

12. In an e-mail—addressed to your instructor—describe any problems that bothered you at various phases of working on your report. Also point out what planning, drafting, and revising strategies worked especially well for you.

13. The following paragraphs are wordy and full of awkward, hard-to-read sentences. Edit these paragraphs to make them more readable by using clear and concise words and sentences.
   a. It has been verified conclusively by this writer that our institution must of necessity install more bicycle holding racks for the convenience of students, faculty, and staff. These parking modules should be fastened securely to walls outside strategic locations on the campus. They could be positioned there by work crews or even by the security forces who vigilantly and constantly patrol the campus grounds. There are many students in particular who would value the installation of these racks. Their bicycles could be stationed there by them, and they would know that safety measures have been taken to ensure that none of their bicycles would be apprehended or confiscated illegally. Besides the precaution factor, these racks would afford users maximized convenience in utilizing their means of transportation when

they have academic business to conduct, whether at the learning resource center or in the instructional facilities.

b. On the basis of preliminary investigations, it would seem reasonable to hypothesize that among the situational factors predisposing the Smith family toward showing pronounced psychological identification with the San Francisco Giants is the fact that the Smiths make their domicile in the San Francisco area. In the absence of contrariwise considerations, the Smiths' attitudinal preferences would in this respect interface with earlier behavioral studies. These studies, within acceptable parameters, correlate the fan's domicile with athletic allegiance. Yet it would be counterproductive to establish domicility as the sole determining factor for the Smiths' preference. Certain sociometric studies of the Smiths disclose a factor of atypicality which enters into an analysis of their determinations. One of these factors is that a younger Smith sibling is a participant in the athletic organization in question.

14. Following are very early drafts of memos that business people have sent to their bosses or co-workers. Revise and edit each draft, referring to the revising and editing checklist on page 70. Turn in your revision and the final, reader-ready copy. As you revise, keep in mind that you may have to delete and add information, rearrange the order of information, and make the tone suitable for your readers. As you edit, make sure your sentences are clear and concise and your words well chosen.

   a. TO: All workers
        FROM: B.J. Blackwell
        DATE: February 3, 2011
        RE: Parking

The parking violations around here have gotten very, very bad. And the administration is provoked and wants some action taken. I don't blame them. I have been late for meetings several times in the last month because inconsiderate folks from other divisions have parked their cars in our zone. That just is not fair, and so I must not be the only one who is upset. No wonder the management finds things so bad they have asked me to prepare this memo.

A big part of the problem it seems to me is that employees just cannot read signs. They park in the wrong zones. They also park in visitors' spots. The penalties are going to be stiff. The administration, or so I was led to believe, is thinking of fining any employee who does not obey the parking policies. I know for a fact that I saw someone from the research department pull right into a visitor parking area last week just because it was 8:55 and he did not want to be late for work. That gives our business a bad name. People will not want to do business with us if they cannot even find a parking spot in the area that the company has reserved for them.

Ms. Watson has laid the law down to me about all this and told me to let each and every one of you know that things have to improve. One of the other big problems around here is that some employees have even parked their cars in loading zones, and security had to track them down to move.

As part of the administration's new policy, each employee is going to be issued a company parking policy and will have to come in and sign for it verifying that he received it. I think things really have gotten out of hand and that some drastic action has to be taken. We will all have to shape up around here.

b. TO: Betty Cannales-Worth, Director
FROM: Tom Cranford
DATE: April 21, 2010
RE: Vacation request for vacation from June 12–23

I have been a highly productive employee and so I do not think that it is out of line at all for me to make this request. I have put in overtime and even done others' work in the department while they were away. So, I think that it is fair and just, and I can see no reason why I should not be allowed to take my vacation during the last two weeks of June.

Let me explain some of the reasons. I could have others watch my desk and do the work. I have helped them out, too, and they know it. I have been remarkably dependable. I have been readily accessible whenever their vacations have come around, and so I know it can be done.

I fully realize that this is the busiest time of the year for our company and that vacations are not usually granted at all during this season. But I do have valid personal reasons which I think should be honored/respected. Peak business times are major. I understand this, and I do hope an exception will be made in my case. After all, I do have the on-the-job training that other companies would reward. Thanks very much.

c. TO: All Employees
FROM: George Holmes
DATE: October 20, 2010
RE: Travel

Every company has its policies regarding travel and vouchers. Ours strike me as important and fairly straightforward. Yet for the life of me I cannot fathom why they are being ignored. It is in everyone's best interest. When you travel, you are on company time, company business. Respect that, won't you. Explain your purpose, keep your receipts, document your visits, keep track of meals. Do the math.

If you see more than one client per day, it should not be too hard or too much to ask you to keep a log of each, separate, individual visit. After all, our business does depend on these people, and we will never know your true contributions on company trips unless you inform us (please!) of whom you see, where, why, and how much it costs you. That way we can keep our books straight and know that everything is going according to company policy.

Please review the appropriate pages (I think they are pages 23–25) about travel procedures. Thanks. If you have questions, give me a call, but check your employee handbook or with your office/section manager, first. That will save everyone more time. Good luck.

15. Find a piece of writing—e-mail, memo, letter, brochure, short report, or website—that you believe was not carefully drafted or revised. In a short memo or e-mail, point out to your instructor what is wrong with the piece of writing—for example, not logically organized, inappropriate tone, incomplete or too technical information. Attach a copy of the poor example to your e-mail or memo.

16. Revise the piece of poor writing you analyzed in Exercise 15. Submit your improved version to your instructor.

17. The following sentences contain sexist and other biased language. Edit them to remove these errors.
    a. Every intern had to record his readings daily for the spokesman.
    b. Although Marcel was an amputee, he still could hunt and peck at the keyboard.
    c. She saw a woman doctor, who told her to take an aspirin every day.
    d. Our agency was founded to help mankind.
    e. John, who is a diabetic, has an excellent attendance record.
    f. Every social worker found her schedule taxing—not enough days in the week to help out man to man.
    g. The company golf league never had an African American player before this month.
    h. To be a policeman, each applicant had to pass a rigorous physical and prove himself in the manly art of self-defense.
    i. It's a wise man who can rise to the top in this cutthroat, volatile stock market.
    j. Sandy Frain, a middle-aged woman, came by this morning wanting an appointment about the new policies on energy efficiency.
    k. As long as his medication is adjusted just right, George Smith performs as well as the next man in this company.
    l. He made his PowerPoint presentation as emphatically as an Italian opera singer on stage.
    m. Bill thought the new e-mailing guidelines were really retarded.
    n. Team B tried to disable our proposal by introducing irrelevant references to the many foreigners living on the north side.

CHAPTER 3

Collaborative Writing and Meetings at Work

Visit www .cengage.com/ english/kolin for this chapter's online exercises, ACE quizzes, and Web links.

In the workplace you will not always have to write alone, isolated from co-workers or managers. In fact, much of your business writing time may be spent working as part of a team. With teleconferencing and telecommunication as an important means of communication in the world of work, collaboration is becoming one of an employee's most valuable contributions to corporate culture. You can expect to prepare a document with other employees or managers who will collaborate or, at the very least, review your work and revise it. One survey estimates that in the world of work 90 percent of all business people spend some time writing as part of a collaborative team. Collaborative writing occurs when a group of individuals—as few as two to as many as seven or more—

- combine their efforts to prepare a single document
- share authorship
- work together for the common good, maybe even the survival, of a department, company, or organization

Keep in mind that teamwork must always take into consideration your employer's corporate image, goals, and politics. Your employer determines the subject that your group will discuss and the format(s) in which that information appears. Moreover, everything the team does is subject to review and revision by your employer. Viewing collaboration in this broader corporate context will get you off to a good start in the world of work.

Teamwork Is Crucial to Business Success

Teamwork is essential, then, for success within the world of work. Being a part of a writing team is a major responsibility in a world where each employee is connected to co-workers and managers as well as to customers in the worldwide marketplace of the Internet. Collaboration is networking, and collaborative writing is a vital part of the global network in which individuals depend on each other's expertise, experience, and viewpoints.

A collaborating team at work.   **Figure 3.1**

© Royalty-Free/CORBIS

Successful collaboration hinges on being a team player, one of the most highly valued skills in the workplace. Being a team player means you

- interact successfully on an interpersonal level
- network to get information
- provide feedback and participation
- give and take constructive criticism
- raise important and relevant questions
- get assistance from resource experts in other departments and fields
- put the good of your company above your ego
- contribute to customer service and satisfaction

Collaboration builds teamwork as it helps get writing done more easily and more efficiently. Figure 3.1 shows a collaborating team at work, planning and interacting on a company project.

Collaboration takes place when preparing many types of writing, including brochures, technical manuals, proposals, long reports, websites, and more. Collaboration can be done in a variety of ways—face to face, over the telephone, or via computer (see "computer-supported collaboration" on pages 97–107, which discusses collaborative software).

This chapter introduces you to successful ways to collaborate with co-workers in your office as well as people with whom you do business around the globe.

You'll receive practical advice on how to write in and for a group and how to solve communication problems within a group setting. It takes much work and skillful negotiation to be a member of an effective collaborative writing group, but the energy is worth the effort.

Advantages of Collaborative Writing

Collaborative writing teams benefit both employers and employees. Specific advantages of collaboration include the following, many of them highlighted in Lee Booker's report on his team's interaction at Keeton Pharmaceuticals in Figure 3.2.

1. It builds on collective talents. Many heads are better than one. A writing team profits from the diverse backgrounds and skills of its individual members. Collaboration joins individuals from diverse disciplines in the corporate world—scientists, lawyers, designers, and security experts, among others.

2. It allows for productive feedback and critique. Collaborative writing encourages all members of the team to get involved, offering both their agreement as well as their critical viewpoints in a shared effort to produce the highest-quality document possible.

3. It increases productivity and saves time. When a group has planned its strategies carefully, collaboration actually cuts down on the number of meetings and conferences, saving a company time and employees travel. See how videoconferencing assists a company in doing this in Figure 16.2 (pp. 676–678).

4. It ensures overall writing effectiveness. The more people there are involved in developing a document, the greater the chances are for thoroughness and cohesion. Guided by the shared principles of company style, documentation, and mission, a collaborative team can better guarantee uniformity and consistency in a piece of writing.

5. It accelerates decision-making time. A group investigating problems and offering pertinent solutions can cut down considerably on the time it takes to prepare a document.

6. It increases employee morale and confidence while decreasing stress. Working as part of a team relieves an individual of some job stress because he or she is not solely responsible for planning, drafting, and revising a document. A team, therefore, provides a safety net by assuring individual members that they can always talk over problems and they will have help in meeting deadlines.

7. It contributes to customer service and satisfaction. By pooling their knowledge of an audience's needs, including those of a regulatory government agency such as the FDA cited in Figure 3.2, a collaborative writing team is much more likely—through discussions, interactions, even disagreements—to anticipate a customer's or an agency's requests and complaints.

8. It affords a greater opportunity to understand global perspectives. By working with a multinational work force, such as that described in the long report

The advantages of collaboration in the workplace: A manager's notes.   **Figure 3.2**

Keeton Pharmaceuticals

———————— Notes for Human Resources File ————————

Collaborative writing is essential to the success and morale of the Environmental Testing Lab at Keeton Pharmaceuticals. I supervise four microbiologists, each of whom is responsible for testing a specific area of our plant. We routinely test the environment in production areas throughout the plant by sampling the air, water, and surfaces (e.g., belts, floors, vents). When we find unacceptable ranges of bacteria, we have to investigate and report our findings to management and, ultimately, to the FDA. The report that emerges is the collaborative effort of microbiologists and production personnel, as well as management.

The protocols for such collaboration are as follows. When an unacceptable limit is found, the microbiologist for that area first interviews the production supervisor as well as line personnel to find potential causes of contamination. From such interviews the microbiologist can obtain honest, objective feedback about what happened. Then the microbiologist prepares a memo report in collaboration with the other three microbiologists, who offer helpful and constructive suggestions on how such unacceptable ranges could be eliminated or reduced. This collective brainstorming helps our lab not to overlook key information, strengthens professional communication, and improves team spirit.

After drafting the memo, the microbiologist for that area of the plant sends it to me as the Lab Supervisor and to my boss, the Quality Lab Section Manager, for our response. We confer with the microbiologist and production staff, as necessary, and make any revisions to ensure that the report follows all company guidelines.

Continued

Figure 3.2   (Continued)

Page 2

The report is then forwarded to the Quality Control Manager and, finally, to the plant manager for their approval and signatures. After their review and approval, the report is placed in the plant repository—the Document Center—where it is subject to periodical FDA inspections.

The collaborative efforts of production personnel, microbiologists, and management guarantee that the report will be a model of clear, precise, and ethical writing. In the last three years, management has frequently commended the labs and production staff for our helpful spirit of cooperation.

Lee Booker
Lab Supervisor

in Chapter 15 (pp. 652–666), individuals can develop greater sensitivity to and appreciation of the needs and problems of an international audience.

■ Collaborative Writing and the Writing Process

The writing process described in Chapter 2 also applies to collaborative writing. Groups use the same strategies and confront the same problems that individual writers do. Writing teams must brainstorm, plan, research, draft, and edit. Like the individual writer, too, a team must move through the writing process. But, although the essential steps in the writing process are the same, there are some differences you need to know about. Below is a rundown of how collaborative groups tackle the process of preparing a document.

1. **Groups must plan before they write.** As in individual writing, the planning phase includes brainstorming and outlining (see pages 45–48), as well as identifying the audience, purpose, and scope of a document. Unlike individual writers, though, the group must also set ground rules for the collaborative process to work, select a leader, assign individual responsibilities within the group, and create a schedule that respects individual group members' schedules.

2. **Groups must do research.** Insufficient or incorrect research can jeopardize the success of a collaborative document just as easily as it can derail the efforts of an individual writer. In a group context, the responsibility of researching is usually shared by several individuals who must coordinate the results of their research prior to the group drafting process.

3. Groups must prepare drafts. The drafting process in group writing can be done in two ways. The first way is similar to individual writing in that only one complete draft of the document is prepared at a time, benefiting from the feedback and critique of several individuals. The second, more common, way is unique to group writing—individual members of the group draft separate portions of the document, later to be combined and edited by the entire group.

4. Groups must revise and edit. The revising and editing processes in group writing are also similar to what an individual would do. But group writing offers the added benefit of multiple writers being involved, each of whom may spot organizational problems, inconsistencies in format and tone, as well as grammatical and mechanical mistakes.

Case Study: Collaborative Editing

Figure 3.3 contains an e-mail written by Tara Barber, the Documentation Manager of CText, Inc., a large firm in Ann Arbor, Michigan, that develops software for the publishing industry. Barber describes the writing process followed by her team of collaborative writers. Her approach sheds light on the practical, day-to-day process of group writing and editing. Group harmony is essential to completing documents successfully and on time.

Study Barber's approach. She effectively works *with* and *not against* her staff as she manages the collaborative effort without being heavy-handed or suppressing individual creativity. Her strategy offers a good model to follow.

Collaborative editing: Advice from a pro.   Figure 3.3

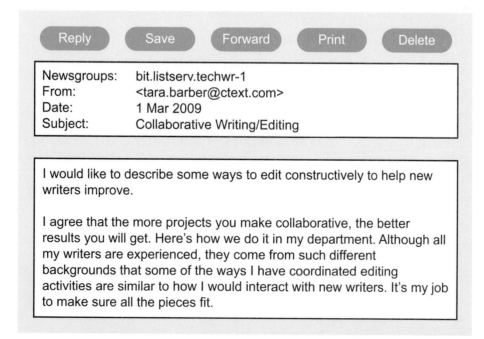

Newsgroups: bit.listserv.techwr-1
From: <tara.barber@ctext.com>
Date: 1 Mar 2009
Subject: Collaborative Writing/Editing

I would like to describe some ways to edit constructively to help new writers improve.

I agree that the more projects you make collaborative, the better results you will get. Here's how we do it in my department. Although all my writers are experienced, they come from such different backgrounds that some of the ways I have coordinated editing activities are similar to how I would interact with new writers. It's my job to make sure all the pieces fit.

Continued

Figure 3.3   (Continued)

When I first took this job, there were several documentation styles being used for the company's manuals. But the department at that time was very small, and so the senior writer and I sat down and prepared a cohesive and consistent style manual. We experienced some tension, but by compromising, we were able to iron out our differences. We developed a style guide and a set of manual conventions that gave new writers precise and helpful guidelines and procedures to follow from the start. The new manual assisted us to eliminate glaring editing problems. That way we do not worry about inconsistencies in spelling, capitalization, formats, headings, or documentation.

But the department has grown since that time, and on at least two occasions all of us have sat down as a department to reassess the style guide and manual. By doing this, we are able to get input from everyone and to allow for new ideas. When a problem arises now, we are comfortable addressing it as a group.

In addition to the usual subject-matter expert reviews, all of us read each other's materials. This review helps us to become familiar with each other's projects, homogenizes our writing styles, and keeps our documentation consistent. It's also a great way to get a good "clean-eyes edit." And, happily, a team member will offer an original idea about how to approach a problem that eliminates a potential editing dilemma. We all win.

In my role as documentation manager, I try to edit early and late, but not in the middle. Our materials usually go through several edits before they're ready. I try to look over early material in the form of outlines and first drafts, to make sure everything looks on track and follows our house style and procedures. I always edit final drafts because I am responsible for the output of my department. But I try not to do too much in the middle unless there is a problem. Peer review works best during this intermediate phase, and I don't want to step on individual creativity.

If I have to make comments, I make them at various levels. And here I'm not talking about catching grammar mistakes, spelling errors, typos, mis-numberings, etc. Instead,

a) I point out problems, or sections that seem confusing, but let the writer suggest fixes.

(Continued)   Figure 3.3

b) I make suggestions and provide examples—more than one if I can.

c) I actively work with the writer to develop ways around a problem. Sometimes this means coming up with a whole new way of approaching the documentation, which is then addressed at our next style meeting. Sometimes such interaction also gets the rest of the department involved in a brainstorming session.

d) If nothing else works, I play the heavy manager and say, "Do it this way because I say so." I try to avoid this, however, if I possibly can.

Working as a constructive team, we rarely have editing problems that we can't solve, with the result that everyone is happy with the comments and the final document. And, even more important, our customers find the documents usable and valuable. You can't really ask for more than that.

Tara Barber
Documentation Manager
CText, Inc.
<tara.barber@ctext.com>

Reprinted with the permission of Tara Barber.

Guidelines for Setting Up a Successful Writing Group

As discussed earlier in this chapter, a group not only has to plan the document but it must also plan how the group can work together most successfully. This planning involves a great deal of thought, organization, and cooperation. Following the guidelines below will increase your group's chances of successfully moving forward to its goal.

1. Get to know one another. To establish group rapport, each member of the group should learn as much as he/she can about the other members. For instance, group members should learn about one another's schedules, professional backgrounds, special competencies, national or cultural perspectives, work habits, and even their pet peeves. By being concerned and friendly, you create an environment that enhances productivity.

2. Set up a preliminary meeting to establish guidelines. At an introductory meeting, the group should discuss the objectives, audience, scope, format, importance of the document, and, most important, the deadline. It is helpful at the first meeting to identify and discuss any directions and/or directives from management and to set up priorities.

3. Agree on the group's leadership. In some situations, the leader of the group will be chosen according to his/her seniority and/or experience. In other situations, the group may vote on which member they want to lead them. Regardless of the selection process, the leader should be chosen to keep the group on track.

If the leader is appointed by the group, the group will also need to decide on his/her responsibilities and level of authority. The leader may take on any number of roles, including setting schedules, presiding over meetings, coordinating the activities of the team, evaluating the work of individual members, writing progress reports, communicating with management, and transmitting the finished document. An effective group leader must be skillful in initiating discussions, providing encouragement, and helping the group reach consensus. As Lee Booker points out in Figure 3.2, management also plays a critical role by giving the group its feedback.

4. Identify each member's responsibilities, but allow for individual talents and skills. A key question to ask and to answer is "Who gets credit for what?" In doing so, members need to acknowledge one another's strengths—in writing, document design, graphics, software, marketing, editing. Assign each member a task appropriate for his or her background.

5. Establish the times, places, and length of group meetings. The group must decide on a calendar of scheduled meetings and ensure that each member has a copy. Various software programs can help groups schedule meetings (see p. 109).

6. Follow an agreed-on timetable, but leave room for flexibility. The group should estimate a realistic time to complete the various stages of their work—when drafts are due or when editing must be concluded, for example. The group's timetable, with major **milestones** (dates when key parts are to be completed), should be sent to each member. **But remember: Projects always take longer than initially planned.** Prepare for a possible delay at any one stage. The group may have to submit progress reports (see pp. 610–617) to management.

7. Use a standard reference guide for matters of style, documentation, and format. Establishing guidelines about style (spelling, abbreviations, dates, capitalizations, symbols), documentation, format, graphics, and so forth makes the group's task easier. Be consistent, write *e-mail* or *email*, not both. Many large companies have policy or style manuals that their teams must use, as Tara Barber notes in Figure 3.3. If such a document is not available, select a manual or other reference work to which members should adhere, such as the American Psychological Association (APA) guide (see pages 354–355).

▌Ten Proven Ways to Be a Valuable Team Player

Being a team player is one of the most highly prized skills in the world of work. By following the ten guidelines below, you can advance the work of your writing team and earn the respect of both your colleagues and your supervisor:

1. Think collectively. Adopt a "we can get this done together" attitude. Do not treat some team members as favorites and ignore others. Both actions nurture resentment.

2. **Participate.** Avoid being a passive observer. Recognize that you have valuable ideas to contribute to the group.

3. **Be clear and precise.** Express your ideas to the group clearly and appropriately. Avoid veering off the point, being vague, or making unsupported generalizations.

4. **Set aside your ego.** Do not view the group as a battleground or contest in which someone wins and someone else loses. Put the team ahead of self.

5. **Be enthusiastic.** Avoid sounding uninterested in or inconvenienced by the group process. Replace cynicism with a "can do" spirit and your team will value you as a vital player.

6. **Listen.** Pay attention to what others say when they are speaking and avoid monopolizing the conversations. Develop a "third ear" to discern the meanings behind group members' words.

7. **Be open-minded.** Give serious consideration to each member's suggestions and ideas. Weigh the pros and cons before deciding on the value of a member's input.

8. **Follow company and group protocols.** Adhere to company and group protocols regarding the organization, style, and format of the document. Follow the group's rules about meeting times, places, and order of business.

9. **Compromise.** Be aware that in a group situation after ample discussion and deliberation, you may need to compromise for the overall benefit of the group.

10. **Go with group consensus.** Don't be a lone ranger. Accept the group's decisions as final. Stubbornness only leads to delay and hard feelings.

◼ Sources of Conflict in Collaborative Groups and How to Solve Them

The success of collaborative writing depends on how well the team interacts. Discussion and criticism are essential to discover ideas, results, and solutions. Members must build on one another's strengths and eliminate or downplay weaknesses. Inevitably, members have different perspectives or viewpoints out of which conflicts may arise.

Conflict in the sense of conflicting opinions—a healthy give-and-take—can be positive if it alerts the group to problems (inconsistencies, redundancies, incompleteness, and down right errors) and provides ways to resolve them. But conflict in the sense of creating problems for the group dynamic is another matter entirely.

Following are some common problems in group dynamics, with suggestions on how to avoid or solve them.

Common Problems, Practical Solutions

1. **Resisting constructive criticism.** Constructive criticism can be vital to the group effort, but when an individual responds to constructive criticism with resistance, or even anger, the group's progress is disrupted.

Solution: When emotions become heated, the leader or other members of the team should move the discussion to another section of the document or to another issue to allow for some cooling-off time. The leader should also discuss (in private) the importance of listening to constructive criticism with the individual or with the entire group of the problem is widespread.

2. Offering only negative criticism. At times, a group member may harshly critique the work of another member of the team or respond to suggestions and ideas with nothing but negativity or sometimes even be condescending. This type of criticism does not help move the group forward. It only embarrasses or irritates the person being critiqued.

Solution: The leader should diplomatically (and in private) remind the individual about the importance of being constructive, perhaps pointing out instances in which he or she has been negative and how these instances could be presented and rephrased more constructively. Individuals should encourage one another, too, for the sake of group harmony.

3. Dominating. The group process is about sharing and responding to ideas, not about taking over. When one group member dominates the discussion and becomes aggressive and territorial, the group process suffers.

Solution: The leader should let the group know that the participation of all members is valued, without singling out any one individual. If one group member continues to dominate the discussion, the leader again needs to ask diplomatically (and in private) that the team member make an effort to defer to other individuals in the group.

4. Declining to participate. Some individuals are afraid to express their opinions because of a lack of self-confidence and so deprive the group of hearing another viewpoint.

Solution: If a group member lacks confidence about an idea or suggestion, he or she should safely test it out on an individual member before presenting it to the group as a whole. When a leader perceives that a group member lacks confidence, that leader should strive to encourage that member to participate at key junctures and assure him/her of a welcome forum to do so. The leader may also want to appoint another member of the team to serve as a listening partner for a member who lacks confidence, giving that individual a chance to voice and test his/her views before the group convenes.

5. Focusing on unimportant issues. Nitpickers can derail any group. Team members can do this by dwelling on insignificant details (for instance, small editing matters during the drafting process) or by steering the group away from larger, more important issues (costs, schedules, personnel, etc.).

Solution: The leader should remind the group (without singling anyone out) about the larger issues at hand and caution them to stay on track both in their discussions and in adhering to the agreed-upon the writing process.

6. Being overly deferential. Group members who agree to everything out of fear of conflict ("yes people") do not help the group. Hiding their feelings, these individuals may actually be depriving the group of valuable input.

Solution: The group leader needs to encourage group members to express their feelings and opinions freely, and to do so courteously and professionally. Again, the leader needs to promote and protect meetings as a safe place to express ideas and air differences for the good of the project.

7. Not respecting cultural differences. You may have individuals from different countries or cultures on your team. Disregarding or misjudging the way they interact with the group can seriously threaten group success and harmony. In addition, discrimination of any type—based on race, age, religious beliefs, nationality—is unacceptable in a collaborative group or workplace.

Solution: Some companies, as the long report example in Chapter 15 illustrates, offer employees seminars on cultural sensitivity that emphasize respecting diversity in the workplace. A group leader must ensure that diversity is honored in the group and if anyone does not, the leader may write that person up, refer him or her to a company handbook, and/or remind that person of antidiscriminatory legislation.

8. Missing deadlines or submitting unsatisfactory work. Meeting deadlines is crucial to collaborative work, as is completing work that meets group standards. Problems can occur when members are not involved in the planning stages, ignore group communications, or do not communicate effectively in writing or at meetings with other group members.

Solution: The leader should not only help set group protocol but also communicate those rules and regulations effectively to the group, check in with them regularly on their progress on writing the collaborative document, and remind them periodically about deadlines.

9. Violating confidentiality. Leaking information about a product, procedure, or operation when the group has been instructed to keep it within the company is a serious violation in the world of work. But realize that sometimes meetings can be closed to everyone in the company except for those who are a part of the group.

Solution: Violating confidentiality is often grounds for dismissal. At a preliminary meeting, and periodically during the writing process, the leader should emphasize the whys, hows, and whens of confidentiality.

▎Models for Collaboration

There are as many types of collaborative writing methods as there are companies. The process can range from relatively simple phone calls or e-mails to a much more extensive use of groupware (see pages 97–107).

The scope, size, and complexity of your document as well as your company's organization will determine what type of (and how much) collaboration is necessary. A shorter assignment (say, a memo) will not require the same type of group participation as would a policy handbook, a proposal, or a long report. At some large companies, for example, a staff of professional editors (such as Tara Barber's team in Figure 3.3) revises the final draft prepared by a departmental team. The more important and more complex a document is, the more extensive collaboration will be.

The following sections describe four models for collaboration used in the world of work.

Cooperative Model

The cooperative model is one of the simplest and most expedient ways to write collaboratively in the business world. An individual writer is given an assignment and then goes through the writing process (see Chapter 2) to complete it. Along the way, he or she may show a draft to a peer to get feedback or to a supervisor for a critique. That is what Randy Taylor did in Figure 3.4. He shared a draft of a letter to a potential client with his boss, Felicia Krumpholtz, who made changes in content, wording, and format and then sent it back to Taylor. Following his boss's suggestions, Taylor then created the revised letter in Figure 3.5.

If Taylor had sent his first draft, the customer would hardly have been impressed with the company's professionalism and might have reconsidered placing an order. But thanks to Krumpholtz's revisions, Taylor made his letter much more effective. Careful writers always benefit from constructive criticism. Even though Felicia Krumpholtz helped Randy Taylor improve his work, strictly speaking it was not a case of group writing. Although the two interacted, they did not share the final responsibility for creating the letter.

Sequential Model

In the sequential model, each individual is assigned a specific, nonoverlapping responsibility—from brainstorming to revising—for a section of a proposal, report, or other document. There is a clear-cut, rigid division of labor. If four people are on the team, each will be responsible for his or her part of the document. For example, one employee may write the introduction, another the body of the report, another the conclusions, and the fourth, the group's recommendation.

Team members may discuss their individual progress and even exchange their work for group review and commentary. They may also choose a coordinator to oversee the progress of their work. When each team member finishes his or her section, the coordinator then assembles the individual parts to form the report.

Functional Model

The division of labor in the functional collaborative model is assigned not according to parts of a document but by skill or job function of the members. For example, a four-person team may be organized as follows:

- The **leader** schedules and conducts meetings, issues progress reports to management, and generally coordinates everyone's efforts to keep the project on schedule.
- The **researcher** collects data, conducts interviews, searches the literature, administers tests, classifies the information, and then prepares notes on the work.

A draft of Randy Taylor's letter edited by his supervisor, Felicia Krumpholtz.   **Figure 3.4**

servitron
4083 randolph street
houston, tx 77016
(713) 555-6761

←————————————————————————— November 15, 2008

Terry Tatum
Manager
i Consolidated Solutions
Houston, TX   *add zip code*

Dear ~~Sir,~~ *Terry Tatum:*

Thank you for asking
~~I am taking the opportunity of answering your request~~ for a price list of Servitron
products. Servitron has been in business in the Houston area for more than ~~twenty-~~
22 ~~two~~ years and we offer unparalleled equipment and service to ~~any customer~~.   *Consolidated*
Using a Servitron product will give you both efficiency and economy.   *Solutions*
                    can   *boldface*   *boldface*
Whatever Servitron model you choose carries with it a full one-year warranty on
all parts and labor. After the expiration date of your warranty ~~you~~ should purchase
our service contract for $75,000 a year.   *might*

~~Here are the models Servitron offers~~

Model Name	Number	Price
Zephyr	81072	$459.95 —— *wrong price*
Colt	86085	$629.95
Meteor	88096	$769.95

Depending on your needs, one of these models should be right for you.

If I might be of further assistance to you, please call on me. I am also enclosing a
brochure giving you more information, including specifications, on those Servitron
products.

~~Truly,~~ *Sincerely yours,*   ⟨*add phone number*
                                           and e-mail⟩   *reverse*
                                                               the order
Servitron ——— *leave 4 spaces*   *of these*
Randy Taylor *sign your name*   *two*
Sales Associate   *sentences*

● www.servitron.com

Figure 3.5   The edited, final copy of Figure 3.4.

November 15, 2008

Terry Tatum
Manager
Consolidated Solutions
Houston, TX 77005-0096

Dear Terry Tatum:

Thank you for asking for a price list of Servitron products. Servitron has been in business in the Houston area for more than 22 years and we can offer unparalleled equipment and service to Consolidated Solutions. Using a Servitron product will give you both **efficiency** and **economy**.

Depending on your needs, one of these models should be right for you.

Model Name	Number	Price
Zephyr	81072	$469.95
Colt	86085	$629.95
Meteor	88096	$769.95

Whatever Servitron model you choose carries with it a full **one-year warranty** on all parts and labor. After the expiration date of your warranty, you might purchase our service contract for $75.00 a year.

I am also enclosing a brochure giving you more information, including specifications, about those Servitron products. If I can help you further, please call me at (713) 555-6761 or e-mail me at rtaylor@servitron.com.

Sincerely yours,

SERVITRON

Randy Taylor

Randy Taylor
Sales Associate

● www.servitron.com

- The **designated writer/editor**, who receives the researcher's notes, prepares outlines and drafts, and circulates them for corrections and revisions.
- The **graphics expert** obtains and prepares all visuals, specifying why, how, and where visuals should be placed, and might even suggest that visuals replace certain sections of text; the graphics expert may also be responsible for the design (layout) and production of the document (see Chapter 11).

This organizational scheme fosters much more group interaction than the sequential model does.

Figure 3.6 illustrates how a functional model works. It describes the behind-the-scenes joint effort that went into Joycelyn Woolfolk's proposal for her boss to authorize a new journal, which would incorporate a newsletter her office currently prepares. A publications coordinator for a large, regional health maintenance organization (HMO), Woolfolk supervises a small staff and reports directly to the public affairs manager, who in turn is responsible to the vice president of the regional office.

As you will see from Woolfolk's functional approach, collaboration can move up and down the chain of command, with participation at all levels. Often, individual employees will pull together information from their separate functional areas (such as finance, information systems, marketing, and sales), and someone else will put that information into a draft that others read and revise until the document is ready to send to the boss. By the time the boss reviews the document, it has been edited and revised many times and by many individuals. This is a common business practice.

Integrated Model

In the integrated model, all members of the team are engaged in planning, researching, and revising. Each shares the responsibility of producing the document. Members participate in every stage of the document's creation and design, and the group goes back to each stage as often as needed. This model offers intense group interaction. Even though individual writers on the team may be asked to draft different sections of the document, all share in revising and editing that document. Depending on the scope of the document and company policy, the group may also go outside the team to solicit reviews and evaluations from experts.

▮ Case Study: Evolution of a Collaboratively Written Document

Figures 3.7, 3.8, and 3.9 show the evolution of a memo that follows an integrated model of collaboration. This memo informed employees that their company was enhancing a recycling program. Alice Schuster, the vice president of Fenton Companies, which manufactures appliances, asked two employees in the human resources department—Abigail Chappel and Manuel Garcia—to prepare a memo ("a few paragraphs" is how Schuster put it) to be sent to all Fenton employees.

Schuster had an initial conference with Chappel and Garcia, at which she stressed that their memo had to convey Fenton's renewed commitment to conservation and

Figure 3.6   Joycelyn Woolfolk's account of how one proposal originated and was collaboratively prepared following a functional model.

Status Report: Coordinating Regional Magazine Project

The idea to start a regional magazine was first expressed in passing by our vice president, who is interested in getting more and higher-level visibility for our regional office. Several other regional offices in our company have created fairly attractive magazines, and one office in particular has earned a lot of good publicity for its online publication.

The public affairs manager (my boss) and I quickly picked up on the vice president's hint and began to formulate ways to investigate the need for such a publication and ways to substantiate our recommendation. For several weeks the public affairs manager and I had a number of conversations addressing specific points, such as the kind of documentation our proposal would need, what our resources for researching the question were, what our capabilities would be for producing such a publication, and so on. We were guided by the twofold goal of getting the vice president's approval and, ideally, meeting a genuine market need.

After discussions with my boss, I met with members of my staff to ask them to do the following tasks:

1. review existing HMO publications and report on whether there was already a regional magazine for the Northwest

2. develop, administer, and analyze a readership survey for current subscribers to our newsletter, which would potentially be incorporated into the new magazine

3. formulate general design concepts for the magazine in print and online that we can implement without increasing staff, while still producing the quality magazine the vice president wants

4. prepare a detailed budget for projected costs

5. consult with experts on our staff (actuaries, physicians, nurses) about topics of interest

6. confer with our Web experts about online designs and problems

---

Continued

(Continued)   Figure 3.6

Page 2

I requested e-mail, Web sources, and other written documentation from my staff members about most of these tasks, and then I used that information to draft the proposal that eventually would go to the vice president, and perhaps even to the president's office. And I communicated with my staff often, through e-mails, blogs, personal meetings, and group sessions.

After I revised my draft several times, I asked my staff to look over each draft for feedback and proofreading. Based on their comments, I made further revisions and did careful editing. My boss reviewed my proposal, revising and editing it in minor ways innumerable times. Ultimately, the proposal will go out as a memo from me to my boss, who will then send it under her name to the vice president.

The copy of my proposal may be further revised in response to the vice president's comments when she gets it. She may use some portion or all of it in another report from her to the president, or she may write her own proposal to the president supporting her argument with the specifics from my proposal. The vice president will word the proposal to fit the expressed values and mission of our regional office as well as provide information that addresses budgetary and policy concerns for which her readers (our company president and board members) have final responsibility.

This is how a proposal started and where it will eventually end.

that, as part of that commitment, the employees had to intensify their recycling efforts. Chappel and Garcia thus shared the responsibility of convincing co-workers of the importance of recycling and educating them about practicing it.

Chappel and Garcia also had the difficult job of writing for several audiences simultaneously—the boss, whose name would not appear on the memo, other managers at Fenton, and the Fenton work force itself.

First Draft

Figure 3.7 is the first draft that Chappel and Garcia collaborated on and then presented to the manager of the human resources department—Wells McCraw—for his comments and revisions. As you can see from McCraw's remarks, written in ink, he was not especially pleased with their first attempt and asked them to make a number of revisions. As a careful reader (conscious of the document's audience), McCraw found Chappel and Garcia's paragraphs to be rambling and repetitious—the writers were unable to stick to the point. Specifically, he pointed out that they included too much information in one paragraph and not enough in others.

Figure 3.7   Early draft of the Chappel and Garcia collaborative memo, with revisions suggested by Wells McCraw.

FENTON COMPANIES

TO: All Employees
FROM: Abigail Chappel; Manuel Garcia
DATE: February 10, 2009
RE: Improving Our Recycling Program

This ¶ is too long. Too many topics—costs, protecting the environment Keep it short—say what we are doing and why

An in-house study has shown that Fenton sends approximately 26,000 pounds of paper to the landfill. The landfill charge for this runs about $2,240, which we could save by recycling. Fenton Companies is conscious of our responsibility to save and protect the environment. Accordingly, starting March 1 we will begin a more intensive paper recycling program. Our program, like many others nationwide, will use the latest degradable technology to safeguard the air, trees, and water in our community. It has been estimated that of the 250 million tons of solid waste, three-quarters of goes to landfills. These landfills across the country are becoming dangerously overcrowded. Such a practice wastes our natural resources and endangers our air and drinking water. For example, it takes 10 trees to make 1 ton of paper, or roughly the amount of paper Fenton uses in four weeks. If we could recycle that amount of paper, we could save those trees. Recycling old paper into new paper involves less energy than making paper from new trees. Moreover, waste sent to landfills can, once broken down, leach, seep into our water supply, and contaminate it. The dangers are great.

Start off with this key idea

Check your facts; I think it is closer to 16–17

Word "it" left out

Delete — not relevant to our purpose

No cap

By enhancing our recycling, we will not be sending so much to the Springfield Landfill and so help alleviate a dangerous condition there. We will keep it from overflowing. Fenton will also be contributing to transforming waste products into valuable reusable materials. Recycling paper in our own office shows that we are concerned about the environmental clutter. By having an improved paper recycling program, we will establish our company's reputation as an environmentally conscious industry and enhance our company's image.

Add the fact about our saving trees in this ¶

Delete — makes us look bad

Start ¶ with this point

Fenton is primarily concerned with recycling paper. The 200 old phone books that otherwise would be tossed away can get our recycling program off to a good start.

When? How? Implications for saving/costs?

Give some examples

We encourage you to start thinking about the additional kinds of paper around your office/workspace that needs to be earmarked for recycling. When you start to think about it, you will see how much paper we as a company use.

Continued

(Continued)   Figure 3.7

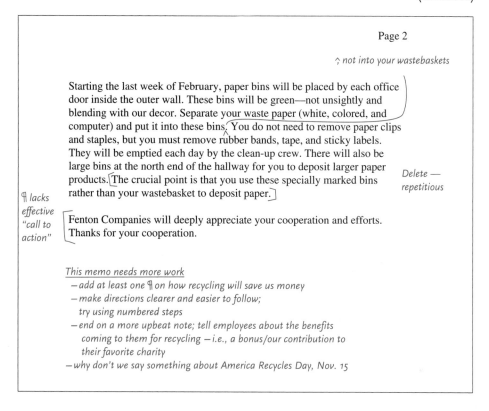

As a good editor/manager, McCraw also directed their attention to factual mistakes, irrelevant and even contradictory comments, and essential information they had omitted. Finally, McCraw offered some advice on using visual devices (see Chapter 1, pp. 21–23) to make their information more accessible to readers.

Subsequent Drafts

Figure 3.8 shows the next stage in the collaboration. In this version of the memo, prepared through several revisions over a two-day period, the authors incorporated McCraw's suggestions as well as several changes of their own. It was this revision that they submitted to Vice President Schuster, who also made some comments on the memo. Schuster's suggestions—all valid—show how different readers can help writing teams meet their objectives.

Note that in the process of revising their memo Chappel and Garcia had to make major changes from the first draft in Figure 3.7. Their work went through several versions to arrive at the document in Figure 3.8. Those changes—shortening and expanding paragraphs, adding and deleting information, and refocusing their approach to meet the needs of their audience—are the essential revisions a collaborative writing team, like an individual writer, would expect to make.

Figure 3.8   Revision of the Chappel and Garcia memo with changes suggested by
Vice President Schuster.

FENTON COMPANIES

TO: All Employees
FROM: Abigail Chappel; Manuel Garcia
DATE: February 10, 2009
RE: Improving Our Recycling Program

To save money and protect our environment, Fenton Companies will begin a
more intensive waste paper recycling program on March 1. Recycling
continues to be an environmental necessity. It has been estimated that three-
quarters of the 250 million tons of solid waste dumped annually in America
could be recycled. Our program will still use the latest recycling technology to
safeguard trees, air, and water.   — *Say strengthen
or continue*

*¶ needs
more
information* This new program will ~~establish~~ Fenton's reputation as an environmentally
conscious company. Fenton now uses one ton of paper every four weeks. This
represents 17 trees that can be saved just by recycling our paper waste.
Recycling also means we will send less waste to the landfill, alleviating
problems of overfill and reducing the potential for contamination of the water
supply. Waste sent to large landfills can leach and seep into water systems. By
recycling paper, Fenton will also reduce the risk of long-term environmental
pollution.   *Add that we
no longer use
Styrofoam
and that our
suppliers
use only
biodegradable
products*

*Indicate
how much
we save* Recycling saves us money. Right now Fenton pays $180 per month to dump
2,000 pounds (one ton) of paper waste at the landfill. However, scrap paper is
worth $100 per ton. Recycling will generate $100 each month in new revenue
while eliminating the $180 dumping expense.

*This
sentence
more
logically
goes in
¶ 3* Ultimately, the success of our project depends <u>on renewed awareness of the
variety of office paper suitable for recycling.</u> Paper products that can be
recycled include newspapers, scrap paper, computer printouts, letters,
envelopes (without windows), shipping cartons, old phone books, and
uncoated paper cups.   *Put in itemized
bulleted list
to stand out
better*

*Do not
start new
¶ here;
keep as
part of
previous ¶* Recycling 200 or so old phone books each year alone will save 22 cubic yards
of landfill space (or close to $100) and will generate $25 in scrap paper
income for our company. The old 2008 phone books, which will be replaced
by new 2009 ones on March 1, will give us an excellent opportunity to
intensify our recycling.

Continued

(Continued)   Figure 3.8

Page 2

Here are some easy-to-follow directions to make our recycling efforts even more effective:

Boldface these words

1. Starting the last week in February, paper bins will be placed inside each office door. Put all waste paper into these bins, which will be emptied by maintenance.
2. Place larger paper products—such as cartons or phone books—in the green bigger paper bins at the end of each main corridor.
3. Put white, colored, computer printout, and newspapers into separate marked bins. Remove all rubber bands, tape, and sticky notes.

Thanks for your cooperation. To show our appreciation for your help, 50 percent of all proceeds from the recycled paper will go toward employee bonuses and the other 50 percent will be given to the office's favorite charity. The benefits of our new recycling program will more than outweigh[the inconvenience it may cause.]We will conduct another study in time for America Recycles Day on November 15 to determine how our company has improved in this area.

Add another sentence to this ¶ on how a safer environment will benefit our company and the employees, too

---

Final Copy

With McCraw's further input and continuing to revise the memo on their own, Chappel and Garcia submitted the final, revised memo found in Figure 3.9 to the vice president a few days later. This final copy received Schuster's approval and was then routed to the Fenton staff.

Thanks to an integrated model of collaboration and careful critiques by McCraw and Schuster, Chappel and Garcia successfully revised their work. Effective team effort and shared responsibility were at the heart of Chappel and Garcia's assignment.

■ Computer-Supported Collaboration

Although collaborating on hard copy documents, as discussed in the previous section, remains a common practice in the workplace, computer-supported collaboration is fast becoming the norm and is being adopted by more and more businesses. Employees today must be proficient at using multiple types of collaborative software systems, otherwise know as **groupware**. You will be expected to know—or at least be adaptable to learning—not only how to use e-mail collaboratively, but also

Figure 3.9   Final copy of the memo prepared by Chappel and Garcia using an integrated model of collaboration.

FENTON COMPANIES

TO: All Employees
FROM: Abigail Chappel; Manuel Garcia
DATE: February 10, 2009
RE: Improving Our Recycling Program

To save additional money and to protect the environment, Fenton will begin a more intensive waste paper recycling program on March 1. For the new program to succeed, we need to increase our paper recycling efforts. Recycling continues to be an environmental necessity. It has been estimated that three-quarters of the 250 million tons of solid waste dumped annually in America is still not being recycled. Our program will continue to rely on the latest recycling technology to safeguard trees, air, and water.

Recycling has already saved us money. Right now Fenton pays $180 per month to dump 2,000 pounds (one ton) of paper waste at the landfill, compared to $360 last year. However, scrap paper is worth $100 per ton. Increased recycling will generate $100 each month in revenue while eliminating the $180 dumping expense.

This new, more intensive program will also strengthen Fenton's reputation as an environmentally conscious company. Three years ago, we stopped using Styrofoam products and asked suppliers to use biodegradable materials for all our shipping containers. Our efforts have proved successful, but Fenton still uses one ton of paper every four weeks. This paper represents 17 trees that can be saved just by recycling our paper waste. We need to send even less waste to the landfill, alleviating problems of overfill and reducing the potential for contamination of the water supply. Waste sent to large landfills can leach and seep into water systems.

Ultimately, the success of this project depends on our being aware of the variety of office paper suitable for recycling. An expanded list of paper products that can be recycled includes:

- newspapers
- letters
- envelopes (without cellophane windows)
- phone books
- uncoated paper cups

- scrap paper
- computer printouts
- junk mail (only black print on white paper)
- shipping cartons

Continued

(Continued)   Figure 3.9

Page 2

The old 2008 phone books, which will be replaced by new 2009 ones on March 1, will give us an excellent opportunity to launch our intensified recycling program.

Here are some easy-to-follow directions to make our recycling efforts even more effective:

1. Starting the last week in February, put all waste paper into the **paper bins** that will be placed **inside each office door.** These bins will be emptied by maintenance.

2. Place **larger paper products**—such as bulky cartons—in the **green bigger paper bins** at the end of the main corridor.

3. Remove all rubber bands, tape, and sticky notes and put white, colored, computer printout, and newspapers into separate marked bins.

Thanks for your cooperation. To show our appreciation for your increased efforts, 50 percent of all proceeds from the recycled paper will go into an annual employee bonus fund and the other 50 percent will be given to the office's favorite charity. We will conduct another study in time for America Recycles Day on November 15 to determine how our company has improved in this area. The benefits of our new recycling program will more than outweigh the inconvenience it may cause. A safer environment—and a more cost-effective way to run our company—benefits us all.

how to use document tracking systems (such as Microsoft Word's Track Changes feature), and Web-based collaboration systems (wikis).

Groupware does not, of course, completely eliminate the need for a group to meet in person to discuss priorities, clarify issues, or build team spirit. But face-to-face communications, although sometimes essential, are frequently accompanied by computer-supported collaboration via groupware.

Advantages of Computer-Supported Collaboration

Collaboration online via software known as groupware offers significant advantages to both employers and workers. Here are some of those benefits:

1. **Increased opportunities to "meet."** Today's meeting-heavy and travel-filled work environments often make collaborating in person difficult and costly. The virtual nature of e-mail and other types of groupware, however, provides less expensive and more frequent opportunities for team members to ask questions, share information, make suggestions and revisions, and troubleshoot.

2. **Reduced stress in updating new group members.** Groupware can efficiently and quickly bring a new group member or members into the ongoing conversation,

saving team members time and effort and reducing the stress of face-to-face meetings for a new member of the team.

3. Expanded options to communicate, free of geographical limitations. Groupware ensures the flexibility to contact group members anywhere—at a home office (telecommuters, freelancers), on the road, at another office, or at another company (vendors, clients). Thanks to groupware, there are no geographical limitations hindering the group.

4. Improved feedback and response rate and more accountability. Because it is accessible and easy to use, groupware helps team members participate actively in the collaboration process. It also helps eliminate problems of members not voicing their concerns (see page 86). Moreover, groupware ensures fair and honest accountability. Because each member's contributions are documented, someone's lack of participation becomes obvious.

5. Enhanced possibility of complete and clear information. Groupware allows everyone involved in the team to truly be "on the same page." It provides a running record, thus saving all information and preventing misinterpretation. Groupware's written environments encourage team members to get to the point and make sure other team members understand that point clearly and quickly.

Types of Groupware

There are essentially three types of groupware commonly used to produce collaboratively written documents in today's workplace: e-mail, document tracking systems, and wikis.

E-Mail

E-mail has a number of functions, the most obvious being to help individuals in the workplace communicate with one another quickly (discussed in Chapter 4). It is also a valuable collaborative tool. Team members can edit documents within the body of an e-mail and forward the e-mail to other members of the team for commentary and further edits.

The main advantage of using e-mail for collaboratively written documents is that it works perfectly well when collaborating on simple documents. As long as the team members are clear about their edits and comments, the document should be relatively simple to revise and then forward to the group.

E-mail does have some drawbacks for collaborative writing. Most important, e-mail does not preserve formatting. Many documents, particularly longer documents, incorporate varying formats—line spacing, paragraph indentations, figures, tables, and so on—that can be garbled when sending e-mails (particularly between different e-mail programs). Another difficulty is that e-mail programs do not offer convenient ways to remove edits and comments. To show their edits and comments clearly, team members can use bold, italic, all capital letters, and, in some e-mail programs, different colors and strikethroughs. However, all of these edits and comments need to be removed to produce a revised draft or final document.

Note how in Figure 3.10 Chantel is very clear about her edits, using blue to highlight any changes or comments, using strikethroughs to indicate suggested cuts,

Visit www
.cengage.com/
english/kolin for
an online exer-
cise, "Collabo-
rating Using
E-Mail."

Collaboration using e-mail.   **Figure 3.10**

Reply	Save	Forward	Print	Delete

To:	<Chantel_Antoine@envirogroup.org>, <rickyyangNYC@yahoo.com>, <Ramon_Calderez@envirogroup.org>, <SeethaRau@aol.com>
From:	<Nicole_Letriere@envirogroup.org>
Date:	Wed. 7 April 2009 9:55 EST
Subject:	West River Park cleanup proposal—concluding paragraph

Hi team—I appreciate your ongoing collaboration with me on this very important proposal. I think we're just about there. Thanks to your efforts, the intro and body just need a little fine-tuning. But before we do that, let's take another close look at the last few sentences of the concluding paragraph. But before we do that, let's take another close look at the concluding paragraph. We really lose focus here and might fail to convince the Neighborhood Parks Commission to take action. The language is flowery and the tone needs to be far more professional and less emotional. We need a firm but more courteous tone.

Chantel, since you are most familiar with the project, would you give the first edit a try, with the above reservations in mind, and then forward your draft to the entire team? Then I would like everyone else to offer further suggestions and edits, and together we can come up with a highly persuasive conclusion.

Many thanks.

Nicole

In light of the above profoundly disturbing toxicology results, we urge you to take IMMEDIATE action to clean up West River Park. The residents in the immediate area of the park have historically been underserved by the NPC, resulting now in a potentially dangerous public health situation, as if this community didn't have enough to worry about. We applaud the NPC's recent efforts to rope off portions of the waterfront and place warning signs in the vicinity. However, instead of simply putting a band aid on the problem, let's heal the wound. It is high time a trained chemical cleanup crew was commissioned to detoxify the waterfront in earnest.

Clear and concise explanation of collaborative needs

Figure 3.10   (Continued)

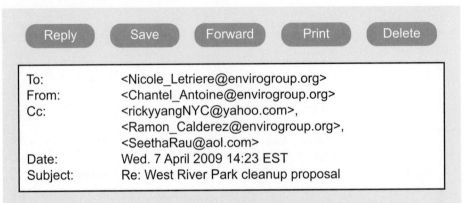

To: <Nicole_Letriere@envirogroup.org>
From: <Chantel_Antoine@envirogroup.org>
Cc: <rickyyangNYC@yahoo.com>,
             <Ramon_Calderez@envirogroup.org>,
             <SeethaRau@aol.com>
Date: Wed. 7 April 2009 14:23 EST
Subject: Re: West River Park cleanup proposal

Brief and straightforward explanations of edits

Hi Nicole (and everyone)—Please review my edits and parenthetical comments in blue below and let me know what you and the team think.

Thanks,
Chantel

Blue color and strikethroughs highlight edit

In light of the above ~~profoundly disturbing~~ *(avoid emotionalism here)* independently researched and verified toxicology results, we strongly urge ~~you~~ the NPC to ~~take IMMEDIATE action~~ *(sounds like an order)* ~~to clean up~~ prioritize the cleanup of West River Park. ~~The residents in the immediate area of the park have historically been underserved by the NPC,~~ *(we can't prove that the NPC has done this).* As river pollution has increased, so has the toxicity level in this stagnant water area, resulting ~~now~~ in a potentially dangerous public health situation~~, as if this particular community didn't already have enough to worry about~~ *(this is off topic).* We applaud the NPC's recent efforts to rope off portions of the waterfront and place warning signs in the vicinity.

Parentheses and italics differentiate comments from original text

However, ~~instead of simply putting a band aid on the problem, let's heal the wound~~ *(this sentence sounds both mocking and flowery).* ~~It is high time~~ we respectfully ask the NPC to hire an experienced chemical cleanup crew ~~were commissioned~~ to detoxify the waterfront ~~in earnest~~ *("in earnest" implies that the NPC has not taken the problem seriously).* We would appreciate your taking prompt action to correct this hazardous condition at West River Park by the end of the month. By doing so, the NPC would assure residents that the cleanup would occur prior to the start of the summer season when the area is used most. We would be happy to assist the NPC by meeting with you and by sharing the results of our field tests. Thank you for moving quickly on this important matter. *(the conclusion was lacking a call to action; I also think we should offer our services)* and using parentheses and italics to differentiate her comments from the text of the edited proposal. Although her method is clear, a team member will need to carefully retype the section to produce a final draft.

Document Tracking Systems

Document tracking systems, such as Microsoft Word's Track Changes feature, are designed specifically for collaborative writing. Figure 3.11 shows an example of collaborative editing using document tracking software. Edited documents using this system are sent as e-mail attachments to team members. These systems preserve all of the original text while automatically adding strikethroughs over deleted text and highlighting in color newly added text. They also come equipped with a Comments feature that allows team members to tag areas of the document they would like to comment on. To do so they can type their responses, which then appear outside the body of the document (they appear at the bottom of the page, or in a separate window). Tracked documents can be shared by multiple team members, each team member using a different color for edits and using different initials to identify his or her comments. The team leader or member working on revising a document can then accept or reject the tracked changes after they are agreed on by the group and then submit a clean, final document free of tracked changes.

Visit www .cengage.com/ english/kolin for an online exercise, "Collaborating Using a Document Tracking System."

The advantages of document tracking systems are numerous. Edits appear clearly as strikethroughs and colored text; comments are easy to add, remove, and differentiate; and everyone can work with a single master document.

But one problem with tracking is that if there are numerous strikethroughs and colored text from various team members, each new editor (or the person revising) may find it difficult to read through the edited document smoothly. It is always a good idea, therefore, to revisit a tracked document in which changes have been fully accepted to make sure it flows smoothly and logically, and that sentences haven't been garbled during the editing process.

Figure 3.11 contains the first draft of a section of a report on increasing the number of parking spaces at a hospital and shows how this draft was edited by four readers. Notice that each reader's initials and his or her numbered comments are "tracked" to hyperlinked boxes on the right. (Keep in mind, though, that depending on the document tracking system you are using and the options you have selected, your tracked documents may look somewhat different.)

Wikis

Wikis are similar to document tracking systems, but have a few crucially different characteristics. First, wikis are not systems found within a software package such as Microsoft Office. They are websites for which team members are given access passwords, enabling them to check documents in and out of the site. Second, wikis typically do not show tracked changes right on the document. Instead, when an edited document is uploaded back to the website, each version is assigned a new version number. Versions of the document can be compared and the differences between versions can be viewed, but these differences will not show up within a single version.

The advantage of wikis over tracked documents is that subsequent versions of a document are as easy to read from start to finish as the first draft. Each wiki version

Figure 3.11   Collaboration using a document tracking system.

Draft One — <u>Expansion Hospital Parking Facilities</u> ~~Proposal on Hospital Expansion~~ — Parking

Community General Hospital needs to expand its parking facilities. Right now there is just too little room for visitors, staff, and patients. These inadequate parking facilities are a ~~detrime~~ <u>detriment</u> to the overall growth of pt. care. They have been an important talking point since the inception of this committee. Maybe they were ok when the hospital opened its doors in 1973, but not today. ~~CGH is a good place to work and t~~ The new parking facilities would definitely benefit the present staff and visitors from walking long distances in the rain and ice.

Comment [LB1] Center and rephrase to read "Expanding Hospital Parking Facilities"

Comment [KT1] Good point but cite source and statistics. See hospital report 2008.

The exact number of new spots is hard to estimate now but I am thinking around 500 might be just right. The problem is the traffic flow around the hospital. While the new parking facilities would alleviate it, it also raises a central question about how to get it done. Perhaps Wentworth Avenue, East to West, might be turned into a one-way street. That way we could add up to 11 new spots in the front of the ER and thus resolve the congestion that has hampered easy ~~egress~~ <u>entrance</u> and ~~ingress~~ <u>exit</u>. Another possibility worth considering is changing Taylor Street — right now it is a two-way street and we could make it one way West to East.

Comment [LB2] We must be precise. It was 417, according to engineering proposal.

Comment [MM1] I think it might be best to start a new section here and title it "Increased Traffic Flow."

Comment [MM2] ER poses a special problem. Let's expand this area into a new section — I've attached a draft.

At any rate, the traffic flow is a key issue we need to solve if we want to expand the parking spots. There are other important engineering problems for us to solve. I met with Eleanor Yi, the professional engineer who has given us lots of details on stress points, pre-cast concrete, and the slope of vehicular access ramps in order for us to erect a two-story parking facility that would accommodate increased traffic flow. I need to review what Yi has sketched out for us in terms of these three issues. Basically, here is a ~~chart~~ of the stats Yi worked up for us. As you can see, Yi believes we do not have the space to locate all new parking spots in one plane area. She recommends a two-story structure and believes the north side of the ER might be the best place.

Comment [MM3] Let's provide a map instead.

Comment [LB3] This is a good start for our initial draft but clearly there's a lot more work to be done on it. From the group's suggestions, I think we need four different sections, tentatively entitled "Overall Growth of Patient Care"; "Increased Traffic Problems"; "Need for More Parking Spaces"; and "Special Parking Problems Our ER Presents."

is a clean document free of complicated tracked edits. Therefore, when team members are revising, they will need to proofread only the final draft, rather than go through the laborious process of accepting/rejecting changes, deleting comments, and troubleshooting lingering inconsistencies.

The disadvantage of wikis, when compared with tracked documents, is that wikis may result in a lack of quality control. Because each group member's changes are not clearly tracked, it may be difficult for group members to keep up with the succession of changes. When using wikis, teams should set up a clear protocol outlining who may make changes to the document and when.

Models for Computer-Supported Collaboration

Employees using groupware will find that the **integrated** model (see below) and the **sequential** model (see page 106) are the most effective collaboration models for computer-supported collaboration.

The Integrated Model

1. All members of the team work on the entire document—developing, expanding, revising, and editing it. The original draft is saved so that it is accessible to all team members who will work on it. When using e-mail, the team should be careful not to lose sight of the original draft. But when employing a document tracking system, writers/revisers need to pass the original draft from hand to hand and continue to edit it. When using a wiki, the original draft will remain available separately as subsequent versions are uploaded.

2. Each member of the team keeps the rest of the team in the loop when he or she makes changes to the document. When using e-mail, the team member making the change should copy all other members of the team on the e-mail. But when using a document tracking system, each team member should inform the others while he or she is making changes so that two writers aren't working on the tracked document at once. With a wiki, however, only the person currently making changes will be able to access the document, since wikis automatically lock out other users when one team member has the document checked out.

3. Each team member looks at each draft of the document as it is edited. When using e-mail, team members should comment on each draft when it is e-mailed to the entire team. Editing through a document tracking system, each subsequent reader/writer should insert comments, noting where he or she agrees or disagrees with the previous editor's comments. With a wiki, though, each member of the team should look at each draft as it is uploaded and provide comments or express concerns to the rest of the group, which can be either fixed by the most recent editor or passed on to the next editor.

4. The team agrees on the final draft. If the steps above for e-mail, document tracking systems, or wikis are followed, the final draft can be reached relatively smoothly. Editing via e-mail or a document tracking system, the team must be careful to consolidate their various drafts by the member(s) assigned to this task. Using a wiki, the team has to agree to the second-to-last draft so that any changes to it will become the final draft version. In any instance, the final draft should receive one final proofread by all members of the team and any loose ends should be resolved.

The Sequential Model

1. **Each member of the team is responsible only for drafting his or her assigned section of the document.** At this point, the writer is simply word processing, not using groupware. The team should choose a team leader, however, who can act as an overall editor (see step 4, below).

2. **Each team member sends his or her completed section to everyone else in the group.** Since the chunks will remain separate for the time being, e-mail should be used (it would be premature to use a document tracking system or wiki).

3. **All members of the group edit the individual sections and send the revised sections to their respective authors.** E-mail continues to be the most appropriate medium to convey these edits.

4. **Each team member revises his or her section based on team comments and sends the revised version to the team leader.** The team leader then combines the reviewed and edited sections into a single document, reads through the entire document for consistency of style and format within each section, and, if additional group feedback is necessary, sends the complete document to the entire group via e-mail (in which case everyone can respond at once), as a tracked document (then each member would need to make edits and add comments in turn), or posted on a wiki (where only one member at a time will be able to check out the document and make changes).

5. **The team leader finalizes the document.** Now that all team members have not only written and revised their sections but also had a chance to edit and comment on the complete document, the team leader has all of the information he or she needs to finalize and submit the document.

Avoiding the Pitfalls of Computer-Supported Collaboration

Before your team begins the process of writing and editing a collaborative document, make sure each of them is aware of the following potential pitfalls of computer-supported collaboration and so can avoid them.

1. **Be sure that all team members have access and authorization.** Team members cannot give input if they cannot open, edit, save, or share a document. E-mail does not present this pitfall. However, when using a document tracking system, all members must have access to the same groupware, and when using a wiki, every team member must have authorization and a password to access the document.

2. **Everyone in the group must be "in the loop."** Each member needs to be included in team e-mails in order to ensure his/her participation. If two team members are editing a tracked document at the same time, one person's edits and

comments will have to be redone later. Finally, if team members are not made aware of changes in the wiki, they won't know if it's their turn to edit next. To avoid these problems, always keep the lines of communication open.

3. Save the original draft and all subsequent drafts. The original document will automatically be saved when the group uses a document tracking system (the one document is essentially the original version and all subsequent versions combined). Wikis automatically create a new version every time the previous version is checked out, saving the previous version. But when you use e-mail, make sure that there is a continuous e-mail thread or, if not, then ask group members to save all previous e-mails that contain drafts.

4. Use a different color for each group member's edits and comments. As we saw, some e-mail programs won't permit you to use different colors, though more advanced e-mail systems allow you to color-code a document. Document tracking systems give each team member a separate color (and initials for the Comments field). Although wikis typically do not allow for color-coding, each version of the document does identify the previous editor.

5. Maintain confidentiality to protect the document from unauthorized users. As discussed earlier, do not send team e-mails or tracked documents to individuals who are not a part of the team. Since wikis require passwords to protect confidentiality, team members should be cautioned not to reveal their passwords to anyone.

6. Require all team members to sign off on and agree to the complete, final document. In some cases, a team leader may be entrusted with this responsibility, but in most cases, the entire team should take ownership of the document. Doing this will create a system of checks and balances and ensures quality control.

Meetings

One of the most frequent ways to collaborate is through meetings, which can take the form of small group discussions or large, formal conferences. Whether it is regularly scheduled (a weekly staff meeting) or a special, unscheduled one, a meeting requires teamwork. Collective energy and goodwill will bear much fruit. The guidelines on collaborative writing (pp. 80–83) also apply to group interactions at meetings. Basically, you need to know how to plan a meeting, create an agenda, and write minutes.

Planning a Meeting

As with a collaboratively written document, meetings have to be carefully planned in order to be successful. And both written documents and meetings need someone to organize them. A group of individuals cannot simply gather and start a conversation; they need to have some focus, some guidelines, some individual to help organize what is to be said, how, why, and when. If you have the responsibility of planning a meeting, you need to carefully consider the following questions before you even convene the meeting:

1. **What is the purpose of the meeting?** Determine why the meeting is necessary, what essential topics need to be discussed during the meeting, and what results/outcomes the meeting should accomplish. Jot down your ideas, which you will later use for the meeting agenda.

2. **Who should attend the meeting?** Identify key people who need to be there, including managers, co-workers, colleagues from other departments, and any individuals outside your company (clients, vendors, etc.). Keep in mind that not every manager needs (or wants) to be present at every meeting. However, as a matter of policy, find out from your boss if management needs to be included. If someone cannot attend, solicit that person's input through e-mail or include him/her via tele-conferencing or videoconferencing (see p. 113 for more on these topics).

3. **When should the meeting take place?** When planning a meeting time, have copies of each team member's schedule to help you avoid conflicts. Also consider that there are good times and bad times to hold a business meeting, as the following schedule shows:

Good Times	Bad Times
1. Midmorning or midafternoon	1. Early in the morning or late in the afternoon
2. Any day of the week that is not Monday morning or Friday afternoon, or immediately before/after a major holiday	2. Monday morning or Friday afternoon
3. Shortly after a major company celebration when morale is high	3. Immediately before or after a major holiday
	4. Same day as a long training session or long meeting

4. **Where should the meeting take place?** Select an appropriate space: A meeting for eight people does not need to take place in an auditorium, while a meeting for ten people in a room with only six chairs will not work. In addition, make sure the room you choose is equipped with the technology you or your speakers plan to use.

5. **What documents need to be presented at the meeting?** Collect any test re-sults, reports, client communications, statistics, models/blueprints, maps, and so on that are relevant for discussion. Make sure that copies of any of these documents you plan to discuss are available for the group to review ahead of time. Always specify when and where they can be read. If you plan to use presentation software, such as PowerPoint slides (see pp. 676–678), make sure you review them ahead of time to avoid/fix any problems.

Creating an Agenda

Out of your planning will come your *agenda,* or the topics to be covered at the meeting. An agenda is a one-, sometimes two-page outline of the main, pertinent points. The agenda should list only those items that your group, based on its work

Tech Note

Scheduling a Meeting with Groupware

Some groupware will automatically schedule meetings for those listed in your group directory. It will notify each group member of the date, time, and place of the meeting; automatically set up an appointment in their calendars; and alert you about any conflicts. The program also provides an alarm to remind group members 10 minutes, 30 minutes, or even a day or two in advance of the meeting. Other options you can choose include specifying whether the meeting is confidential or if visitors have been invited. Such a program eliminates the need for phone calls, memos, or e-mails and speeds up the process of group communication.

and interaction, regards as most crucial. Prioritize your action items so that the most important ones come first. Your agenda might also include short reports or presentations for which one or two members of your group are responsible. Always distribute the agenda ahead of time (at least a day or two) so that your team will be prepared and better able to contribute.

Observing Courtesy at a Group Meeting

To show your team members the courtesy they deserve during a meeting, observe the following guidelines.

1. Be on time. Coming in late can disrupt a meeting and cause further delays by bringing someone up-to-date.

2. Silence your cell phone. Set your phone on vibrate if you are expecting an important business call.

3. Do not send text messages. Sending text messages during a meeting is as rude as playing a video game.

4. Avoid side conversations. Talking to others around your while the meeting is in progress shows poor manners and can distract from the business and progress of the meeting.

5. Avoid interrupting. If you need to interject a comment, raise your hand or wait for the meeting planner to ask for comments from the group.

6. Be an active listener. Pay attention to what is being said at the meeting and also try to discern what is being said behind or underneath those words.

7. Be a focused speaker. If you are called on or choose to speak to the group, get to the point quickly, be clear, and avoid straying from your topic.

8. Do not record the meeting or take photographs. Unless you have permission, do not bring a tape recorder or video camera to a meeting and do not take photographs with a cell phone camera.

Writing the Minutes

The *minutes* are a summary of what happened at the meeting. Transmit copies of the minutes to the team members to help them recall what happened at the meeting and to help them prepare for the next one. Copies of minutes are kept on file—they are the official, permanent record of the group's deliberations and are regarded as legal documents. They may be used in courts to verify that a problem existed, that a solution was suggested, or that a specific action was or was not carried out.

Accordingly, minutes must be clear, accurate, and impartial. Keep them free from your own opinions of how well or poorly the meeting went; for example, "Ms. Saunders customarily offered the right solutions" or "Once more Hicks got off the topic" are not appropriate comments. Because the person chairing a meeting cannot take minutes and preside at the same time, another member of the group designated as the secretary should prepare the minutes. Minutes are transmitted usually within twenty-four to forty-eight hours after the meeting has adjourned.

Minutes of a meeting should include the following information:

- date, time, and place of meeting
- name of the group holding the meeting and why
- name of the person chairing the meeting
- names of those present and those absent; at the start of every meeting, circulate a piece of paper so those present can sign their names
- the approval or amendment of the minutes of the previous meeting
- for each major point—the action items—indicate what was done:
  - who said what
  - what was discussed/suggested/proposed
  - what was decided and the vote, including abstentions
  - what was continued (tabled) for a subsequent study, report, or meeting
  - time of next meeting
  - time the meeting officially concluded

To be effective, minutes must be concise and to the point. Here are a few guidelines to help you:

- Make sure of your facts; spell all names, products, tests correctly.
- Concentrate on the major facts surrounding action items. Save the reader's time and your own by condensing lengthy discussions, debates, and reports given at the meeting.
- Do not report verbatim what everyone said; readers will be more interested in what the group did.
- List each motion (or item voted on) exactly as it is worded and in its final form.
- Avoid words that interpret (negatively or positively) what the group or anyone in the group did or did not do.

Figure 3.12 shows how these parts fit together.

Minutes from a business meeting.   **Figure 3.12**

Solutions

Minutes for Environmental Safety Committee (ESC) meeting on February 2, 2009, in Room 203 of Lab Annex Building at 10:00 a.m. EST.

Members Present:

Thomas Baldanza, Grace Corlee (**President**), Virginia Downey, Victor Johnson, Roberta Koos, Kent Leviche (**Secretary**), Ralph Nowicki, Barbara Poe-Smith, Williard Ralston, Morgan Tachiashi, Asah Rashid, and Carlos Zandrillia

Members Absent:

Paul Gordon (sick leave); Marty Wagner

Old Business:

The minutes from the previous meeting on January 6, 2009 were approved as read.

Reports:

(1) Morgan Tachiashi reported on the progress the Site Inspection Committee was making in getting the plant ready for the March 29 visit of the State Board of Examiners. All preparations are on schedule.

(2) The proposal to study the use of biometric identification in place of employee ID badges was nearly complete, according to Asah Rashid, Chair of the Proposal Committee, and will be presented at next month's meeting for approval.

New Business:

(1) Virginia Downey and Ralph Nowicki expressed concern about a computer virus that may strike the plant—Monkey. Disguised as a familiar e-mail, the virus is contained in an attachment that destroys files. A motion was made by Barbara Poe-Smith, seconded by Virginia Downey, that management upgrade its antivirus protection software. Objecting to this expense, Williard Ralston thought the current software was sufficient. The vote carried by 9 to 3.

Supplies essential information on attendance, date, and place of meeting

Refers to previous meeting to provide continuity.

Concisely summarizes progress on ongoing business

Identifies key speakers

Records only main points of discussion and votes

Continued

Figure 3.12   (Continued)

Page 2

*Sticks to basic
facts and re-
sults of vote*

(2) Thomas Baldanza believed that cross-training should be accelerated,
especially in safety areas, to meet the target date of August 9, which
the ESC had set in December 2006. Agreeing, Roberta Koos stressed
that, without cross-training, some departments would be vulnerable to
safety violations. Victor Johnson, on the other hand, found that such
cross-training could not feasibly be accomplished in the original time
frame since several departments could not spare employees to
participate. He moved that the target date motion be amended and
pushed to November 15. The vote to amend was defeated 10–2.

*Explains what
will happen at
next meeting*

Calling attention to the importance of the target date, Grace Corlee
will ask Zandria Pickens, Plant Training Coordinator, to come to the
next ESC meeting to discuss the current status of the training program
and to offer suggestions for its speedy implementation.

*Includes other
business to be
continued*

(3) Kent Leviche calculated computer downtime in the plant during the
month of January—4 outages totaling 7.5 lost working hours—and
asked the ESC to address this problem. After discussion, the ESC
unanimously agreed to appoint a subcommittee to investigate the
outages and determine solutions. Roberta Koos and Thomas Baldanza
will chair the subcommittee and present a survey report at next
month's meeting.

*Excellent
morale builder*

(4) Personnel in the Environmental Testing Lab were commended for
their extra effort in ensuring that their department maintained the
highest professional standards during the month of January. A letter of
commendation was sent to Emily Lu, Lab Supervisor, and her staff.

*Signals end of
meeting and
date of next
one*

(5) Grace Corlee adjourned the meeting at 3:41 p.m.

Next Meeting:

The next meeting of the ESC will be on March 8 at 1:00 p.m. in Room 203
of the Lab Annex Building.

Virtual Meetings: Teleconferencing, Web Conferencing, and Videoconferencing

While face-to-face meetings are the most common gatherings in the workplace, technology has made virtual meetings an important option. Face-to-face meetings certainly have advantages, such as promoting group interaction, fostering close working relationships, and bypassing technical difficulties. However, sometimes one or several individuals from the group may be traveling, work at a remote site, live overseas, or be unable to attend a face-to-face meeting for another reason (weather, missed flights, etc.). For these reasons, as well as for convenience and to save money on travel costs, meetings in the world of work are often conducted in part or entirely over the telephone or via the Web.

Following is a brief description of three ways in which technology is used to bring members of a group together in today's virtual office:

- **Teleconferencing:** Teleconferencing has been a part of the workplace for as long as speakerphones have expanded conversations beyond two people. Teleconferencing systems allow for conference calls in which multiple participants at your office, across the country, or around the globe can speak to each other. Each individual has to know a particular telephone coded number to be connected via a private branch exchange (PBX). Teleconferencing is by far one of the most common ways to conduct a virtual meeting.
- **Web conferencing:** Web conferencing is different from teleconferencing in that all of the participants attend the conference from their own computers and are connected to one another over the Internet, rather than a PBX. People attending Web conferences participate in the meeting in "real time" by visiting a specific website and logging in or downloading software. The advantage of Web conferencing over teleconferencing is that individuals attending a Web conference can view presentations and documents during a meeting.
- **Videoconferencing:** Videoconferencing is the most sophisticated form of virtual conferencing. It combines the audio component of teleconferencing (you can hear the other participants) with the face-to-face component of a traditional meeting (you can see participants), and the textual component of a Web conference (you can view presentations and documents during the meeting). (See pages 676–678).

Taking Notes

Minutes are crucial to the day-to-day operation and long-range planning of a business. Some meetings are held in a company office for a small group such as the Environmental Safety Committee in Figure 3.12. Others take place online as indicated in Figure 3.10. Still others are held off site where, because of time and location, you may be the only member of your team attending and will have to share your notes with co-workers.

Again, a variety of software packages can help you to share your notes via e-mail. Take your laptop to the meeting to help you. As you take notes, include only major issues that have top priority for your audience. Don't get bogged down in recording minor details. Inform your team or boss about important issues affecting new dates, revised schedules, different prices, amended policies or regulations, changes in operations, service, key decisions, and other important business matters. Anticipate the questions your team or boss may have about the meeting and then supply the answers through your notes. Figure 9.8 (pp. 416–417) contains a short summary from the notes two employees made of the major topics covered at a seminar they attended.

Revision Checklist

- [ ] Succeeded in being a team player by putting the success of my group over the needs of my own ego.
- [ ] Followed necessary steps of the writing process to take advantage of team effort and feedback.
- [ ] Attended all group meetings and understood and agreed to responsibilities of the group and my own obligations.
- [ ] Finished research, planning, and/or drafting expected of me.
- [ ] Conducted necessary interviews and conferences to gather, clarify, and verify information.
- [ ] Shared my research, ideas, and suggestions for revision through constructive criticism.
- [ ] Participated honestly and politely in discussions with colleagues.
- [ ] Treated members of my team with respect and courtesy.
- [ ] Was open to criticism and suggestions for change.
- [ ] Read colleagues' work and gave specific and helpful criticism and suggestions.
- [ ] Kept matters in proper perspective by not being a nitpicker and by not interrupting with extraneous points or unnecessary questions.
- [ ] Sought help when appropriate from relevant subject matter experts and from co-workers.
- [ ] Secured responses and approval from management.
- [ ] Took advantage of e-mail, instant messaging, groupware and technology to communicate with my collaborative team.
- [ ] E-mailed messages in clear, diplomatic, and correct sentences.
- [ ] Attached pertinent documents in e-mails to collaborative team.
- [ ] Answered e-mail questions and responded to requests promptly.

Continued

Continued

☐ Used computer-assisted editing technologies (e-mail, tracking systems, wikis) effectively and ethically. Respected confidentiality.
☐ Prepared agenda for meetings clearly and distributed ahead of time.
☐ Wrote minutes that objectively reported what happened.
☐ Took notes that highlighted main points of a meeting for my collaborative team and boss.
☐ Participated in virtual meetings, through teleconferencing or Web or videoconferencing.

Exercises

1. Assume you belong to a three- or four-person editing team that functions the way Tara Barber's does in Figure 3.3. Each member of your team should bring in four copies of a paper written for this course or for another one. Exchange copies with the other members of your team so that each team member has everyone else's papers to review and revise. For each paper you receive, comment on the style, organization, tone, and discussion of ideas as Wells McCraw did in Figure 3.7.

2. With members of your collaborative team (selected by your teacher or self-appointed) select four different brands of the same leading product (such as a software package, a Web browser, a smartphone, a microwave, a DVD player, a power tool, or other item). Each member of your team should select one of the brands and prepare a two-page memo report (see pp. 122–130), evaluating it for your instructor according to the following criteria:

   - convenience
   - performance
   - technical capabilities/capacities
   - appearance
   - adaptability
   - price
   - weaknesses/strengths compared with competitors' models

   Each team member should then submit a draft to the other members of the team to review. At a subsequent group meeting, the group should evaluate the four brands based on the team's drafts and then together prepare one final recommendation report for your instructor.

3. Your company is planning to construct a new office, and you, together with other employees from your company, have been asked to serve on a committee to make sure that plans for the new building adhere precisely to the Americans with Disabilities Act, passed in 1990. According to that act, it is against the law to discriminate against anyone with disabilities that limit "major life activities" — walking, seeing, speaking, or working.

The law is expressly designed to remove architectural and physical barriers and to make sure that plans are modified to accommodate those protected by the law (for example, wider hallways to accommodate wheelchairs). Other considerations include choosing appropriate nonstick floor surfaces (reducing the danger of slipping), placing water fountains low enough for use by individuals in wheelchairs, and installing doors that require minimal pressure to open and close.

After studying the plans for the new building, you and your team members find several problem areas. Prepare a group-written report advising management of the problems and what must be done to correct them to comply with the law. Divide your written work according to areas that need alteration—doors, floors, water fountains, restroom facilities. Each team member should bring in his or her section, for the group to edit and revise. The group should then prepare the final report for management.

4. A new president will be coming to your college in the next month, and you and five other students have been asked to serve on a committee that will submit a report about campus safety problems and what should be done to solve them. You and your team must establish priorities and propose guidelines that you want the new administration to put into practice. After two very heated meetings, you realize that what you and two other students have considered solutions, the other half of your committee regards as the problems. Here is a rundown of the leading conflicts dividing your committee:

- **Speed bumps.** Half the committee likes the way they slow traffic down on campus, but the other half says they are a menace because they jar car CD players.
- **Sound pollution.** Half your team wants Campus Security to enforce a noise policy preventing students from playing loud music while driving on campus, but the other half insists that would violate students' rights.
- **Van and sport utility vehicle parking.** Half the committee demands that vans and sport utility vehicles park in specially designated places because they block the view of traffic for any vehicle parked next to them; the other members protest that people who drive these vehicles will be singled out for less desirable parking places on campus.

Clearly your committee has reached a deadlock and will be unproductive as long as those conflicts go unresolved. Based on the above scenario, do the following:

a. Have each student on the committee write (or e-mail, if available) the other five students suggesting a specific plan on how to proceed—how the group can resolve their conflicts. Prepare your e-mail message and send it to the other five committee members and to your instructor. What's your plan to get the committee moving toward writing the report to the incoming president?

b. Assume that you have been asked to convince the committee to accept your views on the three areas of speed bumps, noise control, and parking. Send the three opposition students a memo or e-mail persuading them to your way of thinking. Keep in mind that your message must assure them that you respect their point of view.

c. Assume that the committee members reach a compromise after seeing your plan put forth in (a). Collaboratively draft a three-page report to the new president.

d. Collaboratively draft a letter to the editor of your student newspaper defending your recommendations to the student body and explaining how the group resolved its difficulty. This is a public statement that the group felt it was important to write; you will have to choose your words carefully to win campuswide support.

5. You work for a hospital laboratory, and your lab manager, under pressure from management to save money, insists that you and the three other med-techs switch to a different brand of vacuum blood-drawing tubes. You and your colleagues much prefer the brand of tubes you have been using for years. Moreover, the price difference between the two brands is small. As a group project, prepare a memo to the business manager of the hospital explaining why the switch is unnecessary, unwise, and unpopular. Then prepare another collaboratively written memo to your lab manager. Be sensitive to each reader's needs as you diplomatically explain the group's position.

PART II

Correspondence

4 **Writing Routine Business Correspondence**
Memos, Faxes, E-Mails, IMs, and Blogs

5 **Writing Letters**
Some Basics for Audiences Worldwide

6 **Types of Business Letters**

7 **How to Get a Job**
Searching, Preparing, Applying, and Interviewing

Writing Routine Business Correspondence

Memos, Faxes, E-Mails, IMs, and Blogs

Memos, faxes, e-mails, IMs, and blogs are the types of writing you will do most frequently on the job. These five forms of business correspondence are quick, easy, and effective ways for a company to communicate internally as well as externally. You can expect to send one or more of these routine forms of correspondence each day to co-workers in your department, to colleagues in other divisions of your company, to decision makers at all levels, and to clients and customers as well.

Visit www .cengage.com/ english/kolin for this chapter's online exercises, ACE quizzes, and Web links.

What Memos, Faxes, E-Mails, IMs, and Blog Posts Have in Common

Although memos, faxes, e-mails, IMs, and blog posts are very different types of correspondence, they share the following characteristics:

1. **They give busy readers information quickly.** While the messages they contain can be about any topic in the world of work, most often they focus on day-to-day activities and operations at your company—sales and product information, policy and schedule changes, progress reports, orders, troubleshooting problems, and so forth.

2. **Each of these types of correspondence is streamlined for the busy world of work.** Unlike letters, proposals, or reports, which can be long and detailed and contain formal parts and sections, these routine types of correspondence give writers and readers a particularly fast way to communicate. Even though memos do have to be formatted (see pages 127–128), they, like blog posts, IMs, or e-mails, are ready-made to send and receive shorter, quicker messages.

3. **They are informal.** Compared to letters or reports, these kinds of routine correspondence are not as formal. They emphasize a conversational style of writing and can use the jargon and abbreviations in contexts both writer and reader are familiar with.

4. Even though they are routine, they still demand a great deal of thought and time. Although memos, faxes, e-mails, blogs, and IMs are less formal than, for instance, a letter to a client, they all must be written clearly and with correct grammar and punctuation, even when the correspondence is between two employees. Always plan what you are going to write.

5. They represent your company. These routine messages are reflections of your company's image and your professionalism. Be careful about what you write and how you say it. Just because the format of your message is informal does not mean you can be unprofessional, even in a brief e-mail to a co-worker. Your success as an employee can depend as much on your writing an unbiased, ethically proper e-mail or blog post as it does on your technical expertise.

Memos

Memorandum, usually shortened to *memo*, is a Latin word for "something to be remembered." The Latin meaning points to the memo's chief function: to record information of immediate importance and interest in the busy world of work. Memos are brief, informal, but can contain official announcements that serve a variety of functions, including:

- making an announcement
- providing instructions
- clarifying a policy, procedure, or issue
- changing a policy or procedure
- alerting employees to a problem or issue
- offering general information
- providing a brief summary
- making a request
- offering suggestions or recommendations
- providing a record of an important matter
- confirming an outcome
- calling a meeting

Memos are usually written for an in-house audience, although the memo format can be used for documents sent outside a company, such as short reports or proposals (see Chapters 13 and 14) or for cover notes for longer reports (see Chapter 15).

Memos keep track of what jobs are done where, when, and by whom; they also report on any difficulties, delays, or cancellations and what your company or organization needs to do about correcting or eliminating them.

Memo Protocol

As with any other forms of business correspondence, memos reflect a company's image and therefore must follow the company's *protocol* — accepted ways in which in-house communications are formatted, organized, written, and routed. In fact, some companies offer protocol seminars on how employees are to prepare communications. In addition to following your company's protocol, use these common sense guidelines when writing memos:

1. Be timely. Don't send out a memo at the last minute, particularly if the purpose of the memo is to announce a meeting.

2. Be professional. The informal nature of memos doesn't mean that you should compose a poorly organized, poorly written, and/or factually inaccurate memo or one that contains misspellings or punctuation errors. Notice that in Figure 4.1 the informal memo between Roger and Lucy is professionally written and clearly organized and formatted.

Standard memo format.   Figure 4.1

MEMO

TO: Lucy
FROM: Roger
DATE: November 12, 2008
SUBJECT: Review of Successful Website Seminar

As you know, I attended the "How to Build a Successful Website" seminar on November 7 and learned the "rules and tools" we will need to redesign our own site.

Here is a review of the major topics covered by the director, Jackie Chen:

1. Keep your website content-based—hit your target audience.
2. Visualize and "map out" your site's ad links.
3. Design your website to look the way you envision it—make it aesthetically pleasing.
4. Make your site easy to navigate.
5. Maintain your page outline; make changes when necessary.
6. Create hot links and image maps to move users from page to page.
7. Complete your site with appropriate sound and animation.

Could we meet in the next day or two to discuss recreating our own website in light of these guidelines? I would really like your suggestions about this project. Thanks.

Header

Memo parts

Introduction gives background and tells reader what memo will do

Preview

Body: Numbered list helps readers follow information quickly

Conclusion: Asks for comments

3. Be tactful. Be polite and diplomatic, not curt and bossy. For example, in Figure 4.2, notice how Janet Hempstead adopts a firm tone regarding an important safety issue, yet she does not blame or talk down to her readers — the machine shop employees.

4. Send memos to the appropriate personnel. Don't send copies of a memo to people who don't need to read them. It wastes time and energy. Moreover, don't send a memo to high-ranking company personnel in place of your immediate supervisor, who may think you are going over his or her head. For instance, in Figure 4.3, Mike Gonzalez has sent his memo only to the vice president and the public relations officer, the two people who will most likely benefit from the memo's recommendations, and in Figure 4.2, Janet Hempstead has sent her memo only to the machine shop workers, not to the upper management of the Dearborn Company.

Keep in mind, though, that memos are often sent up and down the corporate ladder. Employees send that to their supervisors, and workers send memos to one another. Figure 4.1 shows a memo sent from one worker to another while Figure 4.2 contains a memo sent from the top down. Figure 4.3, however, illustrates a memo sent from an employee to management.

E-Mail or Hard Copy?

A memo can be sent as a printed hard copy, as an e-mail, or in an e-mail attachment. Find out what your company's policy is. Increasingly, e-mail is replacing printed memos but there are times when a hard copy memo is preferred or even essential. Send a notification, such as the one in Figure 4.2, as a hard copy memo for official documentation that cannot be deleted or erased. Also, when your message is confidential (evaluating a co-worker or a vendor's services), making a budget request, or discussing other sensitive financial information), send hard copy. It cannot be forwarded as an e-mail can.

Memo Audience, Style, and Tone

Before you write your memo, think about your audience's needs. Doing this will not only help you determine the content of your memo, but will also help you select the most appropriate style and tone.

In terms of audience, identify who you are writing to and think through the questions your audience will expect your memo to answer. The subject of the memo will answer the question *what*. You also need to determine the answers to the questions *when*, *who*, *where*, and *why*, and consider any questions about cost and technology your audience may have. Here are some key questions your audience may ask and that your memo needs to answer clearly and concisely:

1. When? When did it happen? Is it on, ahead of, or behind schedule? Does it work with workers'/managers' schedules? When does it need to be discussed or implemented? *When* is answered in Figures 4.1 ("November 7," "in the next day or two"), 4.2 ("during the past two weeks," "after each use"), and 4.3 ("in early February," "before the end of the month").

Memo with a clear introduction, discussion, and conclusion.   Figure 4.2

Dearborn Equipment Company

To: Machine Shop Employees
From: Janet Hempstead, Shop Supervisor
Date: September 27, 2009
Subject: Cleaning Brake Machines

During the past two weeks I have received several reports that the brake machines are not being cleaned properly after each use. Through this memo I want to emphasize and explain the importance of keeping these machines clean for the safety of all employees.

When the brake machines are used, the cutter chops off small particles of metal from brake drums. These particles settle on the machines and create a potentially hazardous situation for anyone working on or near the machines. If the machines are not cleaned routinely before being used again, these metal particles could easily fly into an individual's face when the brake drum is spinning.

To prevent accidents like this from happening, please make sure you vacuum the brake machines after each use.

You will find two vacuum cleaners for this purpose in the shop—one of them is located in work area 1-A and the other, a reserve model, is in the storage area. Vacuuming brake machines is quick and easy: It should take no more than a few seconds. This is a small amount of time to make the shop safer for all of us.

Thanks for your cooperation. If you have any questions, please call me at Extension 324 or come by my office.

204 South Mill St., South Orange, NJ 02341-3420 (609) 555-9848 JHEMP@dearco.com

Introduction gives purpose and importance of memo

Discussion states why problem exists and how to solve it

Safety message is boldfaced for emphasis

Conclusion builds goodwill and asks for questions

Figure 4.3   A memo that uses headings to highlight organization.

Company logo

RAMCO INDUSTRIES

Where Technology Shapes Tomorrow

ramco@gem.com   http://www.Ramcogem.com

TO:	Rachel Mohler, Vice President
	Harrison Fontentot, Public Relations
FROM:	Mike Gonzalez
SUBJECT:	Ways to Increase Ramco's Community Involvement
DATE:	March 2, 2009

Introduction supplies background and rationale

At our planning session in early February, our division managers stressed the need to generate favorable publicity for our new Ramco facility in Mayfield. Knowing that such publicity will highlight Ramco's visibility in Mayfield, I think the company's image might be enhanced in the following ways.

Body offers concrete evidence (costs, personnel, location) that plan can work

CREATE A SCHOLARSHIP FUND

Ramco would receive favorable publicity by creating a scholarship at Mayfield Community College for any student interested in a career in technology. A one-year scholarship would cost $4,800. The scholarship could be awarded by a committee composed of Ramco executives and staff. Such a scholarship would emphasize Ramco's support for technical education at a local college.

Headings reflect organization

OFFER SITE TOURS

Guided tours of the Mayfield facility would introduce the community to Ramco's innovative technology. The tours might be organized for community and civic groups. Individuals would see the care we take in protecting the environment in our production and equipment choices and the speed with which we ship our products. Of special interest to visitors would be Ramco's use of industrial robots alongside its employees. Since these tours would be scheduled in advance, they should not conflict with our production schedules.

PROVIDE GUEST SPEAKERS

Many of our employees would be excellent guest speakers at civic and educational meetings in the Mayfield area. Possible topics include the technological advances Ramco has made in designing and engineering and how these advances have helped consumers and the local economy.

Closing requests feedback and authorization

Thanks for giving me your comments as soon as possible. If we are going to put one or more of these suggestions into practice before the facility opens, we'll need to act before the end of the month.

2. Who? Who is involved? Who is affected by it? How many people are involved? How well are they prepared for it? *Who* is answered in Figures 4.1 (Jackie Chen), 4.2 (all machine shop employees), and 4.3 (Ramco Industries as a whole).

3. Where? Where did it take place or will it take place? *Where* is answered in Figures 4.1 (the website seminar), 4.2 (the machine shop), and 4.3 (the Mayfield plant).

4. Why? Why is it an important topic? *Why* is clearly answered in Figures 4.1 (because the website is being redesigned), 4.2 (because it's a safety issue), and 4.3 (because favorable publicity could result).

5. Costs? How much will it cost? Will the cost be lower or higher than a competitor's costs? Will the costs be above or below projections? Is the cost worth it? Not every memo will answer cost questions, but in Figure 4.3, the specific cost of an individual scholarship ($4,800) is provided.

6. Technology? What technology is involved? Why is the technology needed? Is the technology available, current, adaptable, safe for the environment? Again, not every one of your memos will answer questions about technology, but Figures 4.2, 4.3, and especially Figure 4.1 all refer to technological matters or issues.

In terms of style and tone, consider whether your audience will consist of co-workers, managers, people you supervise, individuals outside your company, or a combination of these readers.

When writing a memo to a co-worker you know well, it's best to adopt a casual tone. You want to be seen as friendly and cooperative. To do otherwise would make you look self-important and hard to work with. Consider the friendly tone of the colleague-to-colleague memo in Figure 4.1 and note how Roger ends the memo to Lucy politely but informally. But when writing a memo to a manager, you will want to use a more formal tone. A manager will expect you to show a more respectful, even official, posture. Mike Gonzalez's memo to his bosses in Figure 4.3 is formal yet conversationally persuasive.

Here are two ways of expressing the same message, the first more suitable when writing to a co-worker and the second more appropriate for a memo to a manager:

Co-worker: I think we should go ahead with Marisol's plan for reorganization. It seems like a safe option to me, and I don't think we can lose.

Manager: I think that we should adopt the organizational plan developed by Marisol Vega. Her recommendations are carefully researched and persuasively answer all of the questions our department has about solving the problem.

When writing a memo to people you supervise about policies or procedures, be official and straightforward, but always take your readers' feelings into account. Don't be harsh or condescending. Again, study how in Figure 4.2 Janet Hempstead gets her point across clearly and uncompromisingly yet also offers advice and assistance. She does not blame but emphasizes the need for a safe work environment.

When writing to multiple audiences, it's better to be more formal than informal.

Memo Format

Memos vary in format. Some companies use standard, printed forms (Figure 4.1), while others (as in Figure 4.2) have their names (letterhead) printed on their memos. As we saw, you can also create a memo by including the necessary parts in an e-mail, as in Figures 4.4 and 4.5 on pages 132 and 133.

As you can see from looking at Figures 4.1 through 4.3, memos look different from letters, and they are less formal. Because they are often sent to individuals within your company, memos do not need the formalities necessary in business letters, such as an inside address, salutation, complimentary close, or signature line, as we will see in Chapter 5 (see pp. 157–161).

Basically, the memo consists of two parts: the **header**, or the identifying information at the top, and the **message** itself. The header includes these easily recognized parts: **To, From, Date**, and **Subject** lines.

TO: Aileen Kelly, Chief Computer Analyst
FROM: Stacy Kaufman, Operator, Level II
DATE: January 30, 2009
SUBJECT: Progress report on the fall schedule

Or you can use a memo template in your word processing program that will list these headings, as follows, to save time.

TO:
FROM: Linda Cowan
DATE: October 4, 2010
RE: (Enter subject here.)

On the **To** line, write the name and job title of the individual(s) who will receive your memo or a copy of it. If you are sending your memo to more than one reader, make sure you list your readers in the order of their status in your company or agency, as Mike Gonzalez does in Figure 4.3 (according to company policy the vice president's name appears before that of the public relations director). If you are on a first-name basis with the reader, use just his or her first name, as in Figure 4.1. Otherwise, include the reader's first and last names. Don't leave anyone who needs the information out of the loop.

On the **From** line, insert your name (use your first name only if your reader refers to you by it) and your job title (unless it is unnecessary for your reader). Some companies ask employees to handwrite their initials after their typed name to verify that the message comes from them and that they are certifying its contents, as in Figures 4.2 and 4.3. You do not have to key in your intials in a memo sent as an e-mail.

On the **Date** line, do not simply name the day of the week — Friday. Give the full calendar date — June 1, 2009.

On the **Subject** line, write the purpose of your memo. The subject line serves as the title of your memo; it summarizes your message. Vague subject lines, such as "New Policy," "Operating Difficulties," or "Shareware," do not identify your message precisely and may suggest that you have not restricted or developed it sufficiently. "Shareware," for example, does not tell readers if your memo will discuss new equipment, corporate arrangements, or vendors; offer additional or fewer benefits; or warn employees about abusing the system. Note how Mike Gonzalez's subject line in Figure 4.3 is much more precise than just saying "Ramco's Community Involvement."

Strategies for Organizing a Memo

Don't just dash your memo off. Take a few minutes to outline and draft what you need to say and to decide in what order it needs to be presented. Organize your memos so that readers can find information quickly and act on it promptly. For longer, more complex communications, such as the memos in Figures 4.2 and 4.3, your message might be divided into three parts: (1) introduction, (2) discussion, and (3) conclusion. Regardless of how short or long your memo is, recall the three *P*'s for success — *plan* what you are going to say; *polish* what you wrote before you send it; and *proofread* everything.

Introduction

The introduction of your memo should do the following:

- Tell readers clearly about the problem, procedure, question, or policy that prompted you to write.
- Explain briefly any background information the reader needs to know.
- Be specific about what you are going to accomplish in your memo.

Do not hesitate to come right out and say, "This memo explains new e-mail security procedures" or "This memo summarizes the action taken in Evansville to reduce air pollution." See how clearly this is done in Figure 4.1.

Discussion

In the discussion section (the body) of your memo, help readers in these ways:

- State why a problem or procedure is important, who will be affected by it, and what caused it and why.
- Indicate why and what changes are necessary.
- Give precise dates, times, locations, and costs.

See how Janet Hempstead's memo in Figure 4.2 carefully describes an existing problem and explains the proper procedure for cleaning the brake machines, and how Mike Gonzalez in Figure 4.3 offers carefully researched evidence on how Ramco can increase its favorable publicity in the community.

Conclusion

In your conclusion, state specifically how you want the reader to respond to your memo. To get readers to act appropriately, you can do one or more of the following:

- Ask readers to call you if they have any questions, as in Figure 4.2.
- Request a reply — in writing, over the telephone, via e-mail, or in person — by a specific date, as in Figures 4.2 and 4.3.
- Provide a list of recommendations that the readers are to accept, revise, or reject, as in Figures 4.1 and 4.3.

Routing Memos

As we saw, your memo needs to go to the right people. Don't increase the paper inflation or electronic traffic in your company. It would be presumptuous and a waste of resources to send a memo to individuals (inside or outside your company) who don't need it just to show you are a hard-working employee.

Always verify the names and addresses of the individuals you need to send your memo to. Since memos often carry time-sensitive messages, sending one to the wrong person or to a wrong address could jeopardize the flow of information in a company.

Another important matter to consider is the level of official importance and confidentiality of your memo. If your memo is an official document, you may not want to send it via e-mail, because it could easily be deleted or altered. Moreover, if your memo is confidential (e.g., an evaluation of a co-worker or vendor, a message containing sensitive financial or medical information), you may not want to send it via e-mail, because it could easily be forwarded or fall victim to hackers.

Faxes

Even though you may use e-mail and e-mail attachments extensively, fax (facsimile) machines are still in widespread use in the world of work. A fax is an original document copied and transmitted over telephone or computer lines. Faxes are particularly helpful either when you have only hard copy to send or when you want to send an original signed letter, contract, blueprint, artwork, or other document that you could not send via an e-mail transmission. Faxes demonstrate exactly what original documents look like and allow recipients to obtain a hard copy quickly, without having to wait for mail delivery.

Faxes are among the simplest forms of routine business correspondence because the only writing involved is completing the fax cover sheet. When you send a fax, make sure your cover sheet includes the following information:

1. **The name of the sender and his or her fax and phone numbers.** The phone number is important because it enables the recipient to report an incomplete transmittal.
2. **The name of the recipient and his or her fax and phone number.** The recipient's fax and phone numbers should be included for the sender's reference.
3. **The total number of pages being faxed.** Note that the total number of pages includes the cover sheet itself. The page count helps recipients determine whether the entire fax was received.
4. **A brief explanatory note** that lets the recipient know what the fax is, what its purpose is, and how and when to respond to it.

Guidelines for Sending Faxes

Follow these four guidelines to make sure you fax a clear and complete document:

1. **Be sure the original documents you send are clear.** An unclear faxed document will be difficult for the recipient to read. For example, penciled comments may be too faint to fax clearly, and highlighted comments and those written in blue ink will not transmit (red and black ink are your best choices).
2. **Avoid writing on the top, bottom, or edges of the documents to be faxed.** Any comments written on the outer edges may be cut off or blurred during transmittal.

3. **Do not send overly long faxes.** Be careful about sending anything longer than three or four pages since you will tie up both your own and the recipient's fax machines.
4. **Respect the recipient's confidentiality.** Since faxes may be picked up by other employees in your office, don't assume your message will be confidential, unless the recipient has a private fax machine.

E-Mail

E-mail is the most common form of workplace communication. Professionals in the world of work may receive hundreds of e-mails a day from managers, colleagues, clients/customers, and vendors. E-mail is the lifeblood of every business or organization because it expedites communication in many ways. For instance, using e-mail you can:

1. Send and receive information efficiently — forwarding, storing, and classifying the information by date and sender.
2. Enhance all phases of your collaborative work (see Chapter 3).
3. Send a variety of text/graphics files, including pictures, video clips, sound bites, lists, and financial documents.
4. Eliminate troublesome phone tag.
5. Communicate any time, all the time, 24-7.

E-mail is among the most informal and relaxed type of business correspondence, far more so than a printed memo, a letter, a short report, or a proposal. Think of an e-mail as a polite, informative, yet professional conversation — friendly, to the point, and always accessible. Yet, even though business e-mail is informal, sending it does not mean you can forget your responsibilities as a careful and courteous writer. The sections below will show you how to do that.

Business E-Mail Versus Personal E-Mail

The e-mails you write on the job will require more from you as a writer than your personal e-mails. You cannot write to an employer or customer the way you would to a close friend. The audience for your business e-mails has different expectations. Again, keep in mind that you are representing more than just yourself and your preferences, as in a personal e-mail. You will be speaking on behalf of your employer. Unlike your personal e-mail, therefore, your business e-mail has to consider the impact it will have on your company, your department, and your career.

　　To project the best image of your company, make sure your e-mail is businesslike, carefully researched, and polite. Review each e-mail message before you click on Send. Don't just dash off an e-mail. You need to follow all the rules of proper spelling, capitalization, punctuation, word choice, and courtesy. The tone of your business e-mail should be professional, business-like. Sarcasm and slang do not belong in a business e-mail, and you can be fired for writing an angry or abusive e-mail. Figures 4.4 and 4.5 are examples of effectively written business e-mails.

Visit www .cengage.com/ english/ kolin for an online exercise, "Comparing Your Personal and Professional E-mails."

Figure 4.4   An example of an effectively written business e-mail.

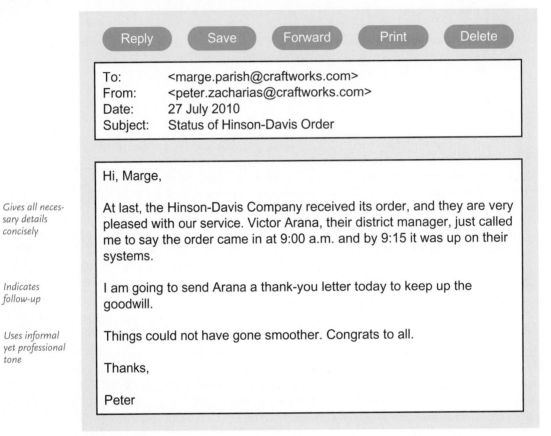

Gives all necessary details concisely

Indicates follow-up

Uses informal yet professional tone

E-Mails Are Legal Records

Employers own their internal e-mail systems and thus have the right to monitor what you write and to whom when you are on the job. Any e-mail on the company's server can be stored, retrieved, forwarded, or intercepted. Keep in mind your e-mail can easily be converted into an electronic paper trail. You never know who will receive and then forward your e-mail — to your boss, an attorney, or a licensing agency. Many companies include disclaimers protecting themselves from legal action against them because of an employee's offensive behavior in a company e-mail.

Here are some safe rules to follow regarding the legal implications of workplace e-mails:

■ Do not use the company e-mail system for personal messages. Use it only for appropriate company business, and make sure you are professional and conscientious.

E-mail sent to a distribution list of co-workers.   Figure 4.5

| Reply | Save | Forward | Print | Delete |

To:	<annulla.cranston@netech.com>; <peter.hwang@netech.com>; <margaret.habermas@netech.com>; <a.pena@netech.com>
From:	<melinda.bell@netech.com>
Date:	19 June 2010
Subject:	Collaboration on annual report

Hello, Team:

To follow up on our conversation yesterday regarding working together on this year's annual report, I'm glad our schedules are flexible. I've checked our calendars and we are all available next Tuesday the 26th at 10:30 a.m. Let's meet in Conference Room 410.

Don't forget we have to draft a two- to three-page overview first that explains Northeast's strategic goals and objectives for Fiscal Year 2011. Not an easy assignment, but we can do it, gang.

It would be a big help if Annulla would bring copies of the reports for the last three years. Would Peter please call Ms. Jhandez in Engineering for a copy of the speech she gave last month to the Powell Chamber of Commerce? If memory serves me correctly, she did a first-rate job summarizing Northeast's accomplishments for 2009.

Thanks for all your splendid work, team. See you Tuesday.

Melinda Bell
Director, Marketing
New Tech
<melinda.bell@netech.com>
FAX: (603) 555-2162
Voice: (603) 555-1505
http://www.netech.com

Starts with context for and confirmation of meeting

Provides clear explanations and instructions

Clearly lists responsibilities

Ends by building morale

Signature block

- Never write an e-mail to discuss a confidential subject — a raise, a grievance, or a complaint about a co-worker. Meet with your supervisor or human resources representative face to face instead.
- Make sure of your facts before sending an e-mail to a customer, vendor, or agency. If you send them the wrong or misleading information about prices,

policies, warranties, or safety features, or if you make a promise your employer can't keep, then your company can be legally liable.

Guidelines for Using E-Mail

When you prepare and organize your e-mail message, always consider your reader's specific needs as well as those of your company. Following the guidelines below will help you to write effective business e-mails:

1. **Always be ethical.**
   - **Never attack your employer, a co-worker, a customer, or a company.** Avoid *flaming*, that is, using strong, angry language that mocks or insults your reader, as in Figure 4.6 (p. 138). Using abusive, obscene, or discriminatory language in an e-mail constitutes grounds for dismissal.
   - **Do not forward a co-worker's or manager's e-mail without approval.** It is unethical to forward someone's e-mail without getting permission first. Keep in mind, too, that e-mails you or other employees write are the property of the company you work for. You violate copyright law by not respecting your employer's ownership.
   - **Avoid changing the wording of a message that you are expected to read and forward.** It is unethical to change someone's wording without checking with him or her, even if your alteration is done for seemingly helpful reasons. Be careful not to change the wording of a message that you are expected simply to read and forward.
   - **Do not forward confidential or sensitive messages to unintended readers.** Always verify that you are sending your message to the right individual(s). A confidential message has legal implications and should not be sent to competitors, individuals in other divisions of your company, or anyone involved in a legal dispute with your firm. Even if a message doesn't have legal implications, it may contain embarrassing news if sent to the wrong party (e.g., telling a co-worker about another worker's personal business).

2. **Make your e-mails easy to read.**
   - **Make the subject of your message clear.** Always provide a clear and concise subject line that precisely summarizes the contents of your message and makes your e-mail easy to identify and store. Consider writing a new subject line when the message changes, even if you are attaching a previous message.
   - **Try to limit your e-mails to one screen.** If you are sending a longer message, send it as an attachment rather than in the body of an e-mail.
   - **Do not send e-mails written in all capital or all lowercase letters.** The use of all capital letters makes you appear to be shouting. Conversely, sending e-mails in all lowercase letters makes you look too lazy to observe the rules of punctuation and capitalization.

- **Break your message into paragraphs.** A screen filled with one dense block of text is intimidating. Make sure each paragraph is no more than a few sentences long, and always double-space between paragraphs.
- **Use plain text.** Avoid overusing certain typefaces (like italic and bold), complex formatting (such as numbered and bulleted lists), and symbols (monetary symbols, accents, etc.) within the text of an e-mail. Different e-mail programs can garble your message, converting typefaces, formatting, and symbols into nonsense. If you use more complex text, make sure your recipient has a compatible e-mail program, or send the text as an attachment.
- **Avoid long strings of e-mail.** Delete long strings of previously answered e-mails when you reply so the reader will not waste time scrolling to find your answer.

3. **Observe the rules of netiquette (Internet + etiquette).**
   - **Respond promptly to e-mail.** Check your messages at least twice a day — once in the a.m. and then again around 2 or 3 p.m. Don't let your e-mails build up, unread. If you will be offline, use the auto-reply feature to tell readers when you expect to return. Prioritize your messages so that you respond to your boss and high-priority matters first.
   - **Give your readers reasonable time to respond.** Consider time zone differences between you and your reader. It may be 2:00 a.m. when your e-mail arrives for an international recipient.
   - **Do not send incessant reminders.** If you don't receive a reply, do not keep sending the same e-mail over and over. You will only antagonize your recipient.
   - **Avoid unfamiliar abbreviations, jargon, and emoticons.** Don't use the types of abbreviations found in your personal e-mails (example: LOL) or those used in text messaging. Use only those abbreviations and jargon that your co-workers will understand. Also, do not use emoticons (smiley faces, sad faces, etc.); they should not be a part of your professional communications.
   - **Don't use red flag words.** Stay away from words like "Urgent," "Crucial," "Top Priority" in your subject line just to get your readers' attention. Your tactic will backfire, potentially upsetting your reader or, worse yet, causing them to ignore genuinely urgent messages you send them in the future.
   - **Include a signature block.** A signature block includes your name, title, and contact information at the end of your message (see Figure 4.5). Always include one when you are communicating with someone outside of your company or agency. Consider using a signature block in your internal e-mails as well to save co-workers at different locations the trouble of locating your contact information.

4. **Adopt a professional style.**
   - **Use a salutation (greeting) and complimentary close.** When e-mailing a co-worker, use an informal salutation and complimentary close (Hi, Hello, Bye, Thanks); but when e-mailing a manager, be more formal (Mr. Botwin,

Tech Note

Using an E-Mail Address Book

The most convenient place to store names and e-mail addresses for easy retrieval is in an e-mail address book. Address books are available in most e-mail services, including Web-based e-mail services, such as Hotmail®, Gmail, and Yahoo!. The benefits of address books are many.

- They allow you to store e-mail addresses, phone numbers, work and home addresses, even IM addresses and birthdays.
- They help you to update contact information so that you have only the most current information in hand.
- By printing all or part of your address book, you have a hard copy to take with you, if necessary.
- Most address books let you import and export from other address books so you can combine information from several address books.

Ms. De Sousa, Thank You). Similarly, when e-mailing clients or customers you do not know, always be formal (Dear Ms. Pietz; Dear Bio-Tech; Sincerely). But always follow your company policy.

- **Keep your messages concise.** Cut wordy phrases and send only the information your reader needs. Exclude nonessential details and chatter.
- **Avoid short, curt e-mails and replies.** Don't turn your e-mail into a telegram with such discourteous, abrupt messages as "Send report immediately; need for meeting" or reply only with "Yes," "No," or "Sure." Save words like "Nope," "Yeah," and "Huh" for your personal e-mails.
- **Respect the cultural traditions of international readers.** Avoid using abbreviations, symbols for monetary units, or measurements that your reader may not use or know. Do not use first names unless the reader approves. Also, stay away from jokes, slang, and political commentary. For more information on communicating with international readers, see "Guidelines for Communicating with International Readers" in Chapter 5 (pp. 173–178).
- **End politely.** In addition to using a complimentary close (Sincerely, Best Regards, Yours Truly), let readers know in your last sentence that you appreciate their help/cooperation and look forward to their reply (see Figure 4.5).

5. **Ensure your e-mails are safe and secure.**
   - **Be careful that you do not contract or spread an e-mail virus.** Do not open attachments or click on hyperlinks from unsolicited e-mails. If you do contract a virus, get help cleaning it up immediately and do not forward the virus.
   - **Don't be a victim of identity theft, or "phishing."** Companies you do business with will never write you an e-mail asking you to provide personal

information, such as your password, address, bank account number, or social security number. If you get a suspicious e-mail (even if the logo looks like that of a legitimate company), delete it.

- **Do not reply to junk e-mail, or "spam."** When you reply to a spam e-mail (even just to tell the sender to take you off their mailing list), you are letting the sender know that your e-mail address is valid, which gives them an opportunity to sell your e-mail address to other spammers.
- **Create an e-mail password that is not easy to guess.** Do not use a password such as "ABC" or your mother's family name (a hacker may be able to locate such public information). In addition, change your password regularly and do not use the same password for all of your accounts.
- **Back up important files, including e-mails.** If you contract a computer virus or your computer crashes, be sure you have saved your most important and current files, including critical e-mails, to disk, a Zip drive, or a flash drive. E-mails may be recoverable via your company server, but you will have to go through all of your past e-mails, deleted or not, to find the ones you need.
- **Use encryption when necessary.** When e-mailing a particularly important or confidential document as an attachment, encrypt it to reduce the risk of a hacker accessing your information.

Tech Note

Sending an E-Mail Attachment

Your e-mail program will tell you whether an attachment has been sent, but attachments may arrive in an unusable or partly usable form if the sender's and the receiver's software are different. You need to use the same or compatible software as the person to whom you are sending the attachment in order to open and work with each other's documents, especially if the files contain graphics, databases, or spreadsheets. To effectively attach a document to an e-mail:

- Send a test document and ask the recipient to send you one in return.
- If this experiment fails with a word processed document, send it again, this time as a text file. Although some formatting may be lost (italics, boldface, indents), text files can be read by most word processing programs.
- When sending image files (such as .jpg or .bmp), try to keep the size of the file as small as possible. Most e-mail programs have specifications as to the size limit attachments cannot exceed.
- Consider adding translation software to your computer system to let you cross platforms (Windows, Macintosh, Unix) or programs.

Figure 4.6   A poorly written e-mail, with annotations.

Figure 4.6 shows an example of a poorly written e-mail that violates many of the preceding guidelines. Figure 4.7 contains an effective revision that reflects the professional and courteous way the writer and his company do business.

A revised, effective version of the poor e-mail in Figure 4.6.   **Figure 4.7**

| Reply | Save | Forward | Print | Delete |

To: <MWood@widedoor.com>
From: <sammy@dataport.com>
Date: 11/19/09
Subject: Upgrades for service contract #4552

Specific e-mail address

Precise subject

Hello, Mary:

I would appreciate your delivering the upgrades for our service contract #4552 by Thursday afternoon, the 22nd of November, if at all possible.

We need to proceed to the next phase of our operation and the upgrades are crucial to that task.

I am attaching a copy of our service agreement with Wide Door for your convenience.

If you run into any problem with the delivery date, please give me a call this afternoon or e-mail me.

Thanks,

Sammy

Samuel Atherton
Operations Assistant
Data Port
4300 Morales Highway
San Padre, CA 95620-0326
Voice mail: 723-555-1298
http://dataport.com

Polite salutation

Gets to the point concisely

Provides explanation and documentation

Ends with clear-cut directions

Professional close

When Not to Use E-Mail

Although e-mail is convenient, easy to use, and appropriate for the routine business correspondence we have been discussing, be careful not to use it in the following situations:

- When you make a new business contact or welcome a new client, write a formal letter, not an e-mail. International readers, in particular, will expect this.
- Always acknowledge a business gift or courtesy by sending a handwritten thank-you note rather than dashing off an e-mail
- Never send an e-mail in place of a letter for any type of legal notification.

Instant Messages (IMs) for Business Use

Visit www .cengage.com/ english/ kolin for an online exercise, "Practicing Your Instant Messaging Skills."

IMs are textual conversations that take place online and in close to real time. They should not be confused with cell phone text messages, which can be more cumbersome and occur in a different environment. Think of IMs as somewhere between phone calls and e-mails. IM conversations are almost as instantaneous as phone conversations, but at the same time they provide written records of communications like e-mails do. Keep in mind that IMs are not just used for communication with your friends; they are also a vital part of workplace correspondence. In fact, researchers estimate that 90 percent of all businesses have or will use IMs for routine workplace correspondence.

IM conversations take place over the Web. While IM conversations could once be conducted only via commercial Web services such as Yahoo!, MSN, AIM, and Jabber, many businesses now have their own IM systems. Two or more individuals can have an IM conversation—or even conduct multiple conversations at once — on their computer screens or via smartphones (see the Tech Note on p. 148). You type your message, and it appears on a small pop-up window on your recipient's computer screen or smartphone pad. See Figure 4.8 for an example of an IM conversation relating to the world of work.

Exchanges through IM reflect the way people in the world of work connect and communicate with each other. IMs allow you to communicate with co-workers and managers in the same office, at remote sites, or around the globe. Crossing time zones, IMs give you access to anyone around the world who is online and connected to the same service.

When to Use IMs Versus E-Mails

Like e-mails, IMs promote collaboration and provide a written record. And like e-mails, too, business IMs further global communication. However, the situations in which these two very different communication technologies should be used are not identical. By answering the following questions you will be better able to decide whether to use IM or e-mail:

1. **How quickly does my message need to be sent/received?** If you need to send or receive information instantly, use an IM rather than e-mail. Because

An IM exchange between two workers.   Figure 4.8

9:14am	Danielle21:	I'm working on the second draft of the goals section of the report today.
9:15am	Sanjit-man:	So am I; let's talk.
10:01am	Danielle21:	I just emailed you my revision—OK or not?
10:09am	Sanjit-man:	Looks good, but too much on Goal 2 and not enough on Goal 3.
10:24am	Danielle21:	I'll take another stab at it later today.
10:31am	Sanjit-man:	FYI, I'll fax a copy of last year's goals for you to look at.
10:35am	Danielle21:	Got it!
10:49am	Sanjit-man:	Let's talk after lunch, OK? Boston will be calling at 11:00.
1:05pm	Danielle21:	Nice job on last year's goals.
1:10pm	Sanjit-man:	Yes, but to get this right we have to be absolutely on target with the specs.
1:12pm	Danielle21:	Agree! Can someone in i-Tech give us the info?
1:15pm	Sanjit-man:	I'll meet with Florence Ng at 2:00.
3:01pm	Danielle21:	How'd the confab with Florence go?
3:11pm	Sanjit-man:	Fantastic. She gave us the most current specs. We are in good shape now.

IM user names are informal but appropriate

Messages are kept to 1–2 lines each

Message exchange sticks to a single topic

Jargon used but avoids "text speak"

of the nature of IMs, recipients generally reply right away to IM messages, whereas netiquette does not require people to read or reply immediately to e-mail messages.

2. **How long or complex is my message?** If you will need to write a long message, ask multiple questions, or convey multiple points, use e-mail. IMs are designed to convey very brief messages, such as asking a quick question, getting a clarification, sending out an alert, or providing brief information, such as the time or location of a meeting.

3. **Do I need to send an attachment?** If you do, realize that some IM programs have attachment capabilities while others may not.

4. **How conversational is my message?** If your message will involve several brief back-and-forth communications, use IM. A conversation over e-mail could last for hours or even days, depending on the time lag between e-mails.

Guidelines on Using IMs in the Workplace

Your IMs may be instantaneous and informal but that does not mean that you can send them with little thought about their content and wording. You have to follow certain guidelines. Keep in mind that your company can monitor, trace, record, and store your IM conversations just as it can with your e-mails. In addition to following all the guidelines for writing workplace e-mails (pp. 134–140), observe these rules for your IMs:

1. **Stay connected.** When you are offline you cannot be reached. When you are out of the IM window, even for a few seconds, tell those on your buddy list when you will be back and/or provide them with alternate contact information.
2. **Keep your messages short.** A line or two at most is enough for an IM. Again, if your message is detailed or involves multiple points, send an e-mail instead.
3. **Write about one topic at a time.** Don't include information about two or three different topics in one IM exchange. IMs should be focused on one topic at a time to prevent the conversational thread from becoming complicated. You can instead conduct several focused IM conversations at once.
4. **Keep your messages professional.** Because IMs are the most informal business messages you can send, be extra careful to maintain professionalism when writing them. Do not send personal messages, tell vulgar jokes, or give yourself a chat name that is in bad taste.
5. **Avoid unfamiliar abbreviations.** Commonly understood abbreviations such as "FYI" are fine — even encouraged — in IMs (as long as you are not writing to a non-native speaker of English). However, always avoid "textspeak" abbreviations, such as "CUL8R" for "see you later."
6. **Turn down the volume.** Quiet the ping sounds that announce an incoming message as a courtesy to your co-workers.
7. **Save your messages.** Save your IM logs for reference purposes and legal ones as well in case you need to prove that a topic was discussed or a decision made.

█ Blogs

Like e-mails and IMs, blogs (an abbreviation of "web logs") are important tools for e-correspondence for employees, managers, and customers. A blog is more than just a casual journal or diary in the world of work. Think of a business blog as an evolving website or daily business newspaper for which employees and management write regular columns, or posts. A blog provides a fast, informal way to give readers information and news on everything from product announcements to project updates to organizational events and procedures. Blogs are also highly interactive — readers can write comments in response to blog posts, allowing for a two- or multiple-way conversation. Customers can give feedback and employees can communicate with one another and with consumers as well via a blog post.

Blogs can be either internal or external, depending on the intended audience. Internal blogs are designed exclusively for employee readers. Employees can interact by posting their comments and asking questions, collaborate on writing documents, and communicate up and down the corporate ladder. Internal blogs can take the place of e-mails and memos (even meetings) in certain situations. In Figure 4.9, a training manager answers employees' questions through her blog.

External blogs are public relations tools allowing companies to express their points of view, their interpretations of events, and their clarifications of issues quickly and publicly — in short, they provide the public with the company's side of the story. Besides being a forum for company viewpoints, external blogs are a fast way to get information to the public about new or updated products and services, research results, changes and improvements in technology, expanded locations, community service and environmental efforts, organizational news and changes — just about anything related to the company's goals and activities. Most important of all, external blogs are highly interactive, encouraging readers to write their comments or ask questions about topics and opinions expressed in the blog. Keep in mind, too, that since blogs are picked up by search engines, you will have an international audience reading and responding to your work. Figure 4.10 shows an example of an external blog.

Guidelines for Writing Internal Blogs

To post a successful internal blog, you need to follow all the guidelines for writing a business e-mail or using workplace IM. You cannot be sloppy or careless; you have to use proper spelling and punctuation. You also have to be professional. A blog written for fellow employees should be friendly and polite. Note the informal tone Maxine Schwartz uses in her blog in Figure 4.9. She is helpful and relaxed but not arrogant, curt, or condescending.

The guidelines below will help you to write an appropriate post for an internal blog:

1. Be ethical. As with any business correspondence, don't post discriminatory comments or reveal confidential information about a product or service. Such information should not be shared across departments within your company or organization.

2. Respect company protocol. Do not post anything that might be considered controversial, restricted, or otherwise off limits without first consulting your boss or the person in charge of maintaining the blog. Often an internal blog will include a usage guidelines statement, which you need to respect.

3. Be professional. Avoid posting personal stories and pictures unrelated to your work on the company's or agency's blog. Most important, do not gossip, criticize an individual (fellow employee, manager, customer, vendor) or a company policy, or make political comments, all of which may be grounds for dismissal.

4. Write to a company-wide audience. Avoid topics or jargon that people outside of your own department may not understand. Keep in mind that an internal blog needs to reach beyond your own department to the company or agency as a whole, as Figure 4.9 illustrates.

5. **Keep your posts short and easy to read.** Your blog should be simple and practical. Most blogs are no more than a few paragraphs. Don't turn your blog into a report or a complicated presentation of technical data. Note how Maxine Schwartz boldfaces and numbers the employee questions she answers.

6. **Write a clear and concise subject line.** Avoid vague subject lines such as "important news," "help," or "any ideas?" You want to attract readers within your company or agency to profit from and possibly respond to your post, not ignore it or be baffled by it. Similarly, avoid excessively long, detailed subject lines that can intimidate readers.

Figure 4.9   An internal blog.

Hong, McCarson, and Steinway, LLC
INTERNAL DISCUSSION BOARD

Blog topics clearly differentiated

Topic: New anti-virus software
(comments: 6)

Topic: Importing HMS logo into memos, letters, etc.
(comments: 7)

Uses clear and concise title

Chooses an informal yet professional tone

Topic: Using the new binding machines
10/6/08, 10:43 a.m. (Mschwartz): Hi, this is Maxine in iTech. I congratulate those of you who have completed our training workshops on the new equipment, and I look forward to working with individuals in the marketing and facilities departments, who are scheduled for a session next week.

Overall, I think the sessions have gone well. However, a number of questions have come up about the new binding machines, and so I thought the best thing to do would be to post the most frequently asked ones here, with answers:

Keeps the blog post brief and chunks text

1. Why are the binding machines only located on the 6th floor?
We're just waiting for three additional machines to arrive from the supplier. When they arrive within the next two weeks, we will put two of the machines in the 5th floor copy room (5-204) for legal services and one in the 3rd floor copy room (3-122) for marketing and facilities. For the time being, thanks for using the two machines in the 6th floor copy room (6-143).

Continued

(Continued)   Figure 4.9

2. Where are the supplies for the binding machines located? They are located in all three copy rooms. iTech would like to apologize about the shelf locations not being more clearly labeled. We fixed that problem. Thanks for letting us know about it. Also, to help our employees further, iTech has posted a detailed supplies and procedures sheet on each copy room door.

Lets readers know i-Tech is responsive to their needs

3. Why can't I get the laminated insert pages to line up properly? Here's a tip to help you with this one: Make sure that the rounded edges of the inserts are facing out and that the squared edges are lined up with the inside edges of the paper. That should help.

4. How do I avoid damaging my legal documents when using the machines? To be on the safe side, insert a few pages of scrap paper into the machine and do a test bind before inserting original documents. You'll then be sure that the machine is aligned properly.

Offers practical help for employees

I hope these answers help everyone out. If you have further questions, please don't hesitate to e-mail me or call me at extension 304.
<u>(comments: 12)</u>

Topic: What is the company policy on maternity/paternity leaves?
<u>(comments: 7)</u>

"Comments" link allows for further discussion

Topic: PowerPoint presentations – is there a specialist in-house to help? <u>(comments: 3)</u>

Guidelines for Writing External Blogs

The purpose of an external blog is enhancing public relations rather than providing routine internal correspondence, and your audience is the outside world, not employees within your own company. But as with successful e-mails and internal blogs, an external blog should honor the confidentiality of company business, avoid abusive or discriminatory remarks, and respect the privacy of co-workers. Your company will likely employ a blog administrator who approves and/or edits external blogs. Keep your company's reputation and mission in mind when you prepare your external blog. The guidelines below will help you post an effective external blog. Refer to Figure 4.10 as you study these guidelines.

1. **Follow company protocol.** As with internal blogs, follow the rules and regulations for blogging outlined by your company. Otherwise, your post may be rejected by the blog administrator.

2. Give each blog a title and date. The blog administrator will likely provide a date, but it will be up to you to give your post a clear and concise title. Readers need to easily locate a particular blog post and know how current each post is.

3. Write to a general, international audience. To ensure that your blog can be understood by general audiences within the United States and across the globe,

Figure 4.10   An external blog.

Search engine and categories fields simplify navigation

"Previous," "next," and "recent," posts allow for further navigation

Provides a clear and concise title and date

Writes to a general audience and avoids jargon

Uses a conversational but professional tone

PH Home | **PH About Us** | **PH Products** | **PH International** | **PH Jobs**

Search: ⦿ POWERHOUSE, Inc. Blogspot   ○ All of POWERHOUSE, Inc.

POWERHOUSE, INC.   **BLOGSPOT**

Categories: | All | Company News | Product News | Distribution | Manufacturing |

\>> Previous Post   >> Next Post

Today's Post (August 15, 2008):	Recent Posts:
\>> **Discontinuation of the PH-450 All-in-One** by Clay Denton-Tyler, District Manager Many of our loyal customers disagree with PowerHouse's decision to discontinue manufacturing the PH-450 All-in-One Power Tool. It is always great to hear from customers and to receive their feedback, even when we've done something wrong. In the hope that customers will better understand our perspective, let me fill in some of the background about why this product will no longer be manufactured. Discontinuing the PH-450 was not an easy decision. After all, this was the product that first brought our company to national attention. Also, I know from many complimentary customer e-mails and responses to earlier blogs, as well as from my personal experience as a proud owner of the PH-450, users have always enjoyed its price, compact design, durability, and all-weather usability. As one customer put it, "Why kill a popular product that has worked so well for 20 years?"	\>> International Tools, SE CFO announces retirement. \>> PowerHouse announces the discontinuation of the PH-450. \>> PowerHouse opens new retail outlets in Sydney, Australia, and Dunedin, New Zealand. \>> International Tools, SE acquires quality Swiss Home Products company. \>> PowerHouse goes international.

Continued

(Continued) **Figure 4.10**

The good news is that even though the PH-450 is being discontinued, our customers can now have a very similar but improved alternative. Recently, our parent company, International Tools, SE, acquired a new multipurpose tool from Swiss Home Products. The SHP-1000 is not only just as compact, durable, and weather-friendly as the PH-450, but it even offers several additional features, such as a nail gun attachment and lifetime limited warranty. And due to an excellent distribution deal negotiated between International Tools, SE, and Swiss Home Products, we can sell it at only 68 percent of the retail cost of the PH-450.

I know it is hard to say goodbye to a reliable helper but like many products in our increasingly technical age, the PH-450 is being replaced by a more efficient alternative. I will miss the old PH-450, but I have found that the SHP-1000 is just as effective in my shop at home. Adapting to a new model has never come easier for me. Why not give it a try? Thanks.

Comments: (21)
- Sign in to add a comment
- First time users, please register first, in order to add a comment

>> More

Attempts to persuade customer to switch to a new model

Sympathizes with readers but offers personal endorsement characteristic of bloggers at same time

Comments link allows for further discussion

avoid jargon as the description of the SHP-1000 does in Figure 4.10. Study the "Guidelines for Communicating with International Readers" in Chapter 5 (pp. 173–178).

4. Speak for yourself. Don't criticize your company, but do offer your own perspective. Keep in mind, too, that a blog is very different from a press release or an advertisement. Readers will easily recognize a blog that sounds as if it were written by the marketing department. Note how in Figure 4.10 Clay Denton-Tyler provides his own views as an owner of the discontinued product.

5. Avoid making your company liable for false or misleading information. Don't make promises, offer guarantees, or provide warranties unless these are approved by the upper management of your company or agency.

6. Request permission and document your sources. If you use someone else's statistics, surveys, illustrations, or ideas, first get permission from the individual or the company that owns the copyright. But since blogs are designed to be posted quickly and provide very current information, it is probably smarter to steer clear of using materials that require permission. Of course, you can quote a few ideas or sentences (see pp. 350–352 in Chapter 8), but always put quotation marks around any words that are not yours and document the source.

Tech Note

Smartphones

Handheld computers, better known as smartphones (and using now-familiar brand names such as BlackBerry, Apple iPhone, and Windows Mobile), are mobile phones with additional built-in e-mail and Web access (allowing users to send IMs and access internal and external blogs). Many smartphones can also be uploaded with extra software, such as office suites like Microsoft Office, making them even more functional for the world of work.

Businesses often provide their employees (especially those who travel frequently) with smartphones to help them communicate via telephone, receive voice mail, and respond. However, since any cellular phone can do that, the real convenience of smartphones is having access to e-mail and the Web.

Yet even though smartphones are very small—they can be hard to manipulate and they can certainly be hard on the eyes with extended use—they can be used in emergency situations to work with longer documents and do extensive Web searching. Stationary and laptop computers, however, remain the more practical alternative when employees work away from their offices.

Revision Checklist

Memos

- ☐ Used appropriate and consistent format.
- ☐ Followed employer's policy of when to send print or e-mail memos.
- ☐ Announced purpose of memo early and clearly.
- ☐ Organized memo according to reader's need for information, putting main ideas up front, giving documentation, and supplying conclusion.
- ☐ Wrote concise and clear memo suitable for audience.
- ☐ Included bullets, lists, underscoring where necessary to reflect logic and organization of memo and make it easier to read.
- ☐ Refrained from overloading reader with unnecessary details.

Faxes

- ☐ Verified reader's fax number and sent cover sheet with phone number to call in the event of transmission trouble.
- ☐ Excluded anything confidential or sensitive if reader's fax machine is not secure.

E-Mail

- ☐ Did not send unsolicited or confidential mail.
- ☐ Sent to reader's correct address.
- ☐ Formatted e-mail with acceptable margins and spacing.
- ☐ Observed netiquette; avoided flaming.
- ☐ Wrote a separate message rather than returning sender's message with a short reply.
- ☐ Kept paragraphs short but used full — not telegraphic — sentences.
- ☐ Avoided unfamiliar abbreviations or terms that would cause reader confusion.
- ☐ Received permission to repeat or incorporate another person's e-mail.
- ☐ Observed all legal obligations in using e-mail.
- ☐ Safeguarded employer's confidentiality and security by excluding sensitive or privileged information.
- ☐ Included enough information and documentation for reader's purpose.
- ☐ Honored reader by observing proper courtesy.
- ☐ Began with friendly greeting; ended politely.
- ☐ Considered needs of international audience.
- ☐ Used antivirus program.
- ☐ Did not forward or reply to spam.

Instant Messaging

- ☐ Used IM only for professional, job-related communications.
- ☐ Kept responses/questions short — not over a line or two.
- ☐ Avoided "textspeak" in business IMs.

Continued

Continued

☐ Notified readers when IM was offline and back online.
☐ Did not send anything confidential through an IM exchange.

Blogs
☐ Posted nothing critical of employers or co-workers, or anything embarrassing, offensive, or confidential.
☐ Made posts conversational and informal, yet professional.
☐ Posted only current and relevant information.
☐ Provided a place for readers to give feedback.

Exercises

1. Write a memo to your boss saying that you will be out of town two days next week and three days the following week for **one** of the following reasons: (a) to inspect some land your firm is thinking of buying, (b) to investigate some claims, (c) to look at some new office space for a branch your firm is thinking of opening in a city five hundred miles away, (d) to attend a conference sponsored by a professional society, or (e) to pay calls on customers. In your memo, be specific about dates, places, times, and reasons.

2. Write a memo to two or three of your co-workers on the same subject you chose for Exercise 1.

3. Send a memo to your public relations department informing it that you are completing a degree or work for a certificate. Indicate how the information could be useful for your firm's publicity campaign.

4. Write a memo to the human resources department notifying it that there is a mistake in an insurance claim you filed. Explain exactly what the error is and give precise figures.

5. You are the manager of a major art museum. Write a memo to various department heads at your museum in which you put the following information into proper memo format.

   Old hours: Mon.–Fri. 9–5; closed Sat. except during July and August, when you are open 9–12
   New hours: Mon.–Th. 8:30–4:30; Fri.–Sat. 9–9
   Old rates: Adults $3.00; senior citizens $1.00; children under 12 free
   New rates: Adults $4.50; senior citizens $1.00; children under 12 free but must be accompanied by an adult

Added features: Paintings by Thora Horne, local artist; sculpture from West Indies in display area all summer; guided tours available for parties of six or more; lounge areas will offer patrons sandwiches and soft drinks during May, June, July, and August

6. Select some change (in policy, schedule, or personnel assignment) you encountered in a job you held in the last two or three years and write an appropriate memo describing that change. Write the memo from the perspective of your former employer explaining the change to employees.

7. How would any of the memos in Exercises 1–6 have to be rewritten to make it suitable as an e-mail message? Rewrite one of them as an e-mail.

8. Bring five or six examples of a company's or an organization's e-mail to class. As a group activity evaluate them for professional style, tone, layout, and preciseness of message. Write a carefully organized e-mail to your instructor evaluating the effectiveness of these e-mails. Include the sample e-mails as an attachment with your report.

9. Send a fax to a company or an organization requesting information about the products or services it offers. Include an appropriate cover sheet.

10. Write an e-mail to a business that provides daily or weekly information to interested customers and submit its response along with your e-mail request to your instructor. Choose one of the following:
    a. an airline: an up-to-date schedule along a certain route and information about any bonus-mile or discount programs
    b. a catalog order company: information about any specials for Internet users
    c. a stock brokerage firm: free quotes or research about a particular stock
    d. a resort: special rates for a given week

11. Write an e-mail with one of the following messages, observing the guidelines discussed in this chapter.
    a. You have just made a big sale and you want to inform your boss.
    b. You have just lost a big sale and you have to inform your boss.
    c. Tell a co-worker about a union or national sales meeting.
    d. Notify a company to cancel your subscription to one of its publications because you find it to be dated and no longer useful in your profession.
    e. Request help from a listserv about research for a major report you are preparing for your employer.
    f. Advise your district manager to discontinue marketing one of the company's products because of poor customer acceptance.
    g. Send a short article (about two hundred words) to your company's online newsletter about some accomplishment your office, department, or section achieved in the last month.
    h. Write to a friend studying finance at a German, Korean, or South American university about the biggest financial news in your town or neighborhood in the last month.

12. Rewrite the following e-mail to your boss to make it more professional.

    Hi—

    This new territory is a pain. Lots of stops; no sales. Ughhhh. People out here resistant to change. Could get hit by a boulder and still no change. Giant companies ought to be up on charges. Will sub. reports asap as long as you care rec.

    The long and short of it is that market is down. No news=bad news.

13. As a collaborative venture, join with three or four classmates to prepare one or more of the e-mail messages for Exercise 11. Send each other drafts of your messages for revision. Submit the final copy of the group's effort.

14. Assume you have received permission to repost in an e-mail all or part of the article on microwaves (pp. 41–42) or virtual reality and law enforcement (pp. 404–408). Prepare an e-mail message to a listserv or Usenet group containing part of the article you have chosen.

15. Send your instructor an e-mail message about a project you are now working on for class, outlining your progress and describing any difficulties you are having.

16. You have just missed work or a class meeting. E-mail your employer or your instructor explaining the reason and telling how you intend to make up the work.

17. As a group activity, instant message two or three other members of your collaborative writing team on a project you are working on. Print out your IM exchanges during this time and submit them to your instructor.

18. Write external blog posts for three or four days about some aspect of your current job or a previous job in which you share with readers news about your company's products/services, technology you are using, professional travel, community service, working with international colleagues, and so forth. Be sure that your posts put your company, department, or agency in a good light.

19. Write an internal blog post for the same three-to-four-day period about one of the topics listed in Exercise 18 or one suggested by your instructor.

CHAPTER 5

Writing Letters

Some Basics for Audiences Worldwide

Letters are among the most important writing you will do on your job. Businesses worldwide take letter writing very seriously, and employers will expect you to prepare and respond to your correspondence effectively. Your signature on a letter tells readers that you are accountable for everything in it. The higher up the corporate ladder you climb, the more letters you will be expected to write.

Because letter writing is so significant to your career, this chapter introduces you to the entire process, provides guidelines and problem-solving strategies, and shows you how to prepare the most frequently written types of business letters. It also shows you how to write for international readers.

Visit www .cengage.com/ english/kolin for this chapter's online exercises, ACE quizzes, and Web links.

Letters in the Age of the Internet

Even in this age of the Internet, letters are still vital in the world of work. A professional-looking letter is one of the most significant symbols in the world of work. It acknowledges the way a company makes important announcements. Letters are important for the following reasons:

1. **Letters represent your company's public image and your competence.** A firm's corporate image is on the line when it sends a letter. Carefully written letters can create goodwill; poorly written letters can anger customers, cost your company business, and project an unfavorable image of you.

2. **Letters are far more formal—in tone and structure—than any other type of business communication.** Memos and e-mails are the least formal communications.

3. **Letters constitute an official legal record of an agreement.** They state, modify, or respond to a business commitment. When sent to a customer, a signed letter constitutes a legally binding contract. Be absolutely sure that what you put in a letter about prices, guarantees, warranties, equipment, delivery dates, and/or other issues is accurate. Your reader can hold you and your company accountable for such written commitments.

4. Unlike e-mails, many businesses require letters to be routed through channels before they are sent out. Because they convey how a company looks and what it offers to customers, letters often must be approved at a variety of corporate levels.

5. Letters are more permanent than e-mails. They provide a documented hard copy. Unlike e-mails that can be erased, letters are often logged in, filed, and bear a written, authorized signature. They are also far more confidential.

6. A letter is the official and expected medium through which important documents and attachments (contracts, specifications, proposals) are sent to readers. Sending such attachments via e-mail or with a memo lacks the formality and respect readers deserve.

7. A letter is still the most formal and approved way to conduct business with many international audiences. These readers see a letter as more polite and honorable than an e-mail for initial contacts and even for subsequent business communications.

8. A hard copy letter is confidential. It is more likely to be delivered to the proper recipient with care in its sealed envelope and less likely to be forwarded to unintended readers, as an e-mail might be.

Letter Formats

Letter format refers to the way in which you print a letter—where you indent and where you place certain kinds of information. Several letter formats exist. Two of the most frequently used business letter formats are full block and modified block.

Full Block Format

In full block format all information is flush against the left margin, double spaced between paragraphs. Figure 5.1 shows a full block letter. Use this format only when your letter is on **letterhead stationery** (specially printed paper giving a company's name and logo, business and Web addresses, fax and telephone numbers, and sometimes the names of its executives).

Modified Block Format

In modified block format the writer's address (if it is not imprinted on a letterhead), the date, the complimentary close, and the signature are positioned at the center point and then keyed toward the right side of the letter. The date aligns with the complimentary close. The inside address, the salutation, and the body of the letter are flush against the left margin. The first lines of paragraphs in the modified style can be indented, as in Figure 5.2, or not.

Full block letter format.   **Figure 5.1**

Nevada Insurance Research Agency
7500 South Maplewood Drive, Las Vegas, NV 89152-0026
(702) 555-9876 **http://www.NIRA.org**

*Letterhead
with company
name*

April 6, 2010

Ms. Molly Georgopolous, C.P.A.
Business Manager
Meyers, Inc.
3400 South Madison Road
Reno, NV 89554-3212

*All text lined
up against the
left-hand
margin*

Dear Ms. Georgopolous:

As I promised in our telephone conversation earlier this afternoon, I am
enclosing a study of the Nevada financial responsibility law. I hope that it will
help you prepare your report.

*Professional,
businesslike
font choice*

Let me emphasize again that probably 95 percent of all individuals who are
involved in an accident obtain reimbursement for medical bills and for
damages to their automobiles. If individuals have insurance, they can receive
reimbursement from their own carrier. If they do not have insurance and the
other driver is uninsured and judged to be at fault, the Nevada Bureau of Motor
Vehicles revokes that party's driver's license until all costs and damages are
paid.

*Text of letter
balanced on
the page*

Please call me again if I can help you.

*Generous
margins on all
sides of the
letter*

Sincerely yours,

Carmen Tredeau

Carmen Tredeau, President

*Signature
written in ink*

Encl.

Bradley Fuller, CPCU	Carmen Tredeau, CPCU	Theodore Kendrick	Iping Li, CPCU	Dora Salinas-Diego, CPCU
Chairperson	President	Vice President	Vice President	Vice President
		Public Affairs	Research	Actuary

Figure 5.2   Modified block letter format with indented paragraphs.

7239 East Daphne Street
Mobile, AL 36608-1012

September 28, 2010

Mr. Travis Boykin, Manager
Scandia Gifts
703 Hardy St.
Hattiesburg, MS 39401-4633

Dear Mr. Boykin:

   I would appreciate knowing if you currently stock the Crescent
pattern of model 5678 and how much you charge per model
number. I would also like to know if you have special prices per
box order.

   The name of your store is listed on the Crescent website as the
closest distributor of Copenhagen products in my area. Would you
please give me directions to your shop from Mobile and the hours
you are open?

   I look forward to hearing from you.

                      Sincerely yours,

                      Arthur T. McCormack
                      Arthur T. McCormack

Continuing Pages

To indicate subsequent pages if your letter runs beyond one page, use one of these two conventions. Note the use of the recipient's name.

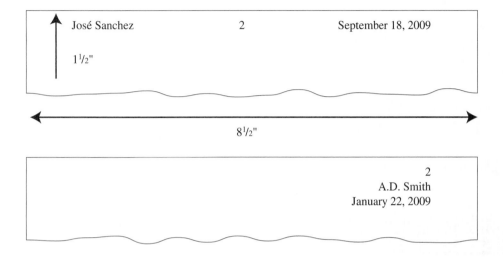

Parts of a Letter

A letter contains many parts, each of which contributes to your overall message. The parts and their placement in your letter form the basic conventions of effective letter writing. Readers look for certain information in key places.

In the following sections, those parts of a letter marked with an asterisk (*) should appear in every letter you write. Figure 5.3 is a sample letter containing all of the parts discussed here. Note where each part is placed in the letter.

*Date Line

Spell out the name of the month in full—"September" or "March" rather than "Sept." or "Mar." The date line is usually keyboarded this way: November 12, 2010. However, see page 176 on listing dates for international readers.

*Inside Address

The inside address, the address of the recipient, is always placed against the left margin, two lines below the date line. It contains the name, title (if any), company, street address, city, state, and Zip Code of the person to whom you are writing.

Make sure that you use the company's exact name consistently—not Hi-Tech, Inc. in one place and Hi Tech Corp. in another. Include the company's full name. If it is Sidow and Lewis, don't write just Sidow Co. Single-space the inside address and do not use any punctuation at the end of the lines.

Figure 5.3   A sample letter, full block format, with all parts labeled.

Letterhead

Madison and Moore, Inc.
Professional Architects

7900 South Manheim Road
Crystal Springs, NE 71003-0092
Phone 402-555-2300   **http://www.MMI.com**

Date line

July 12, 2009

Inside address

Ms. Paula Jordan
Systems Consultant
Broadacres Development Corp.
12 East River Street
Detroit, MI 48001-0422

Salutation

Dear Ms. Jordan:

Thank you for your letter of July 6, 2009. I have discussed your request
with the staff in our planning department and have learned that the design
modules we used for our Vestuvia project are no longer available.

Body of letter

In searching through my files, however, I have come across the enclosed
catalog from a California firm that might be helpful to you. This firm,
California Concepts, offers plans very similar to the ones you are interested
in, as you can tell from the design I checked on page 23 of their catalog.

I hope this will help you and I wish you every success in your project.

Complimentary close

Sincerely yours,

Company name

MADISON AND MOORE, INC.

Signature

William Newhouse

Writer's name and title

William Newhouse
Office Manager

Enclosure
Copy to

Encl.: Catalog
cc: Planning Department

Dr. Mary Petro
Director of Research
Midwest Laboratories
1700 Oak Drive
Rapid City, SD 56213–3406

Always try to write to a specific person rather than just "Sales Manager" or "President." To find out the person's name, check previous correspondence, e-mail lists, the company's or individual's website, or call the company. Make sure you use an abbreviated courtesy title (Ms., Mr., Dr., Prof.) before the recipient's name for the inside address (e.g., Capt. María Torres; Mr. A. T. Ricks; Rev. Siam Tau). Use Ms. when writing to a woman unless she has expressly asked to be called Miss or Mrs.

The last line of the inside address contains the city, state, and ZIP code. (For guidelines on using correct addresses for international readers, see p. 173.)

✱Salutation

Begin with *Dear*, and then follow with a courtesy title, the reader's last name, and a colon (Dear Mr. Brown:). **Never use a comma for a formal letter.** Avoid the sexist "Dear Sir," "Gentlemen," or "Dear Madam" and the stilted "Ladies and Gentlemen" or "Dear Sir/Madam." (For a discussion of sexist language and how to eliminate it, see Chapter 2, pp. 66–68.)

Sometimes you may not be sure of the sex of the reader. There are women named Stacy, Robin, and Lee, and men named Leslie, Kim, and Kelly. If you aren't certain, you can use the reader's full name: "Dear Terry Banks." If you know the person's title, you might write "Dear Credit Manager Banks."

Avoid casual salutations such as "Hello," "Good Morning, "Greetings," or "Happy Tuesday"; these are best reserved for e-mails. And never begin a letter with "To Whom It May Concern," which is old fashioned, impersonal, and trite.

✱Body of the Letter

The body of the letter contains your message. Some of your letters will be only a few lines long, while others may extend to three or more paragraphs. Keep your sentences short and try to hold your paragraphs to under six or seven lines.

✱Complimentary Close

A close is the equivalent of a formal good bye. For most business correspondence, use one of these standard closes:

Sincerely,

Respectfully,

Sincerely yours,

Yours sincerely,

Capitalize only the first word. The entire close is followed by a comma.

If you and your reader know each other well, you can use

> Cordially,
> Best wishes,
> Regards,

But avoid flowery closes, such as

> Forever yours,
> Devotedly yours,
> Faithfully yours,

These belong in a romance novel, not in a business letter.

*Signature

Allow four spaces between the complimentary close and your typed name/title so that your signature will not look squeezed in. Always sign your name in ink. An unsigned letter indicates carelessness or, worse, indifference toward your reader. A stamped signature tells readers you could not give them personal attention.

Some firms prefer using their company name along with the employee's name in the signature section. If so, type the company name in capital letters two line spaces below the complimentary close and then sign your name. Add your title underneath your typed name. Here is an example:

> Sincerely yours,
>
> THE FINELLI COMPANY
>
> *Helen Stravopoulos*
>
> Helen Stravopoulos
> Cover Coordinator

*Enclosure(s) Line

The enclosure line informs the reader that additional materials (such as brochures, diagrams, forms, contracts, a proposal) accompany your letter.

> Enclosure (only one item is enclosed)
> Enclosures (2)
> Encl.: 2006 Sales Report

*Copy Notation

The abbreviation *cc:* (courtesy copy) informs your reader that a copy of your letter has been sent to one or more individuals.

> cc: Service Dept.
> cc: Janice Tukopolous
>   Ivor Vas

Letters are copied and sent to third parties for two reasons: (a) to document a paper trail and (b) to indicate to other readers who else is involved. Professional courtesy dictates that you tell the reader if others will receive a copy of your letter.

The Appearance of Your Letter

The appearance of your letter refers to how it looks after you print it for a reader. The way your letter looks can determine how readers respond to your message. After all, readers get their first impression of you by what you present to them, so if your letter looks sloppy, lopsided, crowded, or misaligned, they will think your ideas are equally careless.

The following tips will help you print professional-looking letters. You may also want to consult the guidelines on document design in Chapter 11 (pp. 496–506) and the Tech Note "Letter Wizards" (p. 162).

- **Use a letter-quality printer.** Letters must look crisp, fresh, and businesslike.
- **Always check your toner cartridge level.** A fuzzy, faint, or messy letter mars your company image and your relationship with your reader.
- **Experiment with different fonts and choose a font that is inviting to the eye.** Avoid script or other fancy fonts (see pp. 501–503), which say you are trying to impress your readers with typeface rather than with content. Also stay away from overly small fonts, which are hard to read, and overly large fonts, which signal to readers that your message may not be very serious.
- **Avoid crowding too much text onto one page.** Squeezing too many characters on a line makes your letter cramped and hard to read. Also, don't cram a long letter onto one page; instead, allow your letter to flow to a second page.
- **Be careful about lopsided letters.** Don't start a brief letter at the top of the page and leave the lower three-fourths of the page blank. Readers will feel as if you've left them hanging. Start a shorter letter near the center of the page.
- **Center your letter, making it look balanced and proportional.** To make your message look clear and businesslike, center your letter on the page so that you leave as much white space above as below. Leave generous left- and right-side margins of approximately 1½ inches. For a shorter letter (such as the letter in Figure 5.6) don't expand your margins to 2 or 3 inches. That will make your letter look unprofessional. And always single space within paragraphs but double space between paragraphs.
- **Use Print Preview to view an image of your letter before you print it.** If your letter does not look professional, you have another chance to make any necessary adjustments to correct it.
- **Always print your letter on high-quality white paper and matching standard size business envelopes.** Heavy, bond paper (20-pound, 8½" × 11" stationery) says that your message is important and substantial. Thin, lightweight paper gives the impression that your message is flimsy. Avoid colored paper, which can look unprofessional.

Tech Note

Visit www
.cengage.com/
english/kolin for
an online exer-
cise, "Using a
Letter Wizard."

Letter Wizards

Most word processing programs, such as Microsoft Word or Corel Word Perfect, have letter wizards to help you format and design your letters. These wizards provide templates that automatically position your address, the current date, the recipient's address, the salutation, paragraphs, and the complimentary close. Using these wizards, all you do is enter the relevant information within a template and select from full or modified block format in addition to other options.

With letter wizards, you can also choose the overall style of your letter—contemporary, traditional, plain, elegant, and so on—by selecting the font and type size. These programs will also help you select a letterhead. You can choose from a list of several predesigned letterheads and then customize any one of them by inserting your, or your company's, name and address. Or you can create an original letterhead.

Although these wizards can create the format and design of a letter for you, do not rely on them too heavily. Keep the following precautions in mind:

- Wizards do not create text for you; they merely set up the layout of the letter and provide space to enter your own text.
- Although letter wizards provide standard formal, informal, and business salutations and closings, be careful about which option you choose. Some may be inappropriate for business correspondence because they look too informal, such as "Hi There."
- Some organizations may prefer not to use standard letter formatting. Always check with your company before using a letter wizard.
- Do not rely exclusively on generic letterhead templates provided by word processing software, since they may be seen as unimaginative by your readers.
- Wizards include parts of a letter you may not need (e.g., copy notation) or that your company does not use (e.g., attention lines).

Organizing a Standard Business Letter

Like the memo in Figure 4.3, a standard business letter can be divided into an introduction, a body, and a conclusion, each section responding to or clarifying a specific issue for your recipient. These three sections can each be one paragraph long, as in Figures 5.1, 5.2, and 5.3, or the body of your letter may be two or more paragraphs, as in Figure 5.4.

To help readers grasp your message clearly and concisely, follow this simple plan for organizing your business letters:

Organization of a business letter.   Figure 5.4

Office Property Management Associates
2400 South Lincoln Highway
Livingston, NJ 07040-9990
(201) 555-3740 www.opma.com

August 15, 2009

Mr. W. T. Albritton
Albritton & Sharp Accounting Services
Suite 400
Suburban Office Complex
Livingston, NJ 07038-2389

Dear Mr. Albritton:

At our July meeting, Office Property Management Associates discussed a number of requests you made about the Suburban Office Complex. I am happy to inform you that the following improvements in services will go into effect within 45 days.

Starting September 7, you will have an on-site manager, Thomas Fiorelli, who will be happy to answer any questions you may have about the Complex and help you with any problems you may encounter. His ten years of experience in managing commercial office parks will benefit you and other businesses at the Complex.

The new outdoor security system you asked for will be installed by September 21. It will offer you and your employees greater protection through additional security cameras around the perimeters of the parking lot, and movement sensors will monitor every outside door.

I want to reassure you that none of these changes will inconvenience the operation of your firm. We are honored to have Albritton & Sharp as residents. I welcome your comments as these changes are implemented as well as any additional suggestions you may have.

Sincerely yours,

Cheryl Hu

Cheryl Hu
Vice President

Introduction comes to point quickly, cordially by referencing reader's earlier request

Body describes changes with specific details

Conclusion builds goodwill by promising reader what will be done and how

- In your first paragraph tell readers why you are writing and why your letter is important to them. Acknowledge any relevant previous meetings, correspondence, or telephone calls early in the paragraph (as in Figures 5.1, 5.3, and 5.4).
- Put the most significant point of each paragraph first to make it easier for the reader to find. Never bury important ideas in the middle or at the end of your paragraph.
- In the second (or subsequent) paragraph of your letter, develop the body of your message with factual support, the key details and descriptions readers need. For instance, note how in Figure 5.4 Cheryl Hu refers to specific changes her reader wanted.
- In your last paragraph, thank readers and be very clear and precise about what you want them to do or what you will do for them. Let them know what will happen next, what they can expect (Figure 5.4), or any combination of these messages. Don't leave your readers hanging. End cordially and professionally.

Making a Good Impression on Your Reader

You have just learned the basics about formatting and organizing your letters. Now we will turn to the content of your letters—what you say and how you say it. Writing letters means communicating to influence your readers, not to alienate or antagonize them. Keep in mind that writers of effective letters are like successful diplomats in that they represent both their company and themselves. You want readers to see you as courteous, credible, and professional.

First, put yourself in the reader's position. What kinds of letters do you like to receive? Letters that are vague, impersonal, sarcastic, pushy, and condescending, or polite, businesslike, and considerate? If you have questions, you want them answered honestly, courteously, and fully. You do not want someone to waste your time with a long, puffy letter when a few sentences would suffice.

To send such effective letters, adopt the **"you attitude"**; in other words, signal to readers that they and their needs are of utmost importance. Incorporating the "you attitude" means you should be able to answer "Yes" to these two questions:

1. Will my readers receive a positive image of me?
2. Have I chosen words that convey both my respect for the readers and my concern for their questions and comments?

The first question deals with your overall view of readers. Do your letters paint them as clever or stupid, practical managers or spendthrifts? The second question concerns the language and tone conveying your view of the reader. Words can burn or soothe. Choose them carefully. As you revise your letters, you will become more aware of and concerned about the ways a reader will respond to you and your message.

Figures 5.5 and 5.6 (pp. 166 and 167) contain two versions of the same letter. Which one would you rather receive?

Achieving the "You Attitude": Four Guidelines

As you draft and revise your work, pay special attention to the following four guidelines for making a good impression on your reader.

1. Never forget that your reader is a real person. Avoid writing cold, impersonal letters that sound as if they were form letters or voice mail instructions. Let the readers know that you are writing to them as individuals. The paragraph below violates every rule of personal and personable communications.

> It has come to our attention that policy number 342q-765r has been delinquent in payment and is in arrears for the sum of $302.35. To keep the policy in force for the duration of its life, a minimum payment of $50.00 must reach this office by the last day of the month. Failure to submit payment will result in the cancellation of the aforementioned policy.

This example shows no sense of one human being writing to another here, of a customer with a name, personal history, or specific needs. The letter uses cold and stilted language ("delinquent in payment," "in arrears for," "aforementioned policy"). Revised, this paragraph contains the necessary personal (and human) touch.

> We have not yet received your payment for your insurance policy (#342q-765r). By sending us your check for $50.00 within the next two weeks, you will keep your policy in force and can continue to enjoy the financial benefits and emotional security it offers you. Please call us if you have any questions. Thank you.

The benefits to an individual reader are stressed, and the reader is addressed directly as a valued customer.

Don't be afraid of using "you," in letters. Readers will feel more friendly toward you and your message. Of course, no amount of "you's" will help if they appear in a condescending context, such as the letter in Figure 5.5.

2. Keep the reader in the forefront of your letter. Make sure the reader's needs control the tone, message, and organization of your letter—the essence of the "you attitude." Stress the "you," not the "I" or the "we." The paragraph below forgets about the reader.

I-Centered Draft
```
I think that our rug shampooer is the best on the market.
Our firm has invested a lot of time and money to ensure
that it is the most economical and efficient shampooer
available today. We have found that our customers are very
satisfied with the results of our machine. We have sold
thousands of these shampooers, and we are proud of our
accomplishment. We hope that we can sell you one of our
fantastic machines.
```

Figure 5.5   A letter lacking the "you attitude."

Brown County · **Office of the Tax Assessor**

County Building, Room 200, Ventura, Missouri 56780-0101
712-555-3000

February 5, 2009

Mr. Ted Ladner
451 West Hawthorne Lane
Morris, MO 64507-3005

Dear Mr. Ladner:

You have written to the wrong office here at the County Building. There is no way we can attempt to verify the kinds of details you are demanding from Brown County.

Simply put, by carefully examining the 2008 tax bill you said you received, you should have realized that it is the Tax Collector's Office, not the Tax Assessor's, that will have to handle the problem you claim exists.

In short, call or write the Tax Collector of Brown County.

Thank you!

Tracey Kowalski
Tracey Kowalski

http://www.browncounty.gov

Tone is sarcastic and uncooperative

Use of "you" alone does not signal a positive image of reader

Insulting and curt ending and close

A you-centered revision of Figure 5.5.   Figure 5.6

Brown County · Office of the Tax Assessor

County Building, Room 200, Ventura, Missouri 56780-0101

712-555-3000

February 5, 2009

Mr. Ted Ladner
451 West Hawthorne Lane
Morris, MO 64507-3005

Dear Mr. Ladner:

Thank you for writing about the difficulties you encountered with your 2008 tax bill. I wish I could help you, but it is the Tax Collector's Office that issues your annual property tax bill. Our office does not prepare individual homeowners' bills.

Thanks reader and gives polite explanation

If you will kindly direct your questions to Paulette Sutton at the Brown County Tax Collector's Office, County Building, Room 100, Ventura, Missouri 56780-0100, I am sure that she will be able to assist you. Should you wish to call her, the number is 458-3455, extension 212. Her e-mail address is **psutton@bctc.gov**.

Helps reader solve problem with specific information

Respectfully,

Tracey Kowalski

Tracey Kowalski

Uses appropriate close

http://www.browncounty.gov

The draft discusses the machine, the company, and its sales success, but readers are interested in how *they* can benefit from the machine, not in how much profit the company makes from selling it.

To win reader confidence, the writer needs to show how and why they will find the product useful, economical, and worthwhile at home or at work. Here is a reader-centered revision.

You-Centered Revision

```
Our rug shampooer would make cleaning your Comfort Rest Mo-
tel rooms easier for you. It is equipped with a heavy-duty
motor that will handle your 200 rooms with ease. Moreover,
that motor will give frequently used areas, such as the
lobby or hallways, the fresh and clean look you want for
your motel.
```

Tech Note

Envelopes, Labels, Mail Merge, and Watermarks

Envelopes and Labels
Many word processing programs provide wizards to help you print mailing labels and envelopes. You can enter your recipient's address, your return address, and the size of your envelope or type of labels, and the wizard helps you print them correctly. Envelopes with printed labels are always more professional looking than those with a handwritten address. Later, you can enter and retrieve one or thousands of addresses using the program's address book, saving you from much tedious keyboarding.

Mail Merge
Use mail merge when you are sending the same letter or e-mail to multiple recipients (for instance, when sending out a survey—see pp. 318–325). Mail merge will allow you to enter the addresses one by one or to select them from your address book before printing, saving you time. If you are sending the same letter to multiple recipients, you can enter their addresses into mail merge. Then, mail merge will automatically print each letter with a new address at the top. You can then use the addresses to create labels or envelopes with your program's wizard.

Watermarks
Software programs will help you to insert a watermark on your letter. A watermark is any graphic, logo, letterhead, or image that is printed very lightly behind the text typed on the page. Watermarks give your letter the appearance of being printed on expensive and official stationery, as in the image of the county seal imprinted on the letters in Figures 5.5 and 5.6.

3. Be courteous and tactful. Refrain from turning your letter into a punch through the mail. Don't inflame your letter or e-mail readers; review Figures 4.6 and 4.7. Again, when you capture the reader's goodwill, your rewards will be greater. The following words can leave a bad taste in the reader's mouth.

it's defective	unprofessional (job, attitude, etc.)
I demand	your failure
I insist	you contend
we reject	you allege
that's no excuse for	you should have known
totally unacceptable	your outlandish claim

Compare the following discourteous sentences with the courteous revisions.

Discourteous	Courteous Revision
We must discontinue your service unless payment is received by the date shown.	Please send us your payment by November 4 so your service will not be interrupted.
You are sorely mistaken about the contract.	We are sorry to learn about the difficulty you experienced over the service terms in our contract.
The new notebook you sold me is third-rate and you charged first-rate prices.	Since the notebook is still under warranty, I hope you can make the repairs easily and quickly.
It goes without saying that your suggestion is not worth considering.	It was thoughtful of you to send me your suggestion, but unfortunately we are unable to implement it right now.

The last discourteous example begins with a phrase that frequently sets readers on edge. Avoid using *it goes without saying*—it can quickly set up a hostile barrier between you and your reader.

4. Don't sound pompous or bureaucratic. Write to your reader as if you were carrying on a professional conversation. Your tone should be polite, but natural, to the point. Make your letters reader-friendly and believable, not stuffy and overbearing. To do that don't resort to using phrases that remind readers of *legalese*—language that smells of contracts, deeds, and stuffy rooms.

In the following list, the words and phrases on the left are pompous expressions that have crept into letters for years; the ones on the right are contemporary equivalents.

Pompous	Contemporary
aforementioned	previously mentioned
as per your request	as you requested
at this present writing	now
I am in receipt of	I have received
contingent upon the receipt of	as soon as we receive
attached herewith	enclosed

I am cognizant of	I know
be advised that	for your information
due to the fact that	because
endeavor	try
forthwith	at once
henceforth	after this
herewith; heretofore; hereby	(drop these three *h*'s entirely)
immediate future	soon
in lieu of	instead of
pursuant	concerning
remittance	payment
under separate cover	I'm also sending you
this writer	I
we regret to inform you that	we are sorry that

Figure 5.7 shows a letter in stilted language written by Brendan T. Mundell to Patricia Lipinski, an executive whose firm has been overcharged for airplane tickets. Mundell's letter overflows with flowery, old-fashioned expressions. The effect is that his message—offering an apology, a credit, and a promise to correct the situation—is long-winded and pompous. It sounds insincere and takes a condescending view of the reader. Note how the revision in Figure 5.8, free of such stilted expressions, is shorter, clearer, and far more personable.

International Business Correspondence

Visit www
.cengage.com/
english/kolin for
an online exer-
cise, "Research-
ing a Country's
Business Culture
Online."

After e-mails, letters are the most frequent type of communication you are likely to have with international readers. Formal letter writing is a very important skill in the global marketplace. But, as we saw in Chapter 1, you cannot assume that people in every culture write letters the way we do in the United States. The conventions of letter writing—formats, inside addresses, salutations, dates, complimentary closes, signature lines—vary from country to country. You may also need to communicate from an international perspective with people at companies located in the United States. For instance, in Figure 6.3 (pp. 202–203) restaurant owner Patrice St. Jacques appeals to the ethnic heritage and pride of a potential customer in his effective sales letter.

It would be impossible to provide information about how to write letters to each international audience. There are at least 5,000 major languages representing diverse ethnic and cultural communities around the globe. But here are some of the most important culturally sensitive questions you need to ask about writing to readers whose cultures are different from yours:

- What is your status in relationship to the reader (client, vendor, salesperson, or international colleague)?
- How should you format and address your letter?
- What is an appropriate salutation?

A letter written in stilted, old-fashioned language.　　**Figure 5.7**

NORTHERN AIRWAYS

July 23, 2009

Ms. Patricia Lipinski
Manager
Lindsay Electronics
4500 South Mahoney Drive
Buffalo, NY 14214-4514

Dear Manager Lipinski:

Please be advised that I am in receipt of yours of July 17th. We are a global air carrier with the interest of our passengers in mind. I would like to take this opportunity to say that we are cognizant of our commitment to good corporate customers like Lindsay and extend our deepest disappointment for the problems your firm has experienced with Northern.

Wordy and pompous

Payment is due your firm, and I hasten to rectify the situation with regard to our error. Forthwith we are adjusting your account #7530, crediting it with the $706.82 you were surcharged. Inasmuch as Northern is such a wonderful company, we are going to enhance the Travel Pass we issue to customers.

Legalese

Condescending tone

I would also like to bring to your attention that in an endeavor to correct such billing errors in the immediate future, I have routed copies of your communication to the manager of the billing department, A. T. Padua. I am confident he will take necessary action at his earliest possible convenience to ascertain the full details of your predicament and make the necessary adjustments in our procedures.

Once again, I want to take the liberty to assure you that Lindsay Electronics is one of our most valued clients. Rest assured that we will take every step imaginable not to jeopardize our long-standing relationship with you. I hope that all the aforementioned problems have now been satisfactorily resolved. Thanking you, I am

Flowery language reeks of insincerity

Inappropriate complimentary close

Faithfully yours,

Brendan T. Mundell

Brendan T. Mundell
General Manager

3000 Airline Highway Tyler, ME 04462-3000 (207) 555-6300 E-MAIL: norair@abc.com
Fly over to our website at www.norair.com

Figure 5.8   A revised version of the stilted and bureaucratic letter in Figure 5.7.

NORTHERN AIRWAYS ≣

July 23, 2009

Ms. Patricia Lipinski
Manager
Lindsay Electronics
4500 South Mahoney Drive
Buffalo, NY 14214-4514

Dear Ms. Lipinski:

Opens with a sincere apology

Thank you for your letter of July 17. I am sorry to learn about the billing problems your employees encountered while traveling on Northern earlier this month. We care very much when we have inconvenienced a good customer like Lindsay Electronics. Please accept my apology.

Gives clear and concise explanations

Lindsay is unquestionably entitled to compensation for our billing error. I am crediting your account #7530 with $706.82, the amount you were overcharged. Furthermore, in appreciation of Lindsay's business, I am also crediting your Travel-Pass account with a bonus 3,000 miles.

Assures reader problem will be corrected in the future

To help us avoid similar incidents, I have sent a copy of your letter to the manager of our billing department, A. T. Padua. I know he will want to learn about Lindsay's experience and will use your comments constructively in an effort to revise our billing procedures.

Ends with offer to help again and thanks reader

As a Frequent Flyer account member, Lindsay is one of our most important customers, and we will continue to work hard to deserve your support. Please call me if I can help you in the future.

Appropriate complimentary close

Sincerely yours,

Brendan T. Mundell

Brendan T. Mundell
General Manager

3000 Airline Highway Tyler, ME 04462-3000 (207) 555-6300 E-MAIL: norair@abc.com
Fly over to our website at www.norair.com

- How should you begin and conclude your letter?
- What types and amount of information will you have to give?
- What is the most appropriate tone to use?

To answer these and similar questions about proper letter protocol for your international readers, you need to learn about their culture by consulting the following resources:

- Country-specific websites prepared by a country's department of tourism or other English-language websites devoted to the culture of a particular country. For instance, Japan Zone (*http://www.japanzone.com*) provides information on contemporary Japanese culture, along with a forum in which you can ask questions about customs, dress, business etiquette, national holidays, and so forth.
- Travel websites such as Yahoo! Travel (*http://www.travel.yahoo.com*), which lets you select the continent, then country of your choice and offers "Expert Advice" links to the country you want to learn about.
- Business travel websites such as Cyborlink.com (*http://www.cyborlink.com/besite/resource.htm*), which provides clickable maps linking to business appearance, behavior, and communication guidelines for a host of countries across the world.
- Either the print or online *World Factbook*, published by the Central Intelligence Agency. The online version is available at *https://www.cia.gov/library/publications/the-world-factbook/index.html* and supplies informational pull-down pages on People, Government, Economy, Communications, and Transnational Issues.
- Cultural information officers at a country's embassy or consulate. Embassy-World.com (*http://www.embassyworld.com*) features a comprehensive database of every embassy and consulate in the world.
- A multinational colleague who is a native speaker and who can advise you about proper letter protocols in his/her country. Susan DiFusco relied on the assistance of a Korean colleague to prepare her letter in Figure 5.12.
- A foreign language instructor who is familiar with the customs of the country or countries where a particular language is spoken.

Guidelines for Communicating with International Readers

The following eleven guidelines will help you communicate more successfully with an international audience and significantly reduce the chances of their misunderstanding you.

1. Use common, easily understood vocabulary. Write basic, simplified English. Choose words that are widely understood as opposed to those that are not used or understood by many speakers. Whenever you have a choice, use the simpler word; for example, use *stop*, not *refrain*; *prevent*, not *forestall*; *happy*, not *exultant*.

2. Avoid ambiguity. Words that have double meanings force non-native readers to wonder which one you mean. For example, "We fired the engine" would baffle your readers if they were not aware of the multiple meanings of *fire.* Unfamiliar with the context in which *fire* means "start up," a non-native speaker of English might think you're referring to "setting on fire or inflaming," which is not what you intend. Such misinterpretation is likely because most bilingual dictionaries would probably list only those two meanings. Or because *fire* can also mean "dismiss" or "let go," a non-native speaker of English might even suspect the engine was replaced by another model.

Be especially careful of using synonyms just to vary your word choice. For example, do not write *quick* in one sentence and then, referring to the same action, describe it as *rapid.* Your reader may assume you have two different things in mind instead of just one.

3. Be careful about technical vocabulary. While a reader who is a non-native speaker may be more familiar with technical terms than with other English words, make sure the technical word or phrase you include is widely known and not a word or meaning used only at your plant or office. Double-check by consulting the most up-to-date manuals and guides in your field, but steer clear of technical terms in fields other than the one with which your reader is familiar. Be especially careful about using business words and phrases an international reader may not know, such as *lean manufacturing, revolving credit, reverse mortgages, amortization,* and so forth.

4. Avoid idiomatic expressions. Idioms are the most difficult part of a language for an audience of non-native speakers to master. As with the example of *fire,* the following colorful idiomatic expressions will confuse and may even startle a non-native reader:

I'm all ears	think outside the box
throw cold water on it	sleep on it
hit the nail on the head	give a heads-up to
new blood	land in hot water
easy come, easy go	touch and go
get a handle on it	pushed the envelope
right under your nose	it was a rough go

The meanings of those and similar phrases are not literal but figurative, a reflection of our culture, not necessarily your reader's. A non-native speaker of English will approach such phrases as combinations of the separate meanings of the individual words, not as a collective unit of meaning.

A non-native speaker of English—a potential customer in Asia or Africa, for example—might be shocked if you wrote about a sale concluded at a branch office this way: "Last week we made a killing in our office." Substitute the idiomatic expression with a clear, unambiguous translation easily understood in international English. "We made a big sale last week." For "Sleep on it," you might say, "Please take a week or two to make your decision."

5. Delete sports and gambling metaphors. These metaphors, which are often rooted in U.S. popular culture, do not translate word for word for non-native speakers and so again can interfere with your communication with your readers. Here are a few examples to avoid:

out in left field	a ballpark figure
struck out	fumbled the ball
drop the ball	out of bounds
down for the count	made a pass
long shot	beat the odds
be in left field	won by a nose

Use a basic English dictionary and your common sense to find nonfigurative translations for those and similar expressions.

6. Don't use abbreviations, acronyms, or contractions. While these shortened forms of words and phrases are a part of U.S. business culture, they might easily be misunderstood by a non-native speaker who is trying to make sense of them in context or by looking them up in a foreign language dictionary. Avoid abbreviations such as pharm., gov., org., pkwy, rec., hdg., hr., mfg., or w/o. The following acronyms can also cause your international reader trouble: ASAP, PDQ, p's and q's, IRA, SUV, RV, DOB, DOT, SSN. If you have to use acronyms, define them. Finally contractions such as the following might lead readers to mistake them for the English word they look like: I've (ivy), he'll (hell), he'ld (held), I'll (ill), we'll (well), can't (cant), won't (wont, want).

7. Watch units of measure. Do not fall into the cultural trap of assuming that your reader measures distances in miles and feet (instead of kilometers and meters as most of the world does, measures temperatures on the Fahrenheit scale (instead of Celsius) buys gallons of gasoline (instead of liters), spends dollars (rather than euros, pesos, marks, rupees, or yen), tells time by a twelve-hour clock (many countries follow a twenty-four-hour clock), and records dates by month/day/year (most countries record dates by day/month/year). See "International Measurements" on pages 176–177 for more detailed information.

8. Avoid culture-bound descriptions of place and space. For example, when you tell a reader in Hong Kong about the Sunbelt or a potential client in Africa about the Big Easy, will he or she know what you mean? When you write from California to a non-native speaker in India about the eastern seaboard, meaning the East Coast of the United States, the directional reference may not mean the same thing to your audience as it does to you. Calling February a winter month does not make sense to someone in New Zealand, for whom it is a summer one.

9. Keep your sentences simple and easy to understand. Short, direct sentences will cause a reader whose native language is not English the least amount of trouble. A good rule of thumb is that the shorter and less complicated your sentences, the easier and clearer they will be for a reader to process. Long (more than fifteen words) and complex (multiclause) sentences can be so difficult for readers to unravel that they may skip over them or simply guess at your message. Do not,

International Measurements

The following guidelines will help you use the proper units of measurement for your international audience.

Distance, Weight, and Volume
Most of the international community uses the metric system rather than English units (meters rather than feet, grams other than ounces, liters rather than quarts, etc.) In fact, the United States is the only country that has not officially adopted the metric system (except in science and technology). When corresponding with an international reader, use metric units. But when writing to multiple audiences—those in the United States and overseas—display both units. For help converting length, pressure, and volume measurements, consult *http://worldwidemetric.com/measurements.html.*

Dates
Different cultures express dates in different ways. Most countries, including those in Europe, list the day, month, and then the year; for example, 1–3-09 refers to March 1, 2009. But in the United States, where the month comes first and then day and year, the date above would be read as January 3, 2009. In Japan, a different system is used; the year is listed first, followed by the month and day, 2009.3.1 (or 09/3/1). To avoid any confusion, write out the month and use all the numbers for the year, for example, 2008 rather than 08.

Currency
Even though many European countries now use one currency, the euro, world currencies still vary widely. Always respect the monetary unit your reader uses and include the local symbol for it. But be careful. Don't write $ or "dollars" and think readers will automatically think of the United States. Other areas of the world—Singapore, Hong Kong, Canada, Australia, New Zealand—also use dollars. Specify what country or region those dollars belong to: $100 dollars (Hong Kong). To convert U.S. dollar amounts into any international currency based on the present rate of exchange, see *http://www.xe.com/ucc/.* This site also identifies world currencies.

Numbers

In many countries, a comma is used instead of a decimal point to separate numbers (expressed in the thousands) and fractions. For example, in the United States 12,001 signifies just over twelve thousand, whereas in Great Britain the number is written as 12.001. In other countries, however, the comma and decimal point are reversed to express percentages: 82.3% in the United States stands for 82 and 3/10 percent while in Germany or Bolivia the percentage is written as 82,3. Many Latin American and European countries, including Scandinavia, use spaces rather than decimals or commas to visually separate figures, for example, 12 001, 12. But be aware that conventions for numbers can change from country to country. Check the scientific guides and manuals in your discipline to find out which practice is internationally endorsed.

Time—Twenty-Four-Hour Clock

Many countries—including most of those in Europe—use the twenty-four-hour clock, or what is sometimes referred to as "military time" in the United States. Unlike the twelve-hour clock in use in the United States, the twenty-four-hour clock does not use **a.m.** or **p.m.** In Paris, for instance, 13:30 is 1:30 p.m. in New York and 9 p.m. in the United States translates to 21:00 in Sweden. Always specify time zones (Eastern Standard Time, Greenwich Mean Time) to help international audiences understand your message more clearly.

Temperature and Season

Most of the world, except the United States and Jamaica, records temperature using the Celsius (represented by C) temperature scale instead of the Fahrenheit scale (represented by F). For this reason, convert Fahrenheit readings to Celsius when writing to international readers. The Celsius scale has a freezing point of 0° and a boiling point of 100°, while on the Fahrenheit scale water freezes at 32° and boils at 212°. Writing to an international reader, accustomed to recording temperatures in degrees Celsius, that it is a pleasant 75 degrees, without noting that the scale is Fahrenheit, may lead to confusion: 75°C is the Fahrenheit equivalent of 120 degrees. To convert Celsius into Fahrenheit, add 40 to Celsius degrees, multiply by 1.8, and subtract 40. Similarly, to convert Fahrenheit to Celsius, add 40, divide by 1.8, and subtract 40.

Be aware of seasonal differences, too. While New York is in the middle of the winter, Australia and Chile are enjoying summer. Be respectful of your readers' cultural (and physical) environment. Thanksgiving is celebrated in the United States in November, but in Canada the holiday is the second Monday in October; elsewhere around the world it may not be a holiday at all.

however, be insultingly childish, as if you were writing for someone in kinder-garten. Also, always try to avoid the passive voice. It is one of the most difficult sen-tence patterns for a non-native speaker to comprehend. Stick to the common subject-verb-object pattern as often as possible. See pages 60–61 and the Writer's Brief Guide to Paragraphs, Sentences, and Words (pp. 693–710).

10. Be cautious about style and tone. Culture plays a major role in how you word your message. Americans expect business letters to be concise and to the point, without flowery compliments and personal details. They want to see conclu-sions and recommendations up front, followed by descriptions of key items leading to the main point. Readers in other cultures, however, would find this approach of-fensive. For instance, German readers expect long letters that unfold slowly through a highly factual and scrupulously documented narrative of events, all of which would lead to a recommendation at the end of the letter. Japanese or Korean readers, unlike either American or German readers, expect the first paragraph or two of a letter to center on the friendship and respect the writer has for the reader, as in Frank Sims's letter in Figure 5.10 and Susan DiFusco's letter in Figure 5.12.

11. Use appropriate salutations, complimentary closes, and signature lines. Find out how individuals in the recipient's culture are formally addressed in a salu-tation (e.g., Señor, Madame, Frau, Monsieur). Unless you are expressly asked to use a first name, always use your reader's surname and include proper titles and other honorifics (e.g., Doctor, Sir, Father). For a complimentary close, use an appropri-ately formal one, such as *Respectfully*, which is acceptable in almost any culture. Fi-nally, find out the country's custom for signature lines. In some countries, the name of the company appears before and above the writer's name, thus giving greater re-spect and authority to the organization than to the employee.

Modifying a Letter for an International Audience

Figures 5.9 and 5.10 show two versions of a letter to an international business exec-utive. The letter in Figure 5.9 violates the guidelines discussed above for writing to such an audience because it:

- uses the incorrect format for the dateline
- misspells the name of the reader's city
- leaves out important postal information
- contains U.S. idioms (e.g., *drop you a line*)
- includes troubling abbreviations
- disregards how the reader's culture records time and temperature

Even more disrespectful, the writer's overall tone is condescending (*south of the border*) and inappropriately casual. At the end of the second paragraph, for instance, the writer tells his reader that his U.S. firm is superior to the Argentine company.

Note how the letter in Figure 5.10, on the other hand, respects the reader's cul-tural conventions because it:

- spells and punctuates the reader's name and address correctly
- incorporates a clear date line
- uses an appropriate salutation and complimentary close
- is written in plain, international English

An inappropriately written letter for an international reader   Figure 5.9

Pro-Tech, Ltd. 452 West Main St. Concord, MA 01742 978-634-2756
www.protech.com

5-8-09   *Misleading*
                                                                  date line

Mr. Antonio Guzman   *Incorrect,*
Canderas   *misspelled*
Mercedes Ave.   *address*
Bunos Aires, ARG.

Dear Tony,   *Too informal*

I wanted to drop you a line before the merger hits and in doing so touch base
and give you the lowdown on how our department works here in the good old
U.S. of A.

None of us had a clue that Pro-Tech was going to go south of the border but   *Culturally*
your recent meeting about the Smartboard T-C spoke volumes to the tech   *condescending*
people who praised your operations to the hilt. So it looks like you and I both
will be getting a new corp. name. Olé. I love moving from Pro-Tech, Ltd. to
Pro-Tech International. We are so glad we can help you guys out.

At any rate, I'm sending you an e-mail with all the ins and outs of our
department struc., layout, employees, and prod. eff. quotas. From this info,   *Filled with*
I'm hoping you'll be able to see ways for us to streamline, cooperate, and soar *American*
in the market. I understand that all of this is in the works that you and I need *idioms*
to have a face-to-face and so I'd appreciate your reciprocating with all the
relevant data stat.

Consequently, I guess I'll be flying down your way next month. Before I take   *Disregards*
off, I would like to give you a ring. How does after lunch next Thursday (say, *time*
1:00–1:30) sound to you? I hope this is doable.   *differences and*
                                                                               reader's
                                                                               24-hour clock

We've had a spell of great weather here (can you believe it's in the low 80's
today!). So, I guess I'll just sign off, and wait 'til I hear from you further. *Ignores, differ-*
                                                                               ences between
                                                                               Fahrenheit/
I send you felicitations and want things to go smoothly before the merger is   *Celsius scales*
upon us.

Adios,   *Close sounds*
                                                                  insincere

Frank Sims

Frank Sims

Figure 5.10   An appropriate revision of Figure 5.9.

Pro-Tech, Ltd. 452 West Main St.   Concord, MA 01742   978-634-2756
www.protech.com

Clear date line

8 May 2009

Complete and correct address

Señor Antonio Mosca-Guzman
Director, Quality Assurance
Tecnología Canderas, S.A.
Av. Martin 1285, 4° P.C.
C1174AAB BUENOS AIRES
ARGENTINA

Courteous salutation

Dear Señor Mosca-Guzman:

Clear and diplomatic opening

As our two companies prepare to merge, I welcome this opportunity to write to you. I am the manager for the Quality Assurance division at Pro-Tech, a title I believe you have at Tecnología Canderas. I am looking forward to working with you both now and after our companies merge in two months.

Respectful view of reader's company

Allow me to say that we are very honored that your company is joining ours. Tecnología Canderas has been widely praised for the research and production of your Smartboard T-C systems. I know we have much to learn from you, and we hope you will allow us to share our systems analyses with you. That way everyone in our new company, Pro-Tech International, will benefit from the merger.

Explanation of business procedures in plain English

Later this week, I will send you a report about how we manage our division. I will describe how our division is structured and the quality assurance inspections we make. I will also give you a brief biography of our staff so that you can learn about their qualifications and responsibilities.

The director of our new company, Dr. Suzanne Nknuma, asked me to meet with you before the merger occurs to discuss how we might help each other. I would very much like to travel to Buenos Aires in the next two months to visit with you and take a tour of your company.

Recognition of reader's time zone

Would you please let me know by e-mail when it may be convenient for us to talk on the telephone so we might discuss the agenda for our meeting? I am always in my office from 11:00 to 17:00 Buenos Aires time.

I look forward to working with and meeting you.

Polite close

Respectfully,

Frank Sims

Frank Sims
Quality Assurance Manager

Gives business title

- recognizes that the reader uses a twenty-four-hour clock
- identifies the writer and makes the reader feel welcome and honored
- clearly informs the reader how the merger will affect his relationship with the writer
- strives to develop a spirit of cooperation
- is courteous throughout
- acknowledges the reader's position of authority in his company

Above all, though, the writer respects his audience's culture and role in the business world and seeks to win the reader's confidence and respect, two invaluable assets in the global marketplace.

Respecting Your Reader's Nationality and Ethnic/Racial Heritage

Do not risk offending any of your readers, whether they are native speakers of English or not, with language that demeans or stereotypes their nationality or ethnic and racial background. Here are six precautions to take.

1. **Respect your reader's cultural traditions.** Be aware of how cultures differ in terms of traditions, customs, and preferences—how people dress, communicate, and celebrate holidays. When writing a letter to someone from a different country, show respect for your reader's background and perspectives. Don't assume that your culture is the only culture. For instance, not everyone celebrates the same holidays you do, or on the same days. As we saw, Thanksgiving is celebrated in October in Canada. Take a look at the cultural calendar that Terri Ruckel incorporates into her long report on recruiting multicultural workers in Figure 15.3.

2. **Respect your reader's nationality.** Always spell your reader's name and country properly, which may mean adding diacritical marks (such as accent marks) not used in English. If your reader has a hyphenated last name (e.g., Arana-Sanchez), it would be rude to address him or her by only part of the name (e.g., only Arana or only Sanchez). In addition, be careful to use the current name of your reader's country or city. For instance, Malaysia (not Malaya), the Czech Republic (not Czechoslovakia), Myanmar (not Burma), and Mumbai (not Bombay).

3. **Honor your reader's place in the world economy.** Phrases like "third-world country," "emerging nation," and "undeveloped/underprivileged area" are derogatory. Using such phrases signals that you regard your reader's country as inferior. Use the name of your reader's country instead of such phrases. Saying that someone lives in the Far East implies that the United States, Canada, or Europe are the center of culture, the hub of the business community. It would be better to simply say "East Asia."

4. **Avoid stereotypes.** Expressions such as "oil-rich Arabs," "time-relaxed Latinos," and "aggressive foreigners" unfairly characterize particular groups. Similarly,

Some Guides to Cultural Diversity for Businesspeople

Bridging the Culture Gap: A Practical Guide to International Business Communication, by Penny Carte and Chris Fox (Kogan Page, 2005).

Cultural Intelligence: A Guide to Working with People from Other Cultures, by Brooks Peterson (Intercultural Press, 2004).

Cultural Intelligence: People Skills for Global Business, by David C. Thomas (Berrett-Koehler, 2004).

Essential Do's and Taboos: The Complete Guide to International Business and Leisure Travel, by Roger E. Axtell (John Wiley & Sons, 2007).

International Business Etiquette Resource Page: *http://www.cyborlink.com/besite/resource.htm.*

Kiss, Bow, and Shake Hands: The Bestselling Guide to Doing Business in More Than 60 Countries, by Terri Morrison and Wayne A. Conway (Adams, 2006).

Multicultural Manners: Essential Rules of Etiquette in the 21st Century, rev. ed., by Norine Dresser (John Wiley & Sons, 2005).

prune from your communications any stereotypical phrase that insults one group or singles it out for praise at the expense of another—"Mexican standoff," "Russian roulette," "Chinaman's change," "Irish wake," "Dutch treat," "Indian giver." The word *Indian* refers to someone from India; use *Native American* to refer to the indigenous people of North America, who want to be known by their tribal affiliations (e.g., the Lakota).

5. Be visually sensitive to the cultural significance of colors. Do not offend your audience by using colors in a context that would be offensive. Purple in Mexico, Brazil, and Argentina symbolizes bad luck, death, and funerals. Green and orange have a strong political context in Ireland. In Egypt and Saudi Arabia, green is the color of Islam and considered sacred. But in China, green can symbolize infertility or adultery. In China, white does not symbolize purity and weddings but mourning and funerals. Similarly, in India if a married woman wears all white, she is inviting widowhood. While red symbolizes good luck in China, it has just the opposite meaning in Korea.

6. Be careful, too, about the symbols you use for international readers. Triangles are associated with anything negative in Hong Kong, Korea, and Taiwan. Political symbols, too, may have controversial implications (e.g., hammer and sickle, a crescent). Avoid using the flag of a country as part of your logo or letterhead for global audiences. Many countries see this as a sign of disrespect, especially in Saudi Arabia, whose flag features the name of Allah.

Case Study: Writing to a Client from a Different Culture

Let's assume that you have to write a sales letter to an Asian business executive. You will have to employ a very different strategy in writing to an Asian executive as opposed to an American executive. For an American reader, the best strategy is to take a direct approach—fast, hard-hitting, to the point, and stressing your product's strengths versus the opposition's weaknesses. Businesses in the United States thrive on the battle of the brands, the tactics of confrontation symbolized in the sports metaphors on page 175. So a sales letter to an American reader would be polite but direct.

But such a strategy would be counterproductive in a sales letter to an Asian reader. Business in East Asia is associated with courtesy and friendship and is wrapped up in a great many social courtesies. In a sales letter to an Asian reader, you have to first establish a friendship before business details are dealt with. The Asian way of doing business, including writing and receiving letters, is far more subtle, indirect, and complimentary than it is in the United States. The U.S. style of directness and forcefulness would be perceived as rude or unfair in, say, Japan, China, Malaysia, or Korea. A hard-sell letter to an Asian reader would be a sign of arrogance, and arrogance suggests inequality for the reader.

Two Versions of a Sales Letter

To better understand the differences between communicating with a U.S. reader and an Asian one, study the two versions of the sales letter in Figures 5.11 and 5.12. The letters, written by Susan DiFusco for Starbrook Electronics, sell the same product, but Figure 5.11 is written to a U.S. executive, while Figure 5.12 adapts the same message for a business person in Seoul, South Korea. See how the two letters differ not only in content but also in the way each is printed.

Format
The full block style of the letter to the U.S. reader signals a no-nonsense, all-business approach. Everything is lined up in neat, orderly fashion. For the Korean reader, however, DiFusco wisely chose the more varied pattern of indenting her paragraphs. For Asian readers the visual effect suggests a much more relaxed and friendly, yet respectful communication. Note the different typefaces, too.

Opening
Pay special attention to how the U.S. letter in Figure 5.11 starts off politely but much more directly, an opening the Korean reader would regard as blunt and discourteous. The sales letter written to the Korean audience (Figure 5.12) starts not with business talk but with a compliment to the reader and his company, praising them for trustworthiness and wishing them much prosperity in the future. Susan DiFusco sought the advice of a Korean co-worker, and from him she included the Korean greeting that opens the letter. She knew that when addressing the Korean administrator she should not get to the bottom line right away but instead used her introductory paragraph to show respect for the company and the reader—the equivalent of East Asian business executives socializing before any mention of business is made.

Figure 5.11   Sales letter to a native English speaker in a U.S. firm.

Starbrook Electronics
Perry, TX 75432-3456
phone (713) 555-2121 E-mail starbrook.com
http://www.starbrook.org

May 5, 2009

Mr. Ellis Fanner
Administrator
Morgan General Hospital
300 Oakland Drive
Morgan, OR 97342-0091

Dear Mr. Fanner:

Begins with an aggressive question and emphasis on hospital's competitive role

How many times have the physicians who work at Morgan General asked when you would be getting an Open MRI? Having the latest, state-of-the-art imaging equipment is important to maintain your reputation as a leading health care provider in the Morgan River Valley.

Hard-hitting appeal to sell reader on MRI features

As the world leader in designing and manufacturing MRI equipment, Starbrook can offer you the latest technology available. This diagnostic technology will save your patients time and improve the care you give them. With the MRI capability of our Open Imaging 500, Morgan General can deliver more accurate and timely diagnoses. Within an hour you can determine whether a patient has had a stroke rather than having to wait a day or longer with more conventional X-ray or scanner technology.

Stresses MRI advantages over competitors' models

Thanks to our Open Imaging 500 model, Morgan General can also improve diagnoses for orthopedic and cardiac problems. Our MRI delivers much more extensive internal imaging than any other of our competitors' equipment.

Continued

Mr. Ellis Fanner   May 5, 2009   2

By obtaining the Open Imaging 500, Morgan General will surpass all other health care providers in the Morgan River Valley. No other hospital within a hundred-mile radius has one. By acting now, you, too, will receive Starbrook's unsurpassed guarantee of service and clarity. You are guaranteed one year's free maintenance by our team of experts.

Urges reader to act to surpass competition

And we will even give you free upgrades to make sure your Open Imaging 500 continues to be state of the art. Since software updates change so often and so radically, no other MRI vendor dares make such an offer. We deliver what we promise. Ask any of our recent satisfied customers—Tennessee General, Grantsville Uptown Clinic, or Nevada Statewide HMO.

Writer praises her own company

We are hosting a demonstration for hospital administrators on the 4th of June in Portland and would like to see you there. Don't hesitate to call me today to arrange for your free showing.

Asks for immediate response from reader

Sincerely yours,

Susan DiFusco
Assistant Manager

The Body

Compare the second and third paragraphs of the letters in Figures 5.11 and 5.12. While the letter to the U.S. reader launches an aggressive campaign to get the reader's business, the letter to the Korean reader avoids the hard sell of U.S. business tactics. Susan DiFusco knew that for her Korean reader she must not promote too strenuously. The more she boasted about Starbrook's work, the less likely it was that she would make a sale. She recognized, again from discussions with other Asian business people over the years, that she had to supply key information—such as the application and the advantages of her product—without overwhelming or pressuring her audience. Yet DiFusco subtly reassures Mr. Kim that her company is honorable and worthy to be recommended to his friends.

View of Competitors

Observe, too, how the letter to the U.S. executive (Figure 5.11) undermines the competition by stating how much better Starbrook's offer is. For most Asian readers, it

Figure 5.12   Sales letter to a non-native speaker of English in a foreign firm.

Starbrook Electronics
Perry, TX 75432-3456

phone (713) 555-2121   E-mail starbrook.com
http://www.starbrook.org

May 5, 2009

Mr. Kim Sun-Lim
Administrator, Tangki Hospital
Nowonku 427–1
Seoul, Korea 100–175

Dear Mr. Kim:

In your beautiful language I say "Gyuihauie bunyanggwa hangbokul kiwonhamnida" on behalf of my firm Starbrook Electronics. I am honored to introduce myself to you through this letter and send you greetings at this beautiful flower season.

Please let us know how we might be of service to you. One of the ways we may be able to serve you is by informing you about our new Open Imaging 500 MRI (magnetic resonance imaging) equipment. This new model can offer you and your patients many advantages. It is far better than conventional X-ray or even scanner models. Your physicians can offer quicker diagnoses for patients with strokes, heart attacks, or orthopedic injuries. MRI pictures will give you clearer and deeper pictures than any X-ray can.

It would be an honor to provide Tangki Hospital with one of our Open Imaging 500s. If you select the Open Imaging 500, we will be happy to give you all maintenance and software updates free for one year. Such service will provide the best in health care for the many people who come to you for help. Our firm is well known for its quality service.

Kindly let me know if I might send you information about the Open Imaging 500. It would be a privilege to meet you and to give you and your staff a demonstration of our MRI.

Respectfully,

Susan DiFusco

Font and format more appropriate for audience— more open and inviting— than in Figure 5.11

Opens with a compliment and subtle reference to business

Gives important equipment information without bashing competitors; appeals to the honor of reader's company

Ends with courteous invitation

would be considered impolite to claim that your product is better than another company's or that your firm is currently doing business with other firms in the reader's country. Asian audiences prefer to avoid anything that hints of impoliteness or assertiveness.

Conclusion.

Finally, contrast the conclusions of the two letters. In her letter to a U.S. audience, DiFusco strongly urges her potential customer to get in touch with her. Such a call to action is customary in a sales letter to a U.S. firm. But in her concluding paragraph to Mr. Kim, DiFusco adopts a more reserved and personal tone. Her use of such appropriate phrases as "kindly let me know" and "it would be a privilege" expresses the friendly sentiments of respect and esteem that would especially appeal to her reader. Note, too, that DiFusco has chosen a complimentary close ("Respectfully") much more in keeping with Mr. Kim's cultural sensitivities than the "Sincerely yours," which was suitable for the letter in Figure 5.11.

As these two letters show, writers need to know and respect their readers' cultures. In addressing an audience of non-native speakers of English, a writer must consider the readers' communication patterns, protocols, and cultural (or subcultural) traditions. The same guidelines apply whether you are writing to non-native speakers abroad or in this country. Your message and vocabulary should always be clear, understandable, and appropriate for the intended audience.

 ## Revision Checklist

Audience Analysis and Research
- ☐ Made sure reader's name and job title are correct.
- ☐ Found out something about my audience—interests and background, well informed or unfamiliar with topic, former clients or new ones.
- ☐ Determined whether audience will be friendly, hostile, or neutral about my message.
- ☐ Did sufficient research—in print, through online sources, in discussions with colleagues—to give audience what they need.
- ☐ Acknowledged previous correspondence.
- ☐ Spent sufficient time drafting and revising letter before printing final copy.

Format/Appearance
- ☐ Followed one letter format (full block, modified block) consistently.
- ☐ Used letter wizards with caution.
- ☐ Left margins wide enough to make my letter look attractive and well proportioned.
- ☐ Included all the necessary parts of a letter for my purpose.
- ☐ Made sure that my letter looks neat and professional.

Continued

Continued

☐ Printed my letter on company letterhead or quality bond paper.
☐ Proofread my letter carefully and made sure each correction was made before final copy was printed.
☐ Eliminated any grammatical and spelling errors.
☐ Signed my letter legibly in blue or black ink.
☐ Sent copies to appropriate parties.
☐ Printed envelopes properly.

Content/Organization

☐ Clearly understood my purpose in writing to reader(s).
☐ Put most important point first in my letter.
☐ Started each paragraph with the central idea of that paragraph.
☐ Answered all the reader's questions and concerns.
☐ Omitted anything offensive, irrelevant, or repetitious.
☐ Stated clearly what I want reader to do.
☐ Used last paragraph to summarize and encourage reader to continue cordial relations with me and my company.

Style: Words, Tone, Sentences, Paragraphs

☐ Emphasized the "you attitude" by seeing things from reader's perspective.
☐ Avoided being too casual or colloquial.
☐ Chose words that are clear, precise, and friendly.
☐ Cut anything sounding flowery, stuffy, or bureaucratic.
☐ Ensured that my sentences are readable, clear, and not too long (less than fifteen to twenty words).
☐ Wrote paragraphs that are easy to read and that flow together.

Writing to International Readers

☐ Did appropriate research about reader's culture, in print, through online sources, and with colleagues who are native speakers (as well as teachers), especially about accepted ways of communicating.
☐ Adopted a respectful, not condescending, tone.
☐ Avoided anything offensive to my reader, especially references to politics, religion, or cultural taboos.
☐ Used plain and clear language that my reader would understand.
☐ Tested my sentences for length and active voice.
☐ Made sure nothing in my letter might be misinterpreted by my reader.
☐ Chose colors and symbols culturally appropriate for the context of my message.
☐ Selected the right format, salutation, and complimentary close for my reader.
☐ Observed the reader's units of measurement for time, temperature, currency, dates, and numbers.

Exercises

1. Find two business letters and bring them to class. Be prepared to identify the various parts of a letter discussed in this chapter.

2. Find a form letter that is addressed to "Dear Customer," "Postal Patron," or "Dear Resident" and rewrite it to make it more personal.

3. Correct the following inside addresses:
   a. Dr. Ann Clark, M.D.
   1730 East Jefferson
   Jackson, MI. 46759
   b. To: Tommy Jones
   Secretary to Mrs. Franks
   Donlevey labs
   Cleveland, O. 45362
   c. Debbie Hinkle
   432 Parkway
   N. Y. C. 10054
   d. Mr. Charles Howe, Acme Pro.
   P.O. Box 675
   1234 S. e. Boulevard
   Gainesville, Flor. 32601
   e. Alex Goings, man.
   Pittfield Industries
   Longview, TEXAS 76450
   f. ATTENTION: G. Yancy (Mrs.)
   Police Academy
   1329 Tucker
   N. O., La. 3410–70122
   g. David and Mahenny
   Lawyers
   Dobbs Build.
   L.A. 94756
   h. Barry Fahwd
   Man., Peninsular, Ltd.
   Arabia

4. Write appropriate inside addresses and salutations to (a) a woman who has not specified her marital status; (b) an officer in the armed forces; (c) a professor at your school; (d) an assistant manager at your local bank; (e) a member of the clergy; (f) your congressperson.

5. Rewrite the following sentences to make them more personal.
   a. It becomes incumbent upon this office to cancel order #2394.
   b. Management has suggested the curtailment of parking privileges.
   c. ALL USERS OF HYDROPLEX: Desist from ordering replacement valves during the period of Dec. 20–30.
   d. The request for a new catalog has been honored; it will be shipped to same address soon.
   e. Perseverance and attention to detail have made this writer important to company in-house work.
   f. The Director of Nurses hereby notifies staff that a general meeting will be held Monday afternoon at 3:00 p.m. sharp. Attendance is mandatory.
   g. Reports will be filed by appropriate personnel no later than the scheduled plans allow.

6. The following sentences taken from letters are discourteous, boastful, excessively humble, vague, or lacking the "you attitude." Rewrite them to correct those mistakes.
   a. Something is obviously wrong in your head office. They have once more sent me the wrong model number. Can they ever get things straight?

b. My instructor wants me to do a term paper on safety regulations at a small factory. Since you are the manager of a small factory, send me all the information I need at once. My grade depends heavily on all this.

c. It is apparent that you are in business to rip off the public.

d. I was wondering if you could possibly see your way into sending me the local chapter president's name and address, if you have the time, that is.

e. I have waited for my confirmation for two weeks now. Do you expect me to wait forever or can I get some action?

f. Although I have never attempted to catalog books before, and really do not know my way around the library, I would very much like to be considered at some later date convenient to you for a part-time afternoon position.

g. It goes without saying that we cannot honor your request.

h. May I take just a moment of your valuable time to point out that our hours for the next three weeks will change and we trust and pray that no one in your agency will be terribly inconvenienced by this.

i. Your application has been received and will be kept on file for six months. If we are interested in you, we will notify you. If you do not hear from us, please do not write us again. The soaring costs of correspondence and the large number of applicants make the burden of answering pointless letters extremely heavy.

j. My past performance as a medical technologist has left nothing to be desired.

k. Credit means a lot to some people. But obviously you do not care about yours. If you did, you would have sent us the $249.95 you rightfully owe us three months ago. What's wrong with you?

7. The following letter, filled with musty expressions and in old-fashioned language (legalese), buries key ideas. Rewrite and reorganize it to make it shorter, clearer, and more reader-centered.

Dear Ms. Granedi:

This is in response to your firm's letter of recent date inquiring about the types of additional services that may be available to business customers of the First National Bank of Bentonville. The question of a possible time frame for the implementation of said services was also raised in the aforementioned letter. Pursuant to these queries, the following answers, this off ice trusts, will prove helpful.

Please be advised that the Board of Directors at First National Bank has a continuing reputation for servicing the needs of the Bentonville community, especially the business community. For the last f ifty years—half of a century— First National Bank has provided the funds necessary for the growth, success, and expansion of many local firms, yours included. This financial support has bestowed many opportunities on a multitude of business owners, residents of Bentonville, and even residents of surrounding local communities.

The Board is at this present writing currently deliberating, with its characteristic caution, over a variety of options suggested to us by our patrons, including your firm. These options, if the Board decides to act upon them, would enhance the business opportunities for financial transactions at First National Bank. Among the two options receiving attention by the Board at this point in time are the creation of a branch office in the rapidly growing north side of Bentonville. This area has many customers who rely on the services of First National Bank. The Board may also place a business loan department in the new branch.

If this office of the First National Bank of Bentonville might be of further helpful assistance, please advise. Remember, banking with First National Bank is a community privilege.

Soundly yours,

M. T. Watkins

Public Relations Director

8. Either individually or in a small group, write a business letter to one of the following individuals and submit an appropriate envelope with your letter.
   a. your mayor, asking for an appointment and explaining why you need one
   b. your college president, stressing the need for more parking spaces or for additional computers in a library
   c. the local water department, asking for information about fluoride supplements
   d. an editor of a weekly magazine, requesting permission to reprint an article in a school newspaper
   e. the author of an article you have read recently, telling why you agree or disagree with the views presented
   f. a disc jockey at a local radio station, asking for more songs by a certain group
   g. a computer dealer, inquiring about costs and availability of software packages and explaining your company's special needs

9. Rewrite the following letters, making them appropriate for a reader whose native language is not English. Identify the intended reader's cultured heritage. As you revise the letters, pay attention to the words, measurements, and sentence constructions you employ. Be sure to consider the reader's cultural traditions by omitting any cultural insensitivity.

a. Dear Chum,

Our stateside boss hit the ceiling earlier today when she learned that our sales quota for this quarter fell precipitously short. Ouch! Were I in her spot, I would have exploded too. Numerous missives to her underlings warned them to get off the dime and on the stick, but they were oblivious to such. These are the breaks in our business, right? We can't all bat 1000.

Let's hope that next quarter's sales take a turn for the best by 12-1-06. If they are as disastrous, we all may be in hot water. Until then, we will have to watch our p's and q's around here. We're freezing here—20° today.

Cheers,

b. Dear Mr. Wong,

It's not every day that you have the chance to get in on the ground floor of a deal so good you can actually taste it. But Off-Wall Street Mutual can make the difference in your financial future. Give me a moment to convince you.

By becoming a member of our international investing group for just under $250, you can just about ensure your success. We know all the ins and outs of long-term investing and can save you a bundle. Our analysts are the hot shots of the business and always look long and hard for the most propitious business deals. The stocks we select with your interests in mind are as safe as a bank and not nearly so costly for you. Unlike any of your undertrained local agents, we can save you money by investing your money. We are penny pinchers with our clients' initial investments, but we are King Midas when it comes to transforming those investments into pure gold.

I am enclosing a brochure for you to study, and I really hope you will examine it carefully. You would be foolish to let a deal like Off-Wall Street Mutual pass you by. Go for it. Call me by 3:00 today.

Hurriedly,

c. Dear Mr. Bafaloukos,

My firm is taking a survey of businesses in your part of the world to see if there is any likelihood of getting you on board our international computer network and so I thought I would see if you might like to take the chance. In today's shaky world, business events can change overnight and

without the proper scoop you could be left out in the cold.
We can alleviate that mess.

Not only do we interface with major exchanges all around
the globe but we make sure that we get the facts to you
pronto. We do not sit on our hands here at Intertel. Check
out our website on who and how we serve and I have no doubts
that you will e-mail or ring us up to find out about joining up.

One last point: Can you really risk going out on a limb
without first knowing that you have all the facts at your
fingertips about worldwide business events? Intertel is
there to save you.

Fondly,

10. Interview a student at your school or a co-worker who was born and raised in
    a non-English-speaking country about the proper etiquette in writing a busi-
    ness letter to someone from his or her country. Collaborate with that student
    to write a letter (for example, a sales letter or a letter asking for information) to
    an executive from that country.

11. In a letter to your instructor, describe the kinds of adaptations you had to make
    for the international reader you wrote to in Exercise 10.

12. Assume you work for a large international corporation that has just opened
    new offices in the following cities:
    a. Dar es Salaam, Tanzania
    b. Istanbul, Turkey
    c. Caracas, Venezuela
    d. Manila, Philippines
    e. Kyiv, Ukraine
    f. Beijing, China
    g. Warsaw, Poland
    h. Mexico City, Mexico
    i. Amman, Jordan
    j. Perth, Australia
    k. Lagos, Nigeria
    l. Prague, Czech Republic

Using the resources mentioned on page 173, write a short report on the main
points of letter etiquette that your boss will have to observe in communicating
with the non-native speaker of English who is the manager at one of the twelve
new offices.

13. As a collaborative project, team up with three other students in your class to
    write separate letters tailored to executives in each of the following cities:
    a. Riyadh, Saudi Arabia
    b. Tokyo, Japan

    c. Munich, Germany
    d. Nairobi, Kenya

Assume you are selling the same product or service to each reader but you will have to adapt your communication to the culture represented by the reader.

Turn in the four letters and explain to your instructor in an accompanying memo how you met the needs of those diverse cultural audiences in terms of style, tone, level of content, format, and sales tactics. Describe the research tools you used to find out about your reader's particular culture (and communication protocols) and how you benefited from using those sources.

CHAPTER 6

Types of Business Letters

As we saw in Chapter 5, letters can be the lifeblood of any company or organization. In this chapter, you will learn to write a variety of letters for different workplace occasions. But regardless of your message, every letter you send needs to

- establish or maintain good rapport with the reader
- protect and promote your company's and your own professional image
- continue or increase business sales, relationships, and opportunities

Letters can help or hurt your company and your career. You want to write letters that lead to promotions and greater professional opportunities.

Visit www
.cengage.com/
english/kolin
for this
chapter's online
exercises, ACE
quizzes,
and Web links.

Formulating Your Message

Writing good letters can be challenging. Business correspondence places a great demand on your ability to formulate and organize a suitable message for your readers. You have to identify your audience, your purpose in writing to them, and their needs in wanting to hear from you. To communicate with them effectively, you have to determine

- what to say
- how to say it
- where to say it

The second and third points are just as important as the first. In fact, they may be even more crucial to your success as a letter writer. As we saw in Chapter 5, the tone and organization of your letter will determine how your message is received. In the process of drafting and revising your letter, consider whether your reader will bristle at or welcome the words you use. Review pages 165–169 on the "you attitude." You will also have to select the best communication strategy (see pp. 208–209), which means taking into account where and how you give readers information. Some messages need to be direct and clear-cut; others require being indirect, holding off a bad news message until you prepare the reader for it. Consider the following two versions of a letter that sends the same message:

Insensitive Version

Your job is being outsourced next quarter. Corporate knows of your twelve years with the firm but downsizing mandates this decision. The firm will try to place you in another job, but expect no guarantees.

Respectful Version

Thank you for your twelve years of excellent work. We appreciate your many contributions. Unfortunately, downsizing means that your position, like several others in the department, will be outsourced. However, I will try to arrange a transfer for you. I may not succeed, but I know you will in your career because of your skills and dedication.

Clearly, the first letter has no respect for the reader's feelings or contributions, while the second is a model of respect and sincerity.

To be a skilled letter writer means that you will have to play several key roles, sometimes all at once. Among them are

- researcher
- problem solver
- decision maker
- honest and ethical spokesperson
- team player

These are the assets of a valued employee and the characteristics that employers want in job applicants (see pp. 251–253). In fulfilling these roles through your business letters, you demonstrate that you can work well with people, are sensitive to their needs, and represent your company with integrity.

Keep in mind the audience for your letters is not just your customer. It includes your employers and co-workers. For instance, you may confer with specialists in engineering to answer a complaint about one of your company's products or work with individuals in the law or marketing department to draft a sales letter. Possibly you may need to get your letter approved by your boss, who may edit your letter (see pp. 91–97) before you send it. Or you may be asked to write a letter for another person's signature, as in Figure 3.6.

This chapter gives you guidelines and strategies to write successful business letters. It will also show you how *not* to write these letters, an equally valuable lesson worth learning about the world of work.

The Four Most Common Types of Business Letters

This chapter discusses the most common types of business correspondence that you will be expected to write on the job:

1. inquiry letters
2. special request letters
3. sales letters
4. customer relations letters

These letter types involve a variety of formats, writing strategies, and techniques. Business letters can be classified as **positive, neutral,** or **negative,** depending on their message and the anticipated reactions of your audience. Inquiry and special request letters are examples of neutral letters that carry neither good nor bad news; they simply inform, responding to routine correspondence.

- Inquiry and special request letters request information about a product or service, place an order, or respond to some action or question.
- Sales letters promoting a product carry positive news, according to the companies that spend millions of dollars a year preparing them.
- Customer relations letters can be positive (responding favorably to a writer's request or complaint) or negative (refusing a request, saying "No" to an adjustment, denying credit, seeking payment, critiquing poor performance, or announcing a product recall).

Inquiry Letters

Inquiry letters ask for information about a product, service, or procedure. Businesses frequently exchange such letters. As a customer, you, too, have occasion to ask in a letter about a special line of products, the price, the size, the color, and delivery arrangements. The clearer your letter, the quicker and more helpful your answers are likely to be.

Figure 6.1 illustrates a letter of inquiry from Michael Ortega to a real estate office managing a large number of apartment complexes. Note that it follows these five rules for an effective inquiry letter:

- states exactly what information the writer wants
- indicates clearly why the writer must have the information
- keeps questions short and to the point
- specifies when the writer must have the information
- thanks the reader

Had Michael Ortega simply written the following very brief letter to Hillsdale Properties, he would not have received the information he needed about size, location, and price of apartments: "Please send me some information on housing in Roanoke. My family and I plan to move there soon."

Special Request Letters

Special request letters make a special demand, not a routine inquiry. For example, these letters can ask a company for information that you as a student or an employee will use in a report, an individual for a copy of a speech, or an agency for facts that your company needs to prepare a proposal or sell a product. The person or company being asked for help stands to gain no financial reward for supplying the information; the only reward is the goodwill that a response creates. Figure 6.2 contains a special request letter from a student asking for information to complete a report.

Figure 6.1   An inquiry letter.

Michael Ortega
403 South Main Street Kingsport, TN 37721-0217
mortega@erols.com

April 7, 2009

Mr. Fred Stonehill
Property Manager
Hillside Properties
701 South Arbor St.
Roanoke, VA 24015-1100

Dear Mr. Stonehill:

States precise request

Please let me know if you will have any two-bedroom furnished apartments available for rent during the months of June, July, and August. I am willing to pay up to $750 a month plus utilities. My wife, one-year-old son, and I will be moving to Roanoke for the summer so I can take classes at Virginia Western Community College.

Explains need for information

Identifies area of interest

If possible, we would like to have an apartment that is within two or three miles of the college. We do not have any pets.

Specifies exact date when a reply is needed

I would appreciate hearing from you within the next two weeks. My e-mail address is **mortega@erols.com**, or you can call me at home (606-555-8957) any evening from 6–10 p.m.

Offers to confer with and thanks the reader

If you have any suitable vacancies, we would be happy to drive to Roanoke to look at them and give you a deposit to hold an apartment. Thanks for your help.

Sincerely yours,

Michael Ortega

Michael Ortega

A special request letter.   **Figure 6.2**

1505 West 19th Street
Syracuse, NY 13206

phone 315-555-1214

October 5, 2009

Ms. Sharonda Aimes-Worthington
Research Director
Creative Marketing Associates
198 Madison Ave.
New York, NY 10016-0092

Dear Ms. Aimes-Worthington:

I am a junior at Monroe College in Syracuse, and am writing a report
on "Internet Marketing Strategies for the Finger Lakes Region of New
York" for my Marketing 340 class. Several of my professors have
spoken very highly of Creative Marketing Associates, and in my own
research I have learned a great deal from reading your article on Web
designs and local economies posted on your website two months ago.

Given your extensive experience in using the Internet to promote
regional businesses and tourism, I would be grateful if you would share
your responses to the following three questions with me:

1. What have been the most effective ways to design websites for
   a regional marketplace such as the Finger Lakes?

2. How can area chambers of commerce best provide links to
   benefit both the local municipalities and businesses in the
   Finger Lakes area?

3. Which other regional area(s) do you see having the same or very
   similar marketing goals and challenges as the Finger Lakes?

jkawatsu@webnet.com

Explains reason for letter

Proves writer has done research

Acknowledges reader's expertise

Lists specific numbered questions on the topic

Continued

Figure 6.2   (Continued)

Page 2

Your answers to these questions would make my report much more authoritative and useful. I would be happy to send you a copy and will, of course, cite you and Creative Marketing Associates in my work.

Because my report is due by December 2, I would greatly appreciate having your answers within the next month so that I can include them. Would you kindly send your responses, or any questions you may have, to my e-mail address, *jkawatsu@webnet.com.* Many thanks for your help.

Sincerely yours,

Julie Kawatsu

Julie Kawatsu

Because your reader is not obliged to respond to your letter, the way you present yourself and your request is crucial to your success. "Why should I help?" is a common question any reader might have. Therefore, try to anticipate and remove any barriers to getting his or her help.

Make your request clear and easy to answer. Supply readers with an addressed, postage-paid envelope, an e-mail address, and fax and telephone numbers in case they have questions. But don't expect a company to fax a long document to you. It is discourteous to ask someone else to pay the fax charges for something you need.

When writing a special request letter, follow these eight points:

1. Make sure you address your letter to the right person.
2. State who you are and why you are writing—a student researching a paper, an employee compiling information for a report, and so on—in your opening paragraph.
3. Indicate clearly your reason for requesting the information. Mention any individuals who may have referred you to write for help and information.
4. Make your questions easy to answer. List, number, and separate them clearly. Keep them to three or four at most.
5. Specify exactly when you need the information. Allow sufficient time—at least three weeks. Be reasonable; don't ask for the impossible.
6. As an incentive for the reader to reply, offer to forward a copy of your report, paper, or survey in gratitude for the help you were given.

7. If you want to reprint or publish the materials you ask for, indicate that you will secure whatever permissions are necessary. State that you will keep the information confidential, if that is appropriate.
8. End by thanking the reader for helping.

Note how Julie Kawatsu's letter in Figure 6.2 adheres to these guidelines.

▌ Sales Letters

Sales letters are written to persuade readers to buy a product, try a service, support some cause, or participate in some activity. A sales letter can also serve as a method of introducing yourself to potential customers. No matter what profession you have chosen or whether you are self-employed, work for a small company, or are part of a large organization, knowing how to write a sales letter is an invaluable skill. There will always be times when you have to sell a product, a service, an idea, a point of view, or yourself!

Preliminary Guidelines

You have undoubtedly received numerous sales letters from large companies, local merchants, charitable organizations, and campus groups. Because of the great volume of sales letters like these in the business world, the ones you write face a lot of competition. To write an effective sales letter that stands out and does its job, you have to do the following:

1. **Identify and limit your audience.** Knowing who your target audience is and how to find them is crucial to your success. You will also have to determine how many people are in your audience. A sales letter may be written to just one person (Figure 6.3) or to hundreds of readers at different companies (Figure 6.4).

2. **Use reader psychology.** Think like your reader and ask: "What am I trying to do for the customer?" Ask that question before you begin writing and you will be using effective reader psychology. Appeal to readers' health, security, convenience, comfort, or pocketbooks by focusing on the right issues (for instance, inform buyers that your product research involves no animal testing). Patrice St. Jacques appeals to the reader's ethnic and community values in Figure 6.3.

3. **Don't be a bore or boast.** Save elaborate explanations about a product for after the sale. Put detailed documentation in instruction booklets, in warranties, or on your company's website. Further, do not turn your sales letter into a glowing commendation of your company, or yourself. Both Figures 6.3 and 6.4 avoid that.

4. **Use words that appeal to the reader's senses.** Use concrete words instead of abstract, vague ones. Choose verbs that are colorful, that put the reader in the picture, so to speak. You will have a greater chance of selling readers if they can hear, see, taste, or touch your product in their mind. That way they can visualize themselves buying or using your product or service. Note how Patrice St. Jacques fills the sales letter in Figure 6.3 with Caribbean sights, sounds, and tastes to appeal to Etienne Abernathy and his company.

Figure 6.3   A sales letter that appeals to a specific international audience.

Distinctive, functional letterhead

ISLAND JACQUES
4700 Cyprus Avenue
Philadelphia, PA 19172

12 May 2009

Mr. Etienne Abernathy, President
Seagrove Enterprises
1800 S. Port Haven Road
Philadelphia, PA 19103-1800

Dear Mr. Abernathy:

Compliments reader on award

Congratulations on winning the Hanover Award for Community Service. We in the Port Haven area are proud that a business with Caribbean roots has received such a distinguished honor.

Appeals to reader's ethnic heritage

To celebrate your and Seagrove's success, as well as all your business entertaining needs (annual banquet, monthly meetings, etc.), I invite you to Island Jacques. We are a family-owned business that for 30 years has offered Philadelphia residents the finest Caribbean atmosphere and food west of the Islands. Our black pepper shrimp, reggae or mango chicken, and Steak St. Lucie—plus our irresistible beef and pork jerk—are the talk from here to Kingston. You and your guests can also savor our original Caribbean art and enjoy our steel drum music.

Features flexible options to make dining easy, convenient, and enjoyable

Island Jacques can offer Seagrove a variety of dining options. With five separate rooms, we are small enough for an intimate party of 4 yet large enough to accommodate a group of 250. We can do early lunches or late dinners, depending on your schedule. And we even cater, if that's your style. Our chefs—Diana Maurier and Emile Danticat—will prepare a special calypso menu just for you. Also a benefit, our prices are generously competitive for the Philadelphia area.

www.islandjacques.netdoor.com
856-555-3295

Continued

Page 2

Please call me soon so you can experience Island Jacques's unique
hospitality. For your convenience, I am enclosing a copy of this week's menu
delights. Check out our website, too, for a taste of Caribbean sound. We
would love to feature Seagrove as Island Jacques's "Guest of the Week"!

Stay Cool, Mon

Patrice St Jacques

Patrice St. Jacques
Manager

*Gives reader
incentive to act
soon*

---

5. Be ethical. Avoid untruths, exaggerations, false comparisons, and unsupported
generalizations. Honesty is the best way to make a sale. Never make false claims
about the cost, safety, or adaptability of your product or service. You could be prose-
cuted for mail fraud or sued for misrepresentation. (Review pp. 27–36 in Chapter 1.)
In addition, never attack a competitor and always get permission before you include
an endorsement.

The Four A's of Sales Letters

Successful sales letters follow a time-honored and workable plan; each sales letter
follows what can be called the four A's:

1. It gets the reader's *attention.*
2. It highlights the product's *appeal.*
3. It shows the customer the product's *application.*
4. It ends with a specific request for *action.*

Those four goals can be achieved in fewer than four or five paragraphs. Look at
Figure 6.4, a one-page letter in which those four parts are labeled. Holding a sales
letter down to one page or less will keep the reader's attention. Television commer-
cials, magazine ads, and Web ads provide useful models of the fourfold approach to
a customer. The next time you see one of these variations of the sales letter, try to
identify the four A's.

Getting the Reader's Attention

Your opening sentence is crucial. That first sentence is bait on a hook. If you lose
readers there, you will have lost them forever. A typical television commercial has
about two to five seconds to catch the viewers' attention and prompt them to keep
watching. Keep your opening short, one or two sentences at most. Show a reader

Figure 6.4 A sales letter sent to a business reader.

Workwell Software
3700 Stewart Avenue   Chicago IL 60637-2210
Phone: (312) 555-3720   **Fax:** (312) 555-7601   **E-mail:** sales@workwell.com
http://www.workwell.com

August 12, 2009

Ali Jen, Manager
Circuit Systems
7 Tyler Place
Oklahoma City, OK 73101

Dear Ali Jen:

Gets reader's attention with a question

Do you know how much money your company loses from repetitive strain injury? Each year employers spend millions of dollars on employee insurance claims because of back pains, fatigue, eye strain, and carpal tunnel syndrome.

Emphasizes the product's appeal

Workwell can solve your problems with its easy-to-use Exercise Program Software that will automatically monitor the time employees spend at their computers and also measure keyboard activity. After each hour (or the specified number of keystrokes), **Workwell** software will take your employees through a series of exercises that will help prevent carpal tunnel syndrome and strains.

Shows specific application of the product and links costs to benefits

Workwell's Exercise Program Software will not interfere with busy schedules. Each of the 15 exercises is demonstrated on screen with audio instructions. The entire program takes less than 3 minutes and is available for Windows XP and Windows Vista. For only $1499.00, you can provide a networked version of this valuable software to all of your employees.

Ends with a call for action

To help your employees stay at peak efficiency in a safe work environment, please call us at 1-800-555-WELL or contact us at **http://www.workwell.com** to order your software today.

Thank you,

Cory Soufas

Cory Soufas
Sales Manager

how your product or service will make his or her life or job easier, save money, or make him or her safer or happier.

The following six techniques are a few of the many interest-grabbing ways to begin a sales letter. Adapt these techniques to your product or service.

1. Ask a question. Ask a question that your reader(s) will want to see answered. Look, for example, at the opening question in Figure 6.4. Avoid such general questions as "Are you happy?" or "Would you like to make money?" Use more specific questions with concrete language. For instance, an ad for Air Force Reserve Nursing asks nurses, "Are you looking for something 30,000 feet out of the ordinary?"

2. Use a how-to statement. This is one of the most frequently used openers in a sales letter. Here are some effective how-to statements: "We can show you how to increase your plant growth up to 91%." "This is how to provide nourishing lunches for less than eighty cents a person." Note that the opening sentence of Cory Soufas's sales letter in Figure 6.4 combines both a how-to and a question approach.

3. Compliment your reader. Appeal to the reader's ego. But remember that readers are not naive; they will be suspicious of false praise. Patrice St. Jacques opens the sales letter in Figure 6.3 with fitting and legitimate praise.

4. Offer a gift. Often you can lure readers further into your letter by telling them they can save money or get a second product free or at half price. For example, a realtor tempts customers to see lots for sale with, "Enclosed is a coupon worth $50 in gas after you tour Deer Run Trails Estates."

5. Introduce a comparison. Compare your product or service with conventional or standard ones. For instance, a clothing firm told police officers that if they purchased a particular jacket, they were really getting three coats in one because the product had a removeable lining for winter or summer use and a visibility coating for nighttime use.

6. Announce a change. Link your sales offer to a current event that will positively affect your prospective buyer. For instance, when the sales tax on hybrids was about to be increased, a dealer sent sales letters to potential buyers, alerting them to the implications of delaying their purchase: "The sales tax on hybrids will jump a WHOPPING 4 percent effective next month. You may not think that 4 percent will mean that much money, but on a new 2009 model that increase could cost you an extra $1,500." The sales letter continued: "Couldn't you use that money for something else, say, those extras you've always wanted, like a GPS system or an extended warranty?"

Highlighting the Product's Appeal

Once you have aroused your reader's attention, introduce your product or service by making it so attractive, so necessary, and so profitable that the reader will want to buy or use it. In Figure 6.3 the Island Jacques letter zeroes in on the reader's ethnic pride and heritage. In Figure 6.4 Workwell Software's mass mailing letter to office managers appeals to their desire for greater productivity and improved safety. Here is an appeal by the Gulf Stream Fruit Company:

> Can you, when you bite into an orange, tell where it was grown? If it tastes better than any you have ever eaten . . . full of rich, golden flavor, brimming with juice, sparkling

with sunshine . . . then you know it was grown here in our famous Indian River Valley where we have handpicked it, at the very peak of its flavor, just for your order.[1]

Showing the Customer the Product's Application

In the third part of your sales letter supply evidence of the value of what you are selling. Don't overwhelm readers with facts, statistics, detailed mechanical descriptions, or elaborate arguments. The emphasis is still on the reader's use of the product and not on the company that manufactures or sells it.

1. **Supply the right evidence.** What evidence best convinces readers about a product's or service's appeal?

- **Descriptions** that emphasize state-of-the-art design and construction, efficiency, convenience, usefulness, and economy.
- **Special features** or changes that make your product or service more attractive. For example, a greenhouse manufacturer stressed that in addition to using its structure just for growing plants, customers would also find it a "perfect sun room enclosure for year-round 'outdoor' activities, gardening, or leisure health spa."
- **Testimonials,** or endorsements, from previous customers as well as from specialists. In Figure 6.4 Cory Soufas cites "a leading orthopedic surgeon" by name.
- **Guarantees, warranties, services, or special considerations** that will make your customer's life easier or happier—loaner cars, free home deliveries, ten-day trial period, a year's free Internet access or upgrades, as in Figure 6.4.

2. **Do I mention costs?** You may be obligated to mention costs in your letter. But postpone them until the reader has been shown how appealing and valuable your product is. Of course, if price is a key selling point, mention it early in the letter. As a general rule, however, do not bluntly state the cost. Note how unobtrusively Patrice St. Jacques handles price at the end of paragraph 3 in Figure 6.3.

Relate prices, charges, or fees to benefits. For example, a dealer who installs steel shutters did not tell readers the exact price of the product but indicated that individuals will save money by buying them: "Virtually maintenance free, your Reel Shutters also offer substantial savings in energy costs by reducing your loss through radiation by as much as 65% . . . and that lowers your utility bills by 35%."

Ending with a Specific Request for Action

The last section of your letter is vital. If the reader ignores your request for action, your letter has been written in vain. Tell readers exactly what you want them to do by when. Make it easy for them to

- authorize a deduction
- visit your business (as in Figure 6.3)
- take a test drive
- participate in a meeting

---

[1]From "Gifts from Gulf Stream," Gulf Stream Fruit Company, Ft. Lauderdale, Fla. Reprinted by permission.

- fill out a pledge card
- respond via the Internet (as in Figure 6.4)
- sign an order blank
- return a form by providing a stamped and addressed envelope

As with price, link the benefits the customers will receive to their responses. "Respond and be rewarded" is the basic message of the last section of your letter. Note that in Figure 6.4 the call to action is made in the last two paragraphs, and urges the reader to act immediately in order to take advantage of the free upgrades and maintenance.

Customer Relations Letters

Much business correspondence deals explicitly with establishing and maintaining friendly working relations. Such correspondence, known as customer relations letters, sends readers good news or bad news, acceptances or refusals. Good news customer relations letters tell customers that

- you have the product or service they want at a reasonable price
- you agree with them about a problem they brought to your attention
- you are solving their problem exactly the way they want
- you are approving their loan
- you are grateful to them for their business

Thank-you letters, congratulations letters, and adjustment letters saying "Yes" with the above messages are all examples of good news messages.

Bad news messages, however, inform readers that

- you do not like their work or the equipment they sold you
- you do not have what they want or you cannot provide it at the price they want to pay
- you are rejecting a proposal they offered
- you are denying someone further use of a facility
- you cannot refund their purchase price or perform a service again
- you want them to pay what they owe you now
- you are downsizing or moving and cannot continue to provide a service

Bad news messages often come to readers through complaint letters, adjustment letters that say "No," and collection letters.

Diplomacy and Reader Psychology

Writing effective customer relations letters requires skill in human relations and reader psychology. Regardless of the news—good or bad—you need to be a diplomatic and persuasive writer. In fact, customer relations letters, like other types of letters, call upon your most effective skills in persuasion (see pp. 24–26). To be at your best persuasively, do some research about your readers—their business needs, their schedules, areas of authority in the chain of command, and even their grievances (if applicable).

Customer relations letters show how you and your company regard the people with whom you do business. The letters should reveal your sensitivity to their needs. The first lesson to learn is that you cannot look at your letter only from your (the writer's) perspective. You have to see the letter from the reader's perspective and anticipate the reader's needs and reactions.

The Customers Always Write

As you read this section on customer relations letters, keep in mind the two basic principles captured in the pun "the customers always write."

1. Customers will write about how they would like to be or have been treated — to thank, to complain, to request an explanation.
2. Customers have certain rights that you must respect in your correspondence with them. They deserve a prompt and courteous reply, whether or not they are correct. If you refuse their request, they deserve to know why; if they owe you money, you should give them an opportunity to explain and a chance, up to a point, to set up a payment schedule. Always be ethical in responding — be fair, honest, undeniably legal, and professional. (Review pp. 27–36.)

Planning Your Customer Relations Letters—Preliminary Guidelines

Whether you are sending good news or bad, determine what to say and how to say it. Construct your letter as if you were building a house — step by step — from laying the foundation (e.g., reader's needs, your purpose, etc.) to painting the trim (polishing and editing). Do some preliminary planning. Outline for a few minutes to find your ideas. Your outline does not have to be formal or even neatly written — a few scribbles sometimes will be enough to get you started. By outlining, you will save (not lose) time, because you can identify your main points, exclude unnecessary or unimportant ones, and avoid the risk of forgetting something essential.

Fortified by your outline, you will feel more confident as you draft your letter. In the process of drafting and revising that letter, consider whether your reader will bristle at or accept the words you use. Your choice of words will determine the success or failure of your letter. You might want to review the discussions on tone and the "you attitude" (pp. 165–169).

After you have written your outline, you will be ready to draft your customer relations letter. As you draft keep in mind two things: the importance of diplomacy and whether you should be direct or indirect.

Being Direct or Indirect

Your message, tone, and knowledge of your reader are essential ingredients in a successful customer relations letter. But success also involves knowing where and how to start, and, especially, where to present your main point. Not every customer relations letter starts by giving the reader the writer's main point, judgment, conclusion, or reaction. *Where you place your main idea is determined by the type of letter you are writing.* Good news messages require one tactic; bad news, another.

Good News Messages.   If you are writing a good news letter, use the direct approach. Start your letter with the welcome, pleasant news that the reader wants to hear. Don't postpone the opportunity to put your reader in the right frame of mind. Then provide any relevant supporting details, explanations, or commentary. Being direct is advantageous when you have good news to convey.

Bad News Messages.   If you have bad news to report, do *not* open your letter with it. Be indirect. Prepare your reader for the bad news; keep the tension level down. If you throw the bad news at your reader right away, you jeopardize the goodwill you want to create and sustain. Consider how you would react to a letter that begins with these slaps:

- Your account is in error.
- Your order cannot be filled.
- Your application for a loan has been denied.
- It is our unfortunate duty to report . . .

Having been denied, disappointed, or even offended in the first sentence or paragraph, the reader is not likely to give you his or her attentive cooperation thereafter.

Notice how A. J. Griffin's bad news letter in Figure 6.5 curtly starts off with the bad news of a rent increase. Receiving such a letter, the owner of Flowers by Dan certainly could not be blamed for looking for a new place of business. Or if he did pay the increase, Griffin's letter would hardly ensure that Mr. Sobol would remain a happy tenant. Griffin was too direct when he should have been diplomatically indirect. He did not consider his reader's reaction; all he was concerned about was delivering his message.

Compare the curt version of Griffin's letter in Figure 6.5 with his revised message in Figure 6.6. In the revised version, Griffin begins tactfully with pleasant, positive words designed to put his reader in a good frame of mind. Then Griffin gives some background information that the owner of Flowers by Dan can relate to. A business person himself, Mr. Sobol doubtless has experienced some recent increases in his own costs. Griffin makes one more attempt to encourage Sobol to recall his good feelings about the Mall—last year they did not raise rents—before introducing the bad news of a rent increase.

Even after giving the bad news, Griffin softens the blow by saying that the Mall knows it is bad news. Griffin's tactic here is to defuse some of the anger that Sobol will inevitably feel. In fact, Griffin words the bad news so that the tenant sees the Mall as acting in the best interest of his flower shop. The Mall will not lower or compromise on the services that the tenant has enjoyed and profited from in the past. Griffin then ends on a positive, upbeat note: a prosperous future for Flowers by Dan.

You will learn more strategies about conveying bad news messages on pages 222–226.

Figure 6.5   An ineffective bad news letter.

River Road Mall

December 1, 2008

Mr. Daniel Sobol
Flowers by Dan
Lower Level 107
River Road Mall

Dear Mr. Sobol:

Blunt opening makes no attempt to help audience understand or accept message

This is to inform you of a rent increase. Starting next month your new rent will be $3,500.00 a month, resulting in a 15 percent increase.

Ends with a demand

Please make sure that your January rent check includes this increase.

Sincerely,

A. J. Griffin
Manager
ajg@rrmall.com

300 First Street
Canton, Ohio 44701
(216) 555-6700
www.rrmall.com

A diplomatic revision of the bad news letter in Figure 6.5.   Figure 6.6

River Road Mall

December 1, 2008

Mr. Daniel Sobol
Flowers by Dan
Lower Level 107
River Road Mall

Dear Mr. Sobol:

It has been a pleasure to have you as a tenant at the Mall for the past two years, and we look forward to serving you in the future.

Opens with positive association

Over these last two years we have experienced a dramatic increase in costs at River Road Mall for security, maintenance, landscaping, pest control, utilities, insurance, and taxes. Last year we absorbed those increases and so did not have to raise your rent. We wish we could do it again but, unfortunately, we must increase your rent by 15 percent, to $3,500.00 a month, effective January 1.

Prepares reader for bad news to follow

States bad news in most concise, upbeat way

Although no one likes a rent increase, we know that you do not want us to compromise on the quality of service that you and your customers expect and deserve here at River Road Mall.

Links bad news to reader benefits

Please let us know how we can assist you in the future. We wish you a very successful and profitable 2009. If you have any questions, please call or visit my office.

Does not apologize but ends with respectful tone

Cordially,

A.J. Griffin

A. J. Griffin
Manager
ajg@rrmall.com

Friendly complimentary close

300 First Street
Canton, Ohio 44701
(216) 555-6700
www.rrmall.com

Follow-Up Letters

A follow-up letter is sent by a company after a sale to thank the customer for buying a product or using a service and to encourage the customer to buy more products and services in the future. A follow-up letter is a combination thank-you note and sales letter that resells the reader. The letter in Figure 6.7 shows how an income tax preparation service attempts to obtain repeat business by doing the following:

1. beginning with a brief and sincere expression of gratitude
2. discussing the benefits already known to the customer and then drawing the customer's attention to a new sales area
3. ending with a specific request for future business

Occasionally, a follow-up letter is sent to a good customer who, for some reason, has stopped doing business with the company. Such a follow-up letter should try to find out why the customer has stopped doing business and to persuade that customer to resume business. Study the letter in Figure 6.8, in which Jim Margolis first politely inquires whether Mr. Janeck has experienced a problem and then urges him to come back to the store.

Complaint Letters

Each of us, either as customers or business people, at some time has been frustrated by a defective product, delayed or wrong shipment, unavailable products or parts, inadequate or rude service, or incorrect billing. When we get no satisfaction from calling an 800 number and are routed through a series of menu options, our frustration level goes up. Usually our first response is to write a letter dripping with juicy insults. But an angry letter, like a piece of flaming e-mail (see Figure 4.6), rarely gets positive results. You never know when the company you are attacking may be one of your firm's clients or a prospective employer of yours. A nasty letter will hurt you and give an unfavorable image of your company. In the world of business you do not want to burn bridges.

Establishing the Right Tone

A complaint letter is a delicate one to write. First, avoid the following:

- name calling
- sarcasm
- insults
- threats
- unflattering clip art
- intimidating type fonts (e.g., all capital letters)

The key thing to keep in mind is that you can disagree without being disagreeable. Be rational, not hostile. Just to let off steam, you might want to write an angry letter but then tear it up and write a new letter with courteous and diplomatic language.

A complaint letter is written for more reasons than just blowing off steam. You want some specific action taken. Register your complaint courteously and tolerantly.

A follow-up letter to encourage repeat business.   Figure 6.7

Taylor Tax Service
Highway 10
North Jennings, TX 78326
phone **(888) 555-9681** e-mail **taylor@aol.com**
http://www.taylor.com

December 2, 2009

Ms. Laurie Pavlovich
345 Jefferson St.
Jennings, TX 78326

Dear Ms. Pavlovich:

Thank you for using our services in February of this year. We were pleased to prepare your 2008 federal and state income tax returns. Our goal is to save you every tax dollar to which you are entitled. If you ever have questions about your return, we are open all year long to help you.

We are looking forward to serving you again next year. Several new federal tax laws will change the types of deductions you can declare. These changes might appreciably increase your refund. Our consultants know the new laws and are ready to apply them to your return.

Another important tax matter influencing your 2009 returns will be any losses you may have suffered because of the hailstorms and tornadoes that hit our area four months ago. Our consultants are specially trained to assist you in filing proper damage claims with your federal and state returns.

To make our services even more convenient, we will file your tax return electronically to speed up any refund. Please call us at (888) 555-9681 or e-mail us at **taylor@aol.com** as soon as you have received all your forms to set up an appointment. We are waiting to serve you seven days a week from 9:00 a.m. to 9:00 p.m.

Sincerely yours,

TAYLOR TAX SERVICE

Demetria Taylor

Demetria Taylor

Links business goal to customer advantage

Stresses reasons/benefits for customer to return for service

Makes it easy and profitable for customer to act soon

Ends with commitment to customer convenience

Figure 6.8   A follow-up letter to maintain customer goodwill.

BROADWAY CLEANERS

April 6, 2009

Mr. Edward Janeck
34 Brompton Lane, Apt. 13
Baltimore, MD 21227-0102

Dear Mr. Janeck:

Thanks customer and asks for feedback

Thank you for allowing us to take care of your cleaning needs for more than three years. It has been our pleasure to see you in the store each week and to clean your shirts, slacks, and coats to your satisfaction. Since you have not come in during the last month, we are concerned that in some way we may have disappointed you. We hope not, because you are a valuable customer whose goodwill we do not want to lose.

Expresses concern

If there is something wrong, please tell us about it. We welcome any suggestions on how we can serve you better. Our goal is to have a spotless reputation in the eyes of our customers.

Shows goodwill

Offers incentive to resume business dealings

The next time you need your garments cleaned, won't you please bring them to us, along with the enclosed coupon worth $10 off your next bill? We look forward to seeing you again—soon.

Cordially,

Jim Margolis

Jim Margolis, Manager

Encl. coupon

Broadway at Davis Drive   Baltimore, Maryland 21228-6210
(443) 555-1962   broadclean@metdoor.com

Companies want to be fair to you to keep you as a satisfied customer and to correct defective products so that other customers will not be inconvenienced. A complaint letter also gives you and your company a written record (always make a copy) that you reported a problem and that you are notifying the reader about it. Never delay in sending your letter. A product or service warranty or guarantee could expire in the meantime. An effective complaint letter can be written by an individual consumer or by a company. Figure 6.9 shows Michael Trigg's complaint about a defective fishing reel; Figure 6.10 expresses a restaurant's dissatisfaction with an industrial dishwasher.

Tech Note

Help in Registering Complaints

Several organizations can assist you when you have a problem with a particular company or agency. These organizations will give you valuable advice on how to proceed with your complaint.

- The Better Business Bureau: *http://www.bbb.org*
- U.S. Chamber of Commerce: *http://www.uschamber.com*
- International Chamber of Commerce—World Business Organization: *http://www.iccwbo.org*
- The Federal Trade Commission: *http://www.ftc.gov/bcp/conline/pubs/credit/crdright.htm* (click on "Your debt and debt collectors")
- CreditCards.com: *http://www.creditcards.com/credit-consumer-rights-article.php*

Writing an Effective Complaint Letter

To increase your chances of receiving a speedy settlement, follow these five steps in writing your letter of complaint. They will help you build your case and answer the key questions.

1. **Begin with a detailed description of the product or service.** Provide the model and serial numbers, size, quantity, color, and cost. Specify check and invoice numbers. Indicate when and where (specific address) and how (through a vendor, the Internet, at a store) you purchased it and also the remaining warranty. If you are returning the product to the company, note how you are sending it—U.S. mail, overnight delivery service, through a sales representative, or another way. If you are complaining about a service, provide the name of the company, the date of the service, the personnel providing it, and their exact duties.

2. **State exactly what is wrong with the product or service.** Precise information will enable the reader to understand and act on your complaint.

Figure 6.9   A complaint letter from a consumer.

17 Westwood Drive

Magnolia, MA 02171

mtrigg@roof.com

October 10, 2008

Mr. Ralph Montoya
Customer Relations Department
Smith Sports Equipment
P.O. Box 1014
Tulsa, OK 74109-1014

Dear Mr. Montoya:

Documents all relevant details about the product

On September 21, 2008, I purchased a Smith reel, model 191, at the Uni-Mart Store on Marsh Avenue in Magnolia. The reel sold for $84.95 plus tax. The reel is not working effectively, and I am returning it to you under separate cover by first-class mail.

Explains politely what is wrong

I had made no more than five casts with the reel when it began to malfunction. The button that releases the spool and allows the line to cast will not spring back into position after casting. In addition, the gears make a grinding noise when I try to retrieve the line. Because of these problems, I was unable to continue my participation in the Gloucester Fishing Tournament last week.

Clearly states what should be done

Specifies an acceptable time frame

I request that a new reel be sent free of charge in place of the defective one I returned. I would also like to know what was wrong with the defective reel.

Thank you for processing my claim within the next two weeks.

Sincerely yours,

Michael Trigg

Michael Trigg

A complaint letter from a business.   Figure 6.10

Camerson and Dale, Sunnyside, California 91793-4116   213-555-7500

June 21, 2009

Priscilla Dubrow
Customer Relations Department
Superflex Products
San Diego, CA 93141-0808

Dear Ms. Dubrow:

On September 15, 2008, we purchased a Superflex industrial dishwasher, model 3203876, at the Hillcrest store at 3400 Broadway Drive in Sunnyside, for $5,000. In the last three weeks, our restaurant has had repeated problems with this machine. Three more months of warranty remain on the unit.

Gives all product details and warranty information

The machine does not complete a full cycle; it stops before the final rinsing and leaves the dishes dirty. The cycle regulators are not working properly because they refuse to shift into the next necessary gear. Attempts to repair the machine by the Hillcrest crew on June 4, 11, and 15 have been unsuccessful.

Describes what's happened and when

The Loft has been greatly inconvenienced. Our kitchen team has been forced to sort, clean, and sanitize utensils, dishes, pans, and pots by hand, resulting in additional overtime. Moreover, our expenses for proper detergents have increased.

Documents reason for adjustment

We want your main office to send another repair crew at once to fix this machine. If your crew is unable to do this, we want a discount worth the amount of the warranty life on this model to be applied to the purchase of a new Superflex dishwasher. This would come to $1,000, or 20 percent of the original purchase price.

Clearly indicates how problem should be solved

So that our business is not further disrupted, we would appreciate your resolving this problem promptly within the next week.

Concludes politely with justification for prompt action

Sincerely yours,

Emily Rashon

Emily Rashon
Co-owner

▼ Browse our menu, which changes daily, at www.theloft.com

- How many times did the machine work before it stopped?
- What parts were malfunctioning?
- What parts of a job were not done or were done poorly?
- When did all this happen?
- Where and how were you inconvenienced?

Stating that "the brake shoes were defective" tells very little about how long they were on your car, how effectively they may have been installed, or what condition they were in when they ceased functioning safely.

3. Briefly describe the inconvenience you have experienced. Show that your problems were directly caused by the defective product or service. To build your case, give precise details about the time and money you lost. Don't be vague and say you had "numerous difficulties." Did you have to pay a mechanic to fix your car when it was stalled on the road? Did you have to buy a new printer or MP3 player? See how Emily Rashon documents The Loft's inconvenience in paragraph 3 of her letter in Figure 6.10.

Where appropriate, refer to any previous telephone calls, e-mail, or letters. Give the names of the people you have written or spoken to and the dates of your calls or correspondence.

4. Indicate precisely what you want done. Do not simply write that you "want something done." Just saying you "want a credit" is not enough. Do you want a complete refund, a replacement, a new model, a credit toward your account, or a credit toward the purchase of another product or service? Always state exactly what you want:

- your purchase price refunded in full
- a credit made to your account
- a credit toward the purchase of another model
- your exact model repaired or replaced
- a new repair crew assigned to the job
- an apology from the company for discourteous treatment

If you are asking for damages, state your request in dollars and cents and always include copies of bills documenting your expenses related to the problem. Perhaps you had to rent a car, were forced to pay a janitorial service to clean up, or had to rent equipment at a higher rate because the company did not make its deliveries as promised.

5. Ask for prompt handling of your claim. In your concluding paragraph, ask the reader to answer questions you may have (such as finding out where calls came from that you were billed for but did not make). You might even specify a reasonable time by which you want to hear from the writer or need the problem fixed. Note how the last paragraphs in Figures 6.9 and 6.10 effectively do this.

Adjustment Letters

Adjustment letters respond to complaint letters by telling customers dissatisfied with a product or service how their claim will be settled. Adjustment letters should reconcile the differences that exist between a customer and a firm and restore the customer's confidence in that firm.

How to (and Not to) Write an Adjustment Letter

An effective adjustment letter requires diplomacy. Be prompt, courteous, and decisive; do not brush the complaint aside in hopes that it will be forgotten. Investigate the complaint quickly and determine its validity by checking previous correspondence, warranty statements, guarantees, and your firm's adjustment policies on merchandise and service. In some cases you may even have to send returned damaged merchandise to your company's laboratory to determine who is at fault.

A noncommittal letter signals to the customer that you have failed to investigate the claim or are stalling for time. Do not resort to vague statements such as the following:

- We will do what we can to solve your problems as soon as possible.
- A company policy prohibits our returning your purchase price in full.
- Your request, while legitimate, will take time to process.
- While we cannot now determine the extent of an adjustment, we will be back in touch with you.

Customers want to be told that they are right; if they cannot get what they request, they will demand to know why. When you comply with a request, a begrudging tone will destroy the goodwill that your refund or replacement would have created. At the other extreme, do not overdo an apology by agreeing that the company is "completely at fault," that "such shoddy merchandise is inexcusable." If you make your company look too bad, you risk losing the customer permanently, and saying your company was negligent or that a product was defective may result in a lawsuit.

Adjustment Letters That Tell the Customer "Yes"

It is easy to write a "Yes" letter if you remember a few useful suggestions. As with a good news message, start with the favorable news the customer wants to hear; that will put him or her in a positive frame of mind to read the rest of your letter. Let the customer know that you sincerely agree with him or her—don't sound as if you are reluctantly honoring the request. For example, if your airline lost or misplaced luggage, apologize before you offer a settlement.

Study the two examples of adjustment letters saying "Yes" in this chapter to see how to write this kind of correspondence. The first example, Figure 6.11, says "Yes" to Michael Trigg's letter in Figure 6.9. You might want to reread the Trigg complaint letter to see what problems Ralph Montoya faced when he had to write to Mr. Trigg. The second example of an adjustment letter that says "Yes" is in Figure 6.12, responding to a customer who has complained about an incorrect billing.

You might also want to review the letter in Figure 5.8, which is an adjustment letter saying "Yes."

Figure 6.11   An adjustment letter saying "Yes."

Smith Sports Equipment
P.O. Box 1014   Tulsa, Oklahoma 74109-1014
(918) 555-0164 ▪ www.smithsport.com

October 19, 2008

Mr. Michael Trigg
17 Westwood Drive
Magnolia, MA 02171

Dear Mr. Trigg:

Apologizes and announces good news

Thank you for alerting us in your letter of October 10 to the problems you had with one of our model 191 spincast reels. I am sorry for the inconvenience the reel caused you. A new Smith reel is on its way to you.

Explains what happened and why problem will not recur

We have examined your reel and found the problem. It seems that a retaining pin on the button spring was improperly installed by one of our new soldering machines on the assembly line. We have thoroughly inspected, repaired, and cleaned this machine to eliminate the problem from happening again.

Expresses respect for customer

Since we began making quality reels in 1955, we have taken pride in helping loyal customers like you who rely on a Smith reel. We hope that your new Smith reel brings you years of pleasure and many good catches, especially next year at the Gloucester Fishing Tournament.

Closes with friendly offer to help again

Thank you for your business. Please let me know if I can assist you again.

Respectfully,

SMITH SPORTS EQUIPMENT

Ralph Montoya

Ralph Montoya, Manager
Customer Relations Department

An adjustment letter saying "Yes."   Figure 6.12

Brunelli Motors
- -
Route 3A, Giddings, Kansas 62034-8100 (913) 555-1521

August 5, 2009

Ms. Kathryn Brumfield
34 East Main
Giddings, KS 62034-1123

Dear Ms. Brumfield:

We appreciate your notifying us, in your letter of July 30, about the problem you experienced regarding warranty coverage on your new Phantom Hawk GT. The bills sent to you were incorrect, and I have cancelled them. Please accept my apologies. You should not have been charged for a shroud or for repairs to the damaged fan and hose, since all those parts, and labor on them, are covered by your warranty.

The problem was the result of an error in the way the charges were listed. Our firm has begun using new software to give customers better service, and the mechanic apparently entered the wrong code for your account. We have since programmed our system to flag any bills for vehicles still under warranty. We hope that this new procedure will help us serve you and our other customers more efficiently.

We value you as a customer of Brunelli Motors. When you are ready for another Phantom, I hope that you will once again visit our dealership.

Sincerely yours,

Susan Chee-Saafir

Susan Chee-Saafir
Service Manager

- -
Experience virtual reality: Drive a new Phantom at
http://www.brunelli.com

Thanks customer and complies with request

Explains why problem occurred and how it has been resolved

Ends courteously and leaves reader with good feeling about the dealership

Guidelines for Writing a "Yes" Letter

The following four steps will help you write a "Yes" adjustment letter.

1. Admit immediately that the customer's complaint is justified and apologize. Briefly state that you are sorry and thank the customer for writing to inform you.

2. State precisely what you are going to do to correct the problem. Let the customer know that you will

- extend warranty coverage
- credit the account with funds, more air miles, or the like
- offer a discount on the next purchase
- cancel a bill or give credit toward another purchase
- repair damaged equipment
- enclose a free pass, coupon, or waiver
- give the customer credit toward another purchase
- upgrade a product or service

Do not postpone the good news the customer wants to hear. The rest of your letter will be much more appreciated and convincing if the customer is told the good news right away. In Figure 6.11 Michael Trigg is told that he will receive a new reel; in Figure 6.12 Kathryn Brumfield learns she will not be charged for parts or service.

3. Tell customers exactly what happened. They deserve an explanation for the inconvenience they suffered. Note that the explanations in Figures 6.11 and 6.12 give only the essential details; they do not bother the reader with side issues or petty remarks about who was to blame. Assure customers that the mishap is not typical of your company's operations. Avoid promising, however, that the problem will never recur. Not only is such a guarantee unnecessary, but keeping it may be beyond your control.

4. End on a friendly—and positive—note. Do not remind customers of the trouble. Leave them with a positive feeling about your company. Say that you are looking forward to seeing them again, that you will gladly work with them on any future orders, or that you can always be reached for questions.

Adjustment Letters That Tell the Customer "No"

Writing to tell customers "No" is obviously more difficult than agreeing with them. You are faced with the sensitive task of conveying bad news, while at the same time convincing the reader that your position is fair, logical, and consistent. Write a letter that is customer-centered. As discussed earlier, do not bluntly start off with a "No." (See p. 209.)

It is not wise to accuse or argue. Avoid remarks that blame, scold, or remind customers of a wrongdoing. Remarks like these are likely to cost you business:

- You obviously did not read the instruction manual.
- Our records show that you purchased the set after the policy went into effect.
- The company policy plainly states that such refunds are not allowed.
- You were negligent in running the machine.

- You claim that our word processor was poorly constructed.
- Your complaint is unjustified.

Guidelines for Saying "No" Diplomatically

The following six suggestions will help you say "No" diplomatically. You can find practical applications of them in Figures 6.13 and 6.14. You might also want to review Figures 6.5 and 6.6 on pages 210 and 211.

1. **Thank customers for writing.** Start by putting your readers in a good frame of mind. Open with a polite, respectful comment, called a buffer, to soften your reader's response before he or she sees your "No." Make sure your buffer is relevant and sincere. Don't put them on the defensive by beginning with "We regret to inform you." The letter writers in Figures 6.13 and 6.14 use buffers to thank the customers for bringing the matter to their attention and sympathize with them about their inconvenience. As with other bad news letters, never begin with a refusal. You need time to calm and convince customers. Telling them "No" in the first sentence or two will negatively color their reactions to the rest of your letter. Use the indirect approach discussed earlier in this chapter (p. 209), and avoid these reader-hostile openings:

- I was surprised to learn that you found our product unsatisfactory.
- We have been in business for years and nothing like this has ever happened.
- There is no way we could give you what you demand.

2. **State the problem so that customers realize that you understand their complaint.** Let customers know that you have read their letter carefully and that you are responding to it fairly. Reassure them that you are aware of the facts of the situation. You thereby prove that you are not trying to misrepresent or distort what the customer told you.

3. **Explain what happened with the product or service before you give the customer a decision.** Provide a factual, respectful explanation to show customers they are being treated fairly. Rather than focusing on the customer's misunderstanding of instructions or failure to observe a service agreement or contract, explain the proper use of the product or the terms of the agreement.

Poor: By reading the instructions on the side of the paint can, you would have avoided the streaking condition that you claim resulted.

Revised: Hi-Gloss Paint requires two applications, four hours apart, for a clear smooth finish.

Note how the explanations in Figures 6.13 and 6.14 emphasize the appropriate ways to use the products.

4. **Give your decision without hedging.** It might seem easier at the time to dangle some hope before the readers that you might say "Yes." But this is not a good strategy for establishing a business relationship. Don't say, "Perhaps some type of restitution could be made later" or "Further proof would have been helpful." Indecision will only infuriate your customers, who believe they have already presented a sound, convincing case.

Figure 6.13   An adjustment letter saying "No."

Smith Sports Equipment

P.O. Box 1014   Tulsa, Oklahoma 74109-1014
(918) 555-0164 ▪ www.smithsport.com

October 19, 2008

Mr. Michael Trigg
17 Westwood Drive
Magnolia, MA 02171

Dear Mr. Trigg:

Buffer—thanks and sympathizes with reader

Thank you for writing to us on October 10 about the trouble you experienced with our model 191 spincast reel. We are sorry to hear about the difficulties you had with the release button and gears.

Explains problem without directly blaming the reader; gives firm decision

We have examined your reel and found the trouble. It seems that a retaining pin in the button spring was pushed into the side of the reel casing, thereby making the gears inoperable. The retaining pin is a vital yet delicate part of your reel. In order to function properly, it has to be pushed gently. Since our replacement guarantee does not cover the use of the pin in this way, we cannot refund your purchase price.

Turns a "No" into a "Yes" for customer

However, we want you to have many more hours of fishing pleasure and so we would be happy to repair your reel for $29.98 and return it to you within 5-7 days. Please let us know your decision.

Ends politely without any reference to the problem

I look forward to hearing from you.

Respectfully,

SMITH SPORTS EQUIPMENT

Ralph Montoya

Ralph Montoya, Manager
Customer Relations Department

Another adjustment letter saying "No."   Figure 6.14

4300 Marshall Drive
Salt Lake City, Utah 84113-1521
(801) 555-6028
www.healthair.com

August 28, 2008

Ms. Denise Southby, Director
Bradley General Hospital
Bradley, IL 60610-4615

Dear Director Southby:

Thank you for your letter of August 20 explaining the problems you
encountered with our Puritan MAII ventilator. We were sorry to learn that you
were unable to get the high-volume PAO_2 alarm circuit to work.

Our ventilator is a high-volume, low-frequency unit that can deliver up to
40 ml. of water pressure. The ventilator runs with a center of gravity attachment
on the right side of the diode. The trouble you had with the high oxygen alarm
system is due to an overload on your piped-in oxygen. Our laboratory inspection of the ventilator you returned indicated that the high-pressure system had
blown a vital adaptor in the MAII. An overload in an oxygen system is not
inculded in the warranty coverage of the ventilator, and so we cannot replace it
free of charge.

We would, however, be pleased to send you another model of the adapter,
which would be more compatible with your system, as soon as we receive your
order. The price of the adapter is $600, but for being a valued customer, our
service representative will install it at no charge.

Please let me know your decision. I look forward to hearing from you.

Sincerely yours,

R. P. Gifford

R. P. Gifford
Customer Service Department

*Professional
you-centered
opening*

*Justifies firm
decision by
explaining
causes of
problem
and conditions
of sale*

*Provides practi-
cal alternative
with financial
incentive*

*Ends with
goodwill*

Arrive at a fair and firm decision, but don't dwell on it. Keep your bad news short and anchor it to an honest explanation, such as those in Figures 6.13 and 6.14. Never apologize for your decision and avoid using words such as *reject, claim,* or *grant. Reject* is harsh and impersonal. *Claim* implies you distrust the customer's complaint. *Grant* signals that you have the power to respond favorably but decline to do so. Instead, use words that reconcile.

5. Turn your "No" into a benefit for readers. Link your firm "No" to an attractive alternative for readers. Your attitude needs to be "Here's what we can do to solve your problem." Find some way to make your "No" more acceptable by supplying a workable alternative to what customers have asked for but did not receive. Surround the bad news with benefits.

- Offer to repair a part for them (Figure 6.13).
- Send them a replacement and install it free of charge or for a lower fee (Figure 6.14).
- Encourage readers to see the "No" of a refusal of credit as good for the writer's business (Figure 6.15).

Never promise to do the impossible or go against company policy, but do continue to convince readers you have their needs in mind.

6. Leave the door open for better and continued business. Close with a diplomatic sentence or two showing respect for readers and expressing your desire to do business with them again. Figure 6.13 ends by letting the reader know that the writer is looking forward to hearing from him, and Figure 6.14 graciously thanks the reader for being a valued and loyal customer.

Refusal-of-Credit Letters

A special set of bad news letters deals with a company's refusing credit to an individual or another company. Writing such a letter requires a great deal of sensitivity. You want to be clear and firm about your decision; at the same time, you do not want to alienate the reader and risk losing his or her business in the future.

How to Say "No"
Follow these guidelines when saying "No" in a refusal-of-credit letter:

1. Begin on a positive—not a negative—note. Find something to thank the reader for; make the bad news easier to take. Compliment the reader's company or previous good credit achievements (if known); certainly express gratitude to the individual for wanting to do business with your company.

2. In a second paragraph provide a clear-cut explanation of why you must refuse the request for credit, but base your explanation on facts, not personal shortcomings or liabilities. Appropriate reasons to cite for a refusal of credit include

- a lack of business experience or prior credit
- being "overextended" or needing more time to pay off existing obligations (Figure 6.15)
- current unfavorable or unstable financial conditions

An effective letter refusing credit.   Figure 6.15

WEST COAST CREDIT INC.   4800 Ridge Road
Los Angeles, CA 91666
Phone (714) 555-3500
FAX (714) 555-4323

March 19, 2009

Mr. Otto L. King
Sunshine Interiors
8235 Mimosa Highway
Vinedale, CA 92004-0318

Dear Mr. King:

We appreciate your interest in doing business with West Coast Credit.
It is always gratifying to see a store like yours open in an expanding
community like Vinedale.

Begins on a positive note

In reviewing your credit application, we checked into the business
history and credit references you supplied. We also called your local
credit bureau. While we found nothing negative in your credit profile,
we did determine that for a business of your size you have already
reached a maximum level of indebtedness. For that reason, we
believe that this would not be the best time to extend your credit line.

Denies credit but explains precisely why

We would, however, encourage you to visit our website and fill out
the credit survey. This site is periodically reviewed and can give you
up-to-date information on credit availability. In the meantime, we wish
you every success in your new business.

Encourages reader to reapply

Cordially,

B. Rimes-Assante

B. Rimes-Assante
Supervisor

www.wcredit.com

- an order that is too large to process without some prepayment
- a lack of equipment or personnel for the company to do the business for which they are seeking credit

3. End on a positive note, too. Encourage the reader to reapply when business conditions have improved or when the reader's firm is in a better financial position. Make an attempt to keep the reader a potential customer, as Figure 6.15 illustrates.

Writing About Credit to a Non-Native Speaker of English

Writing a letter denying credit to a non-native speaker of English requires double tact. As we saw in Chapter 5, you have to consider the cultural expectations of such an audience and use easily understood international English. Compare the inappropriate refusal letter in Figure 6.16 with the far more diplomatic and more acceptably worded letter in Figure 6.17. The letter in Figure 6.16 is rude, uses words a non-native speaker of English may not understand ("expedited," "herewith"), and does not encourage future business dealings with Consolidated Plastics.

The diplomatic letter in Figure 6.17 follows the guidelines for effective communication with non-native speakers and adheres to the suggestions for denying credit. Emma Corson compliments her reader and his firm, expresses an interest in doing business with Mendson SA, and helps her reader to understand how Consolidated's credit policy might even help Mendson in the future.

Collection Letters

Collection letters require the same tact and fairness as do complaint and adjustment letters. Each nonpayment case should be evaluated separately. A nasty collection letter sent to a customer who is a good credit risk after only one month's nonpayment can drive the customer elsewhere. On the other hand, three easygoing letters to a customer who is a poor credit risk may encourage that individual to postpone payment, perhaps indefinitely.

Many businesses send several letters to customers before turning matters over to a collection agency. Each letter in the series employs a different technique, ranging from giving compliments and offering flexible credit terms to issuing demands for immediate payment and threats of legal consequences. One hospital uses the collection letters illustrated in Figures 6.18 (p. 231) and 6.19 (p. 232) to encourage ex-patients to pay their bills. Figure 6.18 is a letter sent early in the collection process when a client is only a month or two late. The collection letter in Figure 6.19 is sent much later to a client who has ignored earlier notices.

The tone of Figure 6.18 is cordial and sincere—now is not the time to say "pay up or else." It stresses how valuable the person is and underscores how pleased the hospital is to have provided the care he needed. The last paragraph makes a request for payment, offering (1) a flexible payment schedule and options, and (2) an escape from the inconvenience (or embarrassment) of receiving past due notices. The bottom of the letter conveniently lists payment options available to the patient.

The later collection letter in Figure 6.19, on the other hand, points out that the time for concessions is over, and reminds the patient of all the efforts that the hospital has expended to collect its bills. Then it announces what consequences will result if he still does not pay.

An inappropriate letter refusing credit to a non-native speaker of English.   Figure 6.16

Consolidated Plastics

May 25, 2009

Mr. Jan Buwalda
Mendson SA
Hoofdstraat 23
Dokkum, The Netherlands 1324 XK

Dear Mr. Buwalda:

I have received herewith your request of the12th and news about your
company. Thanks.

Curt, begrudging tone; uses legalese

Regarding that request to open a credit account with us, it just cannot be
done. I don't know how things are expedited in your country, but in the
United States giving credit to a first-time foreign customer is just not
standard business practice. As you must understand, your credit rating could
be unacceptable as far as we are concerned. You will be expected to pay in
cash for your first transaction with us. We'll evaluate the situation
thereafter.

Insensitive to reader's needs and cultural heritage

Let me know how you anticipate proceeding.

Sincerely,

Emma Corson

Emma Corson
Accounts Executive

Disrespectful close—no attempt to encourage future business

900 Technology Blvd., Bambrake, NH 03243
Phone (603) 555-7000 ■ Fax (603) 555-4321 ■ conplastics@compuserv.com
www.conplastics.com

Figure 6.17   A diplomatic revision of Figure 6.16, a letter refusing credit to a non-native speaker of English.

Consolidated Plastics

May 25, 2009

Mr. Jan Buwalda
Mendson SA
Hoofdstraat 23
Dokkum, The Netherlands 1324 XK

Dear Mr. Buwalda:

Expresses gratitude to and interest in reader's company

Thank you very much for your letter inquiring about opening a credit account with our firm. It is always a pleasure to hear from potential customers in Holland. I was most interested to learn about Mendson's diverse activities.

Welcomes future business contact

We understand and share your company's wish to have a U.S. supplier to work with you. Having Mendson as a customer would be beneficial for Consolidated Plastics, too. Working with you would allow us to enter a new market.

Politely explains procedure to ensure reader's goodwill

However, I am sorry that we cannot open any new account on credit. If you would kindly send us your check for the first month's supplies you need, we would rush your shipment to you. Doing this, you can establish an account with us, and will be able to charge your second month's supplies on that account.

Closes with friendly offer to help

Please write or call me if you have any questions. I look forward to serving you and Mendson in the future.

Cordially,

Emma Corson

Emma Corson
Accounts Executive

900 Technology Blvd., Bambrake, NH 03243
Phone (603) 555-7000 ■ Fax (603) 555-4321 ■ conplastics@compuserv.com
www.conplastics.com

A first, or early, collection letter.   **Figure 6.18**

SABINE COUNTY HOSPITAL
7200 Medical Blvd.
Sabine, TX 77231-0011
512-555-6734
www.sabine.org

May 15, 2009

Mr. Cal Smith
24 Mulberry Street
Valley, TX 77212-3160

Re: Inpatient Services
Date of Hospitalization: March 11–12, 2009
Balance Due: $4,725.48

Dear Mr. Smith:

We are honored that we were able to serve your health care needs during your recent stay at Sabine County Hospital. It is our continuing goal to provide the best possible hospital care for residents of Sabine County and its vicinity. To do so we must keep our finances up-to-date.

Links hospital mission to patient's payment

Our records indicate that your account is now overdue, and that we have not received a payment from you for two months. If you have recently sent one in, kindly disregard this letter and accept our thanks.

Diplomatic reminder to pay now

If for any reason you are unable to pay the full amount at this time, we will be happy to set up a payment schedule that is convenient for you. Just fill in the appropriate blanks below, and return this letter to us. That will enable us to avoid billing you on a "Past Due" basis. Thank you for your cooperation.

Offers options to maintain goodwill

Sincerely,

Morris T. Jukes

Morris T. Jukes
Accounts Department

(   ) I will pay $ _____   (   ) monthly   (   ) quarterly on my account.

(   ) Enclosed is a check for full payment in the amount of $ _____.

Signature

_____

Figure 6.19   A final collection letter.

SABINE COUNTY HOSPITAL
7200 Medical Blvd.
Sabine, TX 77231-0011
512-555-6734
www.sabine.org

September 21, 2009   Re: Inpatient Services
                                     Date of Hospitalization: March 11–12, 2009
Mr. Cal Smith   Balance Due: $4,725.48
24 Mulberry Street
Valley, TX 77212-3160

Dear Mr. Smith:

Direct opening about history and current status of the account

During the past few months we have written to you several times about your balance of $4,725.48 for services you received on March 11 and 12. Your account is more than 190 days overdue, and we cannot allow any further extensions in receiving a payment from you.

Reminds patient of goodwill and requests payment

As you will recall, we have tried to help you meet your obligations by offering several options for paying your bill. You could have arranged for installment payments that would be due each month or quarter, whichever would be more convenient. Because you have not replied, we must ask for full payment now.

States final option

If we do not hear from you within ten days, we will have no alternative but to turn your account over to our collection agency, which will seriously hurt your credit rating. Neither of us would find this a welcome alternative.

Sincerely,

Morris T. Jukes

Morris T. Jukes
Accounts Department

Sending Letter-Quality Messages: Final Advice to Seal Your Success

This chapter has introduced you to different types of letters yo... on the job. Although you will send many more e-mail messa... letters, every job will also expect you to write effective, profes...

Judging and checking your letters in light of the followin... you to draft, tailor, and evaluate the types of business letters... write.

- **Identify your reader:** one individual; a group; a company or agency; a new individual customer or a long-time one; a native or non-native speaker
- **Determine your purpose for writing:** routine or special request; explanation; complaint; apology; sell product/service; build goodwill; express thanks; refuse credit; collect a debt
- **Determine reader's reason for writing:** complaint; request information; make an adjustment; seek credit
- **Organize information:** direct or indirect; begin with good news and save negative message for middle of the letter
- **Include essential information:** schedules; dates; prices and expenses; personnel; explanation of services, warranties, products; background and need for credit
- **Use right style and tone:** professional and courteous; concise and focused; sensitive to reader's needs, including his or her culture and traditions
- **Test your overall goal:** request or provide information; make a new customer or maintain goodwill with an established one; resolve a customer's complaint or explain why not; encourage reader to extend credit

Planning your letter carefully—its purpose, its organization, its content—will help you to make sure your message begins, continues, and ends professionally and successfully.

Revision Checklist

- [ ] Planned what I am going to say to my readers. Did necessary homework and double-checking to answer any questions. Proved to my readers that I am knowledgeable about my topic.
- [ ] Used an appropriate (and consistent) format and page layout for my letters.
- [ ] Followed acceptable company protocol in format, organization, style, and tone of my letters.

Continued

...ued

☐ Adapted style and length of my message for my readers.
☐ Organized and, if necessary, highlighted information in my letters in the most effective way for my message and for my readers.
☐ Emphasized the "you attitude" with my readers, whether employer, customer/client, or co-worker.
☐ Conveyed impression of being courteous, professional, and easy to work with.
☐ Used clear and concise language appropriate for my reader.
☐ Began my correspondence with reader-effective strategies. If reporting good news, told the reader right away. If reporting bad news, was diplomatically indirect and considerate of my reader's reactions.
☐ Followed the four A's of effective sales letters. Identified and convinced my target audience.
☐ Wrote complaint letters in a courteous tone. Informed the reader what is wrong, why it is wrong, and how the problem should be corrected.
☐ Wrote adjustment letters that say "Yes" sincerely and to the point. Made those that say "No" fair but did not begin with a "No." Acknowledged reader's point of view and provided clear explanation for my refusal.
☐ Ensured correspondence was timely. Was prompt and reasonable in answering all my correspondence—both from people in my company and from customers.
☐ Took special care to meet the needs of non-native speakers of English in both tone and message.
☐ Tailored collection letters according to audience, time, and circumstances of overdue account; tried to maintain goodwill.
☐ Followed all legal guidelines in informing readers about past due accounts and refusal-of-credit policies.

Exercises

1. Write a letter of inquiry to a utility company, a safety or health care agency, or a company in your town and ask for a brochure or an annual report describing its services to the community. Be specific about your reasons for requesting the information.

2. In which course(s) are you or will you be writing a report? Write to an agency or company that could supply you with helpful information and request its aid. Indicate why you are writing, precisely what information you need, and why you need it. Offer to share your report with the company.

3. Examine an ad in a magazine, a TV commercial, or on a website and then write a one-page assessment in which you identify the four parts of its sales message.

4. Choose one of the following and write a sales letter addressed to an appropriate audience on why they should
   a. major in the same subject you did
   b. live in your neighborhood
   c. be happy taking a vacation where you did last year
   d. dine at a particular restaurant
   e. shop at a store you have worked for
   f. have their cars repaired at a specific garage
   g. give their real-estate business to a particular agency
   h. visit your website

5. Find at least two sales letters you, your family, or your firm has received, and in an e-mail or a memo to your instructor or employer evaluate how well they follow the four parts of a sales letter discussed in this chapter. Attach a copy of the sales letters to your evaluation. If your e-mail or memo is addressed to your boss, indicate how you would improve your competition's sales letters.

6. As a collaborative project, rewrite the following sales letter to make it more effective. Add any details you think are relevant.

   Dear Pizza Lovers:

   Allow me to introduce myself. My name is Rudy Moore and I am the new manager of Tasty's Pizza Parlor in town. The Parlor is located at the intersection of North Miller Parkway and 95th Street. We are open from 10 a.m. to 11 p.m., except on the weekends, when we are open later.

   I think you will be as happy as I am to learn that Tasty's will now offer free delivery to an extended service area. As a result, you can get your Tasty Pizza hot when you want it.

   Please see your weekly newspapers for our ad. We also are offering customers a coupon. It is a real deal for you.

   I know you will enjoy Tasty's and I hope to see you. I am always interested in hearing from you about our service and our fine product. We want to take your order soon. Please come in.

7. Write a sales letter similar to the one in Figure 6.3 from a manager of one of the following ethnic restaurants or one of your choice. Make sure you include relevant details for the particular audience:
   a. Mexican
   b. Indian
   c. Cuban
   d. Soul food
   e. Czech
   f. Turkish
   g. Vietnamese
   h. Thai
   i. Greek
   j. German
   k. Irish
   l. Chinese
   m. Pakistani
   n. Italian
   o. Argentinean
   p. French
   q. Polish
   r. Hindi

8. Send a follow-up letter to one of the following individuals:
   a. a customer who informs you that she will no longer do business with your firm because your prices are too high
   b. a family of four who stayed at your motel for a week last summer
   c. a wedding party or professional organization that used your catering services last month
   d. a customer who exchanged a coat for the purchase price
   e. a customer who purchased a used car from you and who has not been happy with warranty service
   f. a company that bought software from you nine months ago, alerting them about updates

9. Write a bad news letter based on an experience at your workplace—rejecting an applicant for a job, a warranty claim, an increase in insurance rates, or a request for funding. If you have already written such a letter, rewrite it by applying the principles you learned in this chapter.

10. Write a bad news letter to an appropriate reader about one of the following:
    a. Your company has to discontinue Saturday deliveries because of rising labor and fuel costs.
    b. You are the manager of an insurance company writing to tell one of your customers that, because of reckless driving, his or her rates will increase.
    c. You have to refuse to send a bonus gift to a customer who sent in an order after the expiration date for qualifying for the gift.
    d. You have discontinued a model that a business customer wants to reorder.
    e. You have to notify residents of a community that a bus route or hours of operation are being discontinued.
    f. You represent the water department and have to tell residents of a community that they cannot water their lawns for the next month because of a serious water shortage in your town.
    g. You cannot send customers a catalog—which your company formerly sent free of charge—unless they first send $10 for the cost of that catalog.
    h. You cannot repair a particular piece of equipment because the customer still owes your company for three previous service visits.

11. Write a good news letter about the opposite of one of the situations listed in Exercise 10.

12. You just found out that a business that applied for credit has missed its last mortgage payment. You have to refuse credit to this local firm, which has been in business successfully for eight years. Write a refusal letter without jeopardizing future business dealings.

13. Write a complaint letter about one of the following:
    a. an error in your utility, telephone, credit card, or Internet provider bill
    b. discourteous service you received on an airplane or bus
    c. a frozen food product of poor quality
    d. a shipment that arrived late and damaged

e. an insurance payment to you that is $357.00 less than it should be

f. a public television station's policy of not showing a particular series

g. junk mail or spam that you are receiving

h. equipment that arrives with missing parts

i. misleading representation by a salesperson

j. incorrect information given at a website

14. This exercise might be done as a collaborative project. You are a section manager at e-Tech. Your company has a service contract with Professional Office Cleaners (POC). However, each morning when you arrive at work you are disappointed with what they've done. POC has overlooked some essential tasks and done a poor job on others. Your staff is also disappointed and has e-mailed or spoken to you about problems with POC. Write the following:

a. a memo or e-mail to your boss, the vice president, about POC's shoddy work

b. a complaint letter to POC that the vice president has asked you to write and to sign his name to it

c. a letter to the vice president from the manager of POC who is responsible for your e-Tech section, apologizing for the problem and offering a solution

d. a letter from POC to the vice president taking issue with the complaint made against his cleaning company, offering proof that the work was done according to contract specifications

e. an e-mail you send to your staff about what's happened with POC

15. Write the complaint letter to which the adjustment letter in Figure 6.12 responds.

16. Write the complaint letter to which the adjustment letter in Figure 6.14 responds.

17. Rewrite the following complaint letter to make it more precise, less emotional, and effectively persuasive.

Dear Sir:

We recently purchased a machine from your Albany store and paid a great deal of money for it. This machine, according to your website, is supposedly the best model in your line and has caused us nothing but trouble each time we use it. Really, can't you do any better with your technology?

We expect you to stand by your products. The warranties you give with them should make you accountable for shoddy workmanship. Let us know at once what you intend to do about our problem. If you cannot or are unwilling to correct the situation, we will take our business elsewhere, and then you will be sorry.

Sincerely yours,

18. The following story appeared in a local newspaper.

Residents concerned about relocation of pet food plant

OCEAN SPRINGS (AP) Finicky Pet Food is moving its processing plant from Pascagoula to Ocean Springs, a decision that has some residents concerned about possible odor and other problems.

The plant is moving to an industrial area bordering a subdivision of expensive homes.

"The wind doesn't discriminate," said Jo Souers, who lives in the Bienville Place subdivision. "I don't want this in our neighborhood."

City officials said the plant is moving to an area zoned to accommodate it.

"We don't have a lot of control over it," said city planner Donovan Scruggs. "It is a permitted use for this property."

Scruggs said the property was zoned industrial before the subdivision was built. A body shop, cabinet shop, and boat business are located nearby.

The plant will be built in the small industrial area on U.S. 90, directly across the highway from the Super Wal-Mart.

It is moving into a vacant building, the interior of which has been renovated for its new purpose, city officials said.

The plant will process frozen fish and fish parts for bait and pet food. It will employ 10 workers, with that number doubling during fishing season.

Plant manager Dean Niemann said in a statement that the company no longer needed its Pascagoula location near deep water, which was rented from the county.

Scruggs said the city has investigated the possibility that the plant will emit odors.

"We've told Dean (Niemann) from day one, 'You're locating next to a residential area. If you start stinking, action will be taken,'" Scruggs said.

He said the city has a nuisance ordinance that should handle anything that might arise.

"Residents concerned about relocation of pet food plant," The Associated Press, September 27, 2005. Reprinted by permission.

Based upon information in this story, which you may want to supplement, write the following complaint/adjustment letters:
  a. a complaint letter to the city from resident Jo Souers
  b. a complaint letter to Finicky Pet Food from city officials warning about dangers of pollution to the residential area
  c. a letter from plant manager Dean Niemann to the residents of Bienville Place subdivision
  d. a letter from city officials to the residents of Bienville Place subdivision

19. Write an adjustment letter saying "Yes" to the manager of The Loft, whose letter is in Figure 6.10.

20. Write an adjustment letter saying "No" to the customer who received the "Yes" adjustment letter included in Figure 6.12.

21. Rewrite the following ineffective adjustment letter saying "Yes."

Dear Mr. Smith:

We are extremely sorry to learn that you found the suit you purchased from us unsatisfactory. The problem obviously stems from the fact that you selected it from the rack marked "Factory Seconds." In all honesty, we have had a lot of problems because of this rack. I guess we should know better than to try to feature inferior merchandise along with the name-brand clothing that we sell. But we originally thought that our customers would accept poorer quality merchandise if it saved them some money. That was our mistake.

Please accept our apologies. If you will bring your "Factory Second" suit to us, we will see what we can do about honoring your request.

Sincerely yours,

22. Rewrite the following ineffective adjustment letter saying "No."

Dear Customer:

Our company is unwilling to give you a new toaster or to refund your purchase price. After examining the toaster you sent to us, we found that the fault was not ours, as you insist, but yours.

Let me explain. Our toaster is made to take a lot of punishment. But being dropped on the floor or poked inside with a knife, as you probably did, exceeds all decent treatment. You must be careful if you expect your appliances to last. Your negligence in this case is so bad that the toaster could not be repaired.

In the future, consider using your appliances according to the guidelines set down in warranty books. That's why they are written.

```
Since you are now in the market for a new toaster, let me
suggest that you purchase our new heavy-duty model, number
67342, called the Counter-Whiz. I am taking the liberty of
sending you some information about this model. I do hope
you at least go to see one at your local appliance center.

Sincerely,
```

23. You are the manager of a computer software company, and one of your sales-people has just sold a large order to a new customer whose business you have tried to obtain for years. Unfortunately, the salesperson made a mistake writing out the invoice, undercharging the customer $429. At that price, your company would not break even and so you must write a letter explaining the problem so the customer will not assume all future business dealings with your firm will be offered at such "below market" rates. Decide whether you should ask for the $429 or just "write it off" in the interest of keeping a valuable new customer.
    a. Write a letter to the new customer, asking for the $429 and explaining the problem while still projecting an image of your company as accurate, professional, and very competitive.
    b. Write a letter to the new customer, not asking for the $429 but explaining the mistake and emphasizing that your company is both competitive and professional.
    c. Write a letter to your boss explaining why you wrote the letter in (a).
    d. Write a letter to your boss explaining why you wrote the letter in (b).
    e. Write a letter to the salesperson who made the mistake, asking him or her to take appropriate action with regard to the new customer.

24. Write an appropriate collection letter to one of the following:
    a. a loyal customer who has not responded to a first notice letter
    b. a new business customer who placed a large order with you last quarter and paid for it promptly but who has ignored two notices you have already sent about an order filled this quarter
    c. a customer who has just placed an order over the Internet but has not responded so far to any of your notices for payment for previous purchases
    d. a customer who has been continually late but has always paid eventually
    e. an international customer who has sent in only partial payment

How to Get a Job:

Searches, Dossiers, Portfolios, Résumés, Letters, and Interviews

Obtaining a job involves a lot of hard work. Before your name is added to a company's payroll, you will have to do more than simply walk into the human resources office and fill out an application form. Finding the *right* job takes time in this highly competitive job market. And finding the right person to fill that job also takes time for the employer.

Visit www .cengage.com/ english/kolin for this chapter's online exercises, ACE quizzes, and Web links.

▌Steps the Employer Takes to Hire

From the employer's viewpoint, the stages in the search for a valuable employee include the following:

1. Deciding what duties and responsibilities go with the job and determining the qualifications the future employee should possess
2. Advertising the job on the company website, in newspapers, and in professional publications
3. Scanning and evaluating résumés and letters of application
4. Having candidates complete application forms
5. Requesting further proof of candidates' skills (letters of recommendation, transcripts, portfolios)
6. Interviewing selected candidates
7. Doing further follow-ups and rank ordering those to be interviewed
8. Offering the job to the best-qualified individual

Sometimes the steps are interchangeable, especially steps 4 and 5, but generally speaking, employers go through a long and detailed process to select employees. Step 3, for example, is among the most important for employers (and the most crucial for job candidates). At that stage employers often classify job seekers into one of three groups: those they definitely want to interview, those they may want to interview, and those in whom they have no interest.

▋ Steps to Follow to Get Hired

As a job seeker you will have to know how and when to give the employer the kinds of information the preceding eight steps require. You will also have to follow a certain schedule in your search for a job. The following eight procedures will be required of you:

1. Preparing for your career
2. Analyzing your strengths and restricting your job search
3. Looking in the right places for a job
4. Assembling a dossier and portfolio
5. Preparing a résumé
6. Writing a letter of application and filling out a job application
7. Going to an interview
8. Accepting or declining a job offer

Your timetable should match that of your prospective employer. This chapter shows you how to begin your job search, prepare an appropriate résumé, write a persuasive letter of application, design an effective portfolio, and prepare for an interview.

Job counselors advise students to start planning for their careers several years before they graduate. The more you find out about what career path you want to take ahead of graduation, the better you will be able to target the jobs right for you. But formulating a clear job objective takes lots of preparation. Learn as much as you can about the type of work that characterizes your chosen area and the preparation you need to be successful in it.

Here are some suggestions of things you can do to develop your professional career plans:

- Attend job fairs and workshops on campus as well as those sponsored by state and federal agencies.
- Join student and professional organizations and societies in your area(s) of interest. Membership rates for students are often reduced.
- Apply for relevant internships and training programs.
- Ask your instructors to critique your work in light of your career plans.
- Confer with your academic adviser(s) regularly, not just once a semester.
- Find a mentor—someone in the career field you might want to join.
- Consider developing a competency in a second (or third) language.
- Do volunteer work to gain or enhance experience working in a group setting.
- Consult the U.S. Department of Labor's *Occupational Outlook Handbook* at *http://www.bls.gov/oco/* for valuable information on job prospects, requirements, and salary ranges. The *Handbook* will also list links to sites that provide further information and contacts.

Analyzing Your Strengths and Restricting Your Job Search

Before you apply for jobs, analyze your job skills, career goals, and interests. Here are some points to consider.

1. Make an inventory of your most significant accomplishments in your major and/or on the job. What are your greatest strengths—writing and speaking, working with people in small groups, organizing and problem solving, managing money, speaking a second language, creating graphics?
2. Decide which specialty within your chosen career appeals to you the most. If you are in a nursing program, do you want to work in a large teaching hospital, for a home health or hospice agency, or in a physician's office? What kinds of patients do you prefer to care for—pediatric, geriatric, psychiatric?
3. What are the most rewarding prospects of a job in your profession? What most interests you about a position—travel, technology, international contacts, on-the-job training, helping people, being creative?
4. What are some of the greatest challenges you face in your career today—or will face in five years?
5. Which specific companies or organizations have the best track record in hiring and promoting individuals in your field? What qualifications will such firms insist on from prospective employees?

Once you answer the questions above you can avoid applying for positions for which you are either overqualified or underqualified. If a position requires ten years of related work experience and you are just starting out, you will only waste the employer's time and your own by applying. However, if a job requires a certificate or license and you are in the process of obtaining one, go ahead and apply.

Looking in the Right Places for a Job

One way to search for a job is simply to send out a batch of letters and résumés to companies you want to work for. But how do you know what jobs, if any, those companies have available, what qualifications they are looking for, and what deadlines they might want you to meet? You can avoid these uncertainties by knowing where to look for a job and knowing what a specific job entails. Consult the following resources for a wealth of job-related information.

1. **Networking.** One of the best resources for finding job information is consulting with other people, or networking. Networking pays. It is regarded as the most important strategy to follow. John D. Erdlen and Donald H. Sweet, experts on job searching, cite the following as a primary rule of job hunting: "Don't do anything yourself you can get someone with influence to do for you." Let your professors, friends, classmates, neighbors, relatives, and even your clergy know you are looking for a job. They may hear of something and can notify you or, better yet, recommend you for the position. See how the job seekers in Figures 7.13 (p. 277) and 7.14

(p. 278) have successfully networked with people they know. You can also network with people you don't know personally through professional networking websites such as:

http://www.linkedin.com

http://www.ryze.com

http://www.tribe.net

Also, attend job fairs, professional and organization meetings, and community and civic functions to meet the right contact people whom you can ask for advice and also for possible follow-up help and recommendations.

2. The Internet. Learn about jobs by visiting a company's website to see if it has vacancies and what the qualifications are for them. You can also consult the many online job services that list positions and sometimes give advice, including:

- Monster.com—*http://www.monster.com*
- Riley Guide—*http://www.rileyguide.com*
- Yahoo! Hot Jobs—*http://www.hotjobs.yahoo.com*
- Career Builder—*http://www.careerbuilder.com*
- Job.com—*http://www.job.com*
- Career.com—*http://www.career.com*
- Job Central—*http://www.jobcentral.com*

There are also many specialized job search sites. For a helpful list broken down by job area, visit the University of Delaware Career Services Center's "Specialized Job Sites" page at *http://www.udel.edu/csc/specnet.html.*

3. Newspapers. Look at local newspapers as well as the Sunday editions of large city papers with a wide circulation, such as the *New York Times Job Market* (*http://www.jobs.nytimes.com/js.php*). The *National Business Employment Weekly* (*www.careerjournal.com*), published by the *Wall Street Journal*, also lists jobs in different areas, including technical and managerial positions.

4. Your campus placement office. Counselors keep an online file of current available positions and can also tell you when recruiters will be on campus to conduct interviews. Many placement offices have recruiting databases, allowing students access to a broad range of contacts and interview information. Counselors can also help you locate summer and part-time work, both on and off campus, positions that might lead to full-time jobs. Most important, they will give you sound advice on your job search, including strategies for finding the right job, salary ranges, and interview tips. Many placement offices also sponsor career fairs to bring job seekers and employers together in specific professional fields.

5. Federal and state employment offices. The U.S. government is one of the biggest employers in the country. During 2007 and 2008, for instance, the most active career site on the Web was operated by the federal government, with 1.7 million new hires. Counselors at federal and state employment centers also help job seekers find career opportunities. Figure 7.1 shows the homepage of the website for

USAJOBS, which helps job seekers find employment opportunities with the U.S. government. Consult the following websites for listings of government jobs:

- USA Jobs—*http://www.usajobs.gov*
- Federal Jobs.Net—*http://www.federaljobs.net*
- Studentjobs.gov—*http://www.studentjobs.gov*

6. Professional and trade journals plus associations in your major. Identify the most respected periodicals in your field and search their ads. Each issue of *Food Technology*, for example, features a section called "Professional Placement," a listing of jobs all over the country. Similarly, *CIO Magazine—IT Professional Research Center* (*http://www.cio.com/research/itcareer/*) can help you find jobs in the computer industry, engineering, and technology. Consulting the *Encyclopedia of Associations* (*http://library.dialog.com/bluesheets/html/bl0114.html*) is the quickest way to find out about professional organizations and the journals they publish in your field.

7. The human resources department of a company or agency you would like to work for. Often you will be able to fill out an application even if there is not a current opening. But do not call employers asking about openings; a visit shows a more serious interest.

8. A résumé database service. A number of online services will put your résumé in a database and make it available to prospective employers, who scan the database regularly to find suitable job candidates. Figure 7.2 describes a résumé database

Figure 7.2   A description of one résumé database service.

Employers Benefit from ACM Résumé Database

The ACM Résumé Database is composed of résumés of ACM members—high-caliber, information-technology professionals who can bring expertise to your company. All résumés are up-to-date and can be searched according to criteria you provide, in a short period of time and at low rates. Single searches as well as annual subscriptions to the database can be requested from the database administrator, Resume-Link, at 614-529-0429. ACM institutional members get a 10% discount off the cost of the search.

And for ACM members! Our database also is now searchable for internships and co-op positions, as well as for full-time and consultant placements. Use this free career development service and submit your résumé online at **http://www.Resume-Link.com**.

---

service offered by one professional organization—the Association for Computing Machinery (ACM)—for its members. Check to see if a professional society to which you belong (or might join) offers a similar service.

9. Professional employment agencies. Some agencies list jobs you can apply for free of charge (because the employer pays the fee) and others that charge a stiff fee, usually a percentage of your first year's salary. Be sure to ask who pays the fee for this service. Because employment agencies often find out about jobs through channels already available to you, speak to someone at your campus career center first.

Dossiers and Career Portfolios/Webfolios

Dossiers and portfolios/webfolios play a major role in the job search. These documents work in conjunction with each other in supporting and supplementing your résumé. Increasingly, job seekers are providing both a dossier and portfolio/webfolio to prospective employers. A *dossier*, French for "bundle of documents," provides a file of information about you and your work—recommendations and so on—that others have supplied. But a portfolio (or webfolio if it is submitted electronically) contains documents or multimedia items you have created or produced yourself. The word *portfolio* historically refers to a collection of work that an artist would show to a prospective client. For example, a portfolio might contain reproductions or slides of the artist's paintings or illustrations or samples of photography. Similarly, your career portfolio would contain a collection of samples of your work—written or visual—to show to prospective employers.

Dossiers

Basically, a dossier contains the following documents:

- letters of recommendation
- letters that awarded you a scholarship, gave you an academic honor, or acknowledged your community service
- letters that praised your work on the job
- your academic transcript(s)

Typically, dossiers (or placement files) are stored at your campus placement office, which serves both current students and alumni of your school. You may ask that your dossier be sent to an employer or employers may request it themselves if you have listed the placement office address on your résumé. Job counselors advise that you keep a copy of the documents in your dossier as well.

Obtaining Letters of Recommendation

Should You See Your Letters?

You have a legal right to determine whether you want to see your recommendation letters, but some employers believe that if candidates see what is written about them, the references may be overly complimentary, less frank, and more inclined to withhold information. Also, some of your references may prefer confidentiality and refuse to write a letter of recommendation they know you will see. On the other hand, you may feel more comfortable knowing that your recommendation letters contain only information that will praise your achievements and help you get the job. Before you make any decisions about seeing your recommendation letters, get the advice of your instructors and placement counselors. If you waive your right to see the letters, you must sign an appropriate form, a copy of which is then given to your reference.

Whom Should You Ask?

Whether your recommendation letters are confidential or not, they can sell you or sink your chances, so select your references carefully. Ask the following individuals to be your references and write about your work qualifications and skills:

- previous employers (even for summer jobs or internships) who liked you and your work
- two or three of your professors who know and like your work, have graded your papers, or have supervised you in fieldwork or laboratory activities
- supervisors who evaluated and praised your work in the military
- community leaders or officials with whom you have worked successfully on civic projects

Recommendations from such individuals will be regarded as more objective—and more relevant—than letters from friends, neighbors, or members of the clergy. Of course, if you are asked specifically for a character reference, by all means ask any of these people.

Whomever you do ask, make sure he or she is a strong supporter of yours, someone who has sincerely and consistently complimented your work and encouraged you in your career. Stay clear of individuals who are lukewarm about your work or might be reluctant to recommend you.

Should You Ask Your Current Boss?

Asking your current boss, on the other hand, can be tricky. If your current employer is already aware that you are looking for work elsewhere (for instance, if your job is temporary or if your contract is about to run out) or you are working at a part-time job, by all means ask for a letter of recommendation. However, if you are employed full time and are looking for professional advancement or a better salary elsewhere, you may not want your current employer to know that you are searching for another job.

Always Ask for Permission

Always ask for permission before you list an individual as a reference. You could jeopardize your chances for a job if a prospective employer called someone you named but did not ask to be a reference and that person responds that he or she did not even know you were looking for a job or, worse yet, reveals that you did not have the courtesy to ask to use his or her name.

Career Portfolios/Webfolios

Your portfolio/webfolio needs to represent your best work. Try to see your portfolio/webfolio from the employer's point of view. A portfolio/webfolio is not a personal/family scrapbook but a record of your most professional work. What your portfolio needs to say is "This is a collection of my best work. The documents here illustrate the high professional standards and skills that I will bring with me as an employee of your company."

What to Include in a Career Portfolio/Webfolio

What you include in your portfolio and how you organize it says as much about your qualifications for the job as the documents themselves. It is never too early to start assembling relevant documents for your portfolio. As you prepare it for your job search, get the advice of a job counselor, an instructor, or someone who has been in your field for at least a few years who can evaluate your documents and how you need to arrange them.

Here is a sampling of the kinds of documents you might include in your career portfolio/webfolio, as illustrated in Figure 7.3:

- a mission statement or personal profile (two or three paragraphs) that outlines your career goals and work skills and serves as an introduction to your portfolio
- an additional copy or e-version of your résumé
- copies or scans of documents that prove your academic and/or work history: diplomas, certificates, licenses, internships, concern for the environment, and the like
- copies of awards (academic and job-related) and promotion letters
- relevant examples of written work you did for college courses (include any positive comments provided by instructors)

A sample webfolio with annotated Artifacts/Documents page open.   Figure 7.3

Mission Statement	Résumé	Unofficial College Transcript	Artifacts/ Documents	References

Sample Business Documents

To demonstrate my clear and concise writing style, my ability to design a variety of complex workplace documents, and my understanding of international communication (all of which you stressed in the job description), I have provided links to various business documents I created. None of these contains proprietary or confidential information.

VIEW SAMPLE SALES PROPOSAL ON SAVING ENERGY

VIEW SAMPLE RESEARCH REPORT (TABLE OF CONTENTS, ABSTRACT, (FIRST 3 PAGES)

VIEW SAMPLE LETTER TO INTERNATIONAI CLIENT

VIEW SAMPLE SURVEY QUESTIONNAIRE AND ANALYSIS

Company Newsletter Article

At the end of 2008, I was named "Employee of the Year" at my firm's Cincinnati office. An article about the award appeared in the company newsletter, outlining the selection criteria and describing the job accomplishments that earned me this award.

VIEW NEWSLETTER ARTICLE

PowerPoint Presentation

I understand that the position advertised will entail frequent oral presentations at company meetings. My colleagues and supervisors have praised my abilities as a speaker who can simplify complex ideas for audiences and zero in on the main ideas quickly. To give you a sense of my ability to organize information and provide visual support for my ideas, I attach a sample PowerPoint presentation that accompanied one of my speeches.

VIEW POWERPOINT PRESENTATION

Website

To illustrate both my design skills and my keen understanding of the Web environment, I have provided several screen captures from a company website I created. The construction of the site was handled by my company's Web group, supervised by me.

VIEW WEBSITE SCREEN CAPTURES

- appropriate examples of written work you completed on the job, such as reports, care plans, newsletters or marketing strategies. Rather than including an entire long report, provide just the abstract, table of contents, and a few sample pages. But make sure you do not include in your portfolio any information that would violate a company's confidentiality or compromise its ethics.
- clippings or scans of newspaper or newsletter stories about your academic, community, and/or on-the-job successes (include the name of the periodical, the date, and the page numbers).

- pertinent examples of electronic or graphic work you have done, such as PowerPoint presentations you have created, websites you have designed, layout work you've completed, career-related photographs you have taken
- a list of your references with contact information
- a copy of your college transcript, if available (if you include this in your portfolio, you should realize that it is unofficial—only your college can provide an official transcript via your dossier)

What Not to Include in a Career Portfolio

Be highly selective about what you include in your portfolio/webfolio. Again, be guided by the qualifications a prospective employer is looking for. Never include anything in your portfolio that would contradict or call into question information in your résumé or letter of application. Also, avoid including anything that would put you in a less than favorable light (e.g., a newspaper article that provides a less than complimentary review of the work you did or includes your name along with many others with no mention of your significant, individual contributions or a college paper with negative comments from your instructor).

Exclude the following types of documents from your career portfolio/webfolio:

- any proprietary documents (reports, proposals, letters on company letterhead) or scans of those documents that you wrote on the job and that your employer regards or may regard as confidential
- documents or scans of documents that show your memberships in clubs, fraternities/sororities, sports teams, and so on, unless directly relevant to the job (e.g., you are applying for a job at the national office for Sigma Sigma Kappa or with the Professional Golf Association)
- links to or printouts from personal Web pages, including *Facebook.com* or *MySpace.com* pages (these may contain inappropriate personal information and/or portray you as unprofessional)
- pictures of your family, friends, pets, and the like
- clippings or scans of newspaper or newsletter stories about you that are not directly related to your job search, such as your wining a cruise or other prize or playing on a bowling team
- complaint letters, reviews, or letters to the editor you may have written but that suggest you may be overly critical, harsh, or difficult to get along with

Portfolio/Webfolio Formats

When you provide a prospective employer with your portfolio, you can either mail it as hard copy or send it electronically. If you submit a hard copy portfolio, always make high-quality copies of each document. Also, never include original documents: If an original document were to be lost, it would be difficult, perhaps impossible, to replace it. If you submit a webfolio, make sure that all of your scans are clean and clear. In addition, you can hyperlink parts of your portfolio to your résumé, making it easy for a job recruiter or prospective employer to find evidence of your qualifications.

Whether you submit your portfolio in hard copy or electronically, provide short annotations explaining what each document is and why it is important, as shown in the webfolio in Figure 7.3. Always explain the significance of each portfolio document in terms of the qualifications a prospective employer is looking to fill. As with your résumé, expect to change and adapt the contents and organization of your portfolio depending on the particular job for which you are applying. Finally, remember to bring a hard copy of your portfolio with you to the interview.

Other Types of Business Portfolios

Portfolios have many others functions in the world of work besides being used as a part of your job search. They can be a significant part of an annual review showing your boss the range of activities you have been involved in and documenting your successes. On a larger scale, shareholders often receive portfolios of company accomplishments over a specified period (quarterly, yearly, etc.). And portfolios are also frequently given to prospective clients as part of an overall proposal to document the types of jobs you and your firm have done and to show how well you and your company have carried them out.

█ Preparing a Résumé

The résumé, sometimes called a **curriculum vitae,** may be the most important document you prepare for your job search. It deserves your utmost attention. Regard your résumé as a persuasive ad for your professional qualifications. Prepare your résumé at least six months before you plan to graduate. You'll need the finished document before writing any application letters. A résumé is not your life history or your emotional autobiography, nor is it a transcript of your college work. It is a factual and concise summary of your qualifications, showing a prospective employer that you have what it takes (in education and experience) to do the job you are applying for.

What you include—your key details, the wording, the ordering of information, and formatting—are all vital to your campaign to sell yourself and land the interview. Employers want to see the most crucial details about your qualifications quickly. Accordingly, keep your print résumé short (preferably one page, never longer than two) and hard hitting. Everything on your résumé needs to convince an employer you have the exact skills and background he or she is looking for.

What Employers Like to See in a Résumé

Prospective employers will judge you and your work by your résumé, their first view of you and your qualifications. They want an applicant's résumé to display the following seven characteristics:

- **Honesty.** Be truthful about your qualifications—your education, experience, and skills. Distorting, exaggerating, or falsifying information about yourself in your résumé is unethical and could cost you the job you get. If you were a clerical assistant to an attorney, don't describe yourself as a paralegal. Employers demand trustworthiness.
- **Attractiveness.** The document should be pleasing to the eye and easy to read. It should use appropriate spacing, fonts, and boldface. Your résumé

needs to show you have a sense of proportion and document design and that you are neat. See pages 496–506 of Chapter 11.

- **Careful organization.** The orderly arrangement of information must be easy to follow, logical, and consistent. By doing this, you demonstrate that you have the ability to process information and to summarize. Employers prize analytical thinking.
- **Conciseness.** Generally, keep your résumé to one page, as in Figure 7.4. However, depending on your education or job experience, you may want to include a second page. Résumés are written in short sentences that omit "I" and that use action-packed verbs, such as those listed in Table 7.1.
- **Accuracy.** Grammar, spelling, dates, names, titles, and programs must be correct; your résumé tells readers you can communicate effectively. Typos and inconsistencies only demonstrate that you don't check your facts and figures.
- **Currentness.** All information must be up-to-date and documented, with no gaps or sketchy areas. Missing or incorrect dates or outdated contact information are flags to reject your résumé.
- **Relevance.** The information needs to be appropriate for the job level. It must show that you have the necessary education and experience and confirm that you can be an effective team player.

TABLE 7.1   Action Verbs to Use in Your Résumé

accommodated	customized	instituted	reported
accomplished	dealt in	instructed	researched
achieved	demonstrated	interned	saved
administered	designed	interpreted	scheduled
adopted	determined	led	searched
analyzed	developed	maintained	selected
arranged	directed	managed	served
assembled	drafted	monitored	settled
assisted	earned	motivated	sold
attended	elected	navigated	solved
awarded	established	negotiated	supervised
built	estimated	operated	taught
calculated	evaluated	organized	tested
coached	expanded	oversaw	tracked
collected	expedited	performed	trained
communicated	figured	planned	transformed
compiled	fulfilled	prepared	translated
completed	guided	profiled	troubleshot
composed	handled	programmed	tutored
computed	headed	provided	updated
conducted	implemented	published	upgraded
contributed	improved	ran	verified
converted	increased	reappraised	weighed
coordinated	informed	reconciled	won
created	initiated	reduced	worked
cultivated	installed	re-evaluated	wrote

Your goal is to prepare a résumé that shows the employer you possess the sought-after job skills. A résumé that is unattractive, difficult to follow, poorly written, filled with typos and other errors, and not relevant for the prospective employer will not make the first cut.

It might be to your advantage to prepare several versions of your résumé and then adapt each one you send out to the specific job skills a prospective employer is looking for. It pays to customize your résumé. Following the process in the next section will help you prepare any résumé.

The Process of Writing Your Résumé

To write an effective résumé, ask yourself the following important questions:

1. What classes did you excel in?
2. What papers, reports, or presentations earned you your highest grades?
3. What computer skills have you mastered—languages, software knowledge, navigating and developing Internet resources? Knowledge of e-commerce? E-journaling? Skill at designing a blog and/or website?
4. What jobs have you had? For how long and where? What were your primary duties?
5. How did you open or expand a business market? Increase a customer base?
6. What did you do to earn a raise or a promotion in a previous job?
7. What technical skills (other than computer skills) have you acquired?
8. Do you work well with people? What skills do you possess as a member of a team working toward a common job goal (e.g., finishing a report)?
9. Can you organize complicated tasks or identify and solve problems quickly?
10. Have you had experiences/responsibilities managing money—preparing payrolls, conducting nightly audits, and so on?
11. Have you won any awards or scholarships or received a raise, bonus, commendation, and/or promotion at work?

Pay special attention to your four or five most significant, job-worthy strengths and work especially hard on listing them concisely.

Although not everything you have done relates directly to a particular job, indicate how your achievements are relevant to the employer's overall needs. For example, supervising staff in a convenience store points to your ability to perform the same duties in another business context.

Balancing Education and Experience

If you have years of experience, don't flood your prospective employer with too many details. You cannot possibly include every detail of your job(s) for the last ten or twenty years.

- Emphasize only those skills and positions most likely to earn you the job.
- Eliminate your earliest jobs that do not relate to your present employment search.
- Combine and condense skills acquired over many years and through many jobs.

Figures 7.6 and 7.9 show the résumés of individuals who have a great deal of job experience to offer prospective employers.

Many job candidates who have spent most of their lives in school are faced with the other extreme: not having much job experience to list. The worst thing to do is to write "None" for experience. Any part-time, summer, or other seasonal jobs, as well as work done for a library or science laboratory, show an employer that you are responsible and knowledgeable about the obligations of being an employee. Figure 7.4 shows a résumé from Anthony Jones, a student with very little job experience; Figure 7.5 shows the résumé of María Lopez, a student with a few years of experience.

What to Exclude from a Résumé

Knowing what to exclude from a résumé is as important as knowing what to include. Since federal employment laws prohibit discrimination on the basis of age, sex, race, national origin, religion, marital status, or disability, do not include such information on your résumé. Here are some other details best left out of your résumé:

- salary demands, expectations, or ranges
- preferences for work schedules, days off, or overtime
- comments about fringe benefits
- travel restrictions
- reasons for leaving your previous job
- your photograph (unless you are applying for a modeling or acting job)
- your social security number
- comments about your family, spouse, or children
- height, weight, or hair/eye color
- personal information, hobbies, interests (unless relevant to the job you are seeking)

Save comments about salary and schedules for your interview. (See pages 281–288.) The résumé should be written to earn you an interview.

Parts of a Résumé

As with, memos, letters, and reports, résumés consist of specific parts. These parts—contact information, career objective, credentials (education and experience), related skills and achievements, and references—need to be included in any résumé.

Contact Information

At the top of the page, center your name (do not use a nickname), address, including your ZIP code; telephone number; and e-mail address. Avoid unprofessional e-mail addresses such as *Toughguy@netfield.com* or *barbiegirl@techscape.com.* Also include your website and fax number if you have these for an employer to contact you.

Career Objective

One of the first things a prospective employer reads is your career objective statement that specifies the exact type of job you are looking for and in what ways you are qualified to hold it. Create an objective that precisely dovetails with the prospective employer's requirements. Such a statement should be the result of your

Résumé from a student with little job experience.   Figure 7.4

Anthony H. Jones

WEBSITE DEVELOPER
DESIGNER
GRAPHIC ARTIST

73 Allenwood Boulevard
Santa Rosa, California 95401-1074
707-555-6390
ajones@plat.com
www.plat.com/users/ajones/resume.html

Headlines major achievements and gives contact details

Career Objective

Full-time position as a layout artist with a commercial publishing house using my knowledge of state-of-the-art design technology.

Offers precise, convincing objective

Education

Santa Rosa Junior College, 2006–2008, A.S. degree to be awarded in 2008
Dean's List in 2005; GPA 3.45
Major: Commercial Graphics Illustration, with specialty in design layout
Related courses included:
- Digital Photography
- Graphics Programs: Illustrator, Photoshop
- Desktop Publishing: QuarkXPress, WordPerfect Suite 8

Starts with most important qualification— education

Internship, 2007–2008, McAdam Publishers
Major projects included:
- Assisting layout editors with page composition and importing images.
- Writing detailed reports on digital photography, designs, and artwork used in *Living in Sonoma County* (www.sonomacounty.com) and *Real Estate in Sonoma County* (www.resc.net) magazines.

Stresses job-related activities of apprenticeship

Experience

Salesperson (part-time), **2003–2005,** Buchman's Department Store
Duties included assisting customers in sporting goods and appliance departments and coordinating sport shop by displaying merchandise.

Includes part-time work experience

Computer Skills

Know QuarkXPress, Passport, WordPerfect X3.

Related Activities

Volunteer; designed website and three-fold brochure for the Santa Rosa Humane Society's 2007 fund drive; helped raise $4,300.

Demonstrates skills in Web design

References/Webfolio

References, college transcripts, and a webfolio of Web designs, photographs, and graphics available upon request.

Credentials and portfolio emphasize accomplishments

focused self-evaluation and your evaluation of the job market. It will influence everything else you include. Depending on your background and the types of jobs you are qualified for, you might formulate two or three different career or employment objectives to use with different versions of your résumé as you apply for various positions.

To write an effective career objective statement, ask yourself four basic questions:

1. What kind of job do I want?
2. What kind of job am I qualified for?
3. What capabilities do I possess?
4. What kinds of skills do I want to learn?

Avoid trite or vague goals such as "looking for professional advancement" or "want to join a progressive company." Compare the vague objectives on the left with the more precise ones on the right.

Unfocused	Focused
Job in sales to use my aggressive skills in expanding markets.	Regional sales representative using my proven skills in e-commerce and communication to develop and expand a customer base.
Full-time position as staff nurse.	Full-time position as staff nurse on cardiac step-down unit to offer excellent primary care nursing and patient/family teaching.

Credentials

The order of the next two categories—**Education** and **Experience**—can vary. Generally, if you have lots of work experience, list it first as Anna Cassetti did (see Figure 7.6). However, if you are a recent graduate short on job experience, list education first, as Anthony Jones did (see Figure 7.4). María Lopez (Figure 7.5) also decided to place her education before her job experience because the job she was applying for required the formal training she received at Miami-Dade Community College.

Education.   Begin with your most recent education first, then list everything significant since high school. Give the name(s) of the school(s); dates attended; and degree, diploma, or certificate earned. Don't overlook military schools or major training programs (EMT, court reporter), institutes, internships, or workshops you have completed.

Remember, however, that a résumé is not a transcript. Simply listing a series of courses will not set you apart from hundreds of other applicants taking similar courses across the country. Avoid vague titles such as Science 203 or Nursing IV. Instead, concentrate on describing the kinds of skills you learned.

30 hours in planning and development courses specializing in transportation, land use, and community facilities; 12 hours in field methods of gathering, interpreting, and describing survey data in reports.

Completed 28 hours in major courses in business marketing, management, and materials in addition to 12 hours in information science, including HTML/Web publishing.

Résumé from a student with some job experience.   **Figure 7.5**

MARÍA H. LOPEZ
1725 Brooke Street
Miami, Florida 32701-2121
(305) 555-3429———**mlopez@eagle.com**

Career Objective	Position assisting dentist in providing dental care, counseling, and preventive dental treatments, especially in pedodontics.
Education	A.S. in Dental Hygiene, Miami-Dade Community College, Miami, Florida August 2007–May 2009 Completed courses in oral pathology, dental materials and specialties, periodontics, and community dental health. Experienced with procedures and instruments used with oral prophylaxis techniques. Subject of major project was proper nutrition and dental health for preschoolers.
	Minor: Psychology (twelve hours in child and adolescent psychology) Received excellent evaluations in business writing course. GPA is 3.3. Bilingual: Spanish/English.
	Will take American Dental Assistants' Examination on June 2.
Experience	St. Francis Hospital, Miami Beach, Florida April 2005–July 2007 Full-time clerk on the pediatric floor. Ordered supplies, maintained records, transcribed orders, greeted and assisted visitors.
	Murphy Construction Company, Miami, Florida June 2004–April 2005 Office assistant-receptionist. Did keyboarding, billing, and mailing in small office (four employees).
	City of Hialeah, Florida Summers 2002–2003 Water meter reader
Computer Skills	Microsoft Office, Excel, Health Care spreadsheet

Provides easy-to-find contact information

Lists course and clinical work required for licensure and job

Calls attention to counseling and bilingual skills that benefit employer and patients

Emphasizes professional licensure quali-fications

Highlights previous job responsibilities in health care setting

Continued

Figure 7.5   (Continued)

*Wisely lists
professionals
well known to
prospective/
area employers*

Lopez 2

References   The following individuals have written letters of recommendation
for my placement file, available from the Placement Center,
Miami-Dade Community College, Medical Center Campus,
Miami, FL 33127-2225.

Sister Mary Pela	Professor Mitchell Pelbourne
Pediatric Unit	Department of Dental Hygiene
St. Francis Hospital	Miami-Dade Community College
10003 Collins Avenue	Medical Center Campus
Miami Beach, FL 33141	Miami, FL 33127
(305) 555-5113	(305) 555-3872
Tia Gutiérrez, D.D.S.	Mr. Jack Murphy
9800 Exchange Avenue	1203 Francis Street
Miami, FL 33167	Miami, FL 33157
(305) 555-1039	(305) 555-6767

List your grade point average (GPA) only if it is 3.0 or above; otherwise, indicate your GPA in just your major or during your last term, again if it is above 3.0.

Experience.   Your job history is the key category for many employers. It shows them that you have held jobs before and that you are responsible. Here are some guidelines about listing your experience.

1. Begin with your most recent position and work backward—in reverse chronological order. List the company or agency name, location (city and state), and your title. Do not mention why you left a job.

2. For each job or activity, provide short descriptions of your duties and achievements. If you were a work-study student, don't say that you helped an instructor. Emphasize your responsibilities; for example, you helped to set up a chemistry laboratory, you ordered supplies and kept an inventory of them. Rather than saying you were a secretary, indicate that you wrote business letters and used various software programs, designed a company website, prepared schedules for part-time help in an office of twenty-five people, or assisted the manager in preparing accounts.

3. In describing your position(s), emphasize any responsibilities that involved handling money (for example, billing customers, filing insurance claims, or preparing payrolls); managing other employees; working with customer accounts, services, and programs; or writing letters and reports. Prospective employers are interested in your leadership abilities, financial shrewdness (especially if you saved your company money), tact in dealing with the public, and communications skills. They will also be favorably impressed by promotions you may have earned.

4. If you have been a full-time parent for ten years or a caregiver for a family member or friend, indicate the management skills you developed while running a household and any community or civic service, as Dora Cooper Bolger does in her résumé in Figure 7.7. She skillfully relates her family and community accomplishments to the specific job she seeks. Her volunteer work translates into marketing skills an employer wants to see in a prospective employee's résumé.

Résumé from an individual with ten years' job experience.   Figure 7.6

ANNA C. CASSETTI

6457 Blackstone Avenue
Fort Worth, TX 76321-6733
(817) 555-5657
acassetti@netdor.com

MacMurray Real Estate
1700 Ross Boulevard
Haltom City, TX 77320-1700
(817) 555-7211

Provides home and work contact information

OBJECTIVE
Full-time sales position with large real-estate office in the Phoenix or Tucson areas with opportunities to use proven skills in real-estate appraisal and tax counseling.

EXPERIENCE
MacMurray Real Estate, Haltom City, Texas, 2008–present
Real-estate agent. Excelled in small suburban office (five salespersons plus broker) with limited listings; **sold individually over three million dollars in residential property**; appraised both residential and commercial listings.

Begins with the most important job qualification first— experience

Dallman Federal Savings and Loan, Inc., Fort Worth, Texas, 2001–2007
Chief Teller. Responsible for supervising, training, and coordinating activities of six full-time and two part-time tellers. Promoted to Chief Teller, March 2000, with bonus.

H&R Block, Westover Hills, Texas, 2001 (Sept.–Dec.)
Tax Consultant. Prepared personal and business returns.

Job history focuses on sales achievements, promotion, and responsibilities.

Cruckshank's Hardware Store, Fort Worth, Texas, 1998–2000
Salesperson.

U.S. Navy, 1991–1997
Honorably discharged with rank of Petty Officer, Third Class.
Served as stores manager; earned three commendations

Gives related military experience

EDUCATION
Texas Christian University, Fort Worth, Texas, 2000–2006
Awarded B.S. degree in Real-Estate Management. Completed thirty-three hours in business and real-estate courses with a concentration in finance, appraising, and property management. Also took twelve hours in programming and Web designing. Wrote reports on appraisal procedures as part of supervised training program.

Documents education; cites specific skills

H&R Block, Westover Hills, Texas, 2000 (Sept.–Dec.)
Earned diploma in Basic Income Tax Preparation after completing intensive ten-week course.

Includes computer and communication skills

Continued

Figure 7.6   (Continued)

Cassetti 2

U.S. Naval Base, San Diego, California, 1995–1998
Attended U.S. Navy's Supply Management School.
Applied principles of stores management at Newport Naval Base.

SKILLS AND ACTIVITIES
Licensed Texas Realtor—756a2737
Chair, Financial Committee, Grace Presbyterian Church, Fort Worth.
Adviser, Junior Achievement.

REFERENCES
References available upon request.

Provides necessary license information

Related Skills and Achievements
Not every résumé will have this section but the following are all employer-friendly things to include:

- second or third languages you speak or write
- extensive travel
- certificates or licenses you hold
- memberships in professional associations (e.g., American Society of Safety Engineers, Black Student Association, National Hispanic Business Association, Texas Women Executives, Child Development Organization)
- memberships in community service groups (e.g., Habitat for Humanity, Salvation Army); list any offices you held—recorder, secretary, fund drive chairperson

Computer Skills.   Knowledge of computer hardware, software, word processing programs, and Web design and search engines is extremely valuable in the job market. Note how Anthony Jones and María Lopez inform prospective employers about their relevant technical competencies in Figures 7.4 and 7.5.

Honors/Awards.   List any civic (mayor's award, community service, cultural harmony award) and/or academic honors you have won (dean's list, department awards, scholarships, grants, honorable mentions). Memberships in honor societies in your major and technical/business associations also demonstrate that you are professionally accomplished and active.

References
You can simply say that you will provide references on request or you can list the names, titles, street addresses, e-mail addresses, and telephone numbers of no more than three or four individuals, as María Lopez did in Figure 7.5. Be sure to obtain their permission first. List your references only when they are well known in the community or

belong to the same profession in which you are seeking employment—you profit from your association with a recognizable name or title.

In this section of your résumé, you may also indicate that a portfolio of your work is available for review, as Anthony Jones did in Figure 7.4.

Organizing Your Résumé

There are two primary ways to organize your résumé: chronologically or by function or skill area. You may want to prepare two versions of your résumé—one chronological and one by function or skill area—to see which sells your talents better. Don't hesitate to seek the advice of a placement counselor or instructor about which may work best for you.

Chronologically

The résumés in Figures 7.4, 7.5, and 7.6 are organized chronologically, with most recent listed first. Information about the job applicants is listed year by year under two main categories—education and experience. This is the traditional way to organize a résumé. It is straightforward and easy-to-read, and employers find it acceptable. The chronological sequence works especially well when you can show a clear continuity toward progress in your career through your job(s) and in schoolwork or when you want to apply for a similar job with another company.

A chronological résumé is appropriate for students who want to emphasize recent educational achievements.

By Function or Skill Area

Depending on your experiences and accomplishments, you might organize your résumé according to function or skill area. According to this plan, you would *not* list your information chronologically in the categories "Experience" and "Education." Instead, you would sort your achievements and abilities—whether from course work, jobs, extracurricular activities, or technical skills—into two to four key areas, such as "Sales," "Public Relations," "Training," "Management," "Technical Capabilities," "Counseling," "Group Leadership," "Communications," "Network Operations," "Customer Service," "Working with People," "Opening New Markets," "Multicultural Experiences," "Computer Skills," or "Problem-Solving Skills."

Under each area you would list three to five points illustrating your achievements in that area. Skills or functional résumés are often called **bullet résumés** because they itemize the candidate's main strengths in bulleted lists. Some employers prefer the bullet résumé because they can more easily skim the candidate's list of qualifications in a few seconds.

Think of your function or skill areas as the common denominators that cross job and educational boundaries—the common threads that link your diverse experiences.

Preparing a Skills Résumé

When you prepare a functional or skills résumé, start with your name, address, telephone number, and career objective, just as in a chronological résumé. To find the best two or three functional areas to include, use the prewriting strategies (especially clustering and brainstorming) discussed in Chapter 2.

The following individuals would probably benefit from organizing their résumés by function or skill instead of chronologically:

- Nontraditional students who have had diverse job experiences
- People who are changing professions
- Individuals who have had changed jobs frequently
- Ex-military personnel reentering the civilian marketplace

Note how Donald Kitto-Klein (Figure 7.9) was able to pull together a series of related, marketable skills from the many different jobs he had held over several years.

After you discover and suitably revise the information to be included in your categories, briefly list your educational and work experiences, as Dora Cooper Bolger (Figure 7.7), Anna Cassetti (Figure 7.8), and Donald Kitto-Klein (Figure 7.9) do.

Tech Note

Developing Your Own Website for Your Job Search

Having your own website has definite advantages in today's highly competitive job market. Prospective employers are always impressed by applicants who are HTML design-literate. A website allows you to update information quickly and to give employers easy access to it.

But be careful about what you put on your website. You do not want to make personal information available to a worldwide audience and risk being the victim of identity theft.

Don't equate a personal website with a professional one. Include only professional details that will increase your marketability without compromising your privacy. You might safely include a link to your webfolio (see pages 248–251) or other examples of work you've done.

And always make sure your website is clear, readable, user-friendly, and easy to navigate. The Internet is full of places that can assist you in building a webpage. Try the following:

- Designing an Accessible Webpage (*http://www.trace.wisc.edu/world/web/index.html*)
- Web Page Design for Designers (*http://www.wpdfd.com*)
- Yahoo! Geocities (*http://www.geocities.yahoo.com/home*)

Also, see Chapter 11 (pages 508–520) for specific advice on writing for a Web audience.

Dora Cooper Bolger's résumé organized by skill areas.   Figure 7.7

DORA COOPER BOLGER
1215 Lakeview Avenue
Westhampton, MI 46532
Home: 616-555-4772   Cell: 616-555-4773   dbolger@aol.com

OBJECTIVE   Seek full-time position as public affairs officer in health
care, educational, or charitable organization

SKILLS

Organizational Communication
- Delivered 20 presentations to civic groups on educational issues
- Recorded minutes and helped formulate agenda as president of large, local PTA for past $6^1/_2$ years
- Possess excellent computer skills in PeopleSoft and Microsoft Word
- Updated and maintained computerized mailing lists for Teens in Trouble and Foster Parents' Association

Financial
- Spearheaded 3 major fund-raising drives (total of $200,000 collected)
- Prepared and implemented large family budget (3 children, 8 foster children) for 15 years
- Served as financial secretary, Faith Methodist Church, for 4 years

Administrative
- Organized volunteers for American Kidney Fund (last 5 years)
- Established and oversaw neighborhood carpool (17 drivers; more than 60 children) for 7 years
- Coordinated after-school tutoring program for Teens in Trouble; president since 1996

HONORS   "Volunteer of the Year"(2008), Michigan Child
Placement Agency

EDUCATION   Metropolitan Community College, A.A., 2005
Mid-Michigan College, B.S., expected 2009; major:
Public Administration; minor: Psychology. GPA 3.45

WORK EXPERIENCE Secretary, 1995–2005 (full- and part-time):
Merrymount Plastics; Foley and Wasson;
Westhampton Health Dept.; G & K Electric

**REFERENCES/
PORTFOLIO**   Available on request

Restricted objective

Aptly features skill areas before education

Links achievements from volunteer and home-based activities most important to employer

Chooses strong, active verbs to convey image of a results-oriented professional

Places education after experience; includes major and related minor plus strong GPA

Figure 7.8   Anna Cassetti's résumé organized by function or skill areas.

ANNA C. CASSETTI
www.acas.yahoo.com

Home	**Office**
6457 Blackstone Avenue	MacMurray Real Estate
Fort Worth, TX 76321-6733	1700 Ross Boulevard
(817) 555-5657	Haltom City, TX 77320-1700
acassetti@netdor.com	(817) 555-7211

Begins with a clear and focused objective

OBJECTIVE   Sales position with real-estate office in the Phoenix or Tucson areas with opportunities to use proven skills in appraisals and tax counseling

Groups accomplishments into the three most relevant and marketable skills for prospective employer

SALES/ FINANCIAL
- Sold over $3 million of residential property in 2 years
- Served as a tax consultant with special interest in real-estate sales/market conditions
- Performed general banking procedures as chief teller
- Responsible for maintaining, purchasing, and ordering supplies for ship's store in U.S. Navy

Uses bullets, strong verbs, and specific examples

PUBLIC RELATIONS
- Helped clients select appropriate property for their needs and income
- Counseled commercial and individual clients about taxes
- Supervised, trained, and coordinated the activities of six bank tellers
- Commended for rapport in assisting customers with their banking needs

Emphasizes written and oral skills

COMMUNICATION
- Prepared standard real-estate appraisals
- Wrote in-depth business reports on appraisal procedures, property management problems, and banking policies affecting real-estate transactions
- Achieved proficiency in CorpSheet and other spreadsheet programs
- Conducted small group training and sales sessions

Stresses educational preparation after skills

EDUCATION   B.S. in Real-Estate Management, 2006
Texas Christian University, Fort Worth, Texas
Advanced course work taken in business, finance, and real estate; minor in data processing

Diploma, Basic Income Tax Preparation, 2000
H&R Block, Westover Hills, Texas

Continued

(Continued)   Figure 7.8

Cassetti 2

EMPLOYMENT	MacMurray Real Estate, Haltom City, Texas 2006–present; sales agent	*Demonstrates professional commitment and successes*
	Dallman Federal Savings and Loan, Inc., Fort Worth, Texas 1999–2005; Chief Teller	
	H&R Block, Westover Hills, Texas 1999 Sept.–Dec.; consultant, tax preparer	
	Cruckshank's Hardware Store, Fort Worth, Texas 1996–1998; salesperson	
	U.S. Navy, San Diego, California (last duty station) 1989–1995; stores manager; honorably discharged with rank of Petty Officer, Third Class	
REFERENCES	Available from the Placement Office, Texas Christian University, Fort Worth, TX 76119-6811	*Supplies contact information for references*

Functional résumé by a candidate who has held a variety of jobs.   Figure 7.9

DONALD KITTO-KLEIN

kitto@gar.com   http://www.dkk@opengate.com

56 South Ardmore Way   Garland.com
Petersburg, NY 15438   Grand Banks, NY 15532
(716) 555-9032   (716) 555-4800, Ext. 5398

Objective	Seek supervisory position in computer maintenance and service to provide network continuity to staff and clients.	*Provide professional website* *Connects writer's technical and administrative skills*
Computer Languages	Java, JavaScript, UNIX, HTML	
Computer Platforms	UNIX(Solaris), Windows, Mac OS, Warp	*Establishes technical background/qualifications for job*
Networking	DECnet, Ethernet, Secure ID, Novell, PGH	

Continued

Figure 7.9   (Continued)

Computer Maintenance/ Service Skills	• Serviced PCs and workstations on a regular basis for $3^1/2$ years • Worked extensively on spreadsheets/database software • Modified software billing program and saved company $400,000 • Coordinated maintenance/service activities
People Skills	• Promoted to Support Services Team Leader • Supervised Computer Servicing with three technicians • Worked closely with computer manufacturers and vendors to minimize hardware down time • Elected to employee benefits committee
Communication Skills	• Collaboratively wrote safety manual for power company road crews • Taught in-house training sessions on computer maintenance, networking, data security • Devised routing systems to expedite work orders • Coordinated small group meetings in systems analysis
Employment	Garland.Com, 2005–present Business Graphics and Computers Store, Salesperson, 1999–2003 U.S. Army, Specialist, 4/E, 1994–2000
Education	B.S., Grand Valley Technical Institute, 2005 U.S. Army schools in computer programs, 1998–2000
References/Webfolio	Available on request.

Emphasizes the management skills and experiences prospective employer wants

Demonstrates ability to be a problem solver, a leader, and a spokesperson for company

Includes relevant military background

The Online Résumé

Visit www .cengage.com/ english/kolin for an online exercise, "Using an Online Résumé Builder."

In addition to preparing a hard copy of your résumé, expect to create a version that you can post on the Internet or e-mail to a prospective employer. In fact, some employers conduct the job application process exclusively online, receiving both applications and résumés through the company's website. In doing so, they are able to scan information quickly and screen candidates for a first cut, often in less than twenty seconds. Because prospective employers receive hundreds of résumés in response to a single job posting, your online résumé needs to capture their attention and distinguish you from other job seekers. The following sections will show you how Anthony Jones's hard copy résumé in Figure 7.4 and Dora Cooper Bolger's hard copy résumé in Figure 7.7 have been changed into online résumés in Figures 7.10 and 7.11 respectively.

Places to Post Your Online Résumé

There are several places to post your online résumé, including:

- Databases provided by commercial job search services such as *http://www.hotjobs.yahoo.com*, *http://www.monster.com*, and others listed on page 244 of this chapter.
- Databases sponsored by professional societies such as ACM's service in Figure 7.2.
- Employer websites that accept unsolicited résumés for scanning, indexing, and archiving.
- Your own website (see Tech Note on page 262).

You have to exercise some caution, though, when you use these online sources. Job counselors advise that posting your résumé on your website may provide more information than a prospective employer needs or reveal too much personal information. And as far as online posting services are concerned, usually it's a good idea to limit your résumé postings to just one or two commercial databases, professional society databases, and/or employer websites. Be careful, too, about posting your résumé on unsecured sites that promise to flood the market.

Converting Your Print Résumé to Text Only

If you have written your print résumé effectively, it should contain all the information you need for the online version. However, the format of an online résumé is very different from that of a hard copy résumé, as Anthony Jones's online résumé in Figure 7.10 or Dora Cooper Bolger's in Figure 7.11 show. Don't think that you can just paste your hard copy résumé into an e-mail or post it on a résumé database as is. You first will have to format it as text only so that is readable from a variety of computer programs. Otherwise, your résumé may be blocked or garbled during online posting or e-mailing.

To convert a résumé to text only, follow these guidelines recommended by *http://www.usajobs.gov*.

1. Open your résumé in your word processing program.
2. Save the résumé as text only. (This feature is usually available by choosing Save As from the File menu.)
3. Close the résumé and reopen the new text-only version.
4. You will see that the résumé is now free of formatting that other computers may not be able to read, such as bullets, shading, tabs, boldface, italics, underlining, unusual fonts, different font sizes, and centering. Edit your résumé to make sure that it reads smoothly and clearly and that nothing has become garbled during the conversion to text only.
5. Save the document.
6. To send your résumé in an e-mail or post it online, choose Select All to highlight the text of your résumé, copy it, and paste it into an e-mail or into the space provided by an online résumé database service or on a prospective employer's website.
7. Follow any additional directions provided by the database service or by a prospective employer. Not following directions is the quickest way to have your résumé rejected.
8. Save your original word-processed résumé too, so that you can print it out and present it at interviews or send it by regular mail to employers.

Making Your Online Résumé Search-Engine Ready

Although converting your résumé to text only allows users of various computer programs to read it, you also want prospective employers to be able to find your résumé through a database service or by searching their own résumé archive. The best way they will be able to find appropriate résumés that match a job's specific requirements is to search these databases and archives for job-relevant keywords. A list of keywords (and licenses/degrees) appropriate for an online résumé appears in Table 7.2.

Follow the guidelines below to make sure your online résumé is search-engine ready:

- **Provide a keywords section at the top of your résumé.** After your address, but before your objective, copy all of the keywords used in the body of your résumé and paste them in a separate section. See Anthony Jones's online résumé in Figure 7.10 and Dora Cooper Bolger's résumé in Figure 7.11, for example. Including a keywords section will not only help you maintain an awareness of the keywords you use, but it will also give a prospective employer an immediate snapshot of your skills when he or she locates your résumé via a search engine.
- **Use descriptive keywords.** Keywords should describe a particular skill you have or software program you know well. Be specific and descriptive, not general and vague (e.g., rather than saying "work well with others" say "team player," and rather than saying "graphics software" say "QuarkXPress").
- **Include keywords used by a prospective employer.** If you are looking for a job at a specific company and you send your résumé to them, first have a look at the job descriptions on their website and use the keywords they favor.
- **Put keywords in noun form.** When employers search for résumés housed in databases, they want you to use nouns to describe an experience or skill rather than a verb. So, replace the action verbs that you would include in a conventional, hard copy résumé (see pp. 251–253) with nouns. Here are some examples:

Conventional Résumé	Online Résumé
Wrote technical report	*Technical writer*
Performed laboratory tests	*Laboratory technician*
Responsible for managing accounts	*Accounts manager*
Won three awards	*Award winner*
Edited company newsletter	*Newsletter editor*
Solved software problem	*Software specialist*

Cybersafing Your Online Résumé

Whenever you post your résumé online, protect your identity as well as your job. You don't want to reveal personal information that hackers could use to steal your identity, and you don't want your current employer to find out—or mistakenly believe that—you are looking for another job. Here are some suggestions to safeguard confidential information during your job search:

- **Avoid listing personal information on your résumé.** Exclude information that prospective employers do not need to know about and that could lead to identity theft, such as your social security number, your date of birth, your photograph, and your home address (use a P.O. box instead).

TABLE 7.2   Sample Keywords for an Online Résumé

Job Title	Area of Expertise	Computer Skills	Degrees/ Licenses	Job/Interpersonal Skills
Accountant	Accounting	Access	AA	Analytical
Consumer advocate	Automotive repair	Adobe Illustrator	AS	Bilingual
Editor	Budget	DreamWeaver	ASID	Competency-based
Environmentalist	Computer support	E-commerce	BA	Competitive
Fitness/wellness	Construction	Excel	BS	Cooperative
Health care	Consumer affairs	Filemaker Pro	BSN	Critical thinking
Hospitality management	Counseling	HTML	CPA	Customer oriented
Intern	Customer service	Adobe InDesign	EMT	Ethical
Law enforcement officer	Engineering	JavaScript	LAT	Experienced
Maintenance expert	Financial affairs	Lotus Notes	LTC	Flexible
Officer	Graphics	Lotus 1-2-3	LPN	Fund raising
Paralegal	Health care	Microsoft Word	MAT	Goal oriented
Pharmacy tech	Human resources	Outlook	MBA	High energy
Planner	Information technology	PageMaker	MT	International
Programmer	Legal	PeopleSoft	NAVT	Leadership
Public relations	Management	PhotoShop	OT	Multitasking
Report writer	Marketing	PowerPoint	PTA	Speaker/writer
Resident manager	Networking systems	QuarkXPress	PT	Research oriented
Sales associate	Office support	QuickBooks	RN	Results oriented
Technician	Public relations	SPSS	RT	Risk taking
Tutor	Purchasing	WordPerfect		Safety conscious
Underwriter	Recruitment	XML		Self-motivated
Veterinary tech	Safety/security			Strong work ethic
Waitperson	Sales			Studied abroad
Web designer	Technical support			Team player

Figure 7.10   An online version of Anthony Jones's hard copy résumé in Figure 7.4.

All lines aligned flush with the left margin

Keywords (for search engine readiness) go at the top of the online résumé

Objective appears in the body of the online résumé

All caps rather than bold, italic, fancy fonts used to highlight categories

Uses terminology appropriate for position

Choose nouns rather than action verbs to increase employer matches

Asterisks rather than bullets mark beginning of lines

Anthony H.Jones
jonesdesigner@gmail.com
Phone:(707) 555-6390

KEYWORDS

Web designer, computer graphics, Illustrator, Photoshop, QuarkXPress, fundraiser, budgets, sales, Soapscan, WriteNow, virus protection, team player

OBJECTIVE

Position as layout and Web design editor with small commercial publisher using my education to benefit company

EDUCATION

Santa Rosa Junior College, A.S. degree to be awarded in June 2008. Commercial Graphics Illustration major. Digital photography minor. GPA 3.45

COMPUTER SKILLS

Excellent working knowledge of computer graphics: Illustrator, Photoshop, QuarkXPress Passport, WordPerfect v.3, Soapscan, WriteNow

EXPERIENCE

* Intern in layout and design department. Preparing page composition, importing visuals, manipulating images, McAdam Publishers, 8 Parkway Heights, Santa Rosa, CA 94211

* Salesperson; display merchandise coordinator, Buchman's Department Store, Greenview Mall, Santa Rosa

* Volunteer: designed website, brochures, and other artwork for successful fund drive, Santa Rosa Humane Society

* Web designer, graphic artist, proofreader, Thunder, student magazine

REFERENCES AVAILABLE ON REQUEST

An online version of Dora Cooper Bolger's résumé in Figure 7.7.   **Figure 7.11**

Dora Cooper Bolger
P.O. Box 3216
Westhampton, MI 46532
Home phone: 616-555-4772
Email: dbolger@aol.com

Avoids giving personal information

KEYWORDS
Activity planner, budget coordinator, fundraising, community service campaign, child advocacy, grant writer, public speaking, interpersonal communication, management internal and external groups, team builder, strategic planner

Uses keywords from job description

OBJECTIVE
Position as public affairs officer for nonprofit organization

RELEVANT EXPERIENCE
* Presenter at 20 civic group functions
* Fund raiser for 3 major campaigns over $200,000 collected
* President and coordinator, Teens in Trouble, a tutoring and mentoring program
* Budget planner, Foster Parents Association; large urban church

Uses asterisks and nouns to list accomplish-ments

VOLUNTEER WORK/AWARDS
Volunteer of the Year, Michigan Child Placement Agency, 2008
Volunteer Coordinator/Leader, American Kidney Fund
President and Secretary, local PTA, 2002-2008

Summarizes experience carefully

EDUCATION
A.A., Metropolitan Community College, 2006
B.S., Mid-Michigan College, 2009. Major: Public Administration.
Minor: Psychology.
GPA: 3.45. Dean's List, 2007-2009.

Avoids any symbols, bold-facing, or ital-ics that could garble text

COMPUTER PROFICIENCIES
Microsoft Word; Peoplesoft; PowerPoint; Financial Planning Platform, BusinessPax

REFERENCES/PORTFOLIO
Available on request

- **Post your online résumé only on reliable sites.** Avoid résumé databases that say they will "flood" or "blast" the market. You don't know where your résumé will end up or how long it will be archived. Stick with those database services listed on page 267 or other reliable services.
- **Never use your e-mail address at your current place of work.** It is unethical to use your current employer's resources and/or company time to search for a job at another company.
- **Use an e-mail address specific to your job search.** Consider opening a free e-mail account (such as a Yahoo!, Gmail, or other such account) to be used only for the job search process rather than citing your regular personal e-mail address. This will protect hackers or spammers from gaining access to your usual e-mail address. Create an address that includes a word or phrase that identifies your area of expertise rather than your full name, for example, *josieprogrammer@aol.com.* Or use an e-mail link with a built-in "mail to" option rather than providing your e-mail address on your online résumé.
- **Never put the names of your references or their addresses online.** Although you may include references on a hard copy résumé, avoid doing this on a publicly posted résumé (or anywhere on your own website). Simply say "References available upon request."
- **Delete your résumé from job search databases once you have been hired.** Do not allow your résumé to be archived or your current employer may think you are planning to leave your job and therefore not give you a raise or a promotion.

Testing, Proofing, and Sending Out Your Online Résumé

Keep in mind the following tips as you test, proofread, and send out your online résumé:

- **Test the formatting of your résumé.** Send the text-only version of your résumé to friends or relatives who use different types of computer software and hardware than you use (e.g., use a Mac instead of a PC, use WordPerfect instead of Microsoft Word) to be sure your file is readable on any computer.
- **Proofread your online résumé carefully just as you would your hard copy résumé.** Make sure that you haven't left any glaring gaps in your work history, missed opportunities to add more keywords, or misrepresented any information. Also make sure that your online résumé is free of typographical errors, misspellings, and grammatical errors.
- **Delete any inappropriate wordings for an online résumé that may have carried over from your hard copy one.** For instance, delete "page 2," "continued," or "above," since your resume is now a continuous document.
- **Write individual cover e-mails to each prospective employer.** Take the extra step to write a two- or three-paragraph cover e-mail to accompany your résumé, not just a note that says "Here is my résumé." Follow the same guidelines you would follow to send a hard copy of your letter of application (see pp. 273–280). But don't send the same e-mail to multiple employers at once. Seeing the list of companies you are sending your resume to tells prospective employers that you are not particularly interested in the job they are advertising. Instead, take the time to contact them individually.

- **Always keep track of where you have sent your online résumé.** Either create a log or save the e-mails you have sent, particularly if you send out a large number of applications and résumés.

Letters of Application

Along with your résumé, you must send your prospective employer a letter of application, one of the most important pieces of correspondence you may ever write. Its goal is to get you an interview and ultimately the job. Letters you write in applying for jobs should be *personable*, *professional*, and *persuasive*—the three P's. Knowing how the letter of application and résumé work together and how they differ can give you a better idea of how to compose your letter.

Visit www
.cengage.com/
english/kolin
for an online
exercise,
"Locating and
Applying for an
International
Job."

How Application Letters and Résumés Differ

The résumé is a persuasive record of dates, important achievements, skills, names, places, addresses, and jobs. As noted earlier, you may prepare several different résumés depending on your experience and on the job market.

Your letter of application, however, is much more personal. Because you must write a new, original letter to each prospective employer, you may write (or adapt) many different letters. Each letter of application should be tailored to a specific job. It should respond precisely to the kinds of qualifications the employer seeks.

The letter of application is a sales letter that emphasizes and applies the most relevant details (of education, experience, and talents) in your résumé. In short, the résumé contains the raw material that the letter of application transforms into a finished and highly marketable product—you.

Résumé Facts to Exclude from Letters of Application

The letter of application should not simply repeat the details listed in your résumé. In fact, the following details that you would include in your résumé should *not* be restated in the letter:

- personal data, including license or certificate numbers
- specific names of courses in your major
- names and addresses of all your references

Writing the Letter of Application

The letter of application can make the difference between your getting an interview and being eliminated early from the competition. It should convince a prospective employer that you will use the experience and education listed on your résumé in the job he or she is hoping to fill. You want your letter to be placed in the "definitely interview" category. Limit your letter to one page. As you prepare your letter, use the following general guidelines.

   1. **Follow the standard conventions of letter writing** (see Chapter 5). Print your letter on good-quality, white 8.5" × 11" paper. Proofread meticulously; a spelling

error, typo, or grammatical mistake will make you look careless. Don't rely only on your spell-check.

2. Make sure your letter looks attractive. Use wide margins and don't crowd your page. Keep your paragraphs short and readable—no more than four or five sentences long (see p. 159 in Chapter 5).

3. Send your letter to a specific person. Never address an application letter "To Whom It May Concern", "Dear Sir or Madam," or "Director of Human Resources." Get an individual's name by double-checking the company's website or calling the company's switchboard, and be sure to verify the spelling of the person's name and his or her title.

4. Don't forget the "you attitude" (see Chapter 5, pp. 165–170). See yourself as an employer sees you. Focus on how your qualifications meet the employer's needs, not the other way around. Employers are not impressed by vain boasts ("I am the most efficient and effective safety engineer"). Convince prospective employers that you will be a valuable addition to their organization—a team player, a problem solver, a skilled professional.

5. Don't be tempted to send out your first draft. Write and rewrite your letter of application until you are convinced it presents you in the best possible light. Getting the job may depend on it. A first or even second draft rarely sells your abilities as well as a third, fourth, or even fifth revision does.

The sections that follow give you some suggestions on how to prepare the various parts of an application letter successfully.

Your Opening Paragraph

The first paragraph of your letter of application is your introduction. It must get your reader's attention by answering three questions:

1. Why are you writing?
2. Where or how did you learn of the vacancy or the company or the job?
3. What is your most important qualification for the job?

Begin your letter by stating directly that you are writing to apply for a job. Don't say that you "want to apply for the job"; such an opening raises the question, "Why don't you, then?"

Avoid an unconventional or arrogant opening: "Are you looking for a dynamic, young, and talented accountant?" Do not begin with a question; be more positive and professional.

If you learned about the job through a website or journal, make sure you italicize or underscore its title.

> I am applying for the food service manager position you advertised in the May 10 edition of the *Los Angeles Times* online.

Since many companies announce positions on the Internet, check there first to see if their position is listed online, as Anthony Jones did (Figure 7.12).

If you learned of the job from a professor, a friend, or an employee at the firm, state so. Take advantage of a personal contact who is confident that you are

Letter of application from Anthony Jones, a recent graduate with little job experience.   **Figure 7.12**

Anthony H. Jones
73 Allenwood Boulevard
Santa Rosa, California 95401-1074
707-555-6390
ajones@plat.com
www.plat.com/users/ajones/resume.html

Clear, professional-looking letterhead

May 24, 2008

Ms. Jocelyn Nogasaki
Human Resources Manager
Megalith Publishing Company
1001 Heathcliff Row
San Francisco, CA 94123-7707

Dear Ms. Nogasaki:

I am applying for the layout editor position advertised on your website, which I accessed on 14 May. Early next month, I will receive an A.S. degree in commercial graphics illustration from Santa Rosa Junior College.

Identifies position and source of ad

With a special interest in the publishing industry, I have successfully completed more than 40 credit hours in courses directly related to layout design, where I acquired experience using QuarkXPress as well as Illustrator and Passport. You might like to know that many Megalith publications were used as models of design and layout in my graphics communications and digital photography classes.

Validates necessary education and applies it directly to employer's business

My studies have also given me practical experience at McAdam Publishers as part of my Santa Rosa internship program. While working at McAdam, I was responsible for assisting the design department in page composition and importing images. Other related experience I have had includes creating a website for the Santa Rosa Humane Society and displaying merchandise at Buchman's Department Store. As you will note on the enclosed résumé, my references and webfolio are available upon request.

Convincingly cites related job experience

Refers to résumé/ webfolio

I would appreciate the opportunity to discuss with you my qualifications in graphic design. My phone number is listed above. After June 12, I will be available for an interview at any time that is convenient for you.

Asks for an interview and gives contact information

Sincerely yours,

Anthony H. Jones

Anthony H. Jones
Encl. Résumé

qualified for the position, as María Lopez (Figure 7.13) and Dora Cooper Bolger (Figure 7.14, p. 278) did. But first confirm that your contact gives you permission to use his or her name.

The Body of Your Letter

The body of your letter, comprising one or two paragraphs, provides the evidence from your résumé to prove you are qualified for the job. You might want to spend one paragraph on your education and one on your experience or combine your accomplishments into one paragraph.

Follow these guidelines for the body of your letter:

1. **Keep your paragraphs short and readable—four or five sentences.** Avoid long, complex sentences. Use the active voice to emphasize yourself as a doer. Review the action verbs in Table 7.1 and use keywords found in the employer's ad.
2. **Don't begin each sentence with "I."** Vary your sentence structure. Write reader-centered sentences, even those beginning with "I."
3. **Concentrate on seeing yourself as your employer sees you.** Prove that you can help an employer's sales and service, promote an organization's mission and goals, and be a reliable team player.
4. **Highlight your qualifications by citing specific accomplishments.** Tell your reader exactly how your schoolwork and job experience qualify you to perform and advance in the job advertised. Don't simply say you are a great salesperson. Demonstrate your accomplishments by stressing that you increased the sales volume in your department by 20 percent within the last six months.
5. **Mention you are enclosing your résumé.** Put an "encl" notation at the bottom of your letter.

Education.   Recent graduates with little work experience will, of course, spend more time discussing their education. But regardless, emphasize marketing why and how your most significant educational accomplishments—course work, degrees, training—are relevant for the particular job. Employers want to know which specific skills from your education translate into benefits for their company.

Simply saying you will graduate with a degree in criminal justice does not explain how you, unlike all the other graduates of such programs, are best suited for a particular job. But when you indicate that you have completed 36 credit hours in software security and have another 12 credit hours in global business, you show exactly how and where you are qualified. Note how Anthony Jones in Figure 7.12 and María Lopez in Figure 7.13 establish their educational qualifications with specific details about their training.

Experience.   After you discuss your educational qualifications, turn to your job experience. But if your experience is your most valuable and extensive qualification for the job, put it before education, as Donald Kitto-Klein does in Figure 7.15. If you are switching careers or returning to a career after years away from the work force, start the body of your letter with your experience or your community and civic service, as Dora Cooper Bolger does in Figure 7.14. Her volunteer work convincingly demonstrates she has the organizational and communication skills her prospective employer seeks. Never minimize such contributions.

Letter of application from María Lopez, a recent graduate with some job experience.   Figure 7.13

1725 Brooke Street
Miami, FL 32701-2121
mlopez@eagle.com   305-555-3429

May 15, 2008

Dr. Marvin Henrady
Suite 34
Medical/Dental Plaza
839 Causeway Drive
Miami, FL 32706-2468

Dear Dr. Henrady:

Mr. Mitchell Pelbourne, my clinical instructor at Miami-Dade Community College, informs me you are looking for a dental hygienist to work in your northside office. My education and experience qualify me for that position. This month I will graduate with an A.S. degree in the dental hygienist program, and I will take the American Dental Assistants' Examination in early June.

I have successfully completed all course work and clinical programs in oral hygiene, anatomy, and prophylaxis techniques. During my clinical training, I received intensive practical instruction from several local dentists, including Dr. Tia Gutiérrez. Since your northside office specializes in pedodontal care, you might find the subject of my major project—proper nutrition and dental care for pre-schoolers—especially relevant.

I also have related job experience in working with children in a health care setting. For over two years, I was a unit clerk on the pediatric unit at St. Francis Hospital, and my experience in greeting patients, transcribing orders, and assisting the nursing staff would be valuable to you in your office. An additional job strength I would bring to your office is my bilingual (Spanish/English) communication skills. You will find more detailed information about my accomplishments in the enclosed résumé.

I would welcome the opportunity to talk with you about the position and my interest in pedodontics. I am available for an interview any time after 2:30 until June 11. After that date, I could come to your office any time at your convenience.

Sincerely yours,

María H. Lopez

María H. Lopez

Encl. Résumé

Begins with personal contact

Verifies she will have necessary licensure

Links training to job responsibilities; demonstrates knowledge of employer's office

Relates previous experience to employer's needs; refers to résumé

Ends with a polite request for an interview

Figure 7.14   Letter of application from Dora Cooper Bolger, a recent graduate with community and civic experience.

Dora Cooper Bolger
1215 Lakeview Avenue
Westhampton, MI 46532
Home: 616-555-4772   **Cell:** 616-555-4773   dbolger@aol.com

February 9, 2009

Dr. Lindsay Bafaloukos
Tanselle Mental Health Agency
4400 West Gallagher Drive
Tanselle, MI 46932-3106

Dear Dr. Bafaloukos:

Begins with contact made at professional meeting, highlighting her interests

At a recent meeting of the County Services Council, a member of your staff, Homer Morales, told me that you will be hiring a public affairs coordinator. Because of my extensive experience in and commitment to community affairs, I would appreciate your considering me for this opening. I expect to receive my B.S. in Public Administration from Mid-Michigan College later this year.

Relates proven past successes to employer's needs; gives concrete examples of her skills

For the past ten years, I have organized community groups with outreach programs similar to Tanselle's. I have held administrative positions in the PTA and the Foster Parents' Association and was president of Teens in Trouble, a volunteer group providing assistance to dysfunctional teens. My responsibilities with Teens have included coordinating our activities with various school programs, scheduling tutorials, and representing the organization before local and state government agencies. I have been commended for my organizational and communication skills. My twenty presentations on foster home care and Teens in Trouble demonstrate that I am an effective speaker, a skill I could put to work for Tanselle immediately.

Encourages reader to see her as best-prepared candidate; includes résumé

Because of my work at Mid-Michigan and in Teens and Foster Parents, I have the practical experience in communication and psychology to promote Tanselle's goals. The enclosed résumé provides details of my experience and education.

Requests interview and gives contact information

I would enjoy discussing my accomplishments with Teens and the other organizations with you. I am available for an interview at your convenience. Thank you for reaching me at the phone numbers or e-mail address above. My portfolio is available upon request.

Sincerely yours,

Dora Cooper Bolger

Dora Cooper Bolger

Encl. Résumé

Letter of application from Donald Kitto-Klein, who has strong work experience.   Figure 7.15

DONALD KITTO-KLEIN
▪ ▪ ▪ ▪ ▪ ▪ ▪ ▪ ▪ ▪ ▪

56 South Ardmore Way
Petersburg, NY 15438
716-555-3177

📱 (716) 555-9032
🖥 kitto@gar.com
🌐 http://www.dkk.opengate.com

Designs profes-sional, distinc-tive letterhead

November 4, 2006

Ms. Michelle Washington
Vice President, Operations
Patterson Corporation
Sun Valley, CA 94356

Dear Ms. Washington:

I am very interested in the position of Director of Computer Services posted on your website 30 Oct. My knowledge of computers and my proven ability to work with people in a large organization such as Patterson qualify me to be a productive member of your management team.

Identifies spe-cific position and his major qualification

For the last three years, I have been responsible for all phases of computer maintenance and service at Garland.Com. As a regular part of my duties I have serviced and repaired 15 to 20 PCs a month as well as supervised a mainframe. I have successfully coordinated the activities of my department with Garland's other offices and worked closely with vendors and manufacturers. The training programs and in-house conferences I have conducted repeatedly receive high praise from Garland's management. Because I have worked so effectively with management and staff, I was promoted to support services team leader.

Documents ac-complishments in technology, training, and management pertinent to the job

My educational achievements include a degree in computer technology from Grand Valley and certificates from U.S. Army schools in computers. The enclosed résumé will give you information about these and my other accomplishments.

Puts education after experi-ence

I would like to put my ability to motivate personnel and to manage computer services to work for you and Patterson Corporation. I am available for an interview at your convenience. I look forward to hearing from you.

Emphatically closes by offer-ing to put his career successes to work for employer and requests an interview

Sincerely yours,

Donald Kitto-Klein

Donald Kitto-Klein

Encl. Résumé

Relate Your Experience/Education to the Job.   Link your education and experience as benefits to the particular job you apply for. Persuasively show a prospective employer how your previous accomplishments have prepared you for future success on the job. Relate your course work in computer science to being an efficient programmer. Indicate how your summer jobs for a local park district reinforced your exemplary skills in customer service. Connect your background to the prospective employer's company. Any homework you can do about the company's history, goals, or structure (see p. 281) will pay off.

- By citing Megalith publications as a model in his courses, Anthony Jones in Figure 7.12 stresses he is ready to start successfully from the first day on the job.
- Note how María Lopez in Figure 7.13 links her work at a hospital pediatric unit to Dr. Henrady's specialty.
- Dora Cooper Bolger in Figure 7.14 likewise shows her prospective employer that she is familiar with and can contribute to Tanselle's programs in community mental health though her volunteer work and public speaking experience.

Closing

Keep your closing paragraph short—about two or three sentences—but be sure it fulfills the following three important functions:

1. emphasizes once again briefly your major qualifications
2. asks for an interview or a phone call
3. indicates when you are available for an interview

End gracefully and professionally. Be straightforward. Don't leave the reader with a single weak, vague sentence: "I would like to have an interview at your convenience." That does nothing to sell you. Say that you would appreciate talking with the employer further to discuss your qualifications. Then mention your chief talent. You might also express your willingness to relocate if the job requires it.

After indicating your interest in the job, give the times you are available for an interview and specifically tell the reader where you can be reached. If you are going to a professional meeting that the employer might also attend, or if you are visiting the employer's city soon, say so.

The following samples show how *not* to close your letter and why not.

Pushy:	I would like to set up an interview with you. Please phone me to arrange a convenient time. (That's the employer's prerogative, not yours.)
Too Informal:	I do not live far from your office. Let's meet for coffee sometime next week. (Say instead that since you live nearby, you will be available for an interview.)
Introduces New Subject:	I would like to discuss other qualifications you have in mind for the job. (How do you know what the interviewer might have in mind?)

Note that the closing paragraphs in Figures 7.12 through 7.15 avoid these errors.

Going to an Interview

There are various ways for a prospective employer to conduct an interview. It might be a one-on-one meeting—you and the interviewer. Or you may visit with a group of individuals who are trying to decide if you would fit in. Or you might have your interview over the telephone or through a videoconference. An interview can last thirty minutes or take all day.

Preparing for the Interview

Before you go to the interview be prepared by doing the following:

1. Do your homework about the company. Click on "About Us" and other links on the corporate or agency website to find out who founded the company, who the current CEO is, if it is a local firm or a subsidiary, what the company's chief products or services are, how many years the company has been in business, how many employees it has, where its main office and plants are, and who its major clients and competitors are. If possible, obtain a copy of the company's annual stockholder's report. Also check such business directories as Hoover, Zack's, and Corporate Information, as discussed in Chapter 8 (pp. 333–334).

2. Review the job description carefully. Bring a copy of the job description with you to review just prior to the interview so that you are clear about what the job entails. Review the most relevant technical skills you need for the job.

3. Prepare a one- or two-minute summary of your chief qualifications. You will most likely be asked to summarize your education and experience during the job interview.

4. Take your portfolio, including three or four extra copies of your résumé, with you. Also bring a note pad and a pen (or a BlackBerry or PDA, if you have either) to jot down essential details. Leave your laptop at home.

5. Practice your interview skills with a friend or job counselor. Be sure that this person asks tough questions so that you will get practice answering these questions realistically and convincingly.

6. Brush up on business etiquette. Remember the name(s) of the interviewer(s) and others you may meet. Always be polite and respectful. Words such as "Thank You," "You're Welcome," and so on are essential for job success. Pay special attention to acceptable ways of communicating with international audiences if your interview is with a multinational company or a non-native speaker of English.

7. Bring your photo ID and social security card. Bring any licenses or certificates you may be asked to present to Human Resources. If you are not a U.S. citizen, bring your work visa.

8. Bring your enthusiasm with you. Be upbeat, positive, and eager. Express your desire to know more about the company and your desire to work for it.

Questions to Expect at Your Interview

The following questions are typical of those you can expect from interviewers, with advice on how to answer them.

- **Why do you want to work for us?** (Recall any job goals you have and apply them specifically to the job under discussion.)
- **What qualifications do you have for the job?** (Point to educational achievements and relevant work experience, especially computer skills.)
- **What could you possibly offer us that other candidates do not have?** (Say "enthusiasm," being a team player, having problem-solving abilities.)
- **Why did you attend this school?** (Be honest—location, costs, programs.)
- **Why did you major in "X"?** (Do not simply say financial benefits; concentrate on both practical and professional benefits. Be able to state career objectives.)
- **Why did you get a grade of C in a course?** (Don't say that you could have done better if you'd tried. Explain what the trouble was and mention that you corrected it in a course in which you earned a B or an A.)
- **What extracurricular activities did you participate in while in high school or college?** (Indicate any responsibilities you had—managing money, writing memos, coordinating events. If you were unable to participate in such activities, tell the interviewer that a part-time job, community or church activities, or commuting prevented you from participation. Such answers sound better than saying that you did not like sports or clubs in school.)
- **Did you learn as much as you wanted from your course work?** (This is a loaded question. Indicate that you learned a great deal but now look forward to the opportunity to gain more practical skill, to put into practice the principles you have learned; say that you will never be through learning about your major.)
- **What is your greatest strength?** (Say being a team player, planning and organizing tasks, being sensitive to others' needs, concern about the environment, cooperation, willingness to learn, ability to grasp difficult concepts easily, proficiency at managing time or money, taking criticism easily, and profiting from criticism.)
- **What is your greatest shortcoming?** (Be honest here and mention it, but then turn to ways in which you are improving. Don't say something deadly like, "I can never seem to finish what I start" or "I hate being criticized." You should neither dwell on your weaknesses nor keep silent about them. Saying "None" to this kind of question is as inadvisable as rattling off a list of faults.)
- **How do you handle conflict with a co-worker, boss, or customer?** (Stress your ability to be courteous and honest and to work toward a productive resolution. State that you avoid language, tone of voice, or gestures that interfere with healthy dialogue.)
- **Why did you leave your last job?** ("I returned to school full time" or "I moved from Jackson to Springfield." *Never attack your previous employer.* That only makes you look bad.)

- **Why would you leave your current job?** (Never attack an individual or an organization. Say your current job has prepared you for the position you are now applying for. Emphasize your desire to work for a new company because of its goals, work environment, and opportunities. Say you want a challenge.)

What Do I Say About Salary?

Again, do your homework. Find out what the salary range is for your professional level in your area. Consult the U.S. Bureau of Labor Statistics' *Occupational Outlook Handbook* at *http://www.bls.gov/oco* as well as *http://www.salary.com.* Ask your instructors or individuals you know who work for the company or call a professional organization to which you may belong for information. If the issue of salary comes up, ask if the company has established a salary range for the position and where you stand in relationship to that range. However, since many companies set fixed salaries for entry-level positions, it may be unwise to try to negotiate.

If you are asked about what salary you expect for the job, do not give an exact figure. You may undercut yourself if the employer has a higher figure in mind. By doing your homework on salary ranges, you will have a better feel for the market when the employer does mention salary.

Keep in mind, too, that the figure you are quoted as an annual or monthly salary is only part of the financial picture. Factor other things into your job equation—health insurance, day care, housing, uniform/clothing allowances, product or service discounts, retirement plans, and tuition reimbursement. See Figure 1.1 for an example of such a job perk.

Questions You May Ask the Interviewer(s)

You will have a chance to ask the interviewer(s) questions. Watch for appropriate cues and be prepared to say more than "No, I don't have any questions," which suggests either indifference or unpreparedness on your part. Here are some legitimate questions you can ask interviewers:

1. Will there be any safety, security, proficiency requirements I will need to meet?
2. When is the starting date?
3. Is there a probationary period? If so, how long?
4. How will my work be evaluated (monthly, quarterly, semiannually) and by whom (immediate superior, committee)?
5. What types of on-the-job training are offered?
6. Are there any mentoring programs in place?
7. Is there any support for continuing my education to improve my job performance?

What Interviewer(s) Can't Ask You

There are some questions an interviewer may not legally ask you. Questions about your age, marital status, ethnic background, race, or physical disabilities violate equal opportunity employment laws. Even so, some employers may disguise their interest in those subjects by asking you indirect questions about them. A question such as "Will your husband care if you have to work overtime?" or "How many children do you have?" could probe into your personal life. Confronted with such questions, it is best to answer them positively ("My home life will not interfere with my job," "My family understands that overtime may be required") rather than bristling defensively, "It's none of your business if I have a husband."

Ten Interview Dos and Don'ts

Keep in mind these other interview dos and don'ts.

1. Be on time. In fact, show up about fifteen minutes early in case the interviewer or human resources office wants you to complete some forms.
2. Turn your cell phone off! The last thing you want is for it to ring during your interview.
3. Dress appropriately for the occasion and be well groomed. **Men:** Wear a dark suit and tie. **Women:** Wear a suit (pants or skirt) or equally businesslike attire.
4. Be careful about tattoos. Job counselors warn that visible tattoos can hurt a job seeker's chance for success.
5. Greet the interviewer with a friendly and firm, but not vicelike, handshake. Don't be a wimp, either, with a fishy handshake. Thank your interviewer(s) for inviting you.
6. Don't sit down before the interviewer(s) do. Wait for them to invite you to sit and to indicate where.
7. Speak slowly and distinctly; do not nervously hurry to finish your sentences and never interrupt or finish an interviewer's sentences. Avoid one- or two-word answers, which sound unfriendly or unprepared. Do not use slang (e.g., "Right on," "Way to go," "You go, girl") or overly casual language ("Like . . ." "You know?"). Don't monopolize the discussion by talking too much and always about yourself. Show a keen interest in the company and it products/services/organization. See Figure 7.16, where a candidate reiterates her interest in a prospective employer.
8. Refrain from chewing gum; clicking a ballpoint pen; fidgeting; twirling your hair; or tapping your foot against the floor, a chair, or a desk.
9. Maintain eye contact with the interviewer; do not sheepishly stare at the floor or the desk. If you are interviewed by a group of individuals, make eye contact with each one of them. Body language is equally important. For instance, don't fold your arms—a signal that you are closed to the interviewer's suggestions and comments. Sit up straight; do not slouch.
10. When the interview is over, thank the interviewer for considering you for the job and say you look forward to hearing from him or her.

A follow-up letter.   **Figure 7.16**

2739 EAST STREET
LATROBE, PA 17042-0312

610-555-6373
mlb@springboard.com

September 20, 2008

Mr. Jack Fukurai
Manager of Human Resources
Global Tech
1334 Ridge Road N.E.
Pittsburgh, PA 17122-3107

Dear Mr. Fukurai:

I enjoyed talking with you last Wednesday and learning more about the security officer position available at Global Tech. It was especially helpful to take a tour of the plant's north gate section to see the challenges it presents for the security officer stationed there.

As you noted at the interview, my training in surveillance electronics has prepared me to operate the sophisticated equipment Global Tech has installed at the north gate. I was grateful to Ms. Turner for taking time to demonstrate this technology.

I am looking forward to receiving the handbook about Global Tech's employee services. Would you kindly e-mail me a copy of the newsletter from last quarter that introduced the new security equipment to employees?

Thank you, again, for considering me for the position and for the hospitality you showed me. Please let me know if you have any other questions for me. I look forward to learning from you.

Sincerely yours,

Marcia Le Borde

Marcia Le Borde

Expresses gratitude for interview and singles out main company feature

Reemphasizes qualifications

Asks for newsletter to express further interest

Ends politely by thanking interviewer

The Follow-Up Letter

Within a week after the interview, it is wise to send a follow-up letter, not e-mail, thanking the interviewer for his or her time and interest in you. In your letter, you can reemphasize your qualifications for the job by showing how they apply to conditions described by the interviewer; you might also ask for further information to show your interest in the job and the employer. A sample follow-up letter appears in Figure 7.16.

Accepting or Declining a Job Offer

If you accept a job, send the employer a letter within a week of the offer. Accepting verbally on the phone is not enough. Your letter will make your acceptance official and will probably be included in your permanent personnel file. Accepting a job is easy. Make the communication with your new employer a model of clarity and diplomacy.

A sample acceptance letter appears in Figure 7.17 (p. 287). In the first sentence tell the employer that you are accepting the job and refer to the date of the letter offering you the position, the specific job title, and salary. Indicate when you can begin working. Then mention any pleasant associations from your interview or any specific challenges you are anticipating. That should take no more than a paragraph.

In a second paragraph express your plans to fulfill any further requirements for the job—going to the human resources office, taking a physical examination, having a copy of a certificate or license forwarded, sending a final transcript of your college work. A final one-sentence paragraph might state that you look forward to starting your new job.

Refusing a job requires tact. You are obligated to inform an employer why you are not taking the job. See the sample refusal letter in Figure 7.18 (p. 288).

Do not bluntly begin with the refusal. Instead, prepare the reader for bad news by starting with a complimentary remark about the job, the interview, or the company. Then move to your refusal and supply an honest but not elaborate explanation of why you are not taking the job. Many students cite educational opportunities, work schedules, geographic preference, or more relevant professional opportunities. End on a friendly note, because you may be interested in working for the company in the future and do not want to leave any bad feelings.

Searching for the Right Job Pays

As we saw, finding the right job takes a lot of hard work (researching, organizing, and writing). But all your efforts will pay off with your first and subsequent checks. May all your letters, résumés, portfolios/webfolios, and applications be models of successful writing at work.

Letter accepting a job.   **Figure 7.17**

• KEVIN DUBINSKI •

73 Park Street
Evansville, WI   53536-1016

Home: 608-555-3173   •   Cell: 608-898-4291
KDubinski@global.net

June 29, 2007

Ms. Melinda Haas, Manager
Weise's Department Store
Janesville Mall
Janesville, WI 53545-1014

Dear Ms. Haas:

I am pleased to accept the position of assistant controller that you offered me at a salary of $29,250 in your letter of June 22. Starting on July 18 will be no problem. I look forward to helping Ms. Meyers in the business office. In the next few months I know that I will learn a great deal about Weise's.

As you asked, I will make an appointment early next week with the Human Resources Department to discuss travel policies, salary payment schedules, and insurance coverage.

I am eager to start working for Weise's.

Cordially,

Kevin Dubinski

Kevin Dubinski

Accepts job and shows willingness to be team member

Agrees to conditions

Looks forward to starting

Figure 7.18   Letter refusing a job.

George Alexander

March 8, 2009

Ms. Gail Buckholtz-Adderley
Assistant Editor
The Everett News
Everett, WA 98421-1016

Dear Ms. Buckholtz-Adderley:

Does not begin with refusal

I enjoyed meeting you and the staff photographers at the recent interview for the graphics position at the *News*. Your plans for the special online magazine are exciting, and I know that I would have enjoyed my assignments greatly.

Gives cred-itable reason for not taking job

However, because I have decided to continue my education part time at Western Washington University in Bellingham, I have accepted a position with the *Bellingham American*. Not having to commute to Everett will give me more time for my studies and also for my freelance work.

Thanks reader and ends on a friendly note

Thank you for your generous offer and for the time you and the staff spent explaining your plans to me. I wish you much success with the new online magazine.

Sincerely yours,

George Alexander

George Alexander

---

345 Melba Lane ▪ Bellingham, WA 98225-4912 ▪ galexander@micro.com

Revision Checklist

- [ ] Restricted the types of job(s) for which I am qualified.
- [ ] Prepared a dossier at school placement office, including letters from professors, employers, and community officials.
- [ ] Identified places where relevant jobs are advertised.
- [ ] Notified instructors, friends, relatives, clergy, and individuals who work for the companies I want to join that I am in the job market.
- [ ] Checked with chamber of commerce, state employment office, and relevant government agencies.
- [ ] Researched the companies I am interested in—on the Internet, through printed sources, and in networking with current employees I know.
- [ ] Inventoried my strengths carefully to prepare résumé.
- [ ] Eliminated weak, irrelevant, repetitious, and dated material from inventory.
- [ ] Wrote a focused and persuasive career-objective statement.
- [ ] Determined the most beneficial format of résumé to use—chronological, functional, or both.
- [ ] Prepared an online résumé to send to prospective employers.
- [ ] Investigated creating a website for my job search–related documents.
- [ ] Prepared a portfolio/webfolio that included documents demonstrating my professional skills and achievements relevant for the job.
- [ ] Made résumé attractive and easy to read with logical and persuasive headings and keywords.
- [ ] Made sure résumé contains neither too much nor too little information.
- [ ] Proofread résumé to ensure everything is correct, consistent, and accurate.
- [ ] Created properly formatted online résumé.
- [ ] Wrote letter of application that shows how my specific skills and background apply to and meet an employer's exact needs.
- [ ] Prepared short oral presentation about myself and my accomplishments for an interview.
- [ ] Researched prospective employer's company/organization and salary range(s).
- [ ] Sent prospective employer a follow-up letter within a few days after interview to show interest in position.
- [ ] Sent prospective employer a polite acceptance or rejection letter.

Exercises

1. Using at least four different sources, compile a list of ten employers for whom you would like to work. Get their names, street and e-mail addresses, phone numbers, and the names of the managers or human resources officers. Then select one company and write a profile about it—locations, services, kinds of products or services offered, number of employees, clients served, types of schedules used, and any other pertinent facts.

2. Which of the following would belong on your résumé? Which would not belong? Why?

   a. student ID number
   b. social security number
   c. the ZIP codes of your references' addresses
   d. a list of all your English courses in college
   e. section numbers of the courses in your major
   f. statement that you are recently divorced
   g. subscriptions to journals in your field
   h. the titles of any stories or poems you published in a high school literary magazine or newspaper
   i. your GPA
   j. foreign languages you studied
   k. years you attended college
   l. the date you were discharged from the service
   m. names of the neighbors you are using as references
   n. your religion
   o. job titles you held
   p. your summer job washing dishes
   q. your telephone number
   r. the reason you changed schools
   s. your current status with the National Guard
   t. the URL of your website
   u. your volunteer work for the Red Cross
   v. hours a week you spend reading science fiction
   w. the title of your last term paper in your major
   x. the name of the agency or business where you worked last

3. Indicate what is wrong with the following career statements and rewrite them to make them more precise and professional.

   a. Job in a dentist's office.
   b. Position with a safety emphasis.
   c. Desire growth position in a large department store.
   d. Am looking for entry position in health sciences with an emphasis on caring for older people.
   e. Position in sales with fast promotion rate.
   f. Want a job working with semiconductor circuits.
   g. I would like a position in fashion, especially one working with modern fashion.
   h. Desire a good-paying job, hours: 8–4:30, with double pay for overtime. Would like to stay in the Omaha area.
   i. Insurance work.
   j. Working with computers.
   k. Personal secretary.
   l. Job with preschoolers.
   m. Full-time position with hospitality chain.

n. I want a career in nursing.

o. Police work, particularly in suburb of large city.

p. A job that lets me be me.

q. Desire fun job selling cosmetics.

r. Any position for a qualified dietitian.

s. Although I have not made up my mind about which area of forestry I shall go into, I am looking for a job that offers me training and rewards based upon my potential.

4. As part of a team or on your own, revise the following poor résumé to make it more precise and persuasive. Include additional details where necessary and exclude any details that would hurt the job seeker's chances. Also correct any inconsistencies.

```
RÉSUMÉ OF
Powell T. Harrison
8604 So. Kirkpatrick St.
Ardville, Ohio
345 37 8760
614 234 4587
harrison@gem.com
```

<u>PERSONAL</u>	Confidential
CAREER <u>OBJECTIVE</u>	Seek good paying position with progressive Sunbelt company.
<u>EDUCATION</u> 2006–2008	Will receive degree from Central Tech. Institute in Arch. St. Earned high average last semester. Took necessary courses for major; interested in systems, plans, and design development.
2003–2007	Attended Ardville High School, Ardville, OH; took all courses required. Served on several student committees.
<u>EXPERIENCE</u>	None, except for numerous part-time jobs and student apprenticeship in the Ardville area. As part of student app. worked with local firm for two months.
<u>HOBBIES</u>	Surfing the Net, playing Gameboys. Member of Junior Achievement.
<u>REFERENCES</u>	Please write for names and addresses.

5. Determine what is wrong with the following sentences in a letter of application. Rewrite them to eliminate any mistakes, to focus on the "you attitude," or to make them more precise.
   a. Even though I have very little actual job experience, I can make up for it in enthusiasm.
   b. My qualifications will prove that I am the best person for your job.
   c. I would enjoy working with your other employees.
   d. This e-mail résumé is my application for any job you now have open or expect to fill in the near future.
   e. Next month, my family and I will be moving to Detroit, and I must get a job in the area. Will you have anything open?
   f. If you are interested in me, then I hope that we make some type of arrangements to interview each other soon.
   g. I have not included a résumé since all pertinent information about me is in this letter.
   h. My GPA is only 2.5, but I did make two B's in my last term.
   i. I hope to take state boards soon.
   j. Your company, or so I have heard through the grapevine, has excellent fringe benefits. That is what I care about most, so I am applying for any position that you may advertise.
   k. I am writing to ask you to kindly consider whether I would be a qualified person for the position you announced in the newspaper.
   l. I have made plans to further my education.
   m. My résumé speaks for itself.
   n. I could not possibly accept a position that required weekend work, and night work is out, too.
   o. In my own estimation, I am a go-getter—an eager beaver, so to speak.
   p. My last employer was dead wrong when he let me go. I think he regrets it now.
   q. When you want to arrange an interview time, give me a call. I am home every afternoon after four.

6. Explain why the following letter of application is ineffective. Rewrite it to make it more precise and appropriate.

```
Apartment 32
Jeggler Drive
Talcott, Arizona

Monday

Grandt Corporation
Production Supervisor
Capital City, Arizona
```

Dear Sir:

I am writing to ask you if your company will consider me for the position you announced in the newspaper yesterday. I believe that with my education (I have an associate degree) and experience (I have worked four years as a freight supervisor), I could f ill your job.

My schoolwork was done at two junior colleges, and I took more than enough courses in business management and modern technology. In fact, here is a list of some of my courses: Supervision, Materials Management, Work Experience in Management, Business Machines, Safety Tactics, Introduction to Packaging, Art Design, Modern Business Principles, and Small Business Management. In addition, I have worked as a loading dock supervisor for the last two years, and before that I worked in the military in the Quartermaster Corps.

Please let me know if you are interested in me. I would like to have an interview with you at the earliest possible date, since there are some other f irms also interested in me, too.

Eagerly yours,

George D. Milhous

---

7. From the Sunday edition of your local newspaper or from one of the other sources discussed on pages 243–246, find notices for two or three jobs you believe you are qualified to fill and then write a letter of application for one of them.

8. Write a chronologically organized résumé to accompany the letter you wrote for Exercise 7.

9. Write a functional résumé for your application letter in Exercise 7.

10. Bring the two résumés you prepared for Exercises 8 and 9 to class to be critiqued by a collaborative writing team. After your résumés are reviewed, revise them. Write an e-mail to your instructor about the revisions you made and why they will help you in your job search.

11. Prepare an online version of the résumé in either Exercise 8 or 9. Use persuasive keywords such as those in Table 7.2.

12. Prepare a hard copy portfolio. Collect appropriate documents and arrange them into the most relevant order. Provide an annotation for each.

13. Write an appropriate job application letter to accompany Anna Cassetti's résumés in Figures 7.6 and 7.8.

14. Write a letter to a local business inquiring about summer employment. Indicate that you can work only for one summer and that you will be returning to school by September 1. Include an appropriate résumé.

PART III

Gathering and Summarizing Information

8 Doing Research and
   Documentation on the Job

9 Summarizing Information
   at Work

CHAPTER 8

Doing Research and Documentation on the Job

Being able to do research is crucial for success on the job, whatever company or department you work for and whatever your job title. Companies use research to make major decisions that affect production, sales, service, hiring, promotions, and locations, as the research report at the end of this chapter illustrates (pages 374–387). In the world of work, research helps companies:

Visit www
.cengage.com/
english/kolin for
this chapter's
online exercises,
ACE quizzes,
and Web links.

- stay within budget
- remain up-to-date with market trends and new technologies
- meet the needs of their customers worldwide
- avoid problems with equipment, procedures, and the environment
- explore and open new markets
- increase profits
- report to stockholders, community groups, government agencies

Research follows a process. You have to gather, summarize, and organize information before you can interpret it. Then, in interpreting it, you must be able to answer questions and solve problems. To do effective research, you need to learn the following skills:

- network with people in your department, within your company, outside of your company, and potentially across the globe to gather relevant data
- read a host of print and online sources at your company or a library to find the most relevant studies/opinions on your topic
- do direct observations, perform tests, and make site visits
- interview one person or a carefully selected group of people
- prepare and send out surveys and analyze the results
- organize information into clear and accurate reports that answer the questions and solve the problems
- carefully and completely document your sources to give proper credit, to help readers find these sources

Some Research Scenarios

Research is the life blood of a company. Depending on your organization, department, or job, you can expect to spend as much as 25 to 30 percent of your time at work doing research for your company. Here are a few scenarios in which you may be asked to do research for your company:

- Your division is launching or upgrading a new product or service and the manager wants you to find out exactly what customers expect and want.
- You have been assigned to find out how a product or service your company provides can successfully compete against another company's product or service.
- Your company is lagging behind technologically and you have to write a report recommending the most appropriate and cost-efficient way to upgrade equipment or procedures, taking into account purchasing and installing it, and training employees on how to use the new technology.
- Your department is in danger of exceeding its annual budget and your boss instructs you to write a report showing where and how the department could save time and money.
- A new market trend that affects your company directly has recently received a lot of media attention, and the head of your division wants you to find out if this trend is likely to be long lasting or just a fad.
- Your co-workers have elected you to be the team leader of a five-person research group to uncover ways to attract new customers.
- Your company or organization opens a branch in a country where it has not done business before, and you must investigate this potential market for your firm, focusing on the culture and buying preferences of its people.
- The Bureau of Labor Statistics projects that your city will experience an acute shortage of skilled workers in the next five to ten years. Your supervisor asks you to interpret those statistics in light of your company's future and make recommendations on the types of positions your company will need to fill.

Although you may already have written reports for school assignments, this chapter focuses specifically on the types of research you will be expected to do on the job. Basically, it guides you through the following seven fundamental areas of research:

1. Understanding the differences between research done for school and research done in the workplace
2. Going through the various and interlinked stages involved in the research process
3. Distinguishing between primary and secondary research and how they differ yet relate to each other
4. Knowing how to do primary research, including making observations, doing tests, going on site visits, conducting interviews and focus groups, and taking surveys
5. Being able to do secondary research, including using libraries and the Internet
6. Learning how to take effective notes as you research
7. Documenting your sources and knowing how to cite them correctly both in MLA and APA style

At the end of this chapter (see pages 374–387) you will find an actual business report that illustrates the strategies and tools of primary and secondary research you will need to develop and use to become a successful researcher on the job.

▌The Differences Between School and Workplace Research

The research tools and methods you use in school for a course paper or report are far different from what you can expect to find in the workplace. Understanding the differences in goals, audience, resources, and format/presentation, as well as how credit for your work may be assigned, will help you do your workplace research more effectively.

1. Goals. In school, you do research for educational reasons: to learn about a valuable topic and to prepare for a career. In the world of work, research is aimed at basically one goal: to provide bottom-line information for your employer. Workplace research is pragmatic, almost always addressing financial, operational, and/or personnel issues. It affects the way your company does business, how successful it is, and what the future of its employees, including yourself, will be.

2. Audience. In school, your instructor and your classmates constitute the audience for your research paper. Your instructor will ask you to write your paper following his or her specific guidelines, may help you locate sources, and will grade you on how well you present your findings to him/her and/or your class. The audience for your work-related research, on the other hand, can include numerous and diverse individuals and groups, ranging from stockholders and upper management to your immediate supervisor, co-workers, and customers worldwide. Your boss may give you a question to answer or a problem to solve and may even suggest some sources to consult, but usually you cannot expect to receive the guidance you would when writing a paper for a class assignment.

3. Resources. Most of the resources you use for a school paper or report are print materials (books, journals, encyclopedias, etc.) located at the library, supplemented with Web resources. On the job, however, your research will be far more diverse and less print-centered. In business, research is often based on networking with people, interviewing individuals (customers, experts in the field, vendors, etc.), visiting sites, examining and "shopping" competitors, consulting government and market reports (for population and business statistics), and reading in-house and trade literature. Of course, you will also use libraries and online resources for statistical data, as well as professional journals to keep up on the job, but the central tools of business research involve much more direct contact with individuals.

4. Format/presentation. Your school research report needs to follow the guidelines and format your instructor specifies. Your boss may also ask you to follow specific formats and methods of documentation, but they may be different from those used in school. In the workplace, you will be expected, alone or with your team, to discover the best and most persuasive ways to present and adapt your research. Depending on your audience's needs and the scope of your assignment, you may present your research as a formal report, in an informal e-mail or memo, as a thorough presentation using software such as PowerPoint, or as an informal oral presentation at a meeting or a one-on-one conference.

5. Credit for your work. Your name (as well as the names of any collaborative team members) would certainly be listed on any research report or paper you write in school. But in the world of work, you may not always see your name on the research reports you do. Instead, your manager's name or just the name of your department, division, or company may appear on the document.

Characteristics of Effective Workplace Research

The research you do on the job needs to follow the highest professional and ethical standards. Businesses leave little margin for error and often do not give employees a second chance to get it right. To make sure your research meets your employer's expectations, it must be:

1. Relevant. Job-related research must focus directly on providing specific answers and solutions to the key questions and problems affecting your company and job. Because research in the business world is so costly, your employer will not allow you to waste time and money on side issues, unproductive leads, or personal interests.

2. Current. Your information must be up-to-date. Markets and technologies change rapidly, and employers will insist that your research is on the cutting edge of your profession. By keeping up with the most recent trends and ideas, you can give your employer a bigger picture of the market and help your firm to stay competitive.

3. Accurate. Double- and triple-check all of the facts and figures, dates, addresses, names, regulations, URLs, and so on used in your research. Don't substitute guesswork and unsupported estimates for hard facts. Make sure you record all information accurately. Your boss will expect your reports to be based on quantifiable, measured data that are essential to precise research.

4. Original. Employers want well-thought-out and supported conclusions. Don't just repeat what you have read in a report or in an article, or heard at a conference. Evaluate all data. Question and test what you have read, heard, or observed. That way, you can modify or expand it, reject it, or use it to project and predict. This is the way to arrive at original answers and solutions for your on-the-job research.

5. Thorough. Look at a question or problem from all sides. Network with colleagues to look for any gaps or inconsistencies, as well as business opportunities. Confirm all options and opinions. Never omit important data.

6. Realistic. Base your research on realistic, profitable conclusions. Unsubstantiated recommendations that fly in the face of a company's protocol (e.g., drop a product line, hire/fire 150 people, or move a plant location) may not be logical, profitable, or acceptable. Be sure that your research is consistent with your company's policies.

7. Balanced. Don't just rely on one source or interview and think that you have done sufficient research. You have to discover and acknowledge opposing viewpoints. Keep in mind that even experts can disagree. If you are surveying customer opinions, choose a representative sample of this group and don't leave out those customers who may be critical of the product or service. Negative reactions sometimes lead to productive, profitable changes.

8. Ethical/Legal. Obtain your findings ethically and lawfully, so that you do not infringe on the rights of others. Raiding someone's unpublished research, sharing confidential or privileged information with a third party, or skewing the results of a survey are all unethical acts. So, too, are discounting, or minimizing facts that don't agree with the findings you want or expect. Plagiarism (see pp. 353–354) and omitting evidence are serious violations in the world of work, just as they are in the classroom. Be sure, too, that all of your recommendations are environmentally sound. Many businesses today have adopted a strong green philosophy.

The Research Process

Just as there is a process for writing—brainstorming, drafting, revising, editing— there is a process involved in doing research. In research you go through the steps of finding, assessing, and incorporating information into your written work.

As in the writing process, in doing research you may find yourself repeating certain steps. Say, for example, you are writing a business proposal and are incorporating information from several sources you've researched. At this stage of the process, you might think you've gathered enough information. However, as you work on the proposal, you may realize that it raises new questions. That may lead you back to repeating previous steps. Let's look at the process in more detail:

Step 1. Confirm the purpose and audience of your report. Know who your audience is and why you are writing to them. Will your readers be management, co-workers, clients, government agencies, or a combination of these groups? What are their expectations about length, sources of your information, level of technical detail, conclusions/recommendations, and so on? Again, what is the bottom line— what do you have to answer, argue, propose, solve, uncover, change, describe, or compare—and why?

Step 2. Use a variety of resources. Not only should you include different sources, but they should also be in different media formats: print, online, and possibly audio and video. If you rely on only a single source—an Internet search or one trade journal—your research findings will be based on an incomplete understanding of your topic. Your company could miss out on a sale, or a costly error could result.

Library resources typically include encyclopedias, almanacs, and other reference works; periodicals; bibliographies; and collections of books and other media. But also consult industry publications; databases; information services (e.g., LexisNexis); and in-house publications, newsletters, reports, customers' files, vendor contracts, and institutional archives (see pp. 325–339). You may also have to do fieldwork such as observing an activity firsthand or interviewing a variety of people.

Step 3. Learn to identify and retrieve information. Once you've identified valuable sources of information, you may need to master some techniques for using those sources. Information can be stored in a variety of formats and systems in the world of work. A researcher's life would be much easier if information were stored and retrieved the same way for every source. But, just as the navigation of two websites differs, information from different business areas is organized according to different standards. Other reference materials—almanacs, encyclopedias, government documents, manuals—may be found both in print and online.

Step 4. Know how to evaluate sources, both in print and online. As we've seen, doing research in the world of work means more than just repeating information you find. Given the explosion of information of all kinds found online and the continuing expansion of print materials available, it is crucial that you are able to evaluate the content of what you read. Be prepared to evaluate tests, surveys, interviews, websites, and printed sources critically to see if they have a particular agenda or bias that might slant their opinion on certain topics. Whether the source of information is one person's blog, a local newspaper, or an online journal, here are some questions your boss will expect you to ask:

- Is the information accurate and reliable?
- Who authored this work and what is this person's experience in the field?
- Is the information current? (If the information appears on a website, when was the site last updated?)
- Is the information complete? What is missing? If the source uses statistics, for example, is the amount of data sufficient to draw the conclusions the author reached?
- Is the information valid? If the source uses statistics, how were those statistics gathered and analyzed?

Step 5. Consult appropriate resource people and experts at work, in your profession, and in your community. Doing research in the world of work involves asking interested knowledgeable individuals for information and their perspectives on the topic you are studying. These can be people from different divisions of your company (technical support, human resources, finance), or co-workers and members of your collaborative team who have previously investigated a similar topic, or customers and clients. You might also consult various specialists who work for the local, state, or federal government. Your research could be as simple as sending an e-mail to a customer to ask a brief question or as complex as analyzing the plans to launch a new product line.

Step 6. Continue to ask questions. Research on the job involves more than gathering information. It also means asking the right questions throughout the process of writing your report. As you read, conduct an interview, make a site visit, send e-mails, or search databases, you may encounter dead ends, contradictions, and even new sources or leads you need to investigate. Research does not always go as smoothly as you might expect it to. Don't get discouraged. See such hurdles as temporary and as opportunities to make sure your work is accurate, complete, and relevant. Investigating your original problem may reveal several others you weren't aware of, or you may discover that there is more than one way to solve a problem. Compare and contrast what you find. Be careful to describe and define the information you gather in terms of your company's purpose. In fact, ask these key questions at various times during your research to keep on track:

- Which of my sources will be of the most value to my audience and my purpose?
- What is the best way to incorporate my research within my report?
- Does the information I have found raise more questions? If so, what is the best way to research them? How do they relate to my company's needs?
- Do I need to find more information? If so, where and why?
- Do I agree with the conclusions presented in my sources? Are there major differences of opinion among my sources? How do I determine which, if any, is more appropriate, effective, and economical?

Step 7. Document your sources. One of the most important steps in the research process is documenting—citing the various sources of information (online, in print, from personal interviews, etc.) on which your report or presentation is based. A later section of this chapter will give you specific guidelines on how to document your sources. Remember, to claim another's works as your own is plagiarism. It will undermine you on the job as it does in school. Avoid the mistake of "borrowing" information from an authoritative source or omitting the names of co-workers who assisted you with your report. Readers inevitably discover the truth, and then the author is considered unreliable if not downright dishonest. When you are asked to write on an unfamiliar topic, research it and cite as many authoritative sources as the scope and purpose of your project require. By referring to these experts, your report will carry greater weight.

Step 8. Submit specific recommendations based on your research. Depending on the directions your employer gave you, you may be required to give recommendations. These can be informed projections, predictions, alternative measures, specific solutions, or a plan of action. See, for instance, the business report on pages 375–387.

Step 9. Adhere to a schedule. Research for a business report cannot go on indefinitely. Your employer will demand that you meet deadlines. Businesses are run by deadlines, e.g., weekly summaries, monthly surveys, quarterly reports. If you delay, you will jeopardize not only other ongoing business projects that may depend on your research, but also your company's chance for acquiring new business. Budget your time wisely. Plan ahead, confer with supervisors on a regular basis, and network daily with colleagues through e-mails, IMs, blogs, and, when possible, in person.

Two Types of Research: Primary and Secondary

You can expect to use many sources of information during the research process. But essentially your research will fall into two categories: *primary* and *secondary*. Both kinds of research are important to help you obtain a better understanding of your topic and provide your supervisor or customers with the careful and complete answers and recommendations they expect. You will often do both types of research, as the marketing report at the end of this chapter (pp. 375–387) illustrates. In fact, one type of research sheds light on the other.

Primary research means consulting sources of information not found in printed documents or on the Web. It involves interacting directly with people, places, and things, and it is often done in the office, in the field, or in a laboratory. This type of research often requires gathering information from customers, clients, or other individuals who rely on your company's products or services. For instance, an Internet provider may want to know what customers think about a plan to redesign a company website, a business problem you cannot solve by consulting existing print and online documents. To find out whether or how to redesign its website, the company has to conduct primary research through a customer survey or focus groups (group interviews). To cite another example, a city planning board may want to find out what the residents think about constructing a new park in their neighborhood. To determine their views, the board needs to conduct primary research by interviewing residents and/or making a site visit to the neighborhood to observe the best place to build the park.

Secondary research involves consulting existing print and online sources. When you conduct secondary research, you work with materials that someone else—an expert in your field, a government agency, even a competitor—has published, posted, or distributed. For example, your boss might ask you to write a report on the latest techniques in occupational therapy given to employees injured at a job site so you can develop a similar one for your company. This assignment requires you to locate, read, and summarize the literature in this field. Or an advertising department might need to learn more about how and why a competitor's product is winning a larger share of the market. To find out, they need to consult secondary sources, such as the competitor's website and product literature as well as articles in newspapers and trade journals written about the product. They also need to gather statistical information on customer buying habits (often provided by professional associations or the government), look at relevant websites in which customers provide feedback about the competitor's products, and consult their company's own archived consumer and competitive studies.

Here are some examples of the kinds of primary and secondary research that you can expect to do on the job:

Primary	Secondary
■ making direct observations	■ searching websites
■ performing tests	■ reading books, journals, and magazines
■ going on site visits	■ consulting manuals and reference works
■ conducting interviews	■ examining product reviews
■ coordinating focus groups	■ using government documents
■ sending and analyzing surveys	

Case Study: Primary and Secondary Research in the Real World

The following sections of Chapter 8 describe the various primary and secondary research materials and how you use them in your workplace research. But before you look at these sections, read through Figure 8.1, a real-world explanation of how and why research is done in the workplace, written by Shay Melka, a Vice President for Research and Marketing at B&L Stores, a large retail chain with locations in North America and Asia. The overview demonstrates how essential workplace research is to the day-to-day operations as well as the long-range goals of a company. Melka describes many types of both primary and secondary research and often a wide range of examples and suggestions that apply to almost any business. As you read the discussion of the tools and strategies of primary and secondary research later in this chapter (and the report at the end of this chapter), refer to Figure 8.1 to see how Melka's company has successfully employed them to meet customer needs, increase and improve business, and troubleshoot problems.

Advice from a pro: Why and how research is used in the workplace.   Figure 8.1

The Ways Research Is Conducted at B&L Stores
Shay Melka

I have worked for twelve years with B&L Stores, first as a salesperson, then as an assistant manager, and now in our home office as a vice president in our Research and Marketing Division. B&L has over 200 stores in the United States, Canada, the Caribbean, and Asia. In the highly competitive world of retail sales, our company must keep its hand on the pulse of the consumer. It is essential for us to track the buying habits of our customer base so that we are prepared to act quickly when buying trends change course or when demand is no longer ahead of supply. We need to understand who our customers are by finding out about the following:

- What they purchase
- How they pay for their purchases (cash, credit card, check, debit card, etc.)
- Why they shop with us
- Why they purchase from the competition
- How much they purchase
- Whether they purchase their products in the store or online
- What the in-store experience is like for our customers
- How many times during the year our customers visit our locations
- How advertising campaigns affect current promotions

Continued

Figure 8.1   (Continued)

2

To identify the buying decisions of our current as well as future clientele, we use some basic strategic research tools. Here is a rundown of these tools, or what we call our "outreach programs in customer logistics," with explanations of why, when, and how they work.

Power Think Sessions

This is the name B&L gives to in-house networking, in person and electronically, to do market research, gather vital statistics, and identify key performance indicators. Our weekly Power Think sessions help us obtain essential, current information from sales staff and store managers to highlight the accomplishments of the week, ascertain problems, and entertain proposals for changes (at one such Power Think session a B&L part-time employee gave us the idea for our distinctive gift wrap). We also have larger Power Think sessions several times each year, when B&L brings to the home office the company's top performers to brainstorm. These sessions are usually composed of a cross section of sales staff and managers within our chain, representing urban and rural stores, stores within higher and lower per capita income areas, and stores in large mall and strip mall locations. Each Power Think team submits marketing ideas and business plans. We have also had Power Think sessions with consultants from marketing firms outside the company with experience in information gathering.

Monthly Business Reviews

Sales results are, of course, the driving force for every operational aspect of the B&L Stores' P&L (that is, our profits and losses). Each store manager sends the stats (sales, returns, losses, inventory, etc.) to the home office every day and discusses the store's performance at weekly meetings. While we review numbers on a daily and weekly basis at the home office, B&L, like most corporations, also performs Monthly Business Reviews (MBRs) to answer the following key questions:

- Are sales up, down, at the same level for the last month, quarter, year?
- What is selling and what is not?
- What is our gross margin of profit (what we paid for the item vs. what we sell it for)?
- What are our additional revenues (from warranties, installation, deliveries)?

3

- Are inventory dollars being wasted on product A when product B is the hot seller?
- What merchandise is selling better in various markets?
- Do we need to negotiate new lease terms with any landlords?
- Should older stores be remodeled or closed?
- Is one store strong in cash sales and another in credit sales?
- Does the traffic in one store demand more payroll dollars?
- Do store hours have to be revised, e.g., stay open later, close earlier?
- Are there loss prevention issues at a store (e.g., cash, merchandise)?

The numbers we receive and then analyze in our monthly business reviews help us to answer the questions above. We also gather statistics on individual performance at all levels—from our part-time sales staff to our district managers and vice presidents.

Customer Surveys

We never make a business decision based just on in-house networking and weekly, monthly, and quarterly sales data. In-house networking and statistics are, of course, key sources of information and help us plan. But before we can put any plan into action we must know what our larger customer base thinks of our stores, our Web presence, our merchandise, our employees, our service, and our image. To ensure a positive shopping experience in-store or online, we rely on feedback from both our current and prospective customers. Our goal is to provide a level of service that will make a lasting and positive impression.

Accordingly, we have developed and tested a variety of survey questionnaires, available at our in-store locations and online. For example, several years ago we created a generic survey used across our chain to find out customer opinions about the physical appearance of our stores (are they clean, attractive, and easy to navigate? is there enough parking?) and the professionalism of our sales associates (are they courteous, informative, attentive, and willing to go the extra mile for a customer?).

In addition, because so many of our customers have been concerned about identify theft and other transaction issues, we modified our survey to include five basic questions on this key issue to better assure customers about security. Based on our marketing research and survey, we prepared a policy statement displayed in every store and on our website that lists the steps B&L takes to prevent identify

Continued

Figure 8.1   (Continued)

4

thet. Here is a brief rundown of what B&L promised to do for each customer based upon the feedback we received from our survey:

- Signatures on a credit card are always matched with those on valid IDs.
- Checks are always processed through the Safeguard Approval System.
- Large purchases are always reviewed by district managers.
- Large bills are always verified using a counterfeit pen.
- If a customer uses a private label credit card (Wal-Mart, K-mart), the credit department is always contacted to verify address, phone number, and payment account.

We have also constructed other distinctive surveys for different markets and regions (e.g, those dealing with new store openings, mergers, overseas stores, new Web links, etc.). Customer surveys provide crucial information about how we can provide better customer service. In fact, we have a Customer Support Department (CSD) where any problems or questions raised by our surveys are reported and corrected/answered.

Focus Groups

Focus groups also provide us with informed opinions. We select participants for our focus groups very carefully at B&L. They are made up of our most loyal current customers as well as prospective customers who have filled out a form indicating that they would be willing to participate. We have also purchased the names of individuals who have agreed to be in focus groups for other organizations and may therefore be willing to participate in our groups. We also use paid consultants. Recommendations gathered from these various focus groups are digested, analyzed, and implemented as needed.

Many organizations convene focus groups annually, but at B&L we like to conduct focus groups four or five times a year to help us more confidently forecast trends and to test new products and merchandise in a given market. Our investors (shareholders) and Board of Directors love focus groups. For them, focus groups are the equivalent of getting the opinions of the ideal consumer we want to reach. Ideally, focus groups help B&L gain a better feel for the market and for our customer base. They provide us with a reality check on existing policies and merchandise as well as new directions our company will take.

Focus group meetings are recorded, whether the group is held in person or via teleconferencing. These meetings are presided over

(Continued) Figure 8.1

5

by a moderator who tries to keep the participants from falling into "groupthink," that is, being influenced by a majority of the group rather than voicing an individual opinion that might, in the long run, be very valuable for B&L's future business. Focus groups do B&L a disservice if all they do is give us "yes, great" answers. We want to hear negative criticism in order to remedy a bad situation or solve any problems. The information from a focus group is used to improve customer relations, advertising, online and in-store shopping experiences, merchandise/ordering/stocking, and pricing issues.

Research from Trade Publications, Message Boards, Blogs, Conferences, Seminars, and Workshops

B&L employees and managers also must read the trade journals and professional magazines that are distributed throughout the B&L chain. These publications release sales and real estate news, merchandise trends, new regulations, vendor information, and the like and often include message boards and blogs that help our employees stay current. For instance, individuals in our Retail Architecture department consult online and print trade publications to stay up-to-date on the most current and workable trends in displaying merchandise, whether that involves the arrangement of items, mood lighting, or traffic patterning.

In addition, B&L routinely sends staff and managers in Human Resources, Loss Prevention, Multicultural Diversity, and other departments to conferences, seminars, and workshops. Various departments throughout the year also attend conventions to confer with experts in the field and then make formal reports to the home office, district managers, and relevant store sites. Employees in Human Resources, for instance, regularly attend conferences to find out about the latest federal rulings and guidelines dealing with such matters as overtime, the Americans with Disabilities Act, discrimination, family medical leave, and military leaves.

Research is intensive, ongoing, and the lifeblood of B&L. Gathering, analyzing, and applying information are in a very real sense our most important business.

Primary Research

The key to primary research is planning. You can't just set off to do primary research hoping interesting and relevant information will just pop up. Instead, you will have to follow procedures just as scientists do, formulating hypotheses and deciding how best to research them, systematically conducting the research, and then compiling and interpreting the information you have gathered.

There are several methods of doing primary research, including the following:

- direct observations, site visits, and tests
- interviews and focus groups
- surveys

Direct Observation, Site Visits, and Tests

Direct observation, site visits, and tests are widely used methods of primary research that involve actively observing people, places, and things. Your company or organization will expect your findings/conclusions to be based on keen powers of observation.

Direct observation is seeing what is right in front of you—for instance, watching how an individual performs a task, determining how a piece of equipment works, or studying how a procedure is performed. The key to conducting effective research is observing actively, not passively. For example, if you work in the sales department and you have been asked to write a report on how your company can improve its relationships with its customers via telephone sales calls, you need to

- observe how reps interact with customers
- listen to the reps' tones of voice
- record how clearly they explain company procedures

Site visits require you to use the same keen attention to detail that you use in direct observations, except you will need to go to an off-site location to report what you find there. A site visit could take you to another department in your company, a prospective customer's office, the scene of an incident or accident, or an agricultural location relevant to your business report. See Figure 14.12 for an example of an incident report based on visiting the site where a railroad accident occurred. Regardless of the location, you will have to describe for your boss precisely what you witnessed firsthand. For instance, if your company—say, a chain of dry cleaners—wanted to open a new store and asked you to investigate the most profitable locations, you would need to conduct a great deal of primary research based on site visits. To select the best location, you would have to first identify relevant places for a new store and then find out such things as

- volume of traffic
- availability of parking
- location/size of competing dry cleaners in area
- zoning ordinances/restrictions

Figure 14.8 contains an example of a trip report about opening a new restaurant based on information obtained from a site visit.

Conducting **tests** is another productive way to do primary research involving the observation of people, places, and things. A test can be as simple as examining two pieces of comparable office equipment side-by-side and noting how they compare, or trying out a new e-mail marketing strategy. Or it can be as scientifically demanding as conducting a laboratory test. For example, let's say you work for a company that manufactures waterproof sunscreen lotion and you have been asked to find out why a competitor's product is outselling yours. To answer the question, you will have to arrange for laboratory tests of your competitor's product to determine what ingredients it contains, how effective and safe it is in comparison with your product, and how its packaging may affect customer decisions. Figures 14.10 and 14.11 are examples of reports based on laboratory tests conducted in the world of work.

Sometimes you may have to use all three types of observational research when preparing your report, as Kirk Smith did for his study of Cambridge, Massachusetts, water quality in Figure 8.2. Not only did he observe and record the data collection methods used at the three reservoirs, but he also visited these sites and conducted his own tests. Regardless of the type of observation you make for your research, follow these guidelines:

- **Always plan ahead.** Determine what problems you have to solve, what questions you have to answer, and what data you'll have to gather—procedural guidelines, maps to locations, sample products—to answer those questions and solve those problems.
- **Obtain necessary permissions.** Permission may involve verbal or written consents, permits, or visas. By obtaining approval to make observations, you respect the time, privacy, and working conditions of others.
- **Observe company protocol and ethical standards.** Follow your company's or agency's guidelines and standards for observations. In addition, follow the standards and codes of ethical conduct of your profession. Again, note how Kirk Smith references the U.S. Geological Survey's standards in Figure 8.2.
- **Remain objective and impartial.** Although observations can often be influenced by your subjective point of view, make sure you are open-minded as you record your observations.
- **Keep careful notes and records.** Take a note pad and pen, tape recorder, camera, laptop, smartphone, or whatever other recording device you may need to preserve your observations effectively.
- **Write down precise details.** Be sure to indicate the complete and accurate dates and times you made your observations, any relevant environmental conditions that had on impact on your observations, and any other factors that may have influenced what you observed. See how Kirk Smith recorded the precise methods of water flow/quality measurement in Figure 8.2.

Interviews and Focus Groups

Two other important sources of primary information come from interviews and focus groups. You can do a one-on-one interview with an expert in the field,

Figure 8.2   A report based on direct observation, site visits, and tests.

Water Flow and Quality Evaluation of the Cambridge,
Massachusetts, Drinking Water Source Area

Kirk P. Smith

The drinking water source area for Cambridge, Massachusetts, consists of three primary storage reservoirs (Hobbs Brook Reservoir, Stony Brook Reservoir, and Fresh Pond), two principal streams (Hobbs Brook and Stony Brook), and nine small tributaries. Sites were selected for continuous monitoring to address the water supply regulations followed by the Cambridge Water Department (CWD) and because previous investigations identified specific areas as potentially important sources of contaminants. The purpose of this report is to evaluate the measurement methods used by the Cambridge Water Department.

Reservoir altitude and meteorological measurement were recorded by monitoring stations installed at each reservoir. Water quality measurements of reservoir water were also recorded at USGS stations 01104880 and 42233020. These data were recorded at a frequency of 15 minutes, were uploaded to a U.S. Geological Survey (USGS) database on an hourly basis by phone modem, and are available on the Web at *http://ma.water.usgs.gov*. Stream stage measurements were recorded by monitoring stations on each principal stream and at the outlet of the Stony Brook Reservoir. These data were recorded at a frequency of 15 minutes and were uploaded to a USGS database on an hourly basis by phone modem.

In addition to measurements made on the principal streams, stream stage measurements and water quality measurements were recorded by monitoring stations on 4 of the 9 small tributaries. My visits to these sites and independent water samplings confirm that CWD's measurements comply with USGS standards.

Because the drainage areas of these sites are small and have large percentages of impervious surface, the hydrologic responses, and often the water quality responses, change rapidly. To document these responses effectively, the monitoring stations have recorded stream stage and water quality measurements at variable frequencies as high as 1 minute. These data were uploaded to a USGS database on an hourly basis and are available through *http://ma.water.usgs.gov*. I have found through visits and water sampling that CWD is not only compliant with, but exceeds, USGS standards in measuring drainage area water quality.

Adapted from *Hydrologic, Water-Quality, Bed-Sediment, Soil-Chemistry, and Statistical Summaries of Data for the Cambridge, Massachusetts, Drinking-Water Source Areas, Water Year 2004*, by Kirk P. Smith. U.S. Department of the Interior/U.S. Geological Survey. Open-File Report 2005–1383.

co-worker, client, or other resource person. Or you can hold a focus group, a question and answer session with multiple people—both company representatives and customers—in attendance. See how in Figure 8.1 Shay Melka emphasizes the importance of interviews with individual employees as well as with focus groups to gather essential information from and about a variety of customers.

Interviews

Interviews can be conducted in person, over the telephone, or through e-mail, although a face-to-face meeting is the most productive way to generate relevant information. Figure 8.3 contains an excerpt from an interview on working in a Chinese business culture with a U.S. manager whose company transferred her to the company's Hongzhou office for eighteen months. Note how the interviewer prepared for and structured his questions to help other employees transferred to China.

Follow the process below when you do an interview related to your workplace research:

1. **Set Up the Interview**

   - Determine whether you need to speak with an expert in the field, with a co-worker or manager, or with a client or customer to obtain the relevant information you need.
   - Ask your supervisor or co-workers to help you identify experts or relevant customers you should interview, or consult other sources, such as business directories, client or customer lists, professional organizations, or relevant websites.
   - Politely request an interview with the individual at his or her convenience and set a precise time and place. But be flexible. Your interviewee is giving you his or her time. Always let the individual know ahead of time exactly what you would like to discuss and why you are conducting the interview.
   - Whether for a personal visit or a telephone call, specify how much time you will need for the interview. Be realistic—fifteen minutes may be too short; two hours much too long.

2. **Prepare for the Interview**

   - Gather background information about the person you are interviewing as well as about the organization or professional group that he or she represents, if applicable.
   - Continue to research your topic so that you have sufficient background information and do not waste time asking the obvious or requesting information that is readily available on the Web or from another source.
   - Determine what information you need from the interview to help you solve the problem or answer the questions essential to your report. Be sure to prioritize getting the essential information you need. Be aware, too, that the interviewee may raise some relevant questions you didn't expect but need to follow up on.

Figure 8.3   An excerpt from an interview transcript.

Asks pertinent background question

Q: Did you have any experience working in another country before going to Hongzhou?
A: Not much really, except for a summer-long internship in Lima, Peru, I had as part of my undergraduate education.

Key question on interviewee's responsibilities

Q: What precisely were your duties when I-Systems transferred you to Hongzhou?
A: I was responsible for setting up a local operation for I-Systems and managing a staff of about eighty Chinese employees. As part of all that, I worked closely with local suppliers and developed different e-markets for our new office.

Logical question after asking about background preparation

Q: How did you prepare for your transfer to China?
A: Before I left for my eighteen-month stay, I profited most from participating in teleconferences with our other Chinese offices and attending China trade fairs in the United States and Canada. I also immersed myself in intensive, but admittedly very basic, conversational Chinese. And, of course, I partnered with several of I-Systems Chinese employees and managers here in Pittsburgh.

Turns to problems in new job

Q: What would you say was the biggest obstacle an American manager might face when working in China?
A: Seeing China through Western eyes.

Asks for clarification

Q: When you say "seeing China," what do you mean?
A: By that I mean looking at China from a Chinese business perspective. We tend to think in U.S. terms about expanding and opening markets, that is, what we can do for China. But my Chinese colleagues reminded me that while the United States accounted for only about 5 percent of the world's population, China had about 20 to 25 percent of it and could powerfully influence our company's decisions. Accordingly, we needed to shift our thinking about what China could do for us. To do this, we must have an appreciation of the Chinese way of doing business.

Relevant follow-up question

Q: What characterizes the Chinese way of doing business?
A: For example, the Chinese pharmaceutical industry is one of the most powerful in the world. I think there must be over thirty or forty companies that are developing, manufacturing, and marketing product lines both in China and here. Did you know that China makes nearly 10 percent of the world's aspirin? I found that interesting in

Continued

light of the western perception of Chinese health care—alternative procedures such as acupuncture and homeopathic remedies.

Q: That is fascinating and a topic for a separate interview sometime. But getting back to characterizing Chinese business customs—
A: Oh, yes. Let me illustrate a couple of major differences I had to get used to. Americans have no problems mixing business and pleasure. In fact, we are famous for the business lunch or dinner. Banquets are great occasions to talk shop, to sell our products, services, and websites. But in China a dinner is strictly a social event, one for entertaining and not marketing. It is considered rude in China to inject talk about sales, quotas, operations, or e-markets at a dinner. If you want to give a banquet as a sign of respect for a Chinese executive, leave your specs, stats, briefcases, notebooks, and cell phones at home.

Directs interviewee back to subject of interview

Q: Do you have any other advice for American workers whose companies relocate them to China?
A: Be careful about gestures and gifts.

Asks for further Information

Q: Why do you link the two?
A: To illustrate a major blunder, one of my colleagues kept patting a Chinese executive on the back, a sign in America of friendship and approval. Not so in China. It is seen as discourteous.

Good follow-up question

Q: And the gifts?
A: While some business gifts are appropriate, never give a Chinese executive a clock or stopwatch. It signals doom or death.

3. Draft Your Questions

■ Prepare your questions ahead of time and take them to the interview. Never try to wing it. Your questions should be:
   a. focused on the topic you want to find out about to avoid vague answers
   b. open-ended and designed to prompt thoughtful responses, not just yes or no answers
   c. objectively worded so that the interviewee is not forced to respond to loaded questions

Here are some examples of poorly written questions with effective revisions:

Ineffective Questions	Effective Revisions
Vague Question: How can today's Internet help customers?	*Restricted Question:* In what ways can we improve the navigational signals on our website to help our customers?
Yes or No Question: Do you think big business is opposed to a healthy environment?	*Open-Ended Question:* Would you identify two or three ways we could green our office space?
Loaded Question: Isn't the future of real estate security investments doomed to a bleak future?	*Objectively Worded Question:* What are your thoughts about the future of real estate security investments?

4. **Conduct the Interview**

   - Show up to the interview on time and be dressed appropriately.
   - Always ask if you have permission to record the interview or to take photographs.
   - Start off with a few minutes of small talk to give the interviewee time to get comfortable and to talk about his or her interests and accomplishments.
   - Never begin with your toughest or most controversial question. Save it for last.
   - Stay focused. Don't stray from the topic or delve into personal matters. Should the individual get off the topic, politely return to your questions.
   - Be an attentive and appreciative listener. Let the interviewee do most of the talking.
   - If you discuss anything confidential, politely remind the interviewee about the sensitivity of the topic.
   - If the interviewee does not want to answer a question or has no further information to add, don't press the point. Move to the next question.
   - If the interviewee says that something is "off the record," respect his or her request and do not include it in your transcript or notes, or on tape.
   - At the end of the interview, allow time for your interviewee to clarify any of his or her responses.
   - Thank your interviewee and ask if you might call back with any follow-up questions or further clarifications.

5. **Follow Up After the Interview**

   - Don't wait too long after the interview to review your notes or start your transcription. It's best to read through your notes immediately after the interview, while the conversation is still fresh in your mind. That way you will be better able to understand your notes when you return to them later.
   - Thank the interviewee by letter or e-mail within a day or two following the interview.
   - If the interviewee requested a transcript of the interview, send it to him or her for approval.
   - Determine how much of the interviewee's comments, ideas, statistics, or recommendations you can incorporate into your ongoing research. You cannot

put the entire interview into your report. Your reader(s) want only the bottom-line essentials.

■ Always request permission to quote anything from the interview in your report or presentation to your company or clients.

Focus Groups

Focus groups are typically made up of loyal or prospective customers who have been invited to give a company their reactions and opinions about a specific product, service, or future project. These groups are used to obtain a wider variety of opinions than individual interviews may give and they are more personal and interactive than surveys. Many companies include a moderator to help direct the group's commentary. Businesses rely heavily on focus groups to get honest, well-considered feedback from interested individuals and to incorporate that feedback into their research. Focus groups are usually face-to-face meetings with customers who are within traveling distance of the focus group location, but virtual meeting technologies can allow for online focus groups that bring in the views of people outside of the area, even globally. As you can see in Figure 8.1, B&L Stores uses a variety of focus groups, including those that involve consultants and individuals selected from competitors' lists.

Follow the guidelines below to conduct a successful focus group:

1. **Set Up the Focus Group**

   ■ Determine the location, time, length, topic, and agenda of the focus group prior to contacting potential participants.
   ■ Identify who should be invited to the focus group and how many individuals should make up that group. Effective focus groups usually consist of six to twelve participants in order to get a diversity of opinions but to keep the group from being too crowded and unmanageable. Consult with your supervisor, the company's sales and marketing departments, and customer lists to locate the most helpful participants or consultants.
   ■ Screen potential participants for appropriateness, e.g., knowledge of the product/service, willingness to provide input in a group, and so on.
   ■ Once you decide on the participants, provide them with all of the details they will need to know about location, directions, payment/reimbursement information, and topics to be discussed.

2. **Prepare for the Focus Group**

   ■ Determine the specific questions you will need to ask the focus group. As in a one-on-one interview, prepare these questions ahead of time, avoiding vague, yes or no, or loaded questions. Limit the number of questions to allow for ample discussion time.
   ■ Plan to tape-record or videotape the focus group, and to bring in a co-leader or moderator to take notes. Unlike a one-on-one interview, you will not be able to take effective notes while leading a focus group.
   ■ Create an information form for the participants to fill out at the beginning of the meeting, asking for contact information and which of your company's products/services they use or are familiar with.

3. Conduct the Focus Group

- At the beginning of the meeting, establish reasonable ground rules, such as the importance of staying on topic, speaking in turn, and meeting the goals of the group (see "Sources of Conflict in Collaborative Groups" on pages 85–87).
- Politely remind the group about confidentiality. Many companies have participants fill out confidentiality agreements before the group meets.
- Stick closely to the agenda. Don't stray off topic yourself or allow participants to do so, either.
- Allow sufficient time for each question and ask participants for their recommendations. Then quickly summarize those conclusions for the group to see if participants agree with your assessment.

4. Follow Up After the Focus Group Meets

- Read through your notes soon after the meeting, while the group experience is fresh in your mind. Also, write down any observations about the group dynamic as a whole and about the individual participants, since this information may affect your results.
- If the meeting was recorded on audiotape or videotape, transcribe the recordings.
- Thank the participants again by letter or e-mail within a day or two after the group meets.

Surveys

Visit www
.cengage.com/
english/kolin for
an online exercise,
"Practicing Your
Survey Writing
Skills."

Surveys are among the most frequently used instruments to gather primary research in the world of work. As Shay Melka points out in Figure 8.1., B&L Stores relies heavily on surveys to conducts its business. Think of a survey as an interview with a relatively large number of people. The goal of a survey is simple—to collect and then quantify information about people's attitudes, habits, beliefs, product loyalty, knowledge, or opinions using a survey questionnaire. You can conduct a survey over the phone, online, or by mail.

No doubt you've been asked to participate in a survey recently. For example, many online vendors ask customers, after a purchase, to rate their online shopping experience. Online customer feedback not only helps e-commerce companies learn about the level of their customers' satisfaction, but it also helps them find out about customer preferences to make crucial business decisions.

Note how the WH eComm survey in Figure 8.4 asks both types of questions. Some questions ask about customer preferences (questions 1, 2, 5, 6, 8, and 9 fall into this category), while others ask about customer satisfaction (questions 3, 4, 7, and 10). The questions about customer preferences can be used to help the company decide where to advertise ("How did you hear about our website?") and to determine which products to promote ("What type of product(s) have you purchased from WH eComm?"). The questions about customer satisfaction, meanwhile, elicit information to help the company improve service by assessing such things as the usefulness of its website ("Is the website easy to navigate?") and the quality of its customer service ("How efficient did you find our customer service?").

There are five basic steps you need to follow when using a survey as a part of your research on the job:

1. determining the best way to deliver the survey
2. creating the survey questionnaire
3. choosing the survey recipients
4. reaching the survey recipients
5. compiling and analyzing the survey results

Determining the Best Way to Deliver the Survey

As we saw, surveys can be conducted over the phone, online, or by snail mail. Each medium entails various advantages and disadvantages, as outlined in Table 8.1 below. Decide which medium you think will yield the best results and work within your time frame and budget. For instance, if your top priority is to reach a very large audience, conduct an online or mail survey rather than one over the telephone. If you need to receive detailed answers about a restricted or highly sensitive subject, conduct a telephone survey, which gives you an opportunity to talk directly to the repondents and allows them to clarify their answers. But if your goal is to obtain results quickly and inexpensively, do an online survey.

TABLE 8.1   Advantages and Disadvantages of Different Ways to Send a Survey

Telephone	*Online*	*Mail*
Advantages	**Advantages**	**Advantages**
■ Instant results ■ Ability to ask respondents for clarification ■ Low instance of skipped or partial responses ■ Ability to screen out inappropriate respondents ■ Surveys may be longer than online or mail surveys	■ Ability to reach a large population ■ Fast results ■ Not time consuming ■ Inexpensive ■ Anonymity encourages objective, candid responses ■ Respondents have time to carefully consider their responses	■ Ability to reach a large population ■ Less time consuming than telephone surveys ■ Inexpensive ■ Anonymity encourages objective, candid responses ■ Respondents have time to carefully consider their responses
Disadvantages	**Disadvantages**	**Disadvantages**
■ Inability to reach a large population ■ Time consuming ■ Expensive ■ Possibility of hang-ups (risk of incomplete responses) ■ Lack of anonymity can discourage objective, candid responses ■ Respondents do not have time to carefully consider their responses	■ Low response rate ■ Inability to probe respondents for clarification ■ High instance of skipped or partial responses ■ Inability to screen out inappropriate respondents ■ Surveys must be kept brief	■ Slow results ■ Low response rate ■ Inability to probe respondents for clarification ■ High instance of skipped or partial responses ■ Inability to screen out inappropriate respondents

Figure 8.4   An example of an online survey.

| Reply | Save | Forward | Print | Delete |

To: <Carol_Smith@acme.com>
From: <Gregg_Laos@whecomm.com>
Date: Thurs. 8 April 2009 9:02:00 EST
Subject: Survey

Dear Valued WH eComm Customer,

Cover e-mail explains why topic is important to customers

At WH eComm, we are committed to providing our customers with high-quality e-commerce software through an efficient and user-friendly website. Customer feedback is extremely important in helping us continue to improve our website. So that we may best meet your needs, we ask you to take the time to answer a short survey about your experience at WH eComm. You will find the survey on our website by clicking here. Your answers will help us continue to improve WH eComm online and to offer you the most efficient service you deserve.

Provides incentive to reply

To say thank you for filling out this short questionnaire, we would like to offer you a 5 percent discount on your next purchase. When you have completed the survey, your discount will automatically be recorded in your WH eComm online account.

Many thanks for your time and business,

Gregg Laos
Manager
WH eComm

★WH eComm
revolutionizing e-commerce

1. How many times have you visited our website?
○ First visit   ○ 2–4 times   ○ 5–7 times   ○ more than 7

2. How did you hear about our website?

Fill-in option does not require a lengthy answer

○ Colleague
○ Advertisement in business journal
○ Another website
○ Search engine
○ Other (please specify) [   ]

3. Is the website easy to navigate?

Question is not biased, that is, phrased in a leading manner

○ Very easy   ○ Easy   ○ Somewhat easy   ○ Not easy

4. How effective did you find the following sections of WH eComm?

	Extremely effective	Effective	Could be improved	Ineffective	No response
Web features	○	○	○	○	○
Search	○	○	○	○	○
FAQ	○	○	○	○	○
Online check out	○	○	○	○	○

(Continued)   Figure 8.4

5. How many times have you purchased our products?
○ 1–2 times   ○ 3–4 times   ○ 5–7 times   ○ 8–9 times   ○ more than 9 (please specify)

[   ]

6. What type of product(s) have you purchased from WH eComm? (check as many as apply)
○ E-commerce software
○ Networking software
○ Web design software
○ E-conferencing software

Multiple-choice items clearly differentiated

7. How helpful did you find our customer service?
○ Extremely helpful
○ Helpful
○ Could be improved
○ Not helpful
○ No response

8. How have you most often contacted our customer service office?
○ Phone   ○ E-mail   ○ Fax   ○ Web

Provides non-overlapping choices

9. How soon was your query answered?
○ Same day   ○ Next day   ○ Within 3 days   ○ Within a week   ○ Longer

10. How satisfied were you with the speed and efficiency of our online customer service center?
○ Very satisfied
○ Somewhat satisfied
○ Dissatisfied

11. Please rank, in order of importance, which factors most influence your online purchases.

	Price	Shipping options/time	Returns policy	Website quality
Most important	○	○	○	○
	○	○	○	○
	○	○	○	○
	○	○	○	○
	○	○	○	○
Least important	○	○	○	○

Ranking question supplies all necessary options

12. What would you most like to see changed or improved on our website?

[▲ ▼]

Provides opportunity for respondent to elaborate

Thank you for taking the time to answer our questions.

Home

Creating the Survey Questionnaire

The most crucial part about writing a questionnaire is crafting valid and accurate questions. There are many types of questions you can ask, as illustrated in Figure 8.3, including yes/no, ranking, rating, multiple-choice, or open-ended questions. Researchers advise asking questions that require the least amount of effort on the part of the respondents (yes/no, multiple-choice) to increase their chances of answering your questionnaire. Accordingly, ask open-ended and rating/ranking questions sparingly. In addition, keep your survey to ten to fifteen questions, both to encourage your recipients to reply and to keep sharply focused on those questions that matter most to your company or organization. Again, note how the WH eComm questionnaire in Figure 8.4 includes only twelve key questions, which are most essential for the company to make important decisions. Finally, design your questionnaire to look inviting and streamlined. See pages 491–508 in Chapter 11 for guidelines on document design.

Here are some guidelines for writing specific questions to help you get the results you want—whether you are writing a mail or online questionnaire or preparing a script to follow as you conduct a telephone survey.

1. **Phrase questions precisely.** Vague questions only elicit answers that you cannot use or will be unable to analyze. Use valid, quantifiable questions.

> Ineffective: Are we open enough hours on Saturdays?
> Yes____ No____
> Better: How many hours would you like us to be open on Saturdays?
> 4____ 5____ 6____ 7____ 8____

2. **Ask only one question at a time.** Avoid multiple questions within the same question, since you will not know the exact answer to each question.

> Ineffective: What is your overall impression of our customer support and delivery services?
> poor___ fair____ good____ very good____ excellent____
> Better: (turn the two questions above into two separate queries as follows)
> What is your overall impression of our customer support service?
> poor___ fair____ good____ very good____ excellent____
> How would you rate our delivery service?
> poor___ fair____ good____ very good____ excellent____

3. **Clearly differentiate each option in multiple choice questions.** If respondents are not sure of the differences between/among options, they may answer inappropriately because of question overlap or they may skip the question altogether.

> Ineffective: When is the best time to call you?
> Daytime___ Afternoon___ Weekday___ After work___ Evening___
> Night___
> Better: When is the best time to call you?
> Morning (8:00 a.m.–noon)___ Afternoon (noon–5:00 p.m.)___
> Evening (5:00 p.m.–10:00 p.m.)___

4. Supply all of the necessary options in multiple-choice questions. If you omit an important option, respondents may choose a misleading answer or not answer at all.

> Ineffective: Which types of non-alcoholic beverages would you like us to offer?
> soda___ juice___ coffee/tea___ milk___
>
> Better: Which types of non-alcoholic beverages would you like us to offer?
> soda___ juice___ coffee/tea___ milk___ bottled water___ other (please specify)___
> (the "bottled water" and "other" options give respondents a fuller range of answers)

5. Do not use unfamiliar jargon and abbreviations. Don't assume that respondents will understand the jargon your company or profession uses.

> Ineffective: What was your overall impression of the CGSE in this film?
> poor___ fair___ good___ very good___ excellent___
>
> Better: What was your overall impression of the computer-generated special effects used in this film to create the global village scene?
> poor___ fair___ good___ very good___ excellent___

6. Do not ask inappropriate questions. Refrain from asking questions about income, education level, or other personal matters such as age, ethnicity/race, gender, disability, religion, or sexual orientation unless these questions give you essential demographic information directly relevant to the topic of your survey.

7. Avoid leading or biased questions. Do not give your respondents slanted questions that bias their answer and thus the results of your survey.

> Ineffective: Were you impressed by this award-winning product?
> Yes___ No___
>
> Better: Did you think this was an award-quality product?
> Yes___ No___

7. Limit multiple choice and ranking items to five items. The more complicated your list of multiple-choice or ranked items, the more difficult it will be for your respondents to give a clear and helpful answer and for you to analyze the survey results.

8. Limit rating ranges to a scale of 1 to 5. As with item 7 above, do not complicate your survey by providing a scale with such a wide range of options that respondents are unclear about how they differ or overlap.

Choosing the Survey Recipients

Surveys, of course, depend on targeting the right audience. Sometimes that audience is small and easy to reach. For example, you might survey people within your own company/agency or department (all of the nurses in ICU). But more often the group you want to survey—all of your California customers or vendors, for example—is so large that you could not possibly survey the opinions of every member of that group. In that case, you need to gather information from enough people to make reliable and relevant judgments about the larger population, and you have to select a representative cross section of individuals from the larger group (by age, gender, background, experience, education, etc.).

For example, let's say you wanted to do a survey on the quality of food in a school cafeteria, or the cost and availability of health benefits at a company in order to write a report to a superintendent or a director of human resources. You would need to obtain the opinions of a representative sample of people in that school district or company and choose a cross section of those respondents within that school or company. A cross section is defined as a preselected group that is representative of the population as a whole. Nielson Media Research, *http:// www.nielsenmedia.com/nc/portal/site/Public/*, which rates the popularity of television shows in the United States, conducts its surveys in this way. Obviously, Nielsen cannot survey all 500 million television viewers, so it selects a representative sample of 5,000 households to determine TV ratings.

Reaching the Survey Recipients

Don't expect all of your respondents, or even 40 or 50 percent, for that matter, to reply to your questionnaire. Researchers find that a response rate of 12 percent from a statistically chosen sample group is still valid. But to increase the chances of receiving replies from as many respondents as possible, follow these time-tested procedures:

- Provide a cover letter or e-mail, as in Figure 8.4, asking recipients to reply and thanking them in advance for doing so.
- Offer respondents some incentive to answer the survey, such as the discount WH eComm promises in Figure 8.4.
- Indicate whether respondents should identify themselves or remain anonymous.
- Clearly specify how the respondents are to answer the questions—using a check mark, circling the correct response, writing in a number, or just pointing and clicking.
- If you mail your questionnaire, provide each recipient with a stamped, return-addressed envelope to increase the chance of a reply.

Compiling and Analyzing the Survey Results

The final step in conducting a survey is compiling and analyzing the results. The task of analyzing survey data in the workplace is often left to experts, who employ complex statistical methods (such as those listed at *http://www.spss.com*) to arrive at significant conclusions. However, if you are asked to analyze survey data, especially if your boss wants you to survey a manageable (small) number of responses, here are some helpful tips to follow:

1. **Keep your completed surveys organized and save them.** Assign each completed survey a number or code so that you can easily differentiate often very similar-looking responses from one another. Don't throw away the completed surveys; you may need to refer to specific answers later and/or your company or department may need to archive all surveys.

2. **Create a data sheet.** Originate a data sheet, for instance an Excel spreadsheet, so that you can transpose all of the survey responses into one central document. Break the data sheet into logical categories, for instance, separate rows for each survey question and separate columns for each possible answer.

3. Record responses completely and accurately. Make sure you include the responses to all of the survey questionnaires on your data sheet. Always record responses exactly as you receive them; if the response to a question is blank or illegible (or gives several answers to the same question), discard that response rather than making something up or guessing what the respondent meant. As you record each response to each questionnaire, be careful that you don't risk recording the same answer or survey twice.

4. Present your findings clearly and effectively. To help your boss or other readers understand your findings, create one or more simple tables in which you present key information in an easy-to-read format. Also supply a blank sample questionnaire for reference.

Secondary Research

As we saw earlier (see pages 303–304), secondary research requires you to consult sources that are already available (books, periodicals, reference works, websites, etc.), as opposed to interacting directly with people, places, and things via direct observation, site visits, tests, interviews, focus groups, and surveys for primary research. Secondary research involves gathering documents and reading, summarizing, and incorporating them into your report. As with primary research, your secondary research involves planning and organizing your time to focus directly on the problem/issue your company expects you to investigate.

Libraries are essential to conducting secondary research. To get started in your workplace research, learn about the different types of libraries you can use and the various services they offer, such as online catalogs and periodical databases. You will also need to learn how to use effectively the vast materials available on the Web, such as government documents, sites posted by professional organizations, and a host of company homepages and blogs. Let's begin with an overview of different types of libraries and how and why they can help you in doing your research.

Libraries

Almost every institution—whether a city, state, organization, school, or business—maintains a library or archive. As part of your workplace research, you can expect to use one or more of the following types of libraries:

- corporate libraries
- public and academic libraries
- e-libraries

Corporate Libraries
One of the fastest, easiest, and most profitable ways to locate and collect crucial research data about your company is to consult your company or agency library. Depending on the size of your organization, your corporate library may be only one small room with file cabinets and a computer (with or without a full-time librarian)

Visit www
.cengage.com/
english/kolin for
an online exercise,
"Exploring Online
Libraries and
Other Online
Resources."

or a large, fully staffed library, such as those maintained by multinational pharmaceutical companies, law firms, brokerage houses, and other megacorporations.

Corporate libraries contain a vital history of a company's activities, past and present. Using your company's library, you can obtain much of the sales, historical, competitive, and other data you may need and incorporate into your business report. A company library houses not only books, periodicals, and general reference works, but also business-specific and confidential documents not found in a public or academic library. Corporate libraries include documents often referred to as gray literature (see Tech Note on page 339). If your company or agency does not provide or sponsor its own physical in-house library, most likely it will at least have a password-protected archive that you can access through the company's intranet site (see Tech Note on page 327). You may also need to consult a company's extensive departmental files found in various units of your organization.

Regardless of the size of your company, the following documents and information are likely to be found in a corporate library, on its intranet archive, or in departmental files (e.g., legal, marketing, etc.):

- client and customer records
- corporate reports, studies, surveys, and proposals
- corporate newsletters—from inception to the present
- blueprints/specs
- legal records, including patents and contracts
- financial and operational information organized by month, year, or larger cycles
- books and trade periodicals directly related to your company's research and business
- product and service literature (catalogues, descriptions, technical specifications, training manuals, warranties, etc.)
- competitor information, including comparative analyses, competitor catalogs, and sales information.

Note that, as Shay Melka points out in Figure 8.1, B&L Stores uses many materials from its corporate library or available through numerous divisions of that company. However, while company libraries and archives offer a wealth of valuable research information, keep in mind the following guidelines when you use them:

1. Don't expect to find everything in one central location. As efficient as a corporate library may be, you won't necessarily find every company document you need there. You may have to go to individual departments or branches to locate materials, particularly sensitive legal documents that for ethical reasons are available only through permission.

2. Don't assume the library will be organized like a public or academic library. A corporate library may use the Dewey decimal system or it may not. Become familiar with your company library's location, check-out policies, references, and system of coding and classification.

3. Use corporate library materials with discretion. Recognize that much of your company's financial, statistical, or competitive information may now be irrelevant to your purpose, out of date, or overly optimistic/pessimistic depending on the business climate during the time it was written.

4. Supplement information from your corporate library with information from other sources/locations. Materials found in a corporate library may take you only part of the way toward what you need to learn. Expect to supplement what you find in your corporate library with primary research, Web research, and/or trips to public/academic libraries.

Tech Note

Intranets

To store and share information, diversified companies, government agencies, and many large organizations have created intranets. As the name suggests, intranets are communication networks modeled after the Internet and use the same tools and procedures found on the Web—hyperlinks, directories, passwords, and multimedia content. Unlike the Internet, though, an intranet is designed for internal use by the personnel of a company or organization. Coordinating the websites of various groups and departments, Web administrators and editors oversee a company's intranet. From a centralized directory, information is sent to, from, and about various divisions within the company— management, engineering, sales, safety, or public relations. Information might be designated public or private, depending on audience and content. Some files may be closed to you for reasons of confidentiality. Intranets often use firewalls to secure the site so that other businesses or unauthorized individuals cannot gain access.

Public and Academic Libraries

In the world of work, do not neglect the vast resources and assistance provided at public and academic libraries. While public libraries are available to everyone, however, you may need to be a resident to access particular resources. To use an academic library, you will have to be enrolled at or work for the institution that owns it.

When you use public or academic libraries, you benefit from the expertise of professionally trained librarians who have selected, organized, and recommended authoritative resources. In person or online, a librarian can help you with your research by directing you to appropriate sources (see pages 334–339) and assist you in formulating your online searches (see pages 340–346). A few minutes of conversation with a librarian before you begin your search can save you a great deal of time and effort. Also, a librarian can identify the most relevant databases, subject directories, professional journals, and trade magazines, in print and online, to help you accomplish your workplace research.

To start your library research, access the library's homepage for a full range of its services and for directions on how to conduct a search. Figure 8.5 shows the homepage for one academic library, including how to reach a librarian, access e-resources, and locate relevant documents. Note how the library has made it easy for patrons to connect with the right department to assist them in their research. To make searching even more efficient and convenient, many libraries belong to a network, regional or global, of participating libraries, enabling patrons to access the catalogs of member libraries in the group. For instance, WorldCat, a universal catalog of resources, lets you know what public and academic libraries in your area or the world over own a particular book.

To conduct your research, you'll need to know how the library's online catalog works. At public and academic libraries, the online catalog lists all of the materials the library owns, subscribes to, or has access to. It lists books (listed through the "Anna Catalog" link in Figure 8.5), reference works, databases, websites, audio and video recordings, visuals, and the individual journals and magazines (electronic and print) to which the library subscribes. Starting with a library's online catalog will make your research easier because of the powerful search options available.

Figure 8.5   Library homepage.

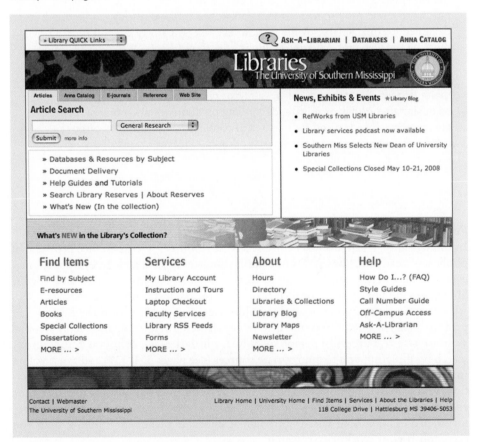

Reprinted by permission of University of Southern Mississippi Library (www.lib.usm.edu).

Here are some guidelines to help you access an online catalog:

- **Know your topic.** To get background information on your topic, start with an encyclopedia or other general reference work so that you are aware of the various issues and subtopics involved.
- **Restrict your subject.** When you search a public/academic library's catalog you need to narrow your topic to find a manageable amount of information. Suppose you are researching the use of lasers in plastic surgery. You might start your search with "medical uses of lasers." To narrow your subject, you might specify "use in cosmetic surgery." You could then further refine it by specifying "in opthamological cosmetic surgery."
- **Take advantage of links.** An online catalog usually provides links to related materials that may be even more helpful to your search. Use the Previous and Next links to navigate through sources on the same topic.

E-Libraries

Unlike public/academic and corporate libraries, which include print and online materials, e-libraries are exclusively Web-based. But do not confuse e-libraries with periodical databases (see pages 330–334), which provide access to online articles only. E-libraries are designed to duplicate the experience of going to a library, as much as that is possible in a digital environment. They provide links to librarian-approved websites in a variety of subject areas, offer links to complete texts of books available online, and alert you to links to online general resources (dictionaries, almanacs, encyclopedias, and more). They also make it easy to reach librarians who are available to answer reference questions online. Most public and academic libraries have also duplicated their library resources online, including access to librarians (see the Tech Note "Library Chat Rooms" on page 332). These sites offer you a variety of helpful links as well.

The most complete e-library is the Internet Public Library (*http://www.ipl.org*), or IPL. The IPL is the most well-known, comprehensive, and academically reliable e-library currently available. It was founded and is run by the University of Michigan School of Information. Figure 8.6 shows the homepage for the IPL. This nonprofit e-library not only provides the e-library resources listed above, but it also includes

- links to online periodical databases
- links to online newspapers around the world
- links to reliable blogs, exhibits of images, and much more

When you look for an e-library for your workplace research, keep these helpful tips in mind:

1. Use a reliable e-library. Consult e-libraries that have clear navigation, provide links to noncommercial websites, and verify links as current. You will lose valuable time if you don't use a high-quality, up-to-date e-library.

2. Consult a nonprofit e-library. Use e-libraries whose Web addresses end in .org or .edu, rather than commercial (.com) e-libraries. Nonprofit e-libraries are designed to help you, while commercial e-libraries are designed to sell something.

Figure 8.6   The Internet Public Library's homepage.

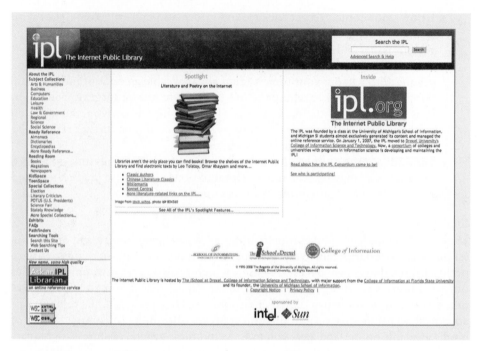

Courtesy of the Internet Public Library, Drexel University.

3. Realize that e-libraries do have limits. Despite the convenience of having a wealth of information at your fingertips, not every document found in a public or academic library will be available for online use. Use e-libraries in conjunction with your primary research and visits to public or academic libraries.

Periodical Databases

Periodical databases are among the most helpful resources for doing research in your library. These online indexes allow you to search for and retrieve a wealth of magazine, journal, and newspaper articles cataloged and classified by various search engines. These databases can be located through a library's online catalog or through a periodical databases link at an e-library such as IPL. Some libraries allow you to access their databases from a remote location, so you can use them when you are traveling. Because most of these databases are available only by subscription, you need a library card to access them at a public library or an access code at a corporate library. To use an academic library, you have to be a registered student or alumnus. With these periodical databases, you can search through thousands of articles in a few minutes. Many databases are updated often—daily, weekly, or monthly. Table 8.2 lists periodical databases that you may often access to do research in the world of work.

TABLE 8.2 Useful Business and Specialty Databases

Database	Contents
ABI/INFORM	Lists business management articles from U.S. and international sources
AGRICOLA	Research materials relating to agriculture
Animal Science	Provides information on animal production and veterinary science
Business & Company Resource Center	Offers a variety of global business topics, company profiles, case studies, and investment information
Business Source Elite	Provides full-text, index, and abstracts of business journals and magazines
Computer Science Index	Abstracts and indexes of over 500 academic journals, professional publications, and other sources on computer science
Contemporary Women's Issues	Women's issues on health and human rights, in over 150 countries
Criminal Justice Abstracts	Covers journals, books, and reports on criminology and related disciplines
EconLit	Includes bibliographic citations, abstracts and journal articles, books, collected essays, and full-text book reviews dealing with economics
e-marketer	Provides market information related to e-marketing and online technologies
Environmental Science & Pollution Management	Journal articles on environmental science and pollution management
FactSearch	Indexes over 1,000 newspapers, periodicals, newsletters, and government documents dealing with criminal justice and legal issues
Facts On File	Provides worldwide news stories from newspapers, periodicals, journals, and online government sources
FirstGov	Official database of the U.S. federal government accessing files, documents, and reports from all government agencies
General BusinessFile ASAP	Covers articles on business and management from over 3,000 publications
Investex Plus	Provides investment analyst reports of over 500 worldwide companies
LexisNexis Business	Full range of sources for business information, financial news, international company information, market research, industry reports
Microcomputer Abstracts	Covers popular magazines and professional journals on microcomputing for business, industry, education, and home use
National Criminal Justice Reference Center	Provides information, documents, and reports from the U.S. Department of Justice
NTIS Homeland Security Information Center	Government resource covering national security concerns, biological and chemical warfare, preparedness and response, and safety training
Regional Business News	Includes full-text articles from over 75 regional business journals
U.S. Bureau of Consular Affairs	Government resource for international travel information, advice, and warnings
U.S. Department of Treasury	Provides government information on accounting and budget, currency, financial markets, small businesses, and taxes
Wall Street Journal	Offers full-text articles from the *Wall Street Journal* from 1984 to the present

Tech Note

Library Chat Rooms

Library chat rooms are often available at no cost to patrons through many public library websites and through the Library of Congress (a U.S. government library that collects almost every publication produced in the United States). Visiting these chat rooms, you can go beyond simply telephoning or e-mailing a question to a librarian and waiting for a response: You can actually carry on a conversation with a librarian for more detailed assistance during library hours, sometimes twenty-four hours a day, seven days a week.

Library chat rooms work like instant messaging (see pages 140–141): You access these chat rooms using your Internet browser, simply key in your questions to the librarian, and receive responses online. Also, like instant messaging, library chat rooms are convenient, because they allow for the ongoing communication of an e-mail conversation combined with the immediate response of a telephone call.

In addition to communicating with you in real time, the librarian can send you attachments and URLs, providing you with the kind of assistance you would get if you went to an academic or public library in person. If you need further help, the e-librarian may refer your question to another librarian in the network who will follow up your request for information. When your chat is finished, the librarian usually sends you a transcript of your communications, which you can file for your records and refer to as you prepare your report.

Information Found in Periodical Databases

Thousands of periodical databases are available from many different information services. They vary in terms of how they list information and what they offer. (See Figure 8.7 for an example of an article page in a periodical database.) But most of them provide the following information, crucial for on-the-job research, when you do a keyword search to locate a particular article:

- **Bibliographic citation**, including author, title, publication date, source information, and perhaps keywords and subject categories.
- An **abstract**, or a brief summary of what the article is about, with enough information to help you determine if the article will be useful for your purposes.
- The **full text of the article**, either retyped for the database or scanned in as a PDF version of the original document. You may need to click on a further link to get past the informational page and open the complete article.
- Additional **factual information** that may be useful, such as the article's International Standard Serial Number (ISSN), information about author, word count, and whether the article contains graphics, statistics, etc.

Article page from a periodical database.   Figure 8.7

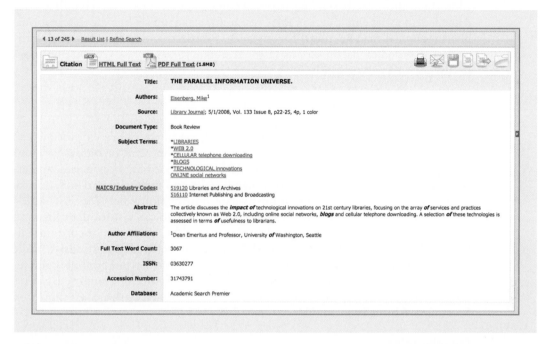

Courtesy of EBSCO Publishing.

Frequently Used Periodical Databases

You need to find out which databases are most relevant to your job-related research. Here are a few full-text databases that your corporate or community library will likely subscribe to:

1. **EBSCOhost.** This database provides full-text articles from thousands of popular magazines, professional journals, and newspapers and claims to offer the largest full-text collection of professional/academic articles in the world.
2. **Lexis/Nexis Academic.** This database is helpful for locating articles, as far back as the 1940s, in business and law.
3. **InfoTrac.** One of the largest periodical databases, InfoTrac contains a variety of articles from both academic journals and general-interest magazines.
4. **NewsBank.** NewsBank's database supplies over 70,000 news articles annually—from over 500 U.S. and Canadian newspapers.
5. **ProQuest.** Indexing more than 7,400 publications, this database includes newspapers and scholarly and general-interest sources in business, news, medicine, humanities, social sciences, hard sciences, and technology.

Business and Other Specialized Databases

Many libraries, including corporate ones, subscribe not only to the most popular databases above, but also to a several business and specialized databases. Almost every professional discipline has its own database. Table 8.2, for example, lists both general business databases and highly specialized subject-specific databases. In addition to these databases, international companies frequently develop their own databases to assist employees in their research. GlaxoSmithKline, the international pharmaceutical manufacturer, for instance, prepared and archived a password-protected periodical database of its research studies for use by its 120,000 employees in labs and offices worldwide.

Reference Materials

In addition to finding articles in periodical databases in your library or through an e-library, expect to use reference materials available in print or online for your workplace research. Reference works include encyclopedias, dictionaries, and atlases, as well as industry directories, handbooks and manuals, statistics, and so forth. These works will give you up-to-date information when you just need a quick and accurate overview of a topic or specific statistical, historical, or financial data. Most reference works are available free at public, academic, or corporate libraries. But use general reference works such as encyclopedias cautiously. While they supply basic information and are easily accessible, do not confine your research to them, because they are limited in scope. A wise rule of thumb is to always double-check the facts you garner from a general reference work against those you find in more specialized works.

Encyclopedias

The word *encyclopedia* comes from a Greek phrase meaning "general education." An encyclopedia is simply a collection of overviews of a wide range topics organized alphabetically, either in book form or online. A general print encyclopedia, such as the *Columbia Encyclopedia* or the *Encyclopaedia Britannica*, is a useful starting point for research because it contains knowledgeable introductions to topics, summarizes events or processes, explains key terms, and includes recent updates, along with lists of further readings. Online encyclopedias include *Wikipedia* at *http://wikipedia.org* (an encyclopedia written and updated by volunteers that contains over 2 million entries), the *Encyclopaedia Britannica Online* at *http://www.britannica.com* (the e-version of the famous encyclopedia), and Encarta at *http://encarta.msn.com/artcenter_/browse.html* (developed and maintained by Microsoft).

Be especially careful when using *Wikipedia*. Unlike the more traditional encyclopedias above, which are authored by scholars and overseen by professional editors, *Wikipedia* articles are written by general readers and are not checked for accuracy or for bias by experts in the area.

For your workplace research, start with a general encyclopedia, but you may need to consult a more specialized business or technology encyclopedia, such as one of the following:

- *Goliath, the Thomson-Gale Encyclopedia of Business, http://goliath.ecnext.com/*
- *Blackwell Encyclopedia of Management, http://www.management encyclopedia.com/*

- *The Encyclopedia of Banking and Finance, http://www.eagletraders.com/ books/int_fin_encyclopedia.htm*
- *Webopedia, http://www.webopedia.com/*

Dictionaries

Rather than relying exclusively on general dictionaries, such as *The American Heritage Dictionary of the English Language* (available online at *http://www.answers .com/library/Dictionary*), you will have to use business and industry-specific dictionaries that define the words and phrases (jargon) of your profession. Most general and specialized dictionaries are available in print and online. Use only those specialized dictionaries that are officially endorsed by your profession. The following are just a few reliable online business dictionaries:

- *The New York Times Glossary of Financial and Business Terms, http://www .nytimes.com/library/financial/glossary/bfglosa.htm*
- *Deardorff's Glossary of International Economics, http://www.personal .umich.edu/~alandear/glossary/*
- *The Washington Post Business Glossary, http://www.washingtonpost.com/ wp-dyn/business/specials/glossary/index.html*

Almanacs

An almanac is a collection of statistical data—charts, tables, graphs, and lists—published annually and carefully organized by general topics such as geography, awards/prizes, and science and technology. Here are some useful almanacs:

- *Information Please (http:// www.infoplease.com)*
- *Plunkett's (http:// www.plunkettresearch.com)*
- *New York Times Almanac*

In addition, you may want to refer to an industry-specific almanac, such as one of Plunkett's various specialist almanacs (advertising, investing, retail, e-commerce, information technology, travel, biotech, insurance, etc.).

Atlases

Atlases are collections of worldwide maps. Many kinds of workplace research involve consulting an atlas for locations as well as for statistical information about specific geographical areas. While print atlases are limited in what they can show because of space considerations, online atlases provide much more detail, including three-dimensional images, shading, and elevations. Among the most widely used print atlases are the *Times Atlas of the World* and the *National Geographic Atlas*. Here are some other handy online atlases:

- *Google Earth, http://earth.google.com/*
- *Encarta, http://encarta.msn.com/encnet/features/mapcenter/map.aspx*
- *National Geographic Map Machine, http://plasma.nationalgeographic.com/ mapmachine/*
- *Mapquest, http://www.mapquest.com/*

Government Documents

The U.S. government is the largest publisher in the country. Virtually every federal agency and department conducts primary and secondary research and then publishes its findings. These publications, collectively referred to as **government documents**, include reports, articles, manuals, speeches, census figures, maps, photos, etc. Government documents are valuable for the statistical information your company may need.

Many government documents are available through the website of the department or agency that created them. But you can access all of these websites via *http://www.USA.gov*, the official government web portal. You can use this link to take advantage of the millions of federal Web resources. Figure 8.8 shows the *USA.gov* Reference Center and General Government page, listing by category a variety of sites. Other major sites to access government documents include:

- *American Fact Finder, http://www.factfinder.census.gov*
- *Library of Congress website, http://www.loc.gov*
- *Google U.S. Government Search, http://www.google.com/ig/usgov*

Directories

If in your job-related research you need to find information about specific companies and their decision makers, turn to one of the many industry-specific directories. They include a wide range of companies along with links to those companies' websites, or,

Figure 8.8   USA.gov Reference Center and General Government page.

more helpfully, staff-written overviews of companies, contact information, lists of key people, and financial statistics. Directories are invaluable in helping you find and build e-mail lists and prospective client lists. Figure 8.9 shows the homepage for the MWBE directory, *www.mwbe.com*, a directory of businesses owned by women and minorities, while Figure 8.10 shows the homepage for Hoover's, *http://www.hoovers.com*, which provides industry overviews, company contact information, business statistics, and lists of competing companies. Some other useful online directories include:

- Corporate Information, *http://www.corporateinformation.com*. Similar to Hoover's, Corporate Information provides snapshots of over 31,000 companies in 59 countries.
- Zacks, *http://www.zacks.com*. Intended for investment purposes, Zack's nonetheless helpfully supplies financial information about many companies.

Homepage for the MWBE directory.   Figure 8.9

MWBE.com. Sponsored and owned by BrainTrust Solutions, Anne Chastain, President.

Figure 8.10   Online version of *Hoover's Handbook of American Business.*

Courtesy of Hoover's, Inc.

Handbooks and Manuals

Similar to industry-specific almanacs, handbooks and manuals collect relevant and reliable information that you can use to write your report. Unlike almanacs, though, handbooks and manuals are less focused on statistical data but instead include explanations of procedures, definitions of terms and concepts, descriptions of industry standards, and overviews of professional issues within the field. The *Occupational Outlook Handbook,* produced by the U.S. Bureau of Labor Statistics, *http://www.bls.gov/oco/,* for instance, "describes what workers do on the job, working conditions, the training and education needed, earnings, and expected job prospects in a wide range of occupations."

Among the most frequently consulted handbooks and manuals for workplace research are *Moody's Manuals* and *Standard and Poor's Corporate Descriptions,* both of which give information on the history of a company, descriptions of products and services, and basic financial details (stocks, earnings, and mergers). Since neither is available online, you'll have to use them at a library.

Statistics

Statistical reference works give you valuable numerical data on a wide range of business-related subjects—employment, housing, immigration, population, pollution, technology, overseas markets, among others. Statistical data is generated from numerous sources, mostly from the U.S. government, private agencies, international organizations, and colleges/universities. Note how Shay Melka's overview in Figure 8.1 stresses the ways B&L Stores uses statistical data to formulate major plans at the company. Here are four statistical references that can assist you in your workplace research:

- Fedstats, *http://fedstats.gov*
- U.S. Bureau of Labor Statistics, *http://www.bls.gov*
- BizStats.com, *http://www.bizstats.com*
- GeoHive, *http://www.geohive.com*

Tech Note

Gray Literature

Gray literature is the name given to documents that are prepared by a specific company, industry, or agency but not readily available through databases or a library's online catalog. Some examples of gray literature include internal reports, newsletters, fact sheets, conference proceedings, instruction booklets, and so forth. Government agencies, special interest groups, universities, research centers, businesses, and professional associations all create or produce gray literature.

Gray literature is an important resource for workplace research, because it often provides information unavailable in commercially published books, journals, or newspapers, or found in scholarly publications such as monographs and academic journals. For instance, public health information leaflets; appliance repair manuals; consumer product ratings; and business and industry reports on such topics as annual stockholder meetings, promotions, and opening new markets are gray literature resources. You would be wise to consult these resources to compare your product/service with a competitor's, assess the corporate health of a firm, or locate the technical information you need.

However, while gray literature can be a rich and diverse source of information, it also poses challenges. Gray literature is typically not indexed by libraries, making it difficult to learn about and locate via bibliographies and databases; and basic information such as author, date, and source are often missing from these materials when you do find them. Moreover, you might want to double-check information you find in, say, a company report, with another source.

The Internet, of course, has transformed the production, distribution, and availability of gray literature. The American Library Association has provided a helpful list of places to locate gray literature on the Internet at *http://www.ala.org/ala/acrl/acrlpubs/crlnews/backissues2004/march04/graylit.cfm*. But when you use gray literature for your business report, be cautious. Try to find out as much as you can about the company or organization from other sources as well to double check and verify information included in a piece of gray literature. Keep in mind that a gray literature document will reflect the viewpoints and preferences of the company or industry that produced it and the financial climate when it was written.

Sherry Laughlin

Internet Searches

Visit www
.cengage.com/
english/kolin for
an online exercise,
"Conducting Inter-
net Searches and
Evaluating
Websites."

There is an enormous amount of information available on the Internet, and this body of information grows larger every day. Acting as a "network of networks," the Internet gives you access to information from millions of websites, databases, libraries, newsgroups, chat rooms, blogs, and other sources. As a result, it is one of the quickest, easiest, and most effective ways to do secondary research.

Searching, Not Surfing, the Web

When you surf the Web, you wander from website to website without any particular goal in mind. However, surfing accomplishes very little when you are researching. Just like conducting library research, your Internet research needs to be carefully planned and focused. You wouldn't wander into a library and browse the stacks until you stumbled across something useful if you were there to do research. Instead, you would consult a database or catalog to locate materials relevant to your topic.

Searching the Web, however, can be more difficult than finding a book in your online library catalog. The Web is much larger and contains much more information. Moreover, there is no single catalog of all the information it holds. To conduct effective research online, you need to use subject directories and search engines, which index webpages like a database, acting as gateways to vast amounts of information.

Subject Directories

Many businesses and groups have created Internet subject directories that easily locate appropriate websites (as opposed to articles, which are located via periodical databases). Subject directories are helpfully organized into logical categories (e.g., Business) and subcategories (e.g., International Business and Trade) for easy navigation. Typically, each subcategory is annotated. These directories are also searchable, but the search area is limited to what the directory creators have selected. Think critically when using subject directories, as some subject directories provide more reliable information than others. Here are a few examples of reliable subject directories:

- **Librarians' Index to the Internet**, *http://lii.org*. Subtitled "Websites you can trust," this index surveys over 20,000 websites, each with a brief explanatory annotation, and is organized into 14 general-interest categories (arts and humanities, business, media, science, etc.).
- **Academic Info**, *http://www.academicinfo.net*. As the name indicates, this is an academic subject directory aimed at both undergraduate and graduate students, and is an excellent way to conduct research.
- **Business.com**, *http://www.business.com*. This site provides an easy way to locate business websites for news and information on companies, products, services, people, and jobs.
- **Google Directory**, *http.//www.google.com/dirhp*. Organized into 15 broad categories with many subcategories, this subject directory provides over 1 million annotated links to websites both academic and commercial.
- **Yahoo! Search**, *http://dir.yahoo.com*. Very similar to the Google Directory, Yahoo! Search provides well over 1 million annotated links to various sites.

Search Engines

When you begin your research, start by using a search engine. Search engines are automated tools that use keywords to locate the webpages containing the most relevant information you need. Using computer robots that "crawl" through and "read" pages on the Web, a search engine indexes the information it finds to create a frequently updated database of results.

Search engine software ranks the results of your search using a computer algorithm that records the popularity of websites, their contents, where the keywords appear on the page, and other sites that link to that page, presenting you with the most relevant matches for your keywords.

Because there is no single index to or abstract of material on the Web, you should expect to use a wide range of search engines. Given the vastness of the Internet, it only makes sense that multiple search engines are available, as Table 8.3 shows. These include general and metasearch engines (which combine multiple search engines) as well as specialized, subject-specific search engines. Like the Web itself, search engines are constantly expanding as they index more pages in cyberspace each second.

Of course, search engines will vary in coverage and relevance for your project. Some are faster than others, while many incorporate results from multiple engines simultaneously. To some extent, all search engines overlap. For example, searching for "CNN" in any search engine will lead you to CNN's homepage at *http://www.cnn.com*. Moreover, no single search engine is comprehensive. Researchers estimate that almost 40 percent of the Web is not even indexed by the most popular search engines. But by using a variety of search engines to conduct keyword searches, you will access a broader scope of sources and thereby increase your chances of finding the information you need.

Listed below are descriptions of four of the most frequently used search engines listed in Table 8.3.

1. **AltaVista,** *http://www.altavista.com,* provides one of the most comprehensive searches in over 25 languages and includes audio and video searches, images, directories, and news. It also offers a free webpage translation service, BabelFish.
2. **Ask.com,** *http://www.ask.com,* provides a variety of personalized search options, including MyStuff, to save and organize your results in folders; view your search history; and share, print, and repeat searches.
3. **Google,** *http://www.google.com,* is currently one of the world's largest and easiest-to-use search engines, indexing over 8 billion pages. Google offers specialized engines such as Google Government, Google University, Google Microsoft (which searches sites only related to Microsoft), Google Apple (which searches Apple/Macintosh-related sites), and Google Local (to search for businesses in your area).
4. **Yahoo!,** *http://www.yahoo.com,* one of the earliest search engines, also includes a human-edited, subject-based directory of the Web. Yahoo! suggests related keywords and allows you to open links in new windows.

Keyword Search Shortcuts

To use search engines effectively, you need to know how to conduct a keyword search, which is crucial to your success. When you enter a keyword into a search

TABLE 8.3   Different Types of Internet Search Engines

Search Engines	*Metasearch Engines*
AltaVista	Excite
Ask.com	DogPile
Google	Mamma
Yahoo!	MetaCrawler
	ProFusion
	Search.com
	WebCrawler

Specialized Search Engines	
General	*News*
AllSearchEngines (directory of subject-specific search engines)	AltaVista News
	Google News
Search Engine Colossus (an international directory of search engines)	MoreOver (updates news hourly)
	NewsDirectory
Media	Newspapers.com (directory of international, national, college, and business papers)
AltaVista MP3 Audio Search	
AltaVista Image Search	World News Network
AltaVista Video Search	Yahoo! News
GMP3	*Business*
Google Image Search	All Business
Google Video	Business.com
ClipArt Searcher	Wow Yellow Pages Business Directory
Yahoo! Image Search	
Yahoo! Video Search	

engine, it searches through its indexed websites and presents you with those that feature your keyword in the title, heading, meta tags, or text of the page. Your choice of keywords is vital since the results you receive depend on the keywords you plug into the search engine.

Following the guidelines below will help you to conduct successful keyword searches.

Be Specific

- **Narrow your subject.** Although it might be tempting to start off with a broad subject like *computer security*, you may find yourself confronted with millions of hits that range from computer theft to message encryptions. Take a look at Figure 8.11, which shows what occurs from a search that floods you with general results. A search on the words *computer security* yielded over 186 million results. Before you start your search, narrow your focus and thereby restrict your search.

- **Use multiple keywords.** Choose at least two or three significant keywords to specify your search. For example, if you key in just the word *virus*, you might receive vast amounts of information on everything from avian flu and the West Nile virus to computer problems. Specify *computer virus* to narrow your search.
- **Don't use prepositions.** Most search engines automatically exclude *of, to, in, from,* and *on.*

Modify Your Keywords

- **Refine your search.** Unlike the vague keywords *computer virus*, specify the type of information you want to receive by refining your topic. Search for *computer virus causes* or *computer virus prevention* to net more precise, pertinent information. Figure 8.12 shows how advanced search options on a search engine can help you narrow your keyword search. Here the search was narrowed by using multiple and more specific search words: *computer security, pop-ups, phishing, viruses,* and *firewalls.*

Example of a general search that yields too many results.   Figure 8.11

Keyword is too broad

Too many hits

Related links

Figure 8.12   Example of a more refined Google search.

Reprinted courtesy of Google Inc.

■ **Find relevant synonyms.** Try to locate possible synonyms or alternative keywords that might give you additional pathways to gain information on your topic. Many search engines even suggest ways to refine your search using other keywords. Take advantage of these suggestions.

Use Boolean Connectors

Boolean connectors, such as AND, OR, and NOT, are essential to limit and guide your search by reducing unrelated search results.

■ Inserting *AND* between keywords will limit your results to those containing all your keywords. For example, *virus AND computer AND prevention* will produce only records that contain all three terms. Note that some search engines, such as Google, automatically search all keywords and therefore don't require the *AND* command.

■ *OR* will produce records containing either or both keywords. *Virus OR computer* will list pages containing either *virus* or *computer* or both words.

■ *NOT* as part of your keyword search excludes unwanted pages. *Telecommuting benefits NOT environmental* indicates that you are not concerned with how telecommuting saves energy or reduces pollution. Instead, you are looking for benefits to employers and employees.

Use Delimiters

To refine your search even further, use the following **delimiters**, sometimes called "wild-card characters":

(margin notes, left side:)

Insert quotation marks to limit search

Use multiple keywords to restrict search

Click on related links

- Quotation marks around a string of keywords limit your term to that particular phrase. For example, Wells Fargo without quotation marks would pull up sites for Wells Engineering; Fargo, North Dakota; and Fargo Wells.
- A plus sign (+) between terms can be used to replace the Boolean AND command, locating results that include all keywords. Similarly, a minus sign (−) can replace the OR command in many search engines.
- An asterisk (*) at the end of a term broadens your search beyond that afforded by a keyword. For example, the term *employ** will return all terms that use *employ* as the stem of a keyword, such as *employment*, *employer*, and *employability*.
- Some search engines, such as Google, allow you to use an asterisk within quotation marks to indicate a missing word. Searching for "*Australian * technology*" will find results including keyword phrases ranging from *Australian medical technology* to *Australian information technology*.

Internet Search Shortcuts

In addition to the keyword search strategies above, there are a number of shortcuts to access frequently requested information quickly and easily on the Internet. Many of these techniques take advantage of advanced search options available on major search engines. Of course, depending on the type of Internet resources available to you, and the constantly changing Web terrain, you will doubtless discover many more shortcuts. But the following ones give you additional ways to access information faster and more efficiently.

- Use Google Alerts by entering your e-mail address and a keyword term into the Alerts system to be notified when your search term appears in the top ten or twenty search results for a Google News or Web search. This service is particularly useful for ongoing business research because it provides constant updates on the latest information relevant to your topic.
- Select the images, audio, or video options from your search engine to obtain thumbnails of all indexed images, MP3 files, or video files that relate to your keywords.
- To find sites in a particular language, file format (Word, PDF, PowerPoint), or date, use advanced search options available on most search engines.
- To obtain frequently sought-after information on weather, time zones, and flight times, just type in such keywords as "Chicago weather," "time Tokyo," or "Delta 346." Or you can key in the name of the city with its ZIP code to obtain a current weather report or one projected over the next 30 days.
- To locate a map, type in the street address of the area you want and a street map will automatically appear.
- To receive a selection of Web definitions taken from a variety of respected sources, type in "define" and then your keyword into Google.
- Choose country-specific search engines, such as those available at Yahoo! or Google, to access information about companies, laws, news, and international relations essential for communicating with international readers.
- To check, stock quotes, key in the stock exchange symbol of the company you are interested in, e.g., TWX for Time Warner Cable.

- To locate information on well-known people, Ask.com provides an instant biography with a photo, along with links to encyclopedia articles and official websites.
- Enter advanced delimiters to specify the exact type of domain you want to search. Searching Google for "*internet security*" *site:gov* will retrieve all *.gov* (government) sites with information on Internet security. Similarly, searching for "*international training*" *site:edu* will pull up all *.edu* (education) sites with information on international training.

Evaluating Websites

The criteria used to evaluate websites are similar to those you follow to assess print documents. Just as not everything you read in print is accurate and unbiased, not everything on the Internet will be correct, up-to-date, objective, or useful. The fact that something is posted on the Web does not make it correct—after all, anyone can post practically anything. Remember that much of the information on the Internet is placed there without a peer review process, that is, facts may not have been checked, sources may not be authoritative, and ideas and opinions may be biased and not be backed up by solid evidence. Overall, the quality of information you find on the Internet can range from unsupported, slanted, or just plain wrong, to authoritative, well supported, and objective.

Keep in mind, too, that as with print sources, websites may be created by special interest groups or individual businesses. The information (e.g., statistical data) you find at a special interest site may be clearly skewed toward/against a particular point of view, and a website created by a company is, of course, going to be prejudiced in the firm's own behalf, because it wants to sell its products and services. You will have to read, assess, and differentiate carefully the Web material you want to include in your report. To do effective workplace research, you have to make critical inquires and judgments. Don't be afraid to seek out and test different viewpoints, and never be afraid to verify what you find on the Web by asking for the opinion of a librarian or a specialist in your company.

Here are some questions to determine whether websites (and print sources, too) are credible, accurate, up-to-date, relevant, and objective.

Credible

- What are the credentials of the author or the organization? Read the "About Us" or "About Me" page of any website critically to learn more about the author's qualifications and awards or the organization's history, mission statement, and track record.
- Does the website refer to the most respected and current research being done in the field?
- Is the author discussed in other sources, including reference works and highly respected websites? Are there links to or from the author's or company's website to those sites?
- Is it possible to contact the author or organization with questions or requests for further help? If the website provides no contact information, contains many dead links, or features undocumented information, be suspicious.

Accurate

- How have the data been gathered, recorded, and interpreted? Are they complete, valid, logical, and consistent with the procedures and protocols of your profession?
- Do the facts, dates, and statistical information match those found in other sources, such as on other websites or in traditional research sources such as almanacs and encyclopedias (see pp. 334–335)? Again, do not rely on only one website for your facts and figures.
- Are all visuals clear, presented without distortion, and consistent with the information presented (see pp. 472–478)?
- Is the website full of typographical, grammatical, and/or spelling errors? If so, it may be unreliable in other matters, such as statistics, conclusions, and recommendations.

Up-to-Date

- When was the content written, placed on the website, and last updated? If the website does not provide such information, its content is likely to be out-of-date.
- Does the website content refer to current events, studies, and experiments, or do all the references go back too many months or years? In some business and technical fields, information needs to be updated weekly, or even daily.
- Is the information given on the website contradicted, supplemented, or labeled out-of-date on other more recent websites and in print reference works?

Relevant

- Is the website's content directly related to the subject of your report, or is it only remotely close to your topic and thus not central to your inquiry? Remember, your boss wants bottom-line information to answer specific questions.
- Does the website offer enough of the right type of data to be useful, or is it sketchy?
- Will others accept or trust the numbers and interpretations you have gathered from the website?
- How does the website help or hurt the case you are trying to build? Even a respected competitor's figures can give you valid and relevant information you can convincingly apply to your report.

Objective

- Does the website openly and clearly state its purpose? Does it claim to give objective information, or is it really just trying to sell something or present a biased point of view? Again, read the "About Me" or "About Us" page carefully.
- Is the website legitimate? Verify whether a company or organization is legitiate by checking with the National Consumer's League, *http://www.nclnet.org*.
- Does the website acknowledge opposing viewpoints and does it freely and clearly admit its own limitations and/or agendas?
- Do the author(s) use a professional, objective tone or is there unwarranted humor or sarcasm?

- Is the site free from sexist, racist, and other demeaning, stereotypical language or visuals?
- Is this website part of a reputable online discussion community (a listerv or chat room) or is it a platform for one individual's or organization's viewpoints?

Note Taking

Once you have consulted appropriate print and online sources, you need some systematic way to record relevant information from them to use in preparing your outline and draft(s) of your business report. But before you can outline or draft, you have to organize and classify the information from your research efficiently. And even after you start to draft your work, you will probably need to continue your research and take additional notes as you investigate new and relevant sources.

The Importance of Note Taking

Note taking is the crucial link between finding sources, reading and responding to them, and writing your business report. At this stage, you are gathering crucial background information to build and support you report. Never trust your memory to keep all of your research facts straight. Taking notes is time well spent.

Take your time. Do not be too quick in getting it done or too eager to begin writing your report. Careless note taking could lead to serious omissions, inconsistencies, or even contradictions between what different parts of your business report say.

Notice how the business research report at the end of this chapter and the one in Chapter 15 on international employees in the work force incorporates a variety of important facts, interpretations, and even visuals from the sources that the writer consulted, including surveys and interviews. Before these research details could be included in the report, the writer had to take careful notes.

How to Take Effective Notes

Effective note taking requires you to (a) identify only the most relevant points, (b) exclude irrelevant or inessential ones, (c) summarize key information concisely and accurately, and (d) document the results. Follow these guidelines to make note taking easier and more efficient:

1. Photocopy or download print or online articles, sections of books, reports, and other sources that you consult and mark relevant quotations, statistics, and other information you may incorporate into your work. Use different colored pens, highlighters, or computer colors to highlight or underline different types of information or to indicate where such information may go in your work (e.g., background material, methods, visuals, etc.).
2. Record all quotations verbatim.
3. Bookmark any sites on the Internet that you know will be important for your work and that you may be likely to return to. (See Tech Note "Using Bookmarks.")
4. Cut and paste information from online sources directly into your word processing program for reference, being careful to include exact source information.

Tech Note

Using Bookmarks

Bookmarks are webpage locations (URLs) that can be easily retrieved. Just as you slip a sheet of paper into a book or turn down the corner of a page of a print magazine to mark your place, you can use a Web browser's bookmark feature to keep track of your most frequently visited sites so that you can quickly return to them as you continue your research. Depending on your Web browser, bookmarks may also be called "favorites." To create a bookmark, simply click on the bookmark or favorites option on your Web browser's pull-down menu while you have the webpage you want to retrieve open, and then give your bookmarked page a name. Most bookmarks will automatically fill in a name based on information supplied by the website, but you may want to supply a more specific name so that you can remember later exactly which site you want to retrieve and why. When you return to the website later, activate the bookmark and your Web browser will automatically take you to that page.

What to Record

To take effective notes you have to know how to summarize a great deal of information accurately. Study the guidelines on writing summaries in Chapter 9 (pp. 398–403). Here are the types of information you need to record in your notes.

1. **Include full bibliographic information for each source.** For *books*, list author, title, city of publication, publisher, date of publication, edition (if not the first), and page numbers. For *journal articles*, include author, title, publisher, volume number, date, and page numbers. For *websites*, write down author, title, name of the site, URL, and the date you accessed the site.

2. **Copy quotations, names, facts, dates, and statistics accurately from the source.** Be sure, too, that you record the author's words correctly. You may want to quote them verbatim in your paper or report. Always compare what you have written down with the original.

3. **Distinguish quotations from paraphrases.** Place quotation marks around any words, sentences, or extended quotations you record directly to avoid the risk of plagiarizing (see pp. 350–352).

4. **Indicate in your notes why the material you quoted or paraphrased is significant.** You may not remember later why the material is significant, so help your memory by leaving yourself a reminder — "use in introduction."

5. **Clearly mark separate works by the same author.** If you are using two or more works by the same author in your research, be sure to indicate which quotations or paraphrases apply to which works.

6. **Identify and record only the most relevant and useful points.** You cannot copy down everything in your notes. Include only what you plan to actually use in your report.

7. **Indicate where graphics might be effective.** If you would like to add a graphic or visual to your report to support your report's findings and/or clarify points for your audience, make a note to yourself so that you remember later.

Tech Note

Electronic Note-Taking Software

Electronic note-taking software offers a fast and convenient way to organize your notes, saving you from worrying about losing note cards or scraps of paper. The four most popular electronic note-taking programs are Microsoft OneNote, EverNote, Google Notebook, and GoBinder. Although each is slightly different, (some allow you to down load or upload audio and video), most have the following features in common to help researchers input, organize, search, save, print, and share their notes:

- Copy and paste text and images from Internet sources. Many programs automatically include the URL.
- Keyboard your own notes or input your handwritten notes using a tablet PC or a pen input device (not all programs provide the handwriting option).
- Flag or color code your notes to highlight and differentiate them.
- Organize your notes into categories, folders, or outlines.
- Reorganize your notes using a drag-and-drop feature.
- Search your notes for keywords or flags to locate relevant information.
- Save your notes to a disk to make sure you don't lose them.
- Print your notes, export them to your office program, or e-mail them to others.
- Collaborate with others who have of the same program via a "user list."

To Quote or Not to Quote

Before recording information from sources, ask yourself three questions:

1. How much do I need to quote directly?
2. Where can I shorten a quote by using ellipses?
3. When should I paraphrase instead of quote?

Incorporating Quotes

A safe rule to follow is this: Quote sparingly. Do not be a human scanner. If you incorporate too many quotations in your report, you will simply be transferring the author's words from the book, article, or website to your paper. That will show that you have read the work but not that you have successfully evaluated it. Do not use direct quotations simply as filler. Save them for when they count most:

- When an author has summarized a great deal of significant information concisely into a few well-chosen sentences

■ When a writer has clarified a difficult concept exceedingly well
■ When an author has made his or her chief statement or thesis

For example, the note in Figure 8.13 contains a brief, well-worded, and significant statement by an author. It has not been necessary to quote verbatim all the evidence leading to that statement. It stands well on its own.

If you are uncertain about exactly how much to quote verbatim, keep in mind that no more than 10 to 15 percent of your report should be made up of direct quotations. Remember that when you quote someone directly, you are telling your readers that these words are the most important part of the author's work as far as you are concerned. Be a selective filter, not a large funnel.

Using Ellipses

Sometimes a sentence or passage is particularly useful, but you may not want to quote it fully. You may want to delete some words that are not really necessary for your purpose. These omissions are indicated by using an *ellipsis* (three spaced dots within the sentence to indicate where words have been omitted). Here are some examples.

> Full Quotation: "Diet and nutrition, which researchers have studied extensively, significantly affect oral health."

> Quotation with Ellipsis: "Diet and nutrition . . . significantly affect oral health."

When the omission occurs at the end of the sentence, you must include the end-of-sentence punctuation after the ellipsis. In the following example, note how the shortened sentence ends with four spaced dots: the three dots for the ellipsis and the closing period.

> Full Quotation: "Decisions on how to operate the company should be based on the most accurate and relevant information available from both within the company and from the specific community that the establishment serves."

> Quotation with Ellipsis: "Decisions on how to operate the company should be based on the most accurate and relevant information available. . . ."

Note containing a direct quotation.   Figure 8.13

Iona Crandell, "Importance of Internet Medicine," <u>Journal of Emergency Medicine</u> 15 (Sept. 2010): 32.

"The Internet is a primary source of medical information for consumers. CybMed, an Internet marketing firm, estimates that more than 80 million people in 2009 consulted the Web for a variety of health-related information. Most users searched popular sites such as Medscape to look up the signs and symptoms of their medical problems. Consumers also flocked to PharmInfo.net, an FDA website, to find information on new drugs and their possible side effects. These websites are the closest thing to a doctor who makes house calls."

Figure 8.14   Note containing a paraphrase.

Acid rain (body of report)
damage to forests

Dampier, www.rainenviron.com

Acid rain is as dangerous to the forests as to the lakes. Victims of "premature senescence," the trees become defoliated and die with no new trees taking their place. Without the trees' protection, wildlife vanishes. Although the exact damage is hard to measure, Swedish scientists have observed that in their country forest products decreased by one percent yearly.

At times you may have to insert your own information within a quotation. This addition, known as an **interpolation**, is made by enclosing your clarifying identification or remark in brackets inside the quotation; for example, "It [the new transportation network] has been thoroughly tested and approved." Anything in brackets is not part of the original quotation.

Paraphrasing

Most of your note taking will be devoted to paraphrasing rather than writing down direct quotations. A **paraphrase** is a restatement in your own words of the author's ideas. Even though you are using your own words to translate or restate, you still must document the paraphrase because you are using the author's facts and interpretations. You do not use quotation marks, though. When you include a paraphrase in your paper, you should be careful to do four things:

1. Be faithful to the author's meaning. Do not alter facts or introduce new ideas.
2. Follow the order in which the author presents the information.
3. Include in your paraphrase only what is relevant for your paper. Delete any details not essential for your work.
4. Use paraphrases in your report selectively. You do not want your report to be merely a restatement of someone else's ideas.

Paraphrased material can be introduced in your paper with an appropriate identifying phrase, such as "According to Dampier's study," "To paraphrase Dampier," or "As Dampier observes." The note shown in Figure 8.14 paraphrases the following quotation:

While the effects of acid rain are felt first in lakes, which act as natural collection points, some scientists fear there may be extensive damage to forests as well. In the process described by one researcher as "premature senescence," trees exposed to acid sprays lose their leaves, wilt, and finally die. New trees may not grow to replace them. Deprived of natural cover, wildlife may flee or die. The extent of the damage to forest lands is extremely difficult to determine, but scientists find the trend worrisome. In Sweden, for example, one estimate calculates that the yield in forest products decreased by about one percent each year. . . .[1]

---

[1]Bill Dampier, "Now Even the Rain Is Dangerous," *International Wildlife* 10, pp. 18–19.

Documenting Sources

Documentation is at the heart of all the research you will do on the job. To document means to furnish readers with information about the print and electronic sources you have used for the factual support of your statements, including books, journals, newspapers, manuals, reports, websites, and other resources such as list-servs and e-mail. Without proper documentation, you will not be able to persuade a customer to buy your product or service and you will not convince your boss that you are doing your best work.

Documentation is an essential part of any research you do for at least four reasons.

1. It demonstrates that you have done your homework by consulting experts on the subject and relying on the most authoritative sources to build your case persuasively.
2. It shows that you are aware of the latest research in your field, thus lending credibility and authority to your conclusions and recommendations.
3. It gives proper credit to those sources and avoids plagiarism (see page 35 in Chapter 1). Citing works by name is not a simple act of courtesy; it is an ethical requirement and, because so much of the material is protected by copyright, a point of law.
4. It informs readers about specific books, articles, or websites you used so they can locate your source and verify your facts or quotations.

The Ethics of Documentation: Determining What to Cite

As a researcher, you have to be sure about what information you must cite and what information you do not need to cite. Before you start consulting sources, you have to be very clear about the ethical standards involved in documentation. The following sections will give you a useful overview to make the documentation process more understandable and easier to follow.

What Must Be Cited

To ensure that your business report avoids any type of plagiarism and maintains high ethical and professional standards, follow these guidelines.

- If you use someone else's exact words, document the source. Document any direct quotation(s), even a single phrase or keyword you could not have come up with on your own.
- Even if you do not use an author's exact words but still get an idea, concept, or point of view from that author, document the work in your paper.
- If any opinions, interpretations, and conclusions expressed verbally or in writing are not your own (e.g., you could not have reached them without the help of another source), document them.

- Stay away from **patchworking**—using bits and pieces of information and passing them off as your own—which is also an act of plagiarism. Always put quotations marks around anything you take verbatim and document it.
- If you use statistical data you have not compiled yourself, document them.
- Always document any visuals—photographs, graphs, tables, charts, images downloaded from the Web (and if you construct a visual based on someone else's data, you must also acknowledge that source).
- Be faithful to the rules established by your collaborative writing team. Never submit as your own work that which was done through collaboration.
- Document any software programs you did not develop yourself.

What Does Not Need to Be Cited

Be careful not to distract readers with unnecessary citations that only demonstrate your lack of understanding of the documentation process and can undercut the professionalism of your report. There is no need to cite the following:

- Common-knowledge scientific facts and formulas, such as "The normal human body temperature is 98.6 degrees Fahrenheit," or "H_2O."
- Readily available geographical data, such as the square mileage of states or countries, elevation of mountains, depth of lakes, rivers etc., populations, and the like.
- Easy-to-obtain historical facts and dates, such as "Ronald Reagan was the 40th president of the United States" or "The first manned moon landing occurred on July 20, 1969."
- Proverbs from folklore, such as "The hand is quicker than the eye."
- Famous quotations, such as "We hold these truths to be self-evident . . . ," though you should mention the name of the person quoted where you include his or her wording in your report.
- Quotations from the *Bible*, the *Koran*, other religious texts, or classic literary works. Although you do not need to provide full citations of these texts at the end of your report, you should identify the specific work and exact place in that work (Deuteronomy 13:5–6, *Merchant of Venice* 2.3) parenthetically in your report.

Documentation Styles

Most professional associations have created their own guidelines for documenting sources. For example, the Council of Biology Editors has outlined a style for biologists in *The CBE Manual for Authors, Editors, and Publishers*, the *Harvard Law Review* provides a style manual for legal professionals in *The Bluebook: A Uniform Style of Citation*, and the National Association of Social Workers uses *Writing for the NASW Press: Information for Authors*. But the two most widely used documentation styles are MLA (Modern Language Association) and APA (American Psychological Association), as detailed in their publications:

- *MLA Handbook for Writers of Research Papers,* Sixth Edition, 2003, *http:// www.mla.org*
- *Publication Manual of the American Psychological Association,* Fifth Edition, 2001, *http://www.apa.org*
- *APA Style Guide to Electronic References,* 2007

MLA is used primarily in the humanities, while APA is used in psychology, nursing, social sciences, and several technological/scientific fields. In business, however, whether you use MLA or APA will be determined by employers. But because MLA and APA are the most well-known and accessible documentation styles, many businesses prefer to rely on one or the other, or adapt or modify these methods to suit the company's needs and those of its clients. The report at the end of this chapter follows MLA style; the report in chapter 15 follows APA style.

What MLA and APA Have in Common

While there are differences between MLA and APA, they share two basic features:

1. Both require in-text citations within your researched report. In-text citations, also known as **parenthetical citations**, are brief citations placed directly in your report, in or at the end of the sentence requiring a reference, to indicate that the information came from a source. These citations include only the most essential information (to avoid interrupting the reader's train of thought), but they let readers know precisely where you borrowed information. An in-text citation also tells readers that they can expect a full bibliographic reference to appear at the end of your report.

2. Both include a complete list of sources cited at the end of the report. MLA uses a **Works Cited** list at the end of the report, while APA calls for a **References** list. These pages provide detailed bibliographic information about all of the sources you cited in your research. Every source you have cited in the text must also be included in these lists, with some exceptions in APA (interviews, surveys, personal conversations, e-mails, and IMs).

Beyond these two shared points, however, MLA and APA follow different citation formats. The next two sections of this chapter will show you how to cite sources (both in the text and at the end of a report) using MLA and APA.

Following MLA Style

This section will help you prepare both in-text citations and a Works Cited list according to MLA guidelines. It will also describe the specific information you have to give readers and where to put it. For each type of citation you will find a helpful example (see also pages 387–388).

Using MLA In-Text Citations

When you insert in-text citations in your report, keep in mind that every citation in your report must correspond to a work you cite in full at the end of your report on the Works Cited page(s). And, correspondingly, every work you cite on the Works Cited list must be included as an in-text citation in your report. Note that when you cite a page number parenthetically, MLA does not use *page, pages, p.* or *pp.* You just list the author's last name(s) and page number(s), without a comma between them (e.g., Jones 32). Here are some guidelines and examples:

- If you do not mention the author(s) by name in your sentence, then you must include the author(s) and the page number(s) in parentheses at the end of the sentence:

  ```
  About 5 percent of the world's population has diabetes
  mellitus, and 25 percent of the world's population
  carries the disease (Walton 56-57).
  ```

- But if you do refer to the author(s) by name in your sentence, supply only the page number(s) consulted in parentheses at the end of the sentence:

  ```
  According to Tamara deTorres, inflation in New Zealand is
  expected to increase rapidly by 2010 (27).
  ```

- When you cite a work written by two authors, list both authors, connected by *and*:

  ```
  While fax machines are still used to transmit copies of
  important documents such as contracts and signed letters,
  e-mail attachments frequently take the place of faxes
  (Perez and Klein 25-27).
  ```

- If the work was written by three authors, list all of the authors, separated by commas, the last two connected by a comma followed by *and*:

  ```
  Tourism has increased by 14 percent this last quarter,
  thanks to individuals passing through our state on their
  way to the national Association of Business Leaders
  convention (Muscovi, Klein, and Philips 57-58).
  ```

- When the work you consulted has more than three authors, list just the first author's name followed by *et al.* (Latin for "and others"):

  ```
  The principles of ergonomics have revolutionized the
  design of office furniture (Brodsky et al. 345-47).
  ```

■ For a work by two authors with the same last name, provide both authors' first and last names, connected by *and*:

```
New York City has the highest percentage of residential
turnover per year (Joyce Yates and Peter Yates 66).
```

■ Sometimes you will have to cite two or more works by the same author. To distinguish one work from the other, provide the author's last name, followed by a comma and a shortened version of the particular work's title (in the example below, the complete title of the work has been shortened from *Compensations and Benefits*):

```
Many firms offer employees 401K stock purchase plans
(Howe, Compensations 16).
```

■ Occasionally, you may have to cite two works in the same sentence. When you do, list both sources, in alphabetical order, separated by a semicolon. As a rule, though, do not overload readers with citations:

```
The use of salt domes to store radioactive wastes has
come under severe attack (Jelinek 56-57; McPherson and
Chin 23-26).
```

■ When the work you consulted was written by a company, organization, or government agency, use a shortened version or abbreviation of the name of the organization (in the example below, *Commission* replaces *Commission on Wage and Price Control*; if abbreviating you would replace *American Medical Association* with *AMA* if it is listed this way in the Works Cited section):

```
Salaries for local electricians were at or above the
national average (Commission 145).
```

■ If the work you've referenced has no author, not even the name of a company, organization, or government agency, use a shortened version of the title and the page number(s).

```
Customers surveyed tended to focus on the product's prob-
lems rather than on its benefits ("Unicom Survey" 22-23).
```

■ Since websites do not have page numbers, all that you have to list parenthetically is the name of the author, or if you use the author's name in your sentence, then simply list a shortened title of the website:

```
D'Arcy claims that by 2012, transportation in Boston will
be even more streamlined to accommodate a projected
increase in riders ("Boston Trans.").
```

■ For a quotation consisting of four or fewer typed lines, include the quotation within your sentence (using quotation marks) and place the in-text citation at the end of the sentence, followed by a period:

```
Pilmer has observed that "renovation of downtown
areas is one of the most pressing issues facing city gov-
ernment" (16).
```

■ For a quotation longer than four typed lines, set the quoted material apart from the text by indenting it ten spaces or one inch on the left side and eliminating the quotation marks. Place the in-text citation after the quotation and the period, as in the following example:

```
L. J. Ronsivalli offers this graphic analogy of how radi-
ation can penetrate solid objects:
          One might wonder how an x-ray, a gamma ray, or a
          cosmic ray can penetrate something as solid as a
          brick wall or a piece of wood. We can't see that
          within the atomic structures of the brick wall
          and the wood there are spaces for the radiation
          to enter. If we look at a cloud, we can see its
          shape, but because distance has made them too
          small, we can't see the droplets of moisture out
          of which the cloud is made. Much too small for
          the eye to see, even with the help of a
          microscope, the atomic structure of solid
          materials is made of very small particles with a
          lot of space between them. In fact, solids are
          mostly empty spaces. (20-21)
```

Preparing an MLA Works Cited List

Place your Works Cited list at the end of your report, on a new page, with the title Works Cited centered at the top. Arrange the list alphabetically by author last names (except when there is no author listed) and begin each citation flush to the left margin (but indent subsequent lines of the citation five spaces or one-half inch). Separate each item within the citation with a period and one space, and double-space within and between entries. When your Works Cited list includes two or more works by the same author, indicate the second work by that author (and subsequent works by that author, if applicable) by replacing the author's name with three hyphens and a period (---.). To see a complete Works Cited list, look at the example at the end of the business research report in this chapter (pages 375–387).

Table 8.4 provides general guidelines on how to format information about author, title, publication data, and page number(s)/URLs on a Works Cited list.

TABLE 8.4   General Guidelines on How to Prepare an MLA Works Cited List

Author	■ Give the author's full name, last name first, followed by a comma, and then first name, and middle name or initial (if applicable). ■ When there are two authors, invert only the first author's name. ■ For a work by three authors, list them in the order they appear in the original source, with the last two authors' names connected by the word *and*. ■ If there are more than three authors, list only the first author's name, followed by *et al.*
Title	■ Cite the full title of source (book, film, monograph), including subtitle, and italicize or boldface it. ■ Capitalize the first letter of every word in titles and subtitles, except prepositions (*with*, *for*, etc.), conjunctions (*and*, *or*), and articles (*a*, *an*, *the*) unless they are the first word of the title. ■ When you cite both an article within a print or online publication and the name of the publication, put the title of the article first (in quotation marks) and then the name of the publication (italicized or underscored, depending on what your employer prefers).
Publication information	■ For books, give the city of publication. If the city may be unfamiliar to readers, follow the name of the city with a comma and a two-letter abbreviation for the state, province, or country (e.g., Gambier, OH, as opposed to Cincinnati). Use a colon between the name of the city and the publisher's name. ■ Use a shortened versions of publisher names (e.g., Houghton, not Houghton Mifflin); for a university press, use the abbreviation UP (e.g., New York UP, not New York University Press); for other publishers that use the word *Press,* just use the letter *P.* ■ Place the date of publication immediately after the publication data, preceded by a comma.
Volume and page number(s)	■ For articles, give the volume number, followed by the date in parentheses—52 (June 2009); for newspapers and popular magazines, do not give a volume number, just the date—12 Aug. 2010. ■ Do not use the words *page* or *pages* or the abbreviations *p.* or *pp.* ■ Include all digits for page numbers from 1 to 99 (e.g., 34–37, not 33–7), and include at least the last two digits for page numbers over 100, three or more digits if necessary (e.g., 166–68, not 166–8 or 596–607, not 596–07).
URLs	■ Place URLs where you would place page numbers, with beginning and end angle brackets, <> ■ If a URL must break to a second or subsequent line, put it after a forward slash or a period. Do not hyphenate to show a break.

The following examples will give you models for a variety of sources you may include in your MLA Works Cited page(s).

Book by One Author

> Spraggins, Marianne. *Getting Ahead: A Survival Guide for Black Women in Business.* Indianapolis: Wiley, 2009.

Book by Two or Three Authors

> Shaw, Timothy M., and Shaun Breslin. *China in the Global Political Economy.* New York: Palgrave Macmillan, 2007.

Book by Four or More Authors

> Brown, M. Katherine, et al. *Managing Virtual Teams: Getting the Most from Wikis, Blogs, and Other Collaborative Tools.* Plano, TX: Wordware, 2010.

Online Book

If the online book was first published in print form, provide a complete citation as you would for a print book, and then add the name of the website (underlined or in italics) where you accessed the book, the date of access, and a URL that will lead your readers directly to the online book, not just to the website's homepage.

> Sheely, M. *The Birth of Wikis.* San Francisco: Computer Press, 2007. *The Computer Press Online.* 2 Mar. 2008 <http://www.cpo.com/books/sheely.htm/>.

If the online book was never published in print form, list the name of the book and the copyright, but omit the publisher's location and name. Then add the name of the website (underlined or in italics) where you accessed the book, the date you accessed it, and a URL that will lead your readers directly to the online book.

> Sheely, M. *The Birth of Wikis.* 2007. *The Computer Press Online.* 2 Mar. 2008 <http://www.cpo.com/books/sheely.htm/>.

Edited Book/Collection

The word ed. (or eds., not capitalized) appears immediately after the editor's name.

> Martin, Jeanne B., ed. *Honest Work: A Business Ethics Reader.* New York: Oxford, 2007.

Article or Chapter Within an Edited Book

Give the name(s) of the author(s) of the article or chapter first, then the title of the article (in quotation marks), the title of the book (underlined or italicized), and the editor name(s) (preceded by *ed.* or *eds.*). Because the article, not the complete book, is cited, you must provide page numbers at the end of the citation.

> Sleet, David A., and Andrea Carlson Gielen. "Injury Preven-
>   tion." *Health Promotion in Practice*. Ed. Sherri S.
>   Gorfield. San Francisco: Jossey-Bass, 2006. 361-91.

Book by a Corporate/Organizational Author

List the name of the organization as author, but abbreviate the corporation/organization after the place of publication.

> Computer Literacy Foundation. *PC's in the Classroom*. 3rd
>   ed. New York: CLF, 2009.

Article in a Print Professional Journal

If the journal has new pagination each issue, you must provide the volume and issue number immediately after the name of the journal, separated by a period. The copyright year appears in parentheses after the volume/issue, followed by a colon and the page numbers.

> Staples, Sandy, and Linda Zhao. "The Effects of Cultural
>   Diversity in Virtual Teams Versus Face-to-Face Teams."
>   *Group Decision and Negotiation* 15.4 (2009): 22-25.

If the journal has continuous pagination throughout the year, you do not need to provide the issue number, because no other issues in that year will have the same page numbers.

> Ruez, Sonia, and Paul Zuckerman. "The Impact of Community
>   Involvement in Shaping Public Policy Decisions."
>   *Contemporary Community Planning* 12(2009): 271-79.

Article in an Online Professional Journal

The format for citing an online journal is the same as citing a print journal, except you must add the date you accessed the article and replace the page number(s) with a URL that leads directly to the article.

> Manojlovich, Marilyn. "Power and Empowerment in Nursing:
>   Looking Backward to Inform the Future." *Online
>   Journal of Issues in Nursing* 12.1 (2007). 12 Oct.
>   2008 <http://www.nursingworld.org/ojin/topic32/
>   pc32_1.htm>.

Article in a Print Magazine

If the magazine is issued weekly, include the day prior to the month and year. If the magazine is published monthly, include only the month (abbreviated) and year. The page numbers appear at the end, preceded by a colon. If the article does not appear on consecutive pages, indicate the additional pages with a plus sign (+).

Ganey, Jayme. "Is the Boss Watching You?" *Essence* Oct.
    2007: 36+.

Article in a Print Newspaper

If the newspaper has sections, you must provide the section letter (preceded by the abbreviation *sec.*) as well as the page number(s). As with a magazine article, if the article does not appear on consecutive pages, indicate the additional pages with a +. Page numbers appear immediately after the section letter.

Hopkins, Jim. "How to Avoid Spam Avalanche." *USA Today*
    21 Feb. 2007, sec. B: 3+.

Article in an Online Magazine or Newspaper

Indicate the date the article was posted, followed immediately by the date you accessed the article. Also, provide a link directly to the article, not to the magazine's or newpaper's homepage.

Demerjian, Dave. "As Skies Grow Crowded, FAA Preps Air
    Traffic Control 2.0." *Wired.com* 25 Oct. 2007. 28 Jan.
    2008 <http://www.wired.com/science/
    discoveries/news/2007/10/faa>.

Article in a Printed Reference Work

Provide both the author's name and the editor's name for the reference work, if either is provided. The latter appears after the title. Also give the edition number of the reference work after the name of the editor.

Valleri, Angelo. "IM's." *The Encyclopedia of the Internet*.
    Ed. Andrea Jackson. 4th ed. Manchester, UK: Computer P,
    2008. 445-46.

Article in an Online Reference Work

Indicate not only the copyright year of the reference work, but also the date you accessed the article. If an author's name is listed, include it too. The URL must lead directly to the article, not to the online reference's homepage.

Miller, Pauline. "Iphone." *MSN Encarta Encyclopedia*. 2007.
    3 Feb. 2008 <http://encarta.msn.com/
    encyclopedia_761552652_2/Apple_Inc.html#p21>.

Government Document

A government document is listed the same way a book is. However, the name of the government agency that commissioned the report takes the place of the author. Note that another government agency may have published the document.

> Commission on Northeast Labor Statistics. *The Impact of Immigration on Queens, New York: Labor in the New Millennium.* Washington, DC: Bureau of Labor Statistics, 2007.

Brochure

Cite a brochure as you would a book. If no author is provided, just list the title. If the name of the publisher or the date is not provided on the brochure, use the abbreviations *n.d.* for "no date" and *n.p.* for "no publisher" where you would ordinarily place that information. A brochure is a type of gray literature (see Tech Note on page 339).

> Gao, Hubert. *Coping with Carpel Tunnel Syndrome.* New York: Beth Israel Hospital, n.d.

Blog Posting

Include the date of access and the URL directly to the posting, not the blog's homepage. Use the phrase *online posting* immediately after the title of the blog posting. Note that the name of the blog is not underlined or italicized.

> Schwartz, Jonathan. "Sun Analyst Conference." Online posting. Schwartz Business Blog. 12 May 2009. <http:// blogs.sun.com/jonathan/entry/sun_analyst_conference>.

Personal Interview

Supply the name of the person interviewed, the phrase *personal interview*, and the date.

> Alvarez, José. Personal interview. 15 Oct. 2009.

Published Interview

List the interview as you would a magazine, newspaper, or journal article.

> Pacarek, Mildred. Interview. "Chicago's Cyberspace Future." *Windy City News.* 2 Apr. 2008: 13-15.

Survey

Provide the name of the person who conducted the survey, the name of the survey (or just the word *survey*), the company or agency for which it was done, and the date of the survey.

> Guttierez, Joseph. Market survey for Duron, Inc. 10 Apr. 2009.

Radio or Television Broadcast

Keep in mind that you have to include both the episode name (in quotations) and the series name (underscored or italicized). Also give the name of the producer (using the abbreviation *Prod.*) after the program. Then cite the network, the name and location of the station (separated by a comma), and the broadcast date.

> "What Are Firewalls?" *This Week in Technology*. Prod. Raul
> Weber. PBS. WNET, New York. 21 Jan. 2008.

Lecture or Speech

If the lecture or speech is unpublished, include the name of the presenter, the title of the speech in quotations (if known, otherwise include the word *Lecture* or *Speech* without quotations), the name of the conference (if applicable) and location, and the date.

> Philips-Ricks, Jonatha. Lecture. National Association of
> Black Business Leaders Conference, New York. 15 Aug.
> 2010.

If the lecture or speech is published in conference proceedings, treat it like an article or chapter within an edited book.

> Philips-Ricks, Jonatha. "The Future of Corporate Bond
> Funds." *Proceedings of the National Association of
> Black Business Leaders Conference, New York, 15-18
> August 2010*. Ed. Arnold Laven. New York: NABBL P,
> 2010. 234-36.

Map or Graphic

Make sure you credit the person or group who created the map or graphic (if available), the title of the map or graphic, the word *Map* or type of graphic (*Table*, *Chart*, *Illustration*, etc.), and the publication data, including copyright year or date of completion.

> Fineberg, Donald. *Sonoma, California*. Map. Sonoma: Profes-
> sional Maps, 2006.

Online Map or Graphic

Cite an online map or graphic as you would a print version, except replace the publication data with the date the map or graphic was posted. Also give the name of the website, the date you accessed the map or graphic, and the URL leading directly to it.

> Fineberg, Donald. *Sonoma, California*. Map. 20 Mar. 2007.
> American Maps Online. 12 Jan. 2008. <http://www
> .americanmaps.com/california2007/sonoma.html>.

Following APA Style

Like MLA, APA uses both in-text citations and a References list at the end of your report that supplies full publication information. As in the earlier discussion of MLA
style, this section will show you what to include and where, according to the APA
guidelines.

Using APA In-Text Citations

As in MLA, any APA in-text citations in your report must also appear on the References list at the end of your report and vice-versa. However, unlike in MLA, personal communications (see pages 372–373) appear only in the text in APA style. In
addition, APA requires the publication year rather than the page number(s) in in-
text citations (with a comma between the author's name and publication year). When
using APA documentation, follow the specific guidelines below:

- When you do not mention the author(s) by name in your sentence, then you
  must include it and the unabbreviated year of publication in parentheses at
  the end of the sentence:

  ```
  About 5 percent of the world's population has diabetes
  mellitus, and 25 percent of the world's population
  carries the disease (Walton, 2007).
  ```

- But if you do refer to the author(s) by name in your sentence, supply only the
  year of publication of the work in parentheses:

  ```
  According to Tamara deTorres (2008), inflation in New
  Zealand is expected to increase rapidly by 2015.
  ```

- When you cite a work written by two authors, list both authors, connected
  by &:

  ```
  While fax machines are still used to transmit copies of
  important documents such as contracts and signed letters,
  e-mail attachments frequently take the place of faxes
  (Perez & Klein, 2008).
  ```

- If the work was written by three to five authors, list all of the authors, separated by commas, the last two connected by &.

  ```
  Tourism has increased by 14 percent this last quarter,
  thanks to individuals passing through our state on their
  way to the National Association of Black Business Leaders
  Conference (Muscovi, Klein, & Philips, 2010).
  ```

- When the work you consulted has six or more authors, list just the first author's name followed by *et al.* (Latin for "and others").

```
The principles of ergonomics have revolutionized the
design of office furniture (Brodsky et al., 2007).
```

- For a work by two authors with the same last name, provide each author's first initials followed by a period and connected by *&*:

```
New York City has the highest percentage of residential
turnover per year (Yates, P., & Yates, J., 2008).
```

- Sometimes you will have to cite two or more works by the same author. To distinguish one work from the other, simply provide the copyright year that matches the appropriate work:

```
Many firms offer employees 401K stock purchase plans
(Howe, 2007). Corporate contributions to employee tax-
sheltered accounts have decreased over the last three
years (Howe, 2010).
```

- Occasionally, you will have to cite two works in the same sentence. When you do, list both sources, in alphabetical order, separated by a semicolon. As a rule, though, do not overload readers with more than two or three citations.

```
The use of salt domes to store radioactive wastes has
come under severe attack (Jelinek, 2007; McPherson &
Chin, 2008).
```

- When the work you consulted was written by a company, an organization, or a government agency, use the full name of the organization in your first reference to it in a citation, with the abbreviated name of the organization in square brackets. Use the abbreviated version of the organization without the square brackets in subsequent citations:

First Citation:
```
Salaries for local electricians were at or above the
national average (Commission on Wage and Price Control
[Commission], 2007).
```

Subsequent Citations:
```
Salaries for janitorial workers, however, remained steady
(Commission, 2007).
```

- If the work you've referenced has no author—not even the name of a company, organization, or government agency—use a shortened version of the work's title and the copyright year:

```
Customers surveyed tended to focus on the product's prob-
lems rather than on its benefits ("Unicom Survey," 2006).
```

- When you cite an online work, include the author and date, but not the name of the website:

```
D'Arcy claims that by 2012 transportation in Boston will
be even more streamlined to accommodate a projected
increase in riders (2008).
```

- APA regards unpublished interviews, surveys, focus groups, e-mails, telephone conversations, text messages, and IMs as "personal communications." Document personal communications in the text, but do not include them in your References list. Give the name of the person with whom you communicated either in your sentence or in the citation, specify the type of communication, and include month, day, and year:

```
Dr. Patavi indicated that this information came from the
preliminary report submitted last month (interview, April
25, 2008).
```

- For a quotation shorter than forty words, include the quotation within your sentence (using quotation marks) and place the in-text citation at the end of the sentence, followed by a period:

```
Pilmer has observed that "renovation of downtown areas is
one of the most pressing issues facing city government"
(2007).
```

- For quotations longer than forty words, set the quotation apart from the text by indenting it five spaces or one-half inch on the left side and eliminating the quotation marks. Place the in-text year after the author's name and the page number(s) after the period at the end of the quotation, as in the following example:

```
L. J. Ronsivalli (2008) offers this graphic analogy of
how radiation can penetrate solid objects:
    One might wonder how an x-ray, a gamma ray, or a cos-
    mic ray can penetrate something as solid as a brick
    wall or a piece of wood. We can't see that within the
    atomic structures of the brick wall and the wood
    there are spaces for the radiation to enter. If we
    look at a cloud, we can see its shape, but because
    distance has made them too small, we can't see the
    droplets of moisture out of which the cloud is made.
    Much too small for the eye to see, even with the help
    of a microscope, the atomic structure of solid
```

```
materials is made of very small particles with a lot of
space between them. In fact, solids are mostly empty
spaces. (pp. 20-21)
```

Preparing an APA References List

Begin your References list at the end of your report, on a new page, with the title *References* centered at the top and then arrange the list alphabetically by authors' last names (except when there is no author listed). Begin each citation flush to the left margin (but indent subsequent lines five spaces or one-half inch) and separate each item within the citation with a period and one space, and double space within and between entries. Unlike MLA style, when your APA References list includes two or more works by the same author, re-list the author's name for each entry, and if the publication year is the same, use lowercase letters *a*, *b*, etc. after the copyright year to distinguish them. A complete APA References list can be found at the end of the long report about multicultural workers in Chapter 15 (pages 652–667).

Table 8.5 provides general guidelines on how to list author, title, publication data, and page number(s)/URLs on an APA References list. Specific guidelines and examples of how to format different types of sources (e.g., books, magazines, blogs, interviews, etc.) appear below.

Book by One Author

```
Spraggins, M. (2009). Getting ahead: A survival guide for
     black women in business. Indianapolis: Wiley.
```

Book by Two to Five Authors

```
Shaw, T., & Breslin, S. (2007). China in the global polit-
     ical economy. New York: Palgrave Macmillan.
```

Book by Six or More Authors

```
Brown, M. K., et. al. (2007). Managing virtual teams: Get-
     ting the most from wikis, blogs, and other collabora-
     tive tools. Plano, TX: Wordware.
```

Online Book

After the author, copyright year, and title, add the words *Retrieved from* and the URL leading directly to the online book. If the URL leads to the site on which the book is available but not directly to the book, use the words *Available from* instead.

```
Sheely, M. (2007). The birth of wikis. Retrieved from
     http://www.cpo.com/books/sheely.htm/
```

TABLE 8.5   General Guidelines for Preparing an APA References List

Author	■ List the author's last name first, followed by a comma, first initial, and middle initial (if applicable). ■ When there are multiple authors, invert all authors' names (e.g., Smith, J. or Jones, T.). ■ List multiple authors in the order they appear in the original source, with the last two authors connected by &. ■ If there are two to five authors, list all of them; if there are more than six authors, list only the first six authors, followed by *et al.*
Publication date	■ Place the date of publication in parentheses immediately after the author's first initial (or middle initial, if there is one)—Chow, N. (2009). ■ Order the date information by year (followed by a comma), month (spelled out), and day.
Title	■ Cite the full title of the source, including subtitle. ■ For a book, capitalize only the first word of the title (except for proper names in the title), and the first word after a colon; however, for a periodical title, capitalize the first letter of every word except for prepositions, articles, and conjunctions. ■ For an article in a print or online publication, give the title of the article first (not in quotation marks), then list the name of the publication in italics.
Other publication information	■ For books, provide the city of publication. If the city is not familiar to readers, follow the name of the city with a comma and a two-letter abbreviation for the state, province, or country (e.g., Gambier, OH, as opposed to Cincinnati). Use a colon between the city of the publication and the publisher's name. ■ Use a shortened version of the publisher's name (e.g., Houghton, not Houghton Mifflin); for a university press, spell out the words *University Press.*
Volume and page numbers	■ For professional articles, put the volume number (in italics) with the issue number (not in italics) in parentheses immediately after the title. Then insert a comma and include page numbers— *12* (3), 87–102. ■ Do not use the abbreviations *p* or *pp.* or the words *page* or *pages.* ■ List complete page numbers, e.g., 37–51, 481–482, not 481–2 or 481–82). ■ Do not include page or volume numbers for a reference work such as an encyclopedia or dictionary.
URLs	■ Insert URLs where you would place page numbers. Do not use angle or other brackets around a URL or follow a URL with a period. ■ When possible, use a DOI (Digital Object Identifier, which is similar to a book's ISBN number) instead of a URL. DOIs are more permanent and consistent in format than URLs. ■ If a URL runs on to a second or subsequent line, break it before a forward slash or period. Do not hyphenate to show a break or otherwise alter any information within a URL.

Edited Book/Collection
Note that the word *Ed.* (or *Eds.,* capitalized) appears after the editor's name, in parentheses. As with books written by more than five authors, if there are more than five editors, list only the first editor, followed by *et. al.*

> Martin, J. (Ed.). (2007). *Honest work: A business ethics reader.* New York: Oxford.

Article or Chapter Within an Edited Book
Include the author and title of the article first. Then insert *In* followed by the editor's name(s) and the title of the book. Because the article, not the complete book, is cited, you must provide the page numbers (in parentheses after the title of the book).

> Sleet, D. A., and Gielen, A. C. (2006). Injury prevention. In S. S. Gorfield (Ed.), *Health promotion in practice* (pp. 361-391). San Francisco: Jossey-Bass.

Book by a Corporate/Organizational Author
Place the name of the organization where the author's name would ordinarily go.

> Computer Literacy Foundation. (2009). *PC's in the classroom* (3rd ed.). New York: Technology Press.

Article in a Professional Print Journal
If the journal has new pagination each issue, you must provide the volume and issue number immediately after the journal name. Note that the issue number appears in italics, but the volume number appears in plain text and in parentheses.

> Staples, S., & Zhao, L. (2006). The effects of cultural diversity in virtual teams versus face-to-face teams. *Group Decision and Negotiation, 15*(4), 22-25.

If the journal has continuous pagination throughout the year, you do not need to provide the issue number, because no other issues in that year will have the same page numbers.

> Ruez, S., & Zuckerman, P. (2009). The impact of community involvement in shaping public policy decisions. *Contemporary Community Planning 12,* 271-279.

Article in an Online Professional Journal
If the article does not have a DOI, you must use the words *Retrieved from* prior to the URL.

```
Manojlovich, M. (2007). Power and empowerment in nursing:
      Looking backward to inform the future. Online Journal
      of Issues in Nursing 12(1). Retrieved from
      http://www.nursingworld.org/ojin/topic32/pc32_1.htm
```

If the article does have a DOI, replace the URL with the DOI and replace *Retrieved from* with the abbreviation *doi*:

```
Manojlovich, M. (2007). Power and empowerment in nursing:
      Looking backward to inform the future. Online Journal
      of Issues in Nursing 12(1). doi: 10.3435/1114-2456.
      86.6.285
```

Article in a Print Magazine

If the magazine is published weekly, include the month and day after the year. But if the magazine is issued monthly, include only the month after the year. You must also include the page number(s) of the article. If the article does not appear on consecutive pages, indicate the additional pages with a +.

```
Ganey, J. (2007, October). Is the boss watching you?
      Essence, 36+.
```

Article in a Print Newspaper

Include the section number of the newspaper (if applicable) prior to page numbers and then list all pages on which the article appears, even if they are not consecutive.

```
Hopkins, J. (2007, February 21). How to avoid spam
      avalanche. USA Today, pp. B3, B12-13, B18.
```

Article in an Online Magazine or Newspaper

Provide a link to the article preceded by the words *Retrieved from*. In most cases, online articles will not include volume number, issue number, number of pages. But if they are provided, include them in your citation.

```
Demerjian, D. (2007, October 25). As skies grow crowded,
      FAA preps air traffic control 2.0. Wired.com 4(2).
      Retrieved from http://www.wired.com/science/
      discoveries/news/2007/10/faa
```

Article in a Print Reference Work

If an author's name and/or reference work editor's name are provided, APA says to include them in your citation. Use the word *in* prior to the name of the reference work. Also indicate the volume number (if applicable) of the reference work and the pages consulted (in parentheses) after the name of the reference work.

```
Valleri, A. (2008). IM's. In The encyclopedia of the
     Internet (Vol. 2, pp. 445-446). Manchester, UK:
     Computer Press.
```

Article in an Online Reference Work

Because online references are frequently updated, include the retrieval date between the words *Retrieved from*. Use the word *in* prior to the name of the reference work.

```
Miller, P. (2007). Iphone. In MSN Encarta Encyclopedia.
     Retrieved September 12, 2007, from http://encarta.
     msn.com/encyclopedia_761552652_2/Apple_Inc.html#p21
```

Government Document

A government document is treated the same way as a book is. However, cite the name of the government agency that commissioned the report as the author, cite the government agency that published the document as the publisher, and include the document number, if available, in parentheses after the title of the document.

```
Commission on Northeast Labor Statistics. (2007). The
     impact of immigration on Queens, New York, labor in
     the new millennium (43-600081-L). Washington, DC:
     Bureau of Labor Statistics.
```

Brochure

Cite a brochure as you would a book, but include the word *Brochure* in brackets after the title. If no author is given, list just the title. If the name of the publisher or the date is not on the brochure, use the abbreviations *n.d.* for "no date" and *n.p.* for "no publisher." A brochure is a type of gray literature (see Tech Note on page 339).

```
Gao, Hubert (n.d.). Coping with carpel tunnel syndrome
     [Brochure]. New York: Beth Israel Hospital.
```

Blog Posting

You must provide a URL to the blog's homepage and include the words *Message posted to* where indicated.

```
Schwartz, J. (2009, February 8). Sun Analyst Conference.
     Message posted to http://blogs.sun.com/schwartzbusiness
```

Personal Interview

Personal interviews are regarded as unpublished communications. Do not include them on your References list but document them in the text.

Published Interview

Cite a published interview as you would a magazine, newspaper, or journal article. If the article does not have the word *interview* in its title, place *Interview with* and the name of the person interviewed after the title of the article in square brackets. Include the page numbers in parentheses after the name of the publication.

```
Pacarek, M. (2008 April 2). Chicago's cyberspace future
        [Interview with Del O. Bohem]. Windy City News,
        pp. 13-14.
```

Survey

Surveys are unpublished personal communications and so are not included on your References list. However, document them in the text.

Radio or Television Broadcast

Include the following in your citation: the name of the producer (followed by the word *Producer* in parenthesis), the date of the broadcast (in parentheses), the title of the broadcast or episode, the phrase *[Radio broadcast]* or *[Television series episode]* (in brackets), the title of the series (preceded by the word *In*), the location of the station or network, and the name of the station or network.

```
Weber, R. (Producer). (2008, January 21). What are
        firewalls? [Television series episode]. In This Week
        in Technology. New York: WNET.
```

Lecture or Speech

If the lecture or speech is unpublished, include the name of the presenter, the date of the speech, the title of the speech (if known), and the location of the speech, using the words *Speech presented at* or *Lecture presented at*.

```
Philips-Ricks, J. (2009, August 15). Lecture presented at
        the National Association of Black Business Leaders
        Conference, New York, NY.
```

If the lecture or speech is published in conference proceedings, treat it like an article or chapter within an edited book.

```
Philips-Ricks, J. (2009). The future of corporate bond
        funds. In A. Laven (Ed.), Proceedings of the National
        Association of Black Business Leaders Conference, New
        York, 15-18 August 2009 (pp. 234-236). New York:
        National Association of Black Business Leaders
        Conference.
```

Map or Graphic

Include the name of the person or group who created the map or graphic (if available), the title of the map or graphic, the year of publication or completion date (in parentheses), the word *Map* or type of graphic in brackets (*Table, Chart, Illustration,* etc.), and the publication data.

```
Fineberg, D. T. (2006). Sonoma, California [Map]. Sonoma:
     Professional Maps.
```

Online Map or Graphic

Indicate the word *Map* or type of graphic in square brackets after the copyright year. Then provide the name of the map or graphic, and the URL, preceded by the words *Retrieved from.*

```
Fineberg, D. T. (2006). [Map]. Sonoma, California.
     Retrieved from http://www.americanmaps.com/
     california2007/sonoma.html
```

A Business Research Report

The rest of this chapter consists of a real-word business report written by members of the marketing team at New Horizons Development, Inc., a real estate group. The marketing team was asked to create a plan to attract tenants to Sawmill Ridge, a new apartment complex in the Dallas/Fort Worth area. Study the report to see how the team has successfully combined primary and secondary sources to research and write their business report. Also note how the report documents these sources in the text and references them on the Works Cited list using the MLA format. Compare the documentation style in this report to the researched report in Chapter 15 on pages 652–667, which uses APA in-text documentation and includes a References list. While these two reports illustrate MLA and APA guidelines, keep in mind your boss may very well ask you to adapt these styles to conform to your company's policies on documentation.

12 March 2008

New Horizons

2800 Taylor Blvd.
Fort Worth, TX 76003
817-555-3300
www.newhorizons.com

Talia Martinez-Ryals
Project Director
New Horizons-Southeast
Tucson, AZ 85749-3001

Dear Director Martinez-Ryals:

We are happy to send you our attached marketing plan for Sawmill Ridge, scheduled to open in three months in the rapidly growing community of Arlington, Texas.

Our plan projects launching a campaign to market Sawmill Ridge for its community feel and benefits as well as its proximity to expanding major employers and retail centers. We have determined that the most appropriate target audience will be young professionals.

This report is based on intensive demographic and market research. For the last two months, we have thoroughly investigated the South Arlington area, interviewed numerous business and community leaders, and explored current and projected demographics.

Thank you for asking us to prepare this report for New Horizons and you. If you have any questions about this report or need further information, please let us know.

Respectfully,

Adrienne Hong

Adrienne Hong

Tyrell Carpenter

Tyrell Carpenter

Margarita Gonzales

Margarita Gonzales

Building Apartment Homes Since 1992

*Title page
provides report
title, company
name, authors,
and date of
submission*

A Marketing Plan
for Sawmill Ridge

Prepared by

Adrienne Hong, Tyrell Carpenter,
and Margarita Gonzales

March 12, 2008

Prepared for

Talia Martinez-Ryals
Project Director
New Horizons Development, Inc.

1

Introduction: Strategic Plan

Two years ago, New Horizons Development, Inc., purchased a large tract of land (21.2 acres) in the Dallas/Fort Worth area to develop Sawmill Ridge, a Class A apartment community with a 54-building complex. The opening of Sawmill Ridge is scheduled for completion in three months. As a result of our overall objectives for this project, we have created a marketing plan for the next phase of Sawmill Ridge, that is, to attract tenants to the development. Our plan, described in this report, is based on a key marketing strategy linking Sawmill Ridge to community development. We provide background information about market conditions and location, describe the target audience, and, finally, outline plans for advertising the development. Our report also covers how we need to follow up with inquiries and how we can best retain residents.

Introduction clearly states the purpose and organization of the report

Building Community

To attract and retain tenants to Sawmill Ridge, our plan will be based on the marketing approach "Building Community by Offering Neighborhood Convenience." Selling the community is as important to our marketing plan as selling the development itself. Having access to neighborhood conveniences is especially suitable for our target audience, which we analyze below. The theme of community involvement is woven throughout our plan.

Authors succinctly spell out marketing plan and how it will be developed in report

Background

The rental market conditions in the greater Dallas/Fort Worth area, as well as Sawmill Ridge's proximity to three major employers and many new retail establishments, have shaped the focus and goals of our marketing plan:

Rental Market Conditions: Supply and Demand

Sawmill Ridge is opening at a very favorable time. The Dallas/Fort Worth rental market has continued to thrive for over two decades (D'Argento, *Contemporary History* 211–213), and apartment net leasing activity in Dallas/Fort Worth has hit a five-year high in the past twelve months, with absorption reaching 24,390 units (Evinson 211). As Figure 1 on page 2 shows, the annual demand included a strong fourth quarter performance as apartment occupancy in the metro area increased by approximately 5,620 units. During the fourth quarter of 2007, occupancy in the Dallas/Fort Worth area tightened significantly, standing at 92.7%. These figures point to an impressive 3.8% increase over the third quarter, a 5.2% increase over the second quarter, and a 4.7% increase over the first quarter (Commission 22–23). Further contributing to this strong demand, apartment communities lost fewer residents to first-time home purchases (Jiminez).

Typical in-text citation includes author and page

Statistics backed up by current secondary sources

No page number for Web source—Jiminez

2

Bar graph is appropriate visual to summarize statistics

Figure 1   Quarterly Dallas/Fort Worth Occupancy Rates (2007)

**Increase in apartment occupancy,
Dallas/Fort Worth metro area, 2007**

Government source lends authority to job statistics

Cites television program

Most important, though, in fueling this sizable demand for apartments, Dallas/Fort Worth has added a large block of new jobs. According to the preliminary data from the United States Bureau of Labor Statistics, 92,300 jobs were generated during the year ending December 2006, boosting employment by 3.4% (United States Bureau 44–48). According to Barbara Corcoran, a real estate expert appearing on *The Today Show*, "The single most important factor in selecting housing is the location of jobs in the area."

Location: Selling the Community

Citation lists name of map

As the map in Figure 2 on page 3 illustrates, Sawmill Ridge is located in South Arlington, near the intersection of Pike and Interstate 20, a major east-west highway in the metro area (South Arlington). Dallas is 9 miles to the east while Fort Worth is 11 miles to the west. Moreover, Sawmill Ridge is close to major retail centers and several large corporations/businesses.

Proximity to Retail Centers

Successful blend of primary research (interview) and secondary research

Essential to our marketing campaign, Sawmill Ridge is located in a highly desirable retail corridor in Arlington. The Parks Mall, which houses more than 2,000,000 square feet of retail shopping, is only 1.2 miles north of Sawmill Ridge. A new upscale retail center, the Highlands Center, with an additional 900,000 square feet of retail space, is currently being built only 1.8 miles from Sawmill Ridge. Based on our interview with Highland's Associate Director, companies like Ethan Allen, Golf Galaxy, Talbot's, and Studio Movie Grill have already purchased retail space in the

3

Figure 2   Map of Sawmill Ridge and the South Arlington area.

Map provides
readers with a
clear visual
of the site
location

new center (Weir-Wise). Other retailers— possibly the Body Shop, Gemstone Jewelers, Filene's Basement, and Best Buy—have been negotiating retail space and are leasing units quickly (Stratton). Having so many diverse and popular retailers nearby substantially increases the attractiveness of Sawmill Ridge.

Our Arlington Area Quality of Life survey of over 700 area residents, which was conducted between November 15 of last year and January 31 of this year, highlights this fact. Response rate was high—over 30% of those surveyed responded. Ninety-five percent of the area residents who completed the survey indicated that they visit the Parks Mall on a weekly basis, and an additional 86% replied that they plan to visit the Highland Center just as frequently after it opens. Only a small percentage (12%) of respondents was concerned that the Highlands may not have enough stores to cater to their needs (Hong et al., survey).

Primary
research (survey) provides
useful statistics
not available
elsewhere

Proximity to Employers

Sawmill Ridge's proximity to major employers is also a tremendous advantage in terms of rental appeal. Our campaign will make maximum use of this benefit. Among the area's largest employers is Washington Mutual, whose central office with 1,500 employees is within easy walking distance of Sawmill Ridge. Another major group of employers can be found at East Park Office Complex. A recent study projects that over twenty-five tenants will employ more than 350 individuals there, and the Complex is less than a mile from Sawmill Ridge (D'Argento, "Booming Market").

Second work
by previously
cited author
requires title

We can also target our campaign to market our development to current employees at these firms. Our semiannual focus group, which surveyed eleven employees from Washington Mutual and eight employees at the three largest firms at the East Park Office Complex, concluded that more than half of the employees at these companies are dissatisfied with long commutes to work and would strongly consider relocating to Sawmill Ridge. Only 20% of focus group participants claimed they would be "somewhat likely" to relocate (Hong et al., focus group).

Focus group
provides
otherwise
unavailable
information

4

Target Demographic

Based on the high numbers of varied retailers and businesses found in the South Arlington area, we will tailor our marketing strategy to attract young professionals (ages 24–34) with per capita incomes between $27,600 and $38,500 per year, individuals already working at these businesses or most likely to be employed there. Because many individuals in this population range have families with small children or are young single professionals, Sawmill Ridge will emphasize the security and satisfaction with the community that our development offers. (Three top-rated elementary schools and one junior high school are within a 2.4-mile radius of Sawmill Ridge.) Our plan is firmly based on the demographics of the area. The 2007 U.S. Census Bureau report (356–372) and local real estate reports (Kearney-Schwartz; Evinson) provide the following statistics, relevant to the five-mile area surrounding Sawmill Ridge, that reinforce why our plan needs to reach this target audience:

Goverment agency listed like a corporate author

- The total population within this radius is 217,682 people, and the average age is 28.3 years old.
- 59.2% of the population is composed of single-earner or dual-earner family households with children under 12 years old.
- The approximate size of renter-occupied housing units for the average household is 2.56 people.
- The average household income is $32,126, higher than the national average by 12%. Of this population, 33% are in management-level positions, 25.7% occupy non-management but well-paying office positions, and 17.7% hold entry-level jobs.

Target demographic solidly backed up by federal and local statistics

Based on these South Arlington demographics, we should easily be able to attract this target group to Sawmill Ridge, offering them the comfort, convenience, and security they demand from a large apartment developer such as New Horizons. ("New Complex" D11+).

Work with no author listed

Advertising Strategy

We plan to advertise both on the Internet and through print media to reach the maximum number of potential tenants for Sawmill Ridge, with a special emphasis on Internet advertising.

Internet Advertising

As the major medium in attracting clients, the Internet offers a wide range of benefits for our plan. In fact, our marketing plan depends heavily on New Horizons using Internet advertising. About 72% of Americans use the Internet on a regular

Cites only bottom-line numbers

5

basis (Punji, "E-commerce" 48). Of those, an estimated 60% rely on the Internet almost exclusively to research basic information, make everyday purchases, and research matters of personal and financial importance, including home ownership and apartment rentals (Poulos 43–67).

Most relevant for our plan, research further shows that 85% of professionals seeking rental information rely exclusively on the Internet as their resource, versus 13% who use print resources exclusively. Nineteen percent go to a combination of Internet and print resources (Vallerini). In addition, if an Internet inquiry from a prospective renter is followed up within the first eight hours of being received, a leasing professional has a 75% chance of renting the space to that client, a much higher success rate than with prospective tenants responding to print advertisements (Bruckner 45).

In terms of bottom-line costs, Internet advertising is highly advantageous for New Horizons. An advertisement on the Internet will cost us only a fraction of what we would have to spend on a print source, especially an ad in a hard copy Sunday edition of a local newspaper. As Meredith Mobley, in her blog at forrent.com, claims, Internet advertising helps rental professionals "raise the bar" in "direct traffic to their site." Table 1 below shows the prices of print versus Internet advertisements for a one-quarter page general or community-targeted newspaper ad (or comparable Internet ad on their websites) over a one-month period.

TABLE 1. *Print vs. Internet Advertising Rates (February 2008— one-quarter page/Internet comparable rates)*

	Print	Internet
Dallas Monitor (Sunday)	$950 (4 consecutive Sunday ads)	$525 (one month— continuous)
Dallas Tribune (Sunday)	$825 (4 consecutive Sunday ads)	$450 (one month— continuous)
Dallas Daily News (Sunday)	$825 (4 consecutive Sunday ads)	$450 (one month— continuous)
Dallas Business News (business weekly)	$725 (4 weeks)	$375 (one month— continuous)
El Mundo (Hispanic weekly)	$675 (4 weeks)	$300 (one month— continuous)
Dallas Defender (African American weekly)	$675 (4 weeks)	$300 (one month— continuous)
Dallas Alternative (arts and alternative weekly)	$650 (4 weeks)	$275 (one month— continuous)

Citation includes article title because author has two works listed in Works Cited

All statistics and key points solidly backed by research

Blog citation shows use of reliable online sources

Table clearly spells out differences between print and Internet ad rates

No source listed for table because it is original work

6

The Most Effective Internet Rental Sites

Attending seminar represents an important part of secondary research

To determine the most attractive Internet sites, we met with a consultant at E-pointe, a consulting firm New Horizons used successfully on two previous projects. Our team also attended a seminar regarding the top Internet rental sites in the Dallas/Fort Worth area to determine which were most frequently accessed by renters (Je-mun). Based on these consultations, we believe that the following four real-estate-specific sites are the most effective venues for advertising Sawmill Ridge:

Speech cited by speaker's name

Bulleted list breaks up text and eases readability

- www.rent.com—The number one countrywide rental source with a popular Dallas/Fort Worth area section.
- www.apartments.com—The number two rental source in America, again with a Dallas/Fort Worth area section.
- www.rentclicks.com—Another countrywide site, which, although new, is impressive due to its high placement results on major search engines.
- www.DFWapartments.com—A local rental source, accessed more often than any other site by Dallas/Fort Worth residents.

Creating a Custom Website

Interview with expert backs up author's argument

In addition to advertising on sites such as those identified above, we need to follow through with our "Creating Community" theme by designing our own website targeted to reach the demographic audience most likely to rent at Sawmill Ridge. Creating our own website is essential because it will not only be read by our target audience, but it will also enhance our professional image (Punji, "Image"). Reflecting New Horizon's "green" philosophy, our website will show images of the many open spaces in the development. A well-designed site can help us surpass competition with other developers whose Internet presence is weak in the Arlington area and provide our on-site personnel with a powerful marketing tool (Gilbert). Naturally, all of our print media, signage, and stationery will list the URL for our website to give it maximum exposure.

Does not have to cite authors mentioned in sentence

To further ensure that our custom website appeals to our Internet-savvy target audience, we will include an application prospective tenants can fill out and submit online, considerably simplifying the leasing process for both metro area and out-of-town renters. This policy will simultaneously save New Horizons time and money. When potential customers complete the application using our custom URL, we will waive the normal application fee of $50.00, which is an example of "calling the web shopper to action," as Brittany Rooks and Perry Canales term it (224). Moreover, we will give prospective tenants the option of paying their security deposits and monthly rent using their credit or debit card, another convenience for our target audience.

In designing our website, we hope to profit from and be consistent with similar sites that other divisions in our company have created. Figure 3 shows the homepage, located at http://www.newhorizonstucson.com, for a comparable development in Tucson, Arizona, which is targeted to a similar audience. This website receives over

7

3,100 hits a week and incorporates the following features we think need to also be included on the website for Sawmill Ridge:

- Advertisements for discounts and specials at local businesses
- Information and links to area social, recreational, and educational resources
- A "residents page" with uploaded pictures from parties, comments from residents, and news about forthcoming events
- Design and feature updates on a regular basis

Print Advertising

Even though we will use the Internet as our major medium for attracting new tenants, we will not neglect print media. As popular as the Internet is, our young professional audience in the Dallas/Fort Worth area regularly read and subscribe to several glossy print publications. Studying this print market, we believe that the following are the chief publications in which ads for Sawmill Ridge should be run:

- *The Arlington Advocate*—As Table 1 indicates, the cost for ads in newspapers is very high. Yet, *The Arlington Advocate* will combine the cost of print ads and those on its website. After visiting Darius Tant, manager of *The Arlington Advocate*'s advertising department, we were able to negotiate favorable terms for a three-month contract. We can run two ads a month for $750.00 and receive coverage in print and online. We should also consider advertising in the major Dallas/Fort Worth newspapers to reach individuals seeking to relocate to the metroplex area.
- *Apartment Guide Magazine*—One of the most popular rental publications for prospects in the Dallas/Fort Worth area, with a circulation of 125,000, the *AGM* is readily available at many locations in the Sawmill Ridge area. A free advertisement on their website, located at http://www.apartmentguide.com, is included with the cost of a print ad in *Apartment Guide*. The cost for a one-page ad in *Apartment Guide* comes to $850 per month. Sawmill Ridge is also negotiating with *The College Apartment Guide*, a subsidiary of *AGM*, to target Dallas/Fort Worth students seeking to obtain employment in the Arlington area after graduation.
- **Direct Mailers**—Professionally made direct mailers will be created and distributed to comparable communities to attract residents from these areas. We anticipate printing about 5,000 of these.
- **Brochures**—Well-designed and visually appealing brochures will be given to potential tenants during property tours and can be distributed to local businesses and major employer relocation departments for our outreach marketing. We anticipate printing about 5,000–6,000 of these.

Uses privileged (internal) company information to create similar website in South Arlington

Assures reader that print media will not be neglected

Lists in easy-to-read bulleted segments the top five print sources in which to run ads

Researched price and circulation information backs up argument

8

Figure 3   Homepage for New Horizons Tucson complex.

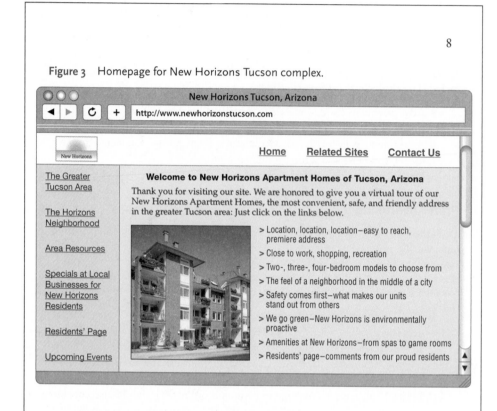

■ **Business Cards**—These cards will feature our attractive design and complement our new Sawmill Ridge brochures. Our office team can use these cards for distance marketing programs in addition to giving them to area residents. Business card advertising is the most direct and effective form of print advertising.

These last three marketing tools can be prepared relatively inexpensively in-house thanks to our new high-speed printers and Photoshop software.

Follow-up Program

Authors plan ahead for second phase

Following up every telephone call, e-mail inquiry, and message board communication from potential tenants is crucial to the success of this marketing plan. When we receive a query, an expression of interest, or a referral, we need to gather all relevant contact information from potential clients, including cell, home, work phone numbers; e-mail address; and street address. In addition, we should try to set up appointments immediately unless a prospective resident is unable to commit, in which case a Sawmill Ridge representative should make a follow-up call to arrange an appointment. Even more important, follow-up contact is a must for prospective

9

tenants who have visited Sawmill Ridge. Accordingly, each visitor to our development will receive both a follow-up phone call (or e-mail) and a thank-you card. As Kelly Talos emphasizes in an article in *Rental Professionals Today*, 47% of rental communities do not follow up with prospective residents effectively, resulting in numerous missed opportunities (90).

Retaining Renters

Our marketing plan also takes into account how to sell residents on Sawmill Ridge beyond their initial leasing agreements. We need to make sure we retain our renters, maintain the overall occupancy of the development, and reduce the cost of turnover expenses. As the remarks below from Roger Cho, in *Resident Developer's Monthly*, indicate, developers need to link keeping residents with initial lease signing:

> Developers and renters often neglect the importance of retaining residents, perhaps the most basic ingredient in the leasing recipe. . . . Retention is particularly important in apartment complexes, where, depending on monthly rental rates, the owners can expect to lose as much as 15% of their annual revenue as a result of turnover and vacant units and as much as 8% of annual revenue due to residents withholding rent when their satisfaction is low and repairs are neglected. Those numbers may seem par for the course, but add them together and you stand to lose nearly 25% annually—a percentage you can drastically reduce with a smart retention plan. (34)

To minimize turnover and to maximize occupancy, our plan will meet the expectations of our audience and be consistent with our marketing strategy of "Creating Community":

- To accommodate our residents' schedules, the office will open early (6:00 a.m.) and close late (7:00 p.m.) three days a week. This policy should help residents unable to visit the office during typical nine to five business hours.
- Sawmill Ridge will host at least five functions each month, including a monthly resident brunch. In months with higher drops in occupancy, the number of activities should increase. Resident functions should appeal to and address all residents. A resident activity team will survey renters for areas of interest and set up events.
- Thanks to an agreement we made with the Pet Motel in South Arlington, Sawmill Ridge residents will receive a 15 percent discount on kennel care.
- Emergency maintenance service will be provided twenty four hours a day, seven days a week. Maintenance technicians will be instructed to leave follow-up cards in each apartment home after they complete the service request.
- All residents will receive greetings on their birthdays, including balloons and candy on their doors and birthday cards put in their mailboxes.

10

Did in-house research to prepare new, related document

■ A monthly newsletter (also available online), such as the one for New Horizons Tucson, can supplement our website, showing appreciation to tenants, providing information about upcoming events and policy changes, and creating a strong community atmosphere.

CONCLUSION

Conclusion succinctly summarizes report

We believe that our twofold approach—targeting the interests of our ideal demographic audience and building a sense of community—is the most effective approach to take in renting the apartments at Sawmill Ridge. As part of this twofold approach, we will market Sawmill Ridge online and in print media to reach and appeal to this large audience of potential renters. We will emphasize customer follow-up during the decision-making process, and we will continue to foster a sense of community and professionalism long after residents have signed their leases to ensure maximum occupancy of our development.

11

Works Cited

Bruckner, Hans. "Don't Forget to Follow Up." *Real Estate Today* Mar. 2007: 45.

Cho, Roger. "Retaining Renters." *Resident Developer's Monthly* Oct. 2007: 34, 41–46.

Commission on Texas Rental Occupancy. *Dallas/Fort Worth Occupancy Rates in 2007*. Washington, DC: United States Census Bureau, 2008.

Corcoran, Barbara. "The Real Estate Market." *The Today Show*. Prod. Victor Talbott. NBC. KVOA, Tucson. 28 Feb. 2008.

D'Argento, Camille. *A Contemporary History of Dallas/Fort Worth Real Estate*. Dallas: Dallas UP, 2007.

---. "South Arlington's Booming Market." *Dallas Real Estate Journal Online* 5.3 (22 Nov. 2007). 1 Feb. 2008 <http://www.dallasrealestatejournal.org/topics/markets/argento2.htm>.

Evinson, Lakita. "Leasing at a Five-Year High in Our Region." *Dallas/Fort Worth Monthly* Jan. 2008: 211.

Gilbert, Jianquin. Personal interview. 4 Oct. 2007.

Hong, Adrienne, Tyrell Carpenter, and Margarita Gonzales. "Arlington Area Quality of Life Survey." 31 Jan. 2008.

---. "Notes." Focus group. 12 Dec. 2007.

Je-mun, Yun. "Selling Dallas on the Web." Dallas Association of Real Estate Brokers conference, Dallas. 14 Sept. 2007.

Jiminez, Alonzo. "Apartment Communities in the Dallas Area—2007 Trends." *The Online Encyclopedia of Real Estate*. 2008. 14 Feb. 2008 <http://www.realestate.org/encyclopedia_23918556.html>.

Kearney-Schwartz, Marianne. "2007 Local Real Estate Statistics." *Dallas Real Estate Journal* 6.1 (23 Jan. 2008): 15–19.

Mobley, Meredith. "MultiFamily Housing Industry Utilizes Streaming and Online Video as Advertising Techniques to Attract Apartment Shoppers." Online posting. Meredith's Blog Page. 29 Oct. 2007 <http://www.meredith.mobley.forrent.com/pressroom/1174058872>.

"New Complex State of the Art." *Dallas Times* 20 Jan. 2008: D11+.

Poulos, Paul. "Internet Real Estate Trends." *Real Estate Today*. Ed. Michael R. Rosenfeld. New York: Jossey-Bass, 2007. 43–67.

Punji, Kamlesh, "E-Commerce Profits from Soaring Rise in Internet Use." *Contemporary Marketing Times* 16 (Jan. 2008): 47–51.

---. "Your Internet Image." Online posting. Punji's Picks Blog. 30 Oct 2007. <http://www.punjispicks.blog.com/internetimage>.

Works Cited list begins on a new page with the title centered at the top

List ordered alphabetically by author

Second and subsequent lines indented five spaces

Two works by same author

12

Talos, Kelly. The Importance of Brokerage Follow-Through." *Rental Professionals Today* 16.5 (2007): 90.

Rooks, Brittany, and Perry Canales. *Selling on the Internet*. New York: Dutton, 2006.

South Arlington, Texas. Map. Dallas: Mapmakers 2006.

Stratton, Daniella-Kay. "Highlands Center On the Move." *Dallas Magazine* Jan. 2008: 341.

United States Bureau of Labor Statistics. *Dallas Employment Statistics for 2006*. Washington: GPO, 2007.

United States Census Bureau. *2007 Dallas Area Report*. Washington: GPO, 2007.

Vallerini, Tomasso. "United States Rental Trends." *Real Estate Today Online*. Oct. 2007. 14 Jan. 2008 <http://www.realestatetoday.com/2007/trends/Vallerini4.htm>.

Weir-Wise, Heidi. Personal interview. 25 Feb. 2008.

Unsigned map listed by map title

Organizations listed alphabetically by name

■ Conclusion

This chapter has introduced you to some very basic yet essential strategies and tools to do primary and secondary research on the job. Clearly, you need to rely on a host of resources—print and online reference works, databases, various search engines, websites, and message boards. Relying on these research tools and strategies, you will have the most up-to-date, most thorough, and most relevant answers to the questions you are asked to investigate and the problems you need to solve on the job. Exploring (and preparing) all of today's potential resources will prepare you to write the types of documents—websites, instructions, short and long reports, proposals—discussed in later chapters.

Revision Checklist

Process of Research

☐ Identified a significant, timely, and restricted topic.

☐ Formulated a mission statement to develop a clear sense of the purpose of my research.

☐ Consulted a wide variety of research materials.

☐ Conducted database, online catalog, and Internet searches.

☐ Used both primary and secondary research methods as needed.

☐ Evaluated the relevance and validity of my sources.

☐ Met with reference librarians or experts in my field.

☐ Documented sources accurately and completely.

☐ Networked with colleagues and boss to meet all deadlines.

Primary Research

☐ Conducted, or observed directly, an experiment or visited a site to describe events, places needed for research.

☐ Set up an appointment with an expert on my topic. Prepared a list of appropriate questions beforehand.

☐ Interviewed experts/customers/goverment officials, using prepared questions, and stayed focused on the subject. Asked important follow-up questions.

☐ Created and distributed questionnaire surveys with only valid, accurate questions to ask and then interpreted my respondents' opinions.

☐ Identified a target audience for my questionnaires.

Secondary Research

☐ Searched online catalog(s) for necessary research materials.

☐ Used databases to find relevant articles and other research materials.

☐ Obtained full-text of appropriate articles and other materials in hard or electronic copy.

Continued

(Continued)

☐ Checked relevant reference materials, such as almanacs, abstracts, encyclopedias.
☐ Located relevant government sources using USA.gov, the Library of Congress, or the U.S. Government Printing Office.
☐ Read and scrutinized periodicals, books, government documents, and websites.

Internet Searches
☐ Searched for pertinent information using at least two search engines.
☐ Used multiple specific, concrete keywords to conduct an Internet search.
☐ Used search strategies such as Boolean connectors or delimiters to narrow and focus my keyword search.
☐ Conducted a directory search or used online-library sources.
☐ Downloaded electronic sources available via my library's online catalog or databases.
☐ Located webpage(s) relating to my topic.
☐ Continued to refine my search using synonyms, alternative keywords, or more specific terms to access only the most useful material for my project.
☐ Joined a newsgroup or surveyed blogs to extend my research.

Taking Notes
☐ Took careful notes and identified precisely the source from which my information came.
☐ Recorded all quotations accurately.
☐ Paraphrased fairly, ethically representing original material.
☐ Distinguished my comments and responses clearly from those of my sources.
☐ Incorporated information from notes into appropriate places in my document.
☐ Used correct punctuation with direct quotations, especially ellipses and brackets for interpolations.

Documentation
☐ Gave full and proper credit to sources consulted and those used for the preparation of my work.
☐ Avoided all forms of plagiarism by supplying complete and accurate documentation of all sources quoted, paraphrased, or consulted for the paper or report.
☐ Recorded all direct quotations accurately; included page references where applicable.
☐ Paraphrased information correctly and acknowledged rightful sources.
☐ Double-checked spelling of authors' and publishers' names and accuracy of all pertinent publication information.
☐ Followed MLA or APA documentation method consistently in preparing Works Cited or Reference lists.
☐ Included all necessary in-text (parenthetical) references; cited each parenthetical reference fully and in correct alphabetical order in Works Cited or Reference lists.
☐ Included only the works referred to in my report in Works Cited or Reference lists.
☐ Alphabetized Works Cited and Reference entries correctly.
☐ Documented all Internet sources properly, giving access dates.

Exercises

1. Choose, define, and restrict a topic based on a problem or issue you might deal with in one of the following divisions of a company:
   a. technology
   b. human resources
   c. security
   d. marketing
   e. accounting
   f. health care
   g. energy/utilities
   h. transportation

   Discuss the steps you took to narrow the topic, the audience you are writing for, and the types of questions that audience may have.

2. Based on the problem you identified in Exercise 1, select an expert relevant to that field to interview for primary research. Confer with co-workers to decide on whom to interview. Prepare a list of ten questions for the interview, remembering to stay focused on your topic.

3. Make a site visit to an appropriate office, plant, agency, or other location relevant to the topic you chose in Exercise 1.

4. Plan a direct observation experiment in a laboratory or other appropriate location to further investigate the topic you selected in Exercise 1. Construct an outline for this experiment, considering the types of information you expect to record and how it will enhance your research.

5. Construct a questionnaire to gather more information about the topic you selected in Exercise 1 or one of the following:
   a. Internet security
   b. online purchasing
   c. business etiquette
   d. safety in the workplace
   e. global business trends

   Come up with a list of ten questions and vary the format to include multiple-choice, dichotomous (yes or no), ranking, and open-ended questions. Think about how you will select your participants, the type of information you want to gather, and how it will enhance your research.

6. Working in a group, assume you are employed by a company that is marketing a new product or service. Your group is conducting market research into possible competitors. Choose three or four different brands of a product or service that is comparable to yours. Prepare a short report in which your group observes, tests, and analyzes these competing products or services, including packaging, contents, pricing, and endorsements, and make recommendations about how the product or service offered by your company can excel in the marketplace.

7. Gather a focus group to plan an advertising campaign based on how best to market your company's product in Exercise 6 above.

8. Prepare a list (providing full bibliographic information) of fifteen articles for the restricted topic you selected in Exercise 1. Use at least one of the online databases discussed in this chapter (pp. 330–334).

9. Using the secondary research strategies explained in this chapter, investigate recent developments in online credit card fraud. Narrow your topic to a specific aspect of credit card fraud, such as prevention, detection, prosecution, and so on. Find the following:
   a. two recent and reliable Web sources using one of the search engines mentioned on pages 340–341
   b. two government sources (e.g., articles, reports, statistics, handbooks) using USA.gov
   c. two online articles using an e-library listed on page 329
   d. two online newspaper articles
   e. one corporate website source using a directory such as those listed on page 340

10. Using the search tools discussed in this chapter, locate the following items related to your major or job. Select titles that are most closely related to your career and explain how they would be useful to you. Prepare a separate bibliographic citation for each title.
    a. titles of three important journals that are available in print and online
    b. an abstract of an article appearing in one of the journals
    c. a term in a specialized dictionary
    d. a description or illustration in a specialized encyclopedia
    e. an article in an international newspaper available online
    f. a training film or recording since 2007
    g. three U.S. government documents released since 2007

11. Assume you have to write a brochure about one of the following topics, introducing it to an audience of consumers. Using the resources and databases discussed in this chapter, prepare a working bibliography of relevant materials that contains at least ten sources. After gathering and reading those sources, prepare the text and a visual for the brochure and submit them with your bibliography to your instructor. This assignment may be done as a collaborative writing exercise.
    a. virtual reality
    b. fiber optics
    c. Internet auctions (e.g., eBay)
    d. interactive television
    e. robotics in medicine
    f. the greenhouse effect
    g. computer dating
    h. laser surgery
    i. globalization
    j. diversity in the workplace
    k. DNA testing

   l. ethics of business blogging
  m. airport security
  n. tablet PCs
  o. any topic your instructor approves

12. Using appropriate references discussed in this chapter, answer any five of the
    following questions. After your answer, list the specific works you consulted.
    Supply complete bibliographic information. For books, indicate author or edi-
    tor, title, edition, place of publication and publisher, date, and volume and page
    numbers. For journals and magazines, include volume and page numbers; for
    newspapers, precise date and page numbers. For Internet sites, provide com-
    plete URL.
    a. What is nanotechnology?
    b. How many calories are there in an orange?
    c. List three interviews that Condoleezza Rice granted between 2007 and 2008.
    d. What is the boiling point of coal tar?
    e. What was the headline in the *New York Times* the day you were born?
    f. List three publications on outdoor recreation issued by the U.S. Depart-
       ment of the Interior from 2006 to the present.
    g. What was the population of Spokane, Washington, in 2000?
    h. List three articles on the advantages of electronic signatures published be-
       tween 2007 and 2008.
    i. Who discovered the neutrino?
    j. What is the first recorded (printed) use of the word *ozone*?
    k. Who edited the second edition of the *Encyclopedia of Psychology*, published
       in 1994?
    l. Approximately how much rainfall does Sierre Leone receive each year?
    m. How many plants does Toyota have in the United States?
    n. Who is the head of public relations for the Red Cross?
    o. Name five plants that have the word *fly* as part of their common name.
    p. What are the names and addresses of all the four-year colleges in the state of
       South Dakota?
    q. Who is India's current head of state?
    r. What is the current membership of the American Dental Association?
    s. What are the names of the justices who currently serve on the U.S. Supreme
       Court?

13. Write a paraphrase of two of the following paragraphs.
    a. Deep-fat frying is a mainstay of any successful fast-food operation and is
       one of the most commonly used procedures for the preparation and pro-
       duction of foods in the world. During the deep-frying process, oxidation
       and hydrolysis take place in the shortening and eventually change its func-
       tional, sensory, and nutritional quality. Current fat tests available to food
       operation managers for determining when used shortening should be dis-
       carded typically require identification of a change in some physical attrib-
       ute of the shortening, such as color, smoke, foam development, etc.

However, by the time these changes become evident, a considerable amount of degradation has usually taken place.[1]

b. E-mail marketing is one of the most effective ways to keep in touch with customers. It is generally cost-effective, and if done properly, can help build brand awareness and loyalty. At a typical cost of only a few cents per message, it's a bargain compared to traditional direct mail at $1 or more per piece. In addition, response rates on e-mail marketing are strong, ranging from 5% to 35% depending on the industry and format. Response rates for traditional mail averages in the 1% to 3% range.

One of the benefits of e-mail marketing is the demographic information that customers provide when signing up for your e-mail newsletter. Discovering who your customers really are—based on age, gender, income, and special interests, for example—can help you target your products and services to their needs. Points to consider when creating your e-mail newsletter:

- HTML vs. plain text: Response rates for HTML newsletters are generally far higher than plain text, and graphics and colors tend to make the publications look far more professional. The downside is that HTML e-mail is slower to download, and some e-mail providers may screen out HTML e-mail.
- Provide incentive to subscribe: Advertise the benefits of receiving your newsletter to get customers to sign up for your newsletter, such as helpful tips, informative content, or early notification of special offers or campaigns.
- Don't just sell: Many studies suggest that e-mail newsletters are read far more carefully when they offer information that is useful to the customers' lives rather than merely selling products and services. Helpful tips, engaging content, and humor are often expected to accompany e-mail newsletters.
- Limit questions: As each demographic question you ask may reduce the number of customers signing up, it's best to limit the amount of information you solicit.[2]

c. Phishing is a form of identity theft in which victims are tricked into turning over their personal information to criminals through bogus e-mail and websites. Phishing schemes, which rely on spam e-mail, emerged in 2004 as a method of capturing personal information to use in identity theft. In phishing schemes, consumers receive e-mail that purports to convey some urgent message about their financial accounts. Recipients are encouraged to respond promptly by clicking a link in the message to what are imitations of legitimate, trusted websites. As a result, the online consumer, believing he or she is connected to a legitimate enterprise, divulges personal financial information, which is diverted to the location of the criminal perpetrator. Essentially,

---

[1]Vincent J. Graziano, "Portable Instrument Rapidly Measures Quality of Frying Fat in Food Service Operations," *Food Technology* 33, page 50. Copyright © by Institute of Food Technologists. Reprinted by permission.

[2]Small Business Administration, "Starting Your Own Business." *E-Marketing*. 2005. http://www.sba.gov/starting_business/marketing/emarketing.html.

phishers have hijacked the trusted brands of well-known banks, credit card issuers, and online retailers, to obtain valuable personal financial information that can be misused or sold to others for the same purpose. Although later phishing attacks have been more generic, they are based on a similar pretext.

Spam e-mail easily reaches thousands if not millions of unsuspecting consumers at a time. With perhaps 5 percent of recipients estimated to divulge personal financial information, the spam scams are lucrative. In November 2004, the Anti-Phishing Working Group (APWG) reported over 1,500 new phishing attacks that month and a 28 percent average monthly growth rate in phishing sites from July through November. The APWG also identified a new form of fraud-based websites that pose as generic e-commerce sites, rather than brand-name sites, and perpetrate loan scams, mortgage frauds, online pharmacy frauds, and other banking frauds. While the United States hosts the largest number of phishing sites, South Korea, China, Russia, Nigeria, Mexico, and Taiwan have also been identified as hosts. Federal law enforcement and others work with foreign counterparts to take down offending sites.

Thus, phishing attacks and "spoofed" e-mail afford criminals an easy and cheap means of obtaining sensitive personal information from consumers which can be very lucrative even if the false website is shut down within 48 to 72 hours.[3]

14. Ask a professor in your major what he or she regards as the most widely respected periodical in your field. Find a copy of the periodical and explain its method of documentation (providing examples). How does it differ from the MLA method?

15. Put the following pieces of bibliographic information in proper form according to the MLA method of documentation for Works Cited. Correct errors in formatting, punctuation, and so on.

   a. *New York Times.* "Say on Pay: A Whisper or a Shout for Shareholders?" Page BU9. April 6, 2008. Claudia Deutsch.

   b. *Business Ethics Quarterly.* 18 (2) 2008. "Good Capitalism, Bad Capitalism, and the Economics of Growth and Prosperity." Denis Collins. Pages 267–271.

   c. *Electronic News.* Access date: 4 April, 2008. Colleen Tayler. "Chip Market Sales Show Growth in March." http://www.edn.comlarticle/CA6437823.html?text=march

   d. Dave Willmer. *Office Pro.* May 2008. "Conference Connections: The Secrets To Successful Networking At Industry Events." Page 6.

   e. "Bill Gates: College Tour 2008." Post date: April 25, 2008. http://www.microsoft.com/presspass/exec/billg/speeches/2008/04-25uw.mspx. University of Washington. Bill Gates. Access Date: 17 May 2008.

---

[3]Anti-Phishing Working Group (APWG), "Phishing Attack Trends Report," (July 2004). See <*http://www.antiphishing.org*>. APWG is an industry association focused on eliminating the identity theft and fraud resulting from phishing. It reports regularly on the form and volume of phishing scams.

f. "The Law Goes Open Source." http://www.forbes.com/technology/forbes/2008/0630/070.html. *Forbes.com.* Daniel Fisher. Access Date: 10 August 2008. Post date: 30 June 2008.

g. "401(k) Plans For Small Businesses." http://www.dol.gov/ebsa/publications/401kplans.html. U.S. Department of Labor. Access Date: 30 June 2008. Post Date: 24 June 2008.

h. "Deconstructing Your Business: Component Business Model." Pages 23–40. Sandy Carter. *The New Language of Business: SOA & Web 2.0.* Published in 2007 by IBM Press, Indianapolis IN.

i. "Managing Channel Profits." *Marketing Science.* Abel P. Jeuland. Pages 29–48. Volume 27, Issue 1. Jan/Feb 2008.

j. Volume 31, Issue 5. "The Trade Policy Jungle: A Survival Guide for Academic Economists" *The World Economy.* Simon J. Evenett. Pages 498–516. May 2008.

k. "Google's Tar Pit." Pages 34–35. Joshua Green. *The Atlantic.* December 2007.

l. *Inc.com.* Access Date: 7 March 2008. Post Date: 1 February 2008. "A Digital Makeover for the Modeling Business." http://www.inc.com/magazine/20080201/a-digital-makeover-for-the-modeling-business.html. David H. Freedman.

m. "Jury Still Out on Labor Rift." *HRMagazine.* Louis J. Gregorio. Pages 23–24. April 2008.

n. *CA Magazine.* May 2008. "Going Green" Page 1. Christian Bellavance.

o. *The International Trade Journal.* "Strategic Environmental Policy with Technology Licensing." Oct-Dec 2007. Volume 21, Issue 4. Jayoti Das. Pages 331–358.

p. July 2008. Volume 86, Issue 7. *Harvard Business Review.* "Making Diverse Teams Click." Jeffrey Polzer. Pages 20–21.

q. "Advocate Aims to Work Himself Out of a Job." Page C7. Cyndia Zwahlen. March 20, 2008. *Los Angeles Times.*

r. Edward Iwata. May 9, 2008. Page B1. *USA Today.* "Tech Tools Bring Big Success To Small Firms."

s. *EH.Net Encyclopedia.* Haupert, Michael J. http://eh.net/encyclopedia/article/haupert.mlb. "The Economic History of Major League Baseball." Edited by Robert Whaples. Post Date: December 3, 2007. Access Date: 18 February 2008.

t. Avis Thomas-Lester. *Washington Post.* "Budding Entrepreneurs Get a Boost; County Chamber Offers Mentors, Support to Business-Minded Students." Page PG01. May 1, 2008.

u. "Giving Time to Give Thanks for Benefits." Kelley Butler. Page 7. *Employee Benefit News.* November 2007.

v. "Job Searchers Face a New Reality." *Business Week.* April 4, 2008. Francesca Di Meglio. 2.

16. Put the bibliographic references you listed in MLA format in Exercise 15 into APA format for a Reference list.

17. Convert the bibliographic information given for one article in the periodical you chose for Exercise 14 into the MLA parenthetical style.

18. The following passage contains mistakes in the MLA method of documentation. Find the mistakes and explain how to correct them.

> More and more companies are allowing employees to "telecommute" (see Smith; Dawson; Brown; Gura and Keith; and Allen). One expert defines telecommuting as "home-based work" (13). Having terminals in their homes "allows employees to work at a variety of jobs" ("New Employment Opportunities"). It has been estimated that currently 900,000 employees work out of their homes (Pennington, p. 56). That number is sure to increase as computer-based businesses multiply in the late 1990s (Brown). In one of her recent articles on telecommuting, Holcomb (167) found that "in the last year alone 43 companies in the metropolitan Phoenix area made this option available to their employees."
>
> Employees who telecommute cite a variety of benefits for such an arrangement (see in particular articles by Gura, Smith, and Kaplan). One employee of a mail order company whose opinion was quoted observed that "I can save about 15–17 hours a week in driving time" (from Allen). Working at home allows the telecommuting employee to work at his or her optimum times ("The Day Does Not Have to Start at 9:00 a.m."). Also, in articles by Kaplan and Keith the benefits of not having to leave home are emphasized: "A telecommuting parent does not have to worry about child care" (39). Telecommuting may "be here to stay" (quoted in a number of different website sources).

19. Submit your preliminary list of references (your tentative Works Cited page) for a long report to your instructor.

CHAPTER 9

Summarizing Information at Work

Visit www
.cengage.com/
english/kolin for
this chapter's
online exercises,
ACE quizzes, and
Web links.

A summary is a brief restatement of the main points of a book, report, website, article, laboratory test, meeting, or convention. A summary saves readers hours of time because they do not have to study the original work or attend a conference. A summary can reduce a report or an article by 85 to 95 percent (or even more) or capture the essential points of a three-day convention in a one-page memo. Moreover, a summary can tell readers whether they should even be concerned about the original; it may be irrelevant for their purposes. Finally, since only the most important points of a work are included in a summary, readers will know they have been given the crucial information they need.

Summaries can be found all around you. For instance, television and radio stations regularly air two-minute "newsbreaks" to summarize in a few sentences the major stories covered in more detail on the evening news and include these summaries on their websites, as, for example, on CBSnews.com. In addition, summaries are widely used on websites to encourage users to go further into a subject. A homepage is in essence a summary of the various links to which it is connected. When search engines retrieve positive "hits," a short summary accompanies each citation to help users determine whether the material is relevant to their needs.

The Importance of Summaries in Business

Summaries are vital in the world of work. They get to the main points—the bottom line—right away for busy readers. New communication technologies such as text messaging and PDAs all require you to summarize information concisely and quickly.

On the job, writing summaries for employers, co-workers, and customers is a regular and important responsibility. Chapter 14 discusses a variety of reports—progress, sales, periodic, trip, test, and incident—whose effectiveness depends on a faithful summary of events. You may be asked to summarize a business trip that lasted a week into one or two pages for your company or agency. You may have to

A summary of a long report on child-care facilities. **Figure 9.1**

HIGH MARKS FOR ON-SITE CARE

The best place to find high-quality child care may be your workplace, according to a new study by Burud & Associates, a California-based work/life benefits consulting firm. The study of 205 work-site child-care programs found that such centers are eight times more likely than other facilities to meet the high standards set by the National Association for the Education of Young Children (NAEYC).

Forty-one percent of work-site centers open at least two years are NAEYC-accredited—compared to only 5 percent of all child-care centers. Other key findings include:

- Ninety-two percent of work-site centers offer infant care.
- Workplace centers provide "substantially better" employee benefits—which help the centers recruit and retain high-quality caregivers.
- One in four centers is open past seven p.m. and one in seven is open before six a.m. and/or offers weekend care.

Source: First appeared in *Working Mother,* May 1999. Reprinted by permission.

condense a proposal to fit a one-page format for an organization. A busy manager may ask you to read and condense a ninety-page report so that she or he will have a knowledgeable overview of its contents. You may be asked to write a **news release**—another type of summary vital for your organization's image, discussed on pages 420–427—for your employer's website.

Figure 9.1 is a summary of a long report evaluating child-care facilities. Note how it concisely identifies the main purpose and conclusions of this report.

Contents of a Summary

The chief problem in writing a summary is deciding what to include and what to omit. As we have just seen, a summary is a streamlined review of *only* the most significant points. You will not save your readers time if you simply rephrase large sections of the original and call the new version a summary. That will simply supply readers with another report, not a summary.

Make your summary lean and useful by briefly telling readers the main points: purpose, scope, conclusions, and recommendations. A summary should concisely answer the readers' two most important questions:

1. What findings does the report or meeting offer?
2. How do those findings apply to my business, research, or job?

Note how the summary in Figure 9.1 successfully answered those two questions by indicating to working mothers where the best child care is likely to be found in the workplace and why.

How long should a summary be? While it is hard to set down precise limits about length, effective summaries are generally 5 to 10 percent, or less, of the length of the original. The complexity of the material being summarized and your audience's exact needs can help you to determine an appropriate length. To help you know what is most important for your summary, the following suggestions will guide you on what to include and what to omit.

What to Include in a Summary

1. **Purpose.** A summary should indicate why the article or report was written or why a convention or meeting was held. Your summary should give the readers a brief introduction (even one sentence will do) indicating the main purpose of the report or conference.

2. **Essential specifics.** Include only the names, costs, titles, places, or dates essential to understanding the original.

3. **Conclusions or results.** Emphasize what the final vote was, the result of the tests, or the proposed solution to the problem.

4. **Recommendations or implications.** Readers will be especially interested in important recommendations—what they are, when they can be carried out, and why they are necessary, or why a plan does not work.

What to Omit from a Summary

1. **Opinion.** Avoid injecting opinions—your own, the author's, or a speaker's. You distract readers from grasping main points by saying that the report was too long or missed the main point, that a salesperson from Detroit monopolized the meetings, or that the author digressed to blame the Land Commission for failing to act properly. A later section of this chapter will deal with evaluative summaries.

2. **New data.** Stick to the original article, report, book, or meeting. Avoid introducing comparisons with other works or conferences; readers will expect a digest of only the material being summarized.

3. **Irrelevant specifics.** Do not include any biographical details about the author of an article that might be included in a section entitled "Notes on Contributors." This information plays no role in the reader's understanding of your summary.

4. **Examples.** Illustrations, explanations, and descriptions are unnecessary in a summary. Readers will want to know outcomes, results, and recommendations, not the illustrative details supporting or elaborating on those results.

5. **Background.** Material in introductions to articles, reports, and conferences can usually be excluded from a summary. Such "lead-ins" prepare the reader for a discussion of the subject by presenting background information, not the big picture readers expect to see.

6. **Jargon.** Technical definitions or jargon in the original document may confuse rather than clarify the essential information for the general reader.

7. **Reference data.** Exclude information found in footnotes, bibliographies, appendixes, tables, or graphs.

Tech Note

Using Software to Summarize Documents

Your word processing software can help you prepare your summary. Download the material (article, report, technical paper) you need to summarize and then, as you read through it the first time on your screen, cut nonessential material. The first time through, it is easier to cut what you don't want than it is to select exactly what you do want. Using this method of editing, you can delete single sentences or several paragraphs at a time. Also, highlight key points as you read through the material you have downloaded. Highlighting during the first pass can guide you on your second reading as you attempt to include only relevant material for your summary.

▮ Preparing a Summary

To write an effective summary, you need to proceed through a series of steps. Basically, you will have to read the material carefully, making sure that you understand it thoroughly, to identify the major points, and, finally, to put the essence of the material into your own words. Follow these steps to prepare a concise, useful summary:

Visit www
.cengage.com/
english/kolin for
an online exercise,
"Practicing Your
Summary-Writing
Skills."

1. **Read the material once in its entirety to get an overall impression of what it is about.** Become familiar with large issues, such as the purpose and organization of the work and the audience for whom it was written. See if there are visual cues—headings, subheadings, words in italic or boldface type, sidebars—that will help you to classify main ideas. Also look for any mini-summaries within the article or report and/or a concluding summary.

2. **Reread the material.** Read it a second time or more if necessary. To locate all and only the main points, underline them. (If the work is a book or in a journal that belongs to your library, work with a photocopy so you can underline.) To spot the main points, pay attention to the key transitional words, which often fall into predictable categories.

- Words that enumerate: *first, second, third, initially, subsequently, finally, next, another*
- Words that express causation: *accordingly, as a result, because, consequently, subsequently, therefore, thus*
- Words that express contrasts and comparisons: *although, by the same token, despite, different from, furthermore, however, in contrast, in comparison, in addition, less than, likewise, more than, more readily, not only . . . but also, on the other hand, the same is true for, similar, unlike*
- Words that signal essentials: *basically, best, central, crucial, foremost, fundamental, indispensable, in general, important, leading, major, obviously, principal, significant*

Pay special attention to the first and last sentences of each paragraph. Often the first sentence of a paragraph contains the topic sentence, and the last sentence summarizes the paragraph or provides a transition to the next paragraph.

Also be alert for words signaling information you do *not* want to include in your summary, such as the following:

- Words announcing opinion or inconclusive findings: *from my personal experience, I feel, I admit, in my opinion, might possibly show, perhaps, personally, may sometimes result in, has little idea about, questionable, presumably, subject to change, open to interpretation*
- Words pointing out examples or explanations: *as noted in, as shown by, circumstances include, explained by, for example, for instance, illustrated by, in terms of, learned through, represented by, such as, specifically in, stated in*

3. Collect your underlined material or notes and organize the information into a draft summary. At this stage do not be concerned about how your sentences read. Use the language of the original, together with any necessary connective words or phrases of your own. Key the draft into your computer. *Expect to have more material here than will appear in the final version.* Do not worry; you are engaged in a process of selection and exclusion. Your purpose at this stage is to extract the principal ideas from the examples, explanations, and opinions surrounding them.

4. Read through and revise your draft(s) and delete whatever information you can. As you revise, see how many of your underlined points can be condensed, combined, or eliminated. You may find that you have repeated a point. Check your draft against the original for accuracy and importance. Be sure to be faithful to the original by preserving its emphases and sequence.

5. Now put the revised version into your own words. Again, make sure that your reworded summary has eliminated nonessential words. Connect your sentences with words that show relationships between ideas in the original (*also, although, because, consequently, however, nevertheless, since*). Compare this version of your summary with the original material to double-check your facts.

6. Do not include remarks that repeatedly call attention to the fact that you are writing a summary. You may want to indicate initially that you are providing a summary, but avoid such remarks as "The author of this article states that water pollution is a major problem in Baytown"; "On page 13 of the article three examples, not discussed here, are found."

7. Edit your summary to make sure it is clear and concise. Check to be certain it is coherent, too. Tell the reader how one point flows into another. Also proofread your summary carefully.

8. Identify the source you have just summarized. Include pertinent bibliographic information in the title of your summary or in a footnote or an endnote. That gives proper credit to the original source and informs your readers where they can find the complete text if they want more details.

Case Study: Summarizing an Original Article

Figure 9.2, a 2,500-word article entitled "Virtual Reality: The Future of Law Enforcement Training," appeared in the *FBI Law Enforcement Bulletin* and hence would be of primary interest to individuals in law enforcement administration. Assume you have been asked to write a summary of the FBI article for your boss, a police chief in a medium-sized city who might be interested in incorporating virtual reality segments into the police academy training program.

By following the steps outlined above, you would first read the article carefully two or three times, underscoring or highlighting the most important points, signaled by key words. Note what has been underscored in the article. Also study the comments in the margins to see why certain information is to be included or excluded from the summary.

After you have identified the main points, extract them from the article and, still using the language of the article, join them into a coherent working draft summary, as in Figure 9.3. Then shorten and rewrite the working draft in your own words to produce the compact final version of your summary, as shown in Figure 9.4. Only 161 words long, the final summary is 12 percent of the length of the original article and records only major conclusions relevant to the audience for the article.

To further understand the effectiveness of the summary in Figure 9.4, review the wordy and misleading summary of the same article in Figure 9.5. The latter summary not only is too long but also dwells on minor details at the expense of major points. It includes unnecessary examples, statistics, and names; it even adds new information while ignoring crucial points about the applications of virtual reality to law enforcement officials. But even more serious, the summary in Figure 9.5 distorts the meaning and the intention of the original article. The reader concludes that the article says virtual reality is not very valuable for law enforcement administrators and officers—just the opposite of the point the article makes. You can avoid such mistakes by de-emphasizing minor points, by making sure that all parts of your summary agree with the original, and by not letting your own opinions distort the message of the original article.

Figure 9.2   An original article with important points underscored for use in a summary.

Virtual Reality:
The Future of Law Enforcement Training

By
Jeffrey S. Hormann

Delete scenario— example of background; an opener

A late night police pursuit of a suspected drunk driver winds through abandoned city streets. The short vehicle chase ends in a warehouse district where the suspect abandons his vehicle and continues his flight on foot. Before backup arrives, the rookie patrol officer exits his vehicle and gives chase. A quick run along a loading dock ends at the open door to an apparently unoccupied building. The suspect stops, brandishes a revolver, and fires in the direction of the pursuing officer before disappearing into the building. The officer, shaken but uninjured, radios in his location and follows the suspect into the building.

Include important observation

Did the officer make a good decision? Probably not by most departments' standards. Whether the officer's decision proves right or wrong, the training gained from this experience is immeasurable, that is, provided the officer lives through it. Fortunately for this officer, the scenario occurred in a realistic, high-tech world called virtual reality, where training can have a real-life impact without the accompanying risk.

Traditional Training Limitations

Major distinction

Experience may be the best teacher, but in real life, police officers may not get a chance to learn from their mistakes. To survive, they must receive training that prepares them for most situations they might encounter on the street. However, because many training programs emphasize repetition to produce desired behaviors, they may not achieve the intended results, especially after students leave the training environment. Thus, the more realistic the training, the greater the lessons learned.

Delete explanation and example

Include significant qualification

Additionally, even some in law enforcement may fall prey to the effects of what has come to be termed "The MTV Generation."[1] As products of this generation, today's young officers purportedly have short attention spans requiring new, nontraditional training methods. The key to teaching this new breed is to provide fast-paced, attention-getting instruction that is clear, concise, and relevant.[2]

Training with Virtual Reality

Emphasize author's main point

Virtual reality can provide the type of training that today's law enforcement officers need. By completely immersing the senses in a computer-generated environment, the artificial world becomes reality to users and greatly enhances their training experiences.

Important reason for its neglect by law enforcement

Although considerable research and development have been conducted in this field, only a limited amount has applied directly to law enforcement. The apparent reason simply is that, for the most part, law enforcement has not asked for it.

Restatement of main point above

Because virtual reality technology is relatively new, most law enforcement administrators know little about it. They know even less about what it can do for

Continued

their agencies. By <u>understanding what virtual reality is, how it works</u>, and <u>how</u> it <u>can benefit them</u>, law enforcement administrators can become significantly involved in the development of this important new technology.

Note parallel items with key words signaling important applications

What Is Virtual Reality?

<u>Simply stated, virtual reality</u> is <u>high-tech illusion</u>. It is a computer-generated, three-dimensional environment that engulfs the senses of sight, sound, and touch. Once entered, it becomes reality to the user.

Include definition

Within this virtual world, users travel among, and interact with, objects that are wholly the products of a computer or representations of other participants in the same environment. Thus, the limits of this virtual environment depend on the sophistication and capabilities of the computer and the software that drives the system.

Important explanation

How Does Virtual Reality Work?

Based on data entered by programmers, computers create virtual environments by generating <u>three-dimensional images</u>. Users usually view these images through a head-mounted device, which, for instance, can be a helmet, goggles, or other apparatus that restricts their vision to two small video monitors, one in front of each eye. Each monitor displays a slightly different view of the environment, which gives users a sense of depth.

Significant phrase

Delete specific pieces of equipment

<u>Another device</u>, called a position tracker, monitors users' physical positions and provides input to the computer. This information instructs the computer to change the environment based upon users' actions. <u>For example</u>, when users look over their shoulders, they see what lies behind them.

Delete example

<u>Because</u> virtual reality users remain stationary, they use a <u>joy stick</u> or trackball to move through the virtual environment. Users <u>also</u> may wear a special glove or use other devices to manipulate objects within the virtual environment. <u>Similarly</u>, they can employ virtual weapons to confront virtual aggressors.

To enhance the sense of reality, some researchers are experimenting with tactile feedback devices (TFDs). TFDs transmit pressure, force, or vibration, providing users with a simulated sense of touch.[3] <u>For example</u>, a user might want to open a door or move an object, which in reality, would require the sense of touch. A TFD would simulate this sensation. At present, <u>however</u>, it is important to remember that these devices are <u>crude</u> and somewhat <u>cumbersome to use</u>.

Delete further examples

Major conclusion

Uses for Virtual Reality

In <u>today's competitive business environment</u>, organizations continuously strive to accomplish tasks faster, better, and inexpensively. This especially holds true in training.

Virtual reality is <u>emerging rapidly</u> as a <u>potentially unlimited</u> method for providing <u>realistic</u>, <u>safe</u>, and <u>cost-effective training</u>. <u>For example</u>, a firefighter can battle the flames of a virtual burning building. A police officer can struggle with virtual shoot/don't shoot dilemmas.[4]

Major value to audience of administrators

Within a virtual environment, <u>students</u> can <u>make decisions</u> and act upon them <u>without risk</u> to themselves or others. By the same token, <u>instructors</u> can critique

Emphasize significant advantages in training

Continued

Figure 9.2   (Continued)

students' actions, enabling students to review and learn from their mistakes. This ability gives virtual reality a great <u>advantage over most conventional training methods</u>.

Use only main points relevant to target audience of administrators

The Department of Defense <u>(DOD) leads public and private industry</u> in <u>developing virtual reality training</u>. <u>Since</u> the early 1980s, DOD has actively researched, developed, and implemented virtual reality to <u>train members</u> of the <u>armed forces</u> to fight effectively in combat.

<u>DOD's current approach</u> to virtual reality training <u>emphasizes team tactics</u>. Groups of military personnel from around the world engage in combat safely on a virtual battlefield. Combatants never come together physically; <u>rather</u>, simulators located at various sites throughout the world transmit data to a central location, where the virtual battle is controlled. Basically, it costs less to move information than people. Consequently this form of training has proven quite cost-effective.

An <u>additional benefit</u> to this <u>type of training</u> is that <u>battles</u> can be <u>fought under varying conditions</u>.

Virtual battlefields <u>re-create real-world locations</u> with <u>interchangeable characteristics</u>. To explore "what if" scenarios, participants can modify enemy capabilities, terrain, weather, and weapon systems.

Note main military advantage

Delete examples

Virtual reality <u>also can re-create actual battles</u>. Based on information from participants, the Institute for Defense Analyses re-created the 2nd Armored Cavalry Regiment Offensive conducted in Iraq during Operation Desert Storm. The success of the virtual re-creation became apparent when, upon viewing the simulations, soldiers who had fought in the actual battle reported the extreme accuracy of the event's depiction and the feeling of reliving the battle.[5] <u>Clearly, virtual reality holds great potential</u> for accurate review and analysis of <u>real-world situations</u>, which would be <u>difficult to accomplish</u> by <u>any other method</u>.

Major conclusion signaled by key word "clearly"

Omit example

Preliminary studies, for instance, show that military units perform better following virtual reality training.[6] <u>Even though</u> virtual environments are only simulations, the complete immersion of the senses literally overwhelms users, totally engrossing them in the action. <u>This realism</u> presumably plays a <u>major role</u> in the <u>program's success</u> and <u>likely</u> will prove positive in future endeavors. <u>In fact</u>, due to its success in training multiple participants in group combat situations, DOD plans to train infantry personnel individually with virtual reality fighting skill simulators.[7]

Key word "major" signals relevant idea for audience

Omit military application

Law Enforcement Training

While virtual reality has proven its value as a training and planning tool for the military, <u>applications for this technology reach far beyond DOD</u>. In varying but key ways, many military uses can <u>transfer to law enforcement</u>, including training in firearms, stealth tactics, and assault skills.

Significant parallel points

Unfortunately, few organizations have dedicated resources to developing virtual reality for law enforcement. <u>According</u> to a recently published resource guide, more than <u>100 companies</u> currently are <u>developing and/or selling virtual reality hardware or software</u>. However, <u>none</u> of these firms <u>mentioned law enforcement uses</u>.[8]

Delete statistics

<u>Further</u>, a review of relevant literature revealed numerous articles on virtual reality technology, but only a few addressed law enforcement applications. <u>Yet</u>, vir-

Subordinate idea

Continued

tual reality <u>clearly offers law enforcement benefits</u> in a number of areas, including pursuit driving, firearms training, high-risk incident management, incident re-creation, and crime scene processing.

Restatement of major point

Pursuit Driving

<u>Pursuit driving</u> represents one area in which <u>virtual reality application</u> has become <u>reality for law enforcement</u>. Law enforcement personnel identified a need and provided input to a well-known private corporation that developed a driving simulator equipped with realistic controls.

Include application but omit example

The simulator provides users with realistic steering wheel feedback, road feel, and other vehicle motions. The screen possesses a 225-degree field of view standard, with 360-degree coverage optional. <u>As noted in demonstrations</u>, simulations can involve one or more drivers, and environments can alternate between city streets, rural back roads, and oval tracks. The vehicle itself can change from a police car to a truck, ambulance, or a number of others.

Delete specific mechanism and explanation of operation of screen mechanism

Virtual reality driving simulators provide police departments invaluable training at a <u>fraction of the long-term cost of using actual vehicles</u>. In fact, the simulator is being used by a number of police departments around the country.

Note cost efficiency again

During the past year, for example, the Los Angeles County Sheriff's Office Emergency Vehicle Operations Center (EVOC) has used a four-station version of the driving simulator to train its officers. The simulators help students develop judgment and decision-making skills, while providing an environment free from risk of injury to students or damage to vehicles. Still, as the EVOC supervisor cautions, <u>virtual reality training</u> should <u>complement, not replace</u>, actual behind-the-wheel instruction.[9]

Include major advantage but exclude specific example

Note major distinction for training purpose

Firearms Training

In another way, virtual reality could <u>greatly enhance</u> shoot/don't shoot <u>training simulators</u> currently in use, such as the Firearms Training System, a primarily two-dimensional approach that possesses limited interactive capabilities. A <u>virtual reality system</u> would <u>allow officers</u> to enter any <u>three-dimensional environment</u> alone or as a member of a team and confront computer-generated aggressors or other virtual reality users.

New subtopic; include advantage but delete examples

Evaluators could specifically observe the <u>training from any perspective</u>, including that of the officers, or the criminal. The <u>training scenarios</u> could involve actual building floor plans or local city streets, and criteria <u>such as</u> weather, number of participants, or types of weapons could be altered easily.

Include significant points on advantages

High-Risk Incident Management

<u>In addition</u> to weapons training, virtual reality <u>could prove invaluable for SWAT team members</u> before <u>high-risk tactical assaults</u>. Floor plans and other known facts about a structure or area could be entered into a computer to create a virtual environment for commanders and team members to analyze prior to action.

Next three reasons to use virtual reality signaled by key words in <u>addition</u>, <u>also</u>, and <u>likewise</u>

Incident Re-creation

Law enforcement agencies could <u>also</u> <u>collect data</u> from victims, witnesses, suspects, and crime scenes <u>to re-create traffic accidents</u>, shootings, and other crimes.

Continued

Figure 9.2   (Continued)

The virtual environment created from the data could be used to refresh the memories of victims and witnesses, to solve crimes, and ultimately, to prosecute offenders.

Crime Scene Processing

Delete examples

Virtual reality crime scenes could <u>likewise</u> be used to <u>train both detectives and patrol officers</u>. First, students could search the site and retrieve and analyze evidence <u>without ever leaving the station</u>. Then, actual crime scenes could be recreated to add realism to training or to evaluate prior police actions.

Is Virtual Reality Virtually Perfect?

Crucial qualification and justification for using virtual reality in law enforcement training

<u>Though</u> virtual reality may appear to be the <u>ideal law enforcement tool</u>, as with any new technology, <u>some drawbacks exist</u>. <u>Currently</u>, areas of concern range from cumbersome equipment to negative physical and psychological effects experienced by some users. Fortunately, <u>however</u>, the <u>field is evolving and improving constantly</u>, and as <u>virtual reality gains widespread use</u>, most <u>major concerns should be dispelled</u>.

Physical Limitations and Effects

Delete examples of limitations/effects

<u>Because</u> computers currently are <u>not fast enough</u> to process large amounts of graphic information in real time, <u>some observers</u> describe virtual environments as "<u>slow-moving</u>."[10] The human eye can process images at a much faster rate than a computer can generate them. In a <u>virtual environment</u>, frames are displayed at a rate of about 7 per second, an extremely slow speed when compared to television, which generates 60 frames per second.[11] Users find the resulting choppy or slow graphics less than appealing.

———————

Adapted from: *FBI Law Enforcement Bulletin* 64, no.7: 7–12.

▌Executive Summaries

An executive summary, found at the beginning of a proposal (Chapter 13) or a long report (Chapter 15), is usually one or two pages long (four to six concise paragraphs) and condenses the most important points from the proposal or report for a busy manager—the executive. This boss wants the big picture, not all the technical details. It is written to help the reader reach a major decision based on the report or proposal. Figure 9.6 is an executive summary of a report on software for a safety training program. Managers use executive summaries so they will *not* have to wade through entire reports. An effective executive summary is like a report itself—self-contained and able to stand on its own.

What Managers Want to See in an Executive Summary

Executive readers are most concerned with managerial and organizational issues—the areas over which they have supervisory control. These readers will look for information on costs, profits, resources, personnel, timetables, and feasibility.

A working draft summary of the "Virtual Reality" article in Figure 9.2.   **Figure 9.3**

Law enforcement officers put their lives on the line every day, yet their training does not fully allow them to anticipate what they will find on the streets. Virtual reality will give them realistic, high-tech benefits of encountering criminals without any risks. Traditional training methods, which work through repetition, cannot equal the advantages of virtual reality when it comes to teaching officers the lessons they must learn to survive in the field. This new breed of officers is demanding the attention-getting, highly realistic training that virtual reality affords them. Virtual reality translates the artificial world of the computer into the real world. Yet even though much research has been done on virtual reality, it is new to law enforcement officials. Moreover, manufacturers have not marketed their technology to them. It is essential that these administrators know how virtual reality works and what it can do for them. Virtual reality has been defined as high-tech illusion through the computer user's perceived interaction with the real world. Working through sophisticated software, virtual reality gives users a three-dimensional (hearing, feeling, and seeing) view of the things and people around them. Virtual reality requires specific equipment including goggles/headsets, a tracker, a trackball, and special gloves. But these devices do have problems; at present, they are crude and can be cumbersome. Even so, virtual reality provides cost-effective and life-saving benefits for law enforcement administrators. Thanks to this technology, students will be able to make quicker and better decisions in the field. Virtual reality has already been tried by the Department of Defense; the armed forces have used it to re-create battlefield conditions, helping the troops better understand the enemy and its position. Yet virtual reality holds great appeal for other real-world applications, especially law enforcement. Unfortunately, the 100 companies that manufacture virtual reality equipment have neglected these law enforcement applications. Yet virtual reality easily accommodates law enforcement instruction. Driving simulators help officers prepare

Continued

Figure 9.3   (Continued)

for high-speed chases. In Los Angeles County, such simulators complement more traditional training. Virtual reality can help officers in a variety of training missions—firearms, high-risk incidents, re-creating crimes, understanding the crime scene. Using virtual reality, officers never have to leave the station. Admittedly, virtual reality has drawbacks, but as this new technology improves, users should face fewer problems.

Figure 9.4   A final, effective summary of the article in Figure 9.2.

Virtual reality offers benefits for law enforcement training that traditional methods cannot provide. This computer-generated technology simulates and re-creates real-life crime scenes without placing officers at risk. Thanks to virtual reality's three-dimensional world of sight, sound, and touch, officers enter the criminals' world to gain invaluable experience interacting with them. Because virtual reality has not been marketed for law enforcement use, administrators may not know about it. Yet it provides a cost-effective, realistic way to enhance training programs. The applications of virtual reality far exceed its military use of simulating battlefield conditions. Virtual reality allows administrators to give trainees hands-on experience in pursuit driving, firearms training, SWAT team assaults, incident re-creation, and crime location processing. Officers can investigate a crime without ever leaving the station. Although virtual reality is an emerging technology with limitations, it is quickly improving and rapidly expanding. Administrators need to incorporate it into their curriculum to give officers field-translatable experiences.

A misleading summary of the article in Figure 9.2.   Figure 9.5

A rookie police officer makes many mistakes in pursuing subjects. Training can cover many realistic situations, but young officers in the MTV Generation have short attention spans. Given the research so far on virtual reality, it holds little promise for law enforcement use. Virtual reality has too many limitations, but it works interestingly through gloves, helmets, and goggles, and with a position tracker users can see over their shoulders. It even has a joy stick (like those in an amusement park) and a crude device—a TFD—that simulates touch (nice to have in a horror movie). Instructors can gain much from virtual reality because they can better criticize their trainees. In the early 1980s, the DOD used virtual reality to duplicate battlefield conditions. The 2nd Armored Cavalry Regiment Offensive won the Iraqi War because of virtual reality. But companies manufacturing virtual reality technology are not interested in law enforcement applications, another indication of its limitations. The Los Angeles Sheriff's EVOC used a driving simulator—offering a 225-degree field of view but it can be ordered with a 360-degree field—but expressed their caution about it. There have been limited interactions in the use of virtual reality for firearms training, though floor plans might have helped SWAT teams. Witnesses may need to refresh their memories with virtual reality. Again drawbacks exist. Computers are not as fast as the human eye in processing information.

Nonessential introductory material

Distorts article, which advocates the benefits of virtual reality for law enforcement

Dwells on specific virtual reality equipment at the expense of the main advantages

Overlooks the significance of virtual reality for trainees

Reverses chronology of events; misrepresents the role of virtual reality

How relevant to law enforcement? Delete unit's name

Again, distorts intention of article

Delete specifications

One-sided; omits success of simulation

Focuses on limits rather than usefulness

Fails to attribute these benefits to virtual reality

Does not subordinate flaws

Your summary must supply key information on the executive's **four E's—evaluation, economy, efficiency,** and **expediency.** Executive readers will expect you to summarize large, complex subjects into easy-to-read, easy-to-understand information that they can act on confidently.

Organization of an Executive Summary

An executive summary must be faithful to the report while giving the readers what they need. First read the report carefully, plan what you want to include, then draft and revise using valuable connective words (p. 402). Clearly, you cannot write an executive summary of your report until after you have written the report itself.

Basically, follow this organizational plan when you write an executive summary:

1. **Begin with the purpose and the scope of the report.** For example, a report might be written to study new marketing strategies, to identify obsolete software, to relocate a branch store.
2. **Relate your purpose to a key problem.** Identify the source (history) and seriousness of the problem.
3. **Identify in nontechnical language the criteria used to solve the problem.** Be careful not to include too much information or too many details.
4. **Condense the findings of your report.** Relate what tests or surveys revealed.
5. **Stress conclusions and possible solutions.** Be precise and clear.
6. **Provide recommendations.** For example, buy, sell, hire more personnel, relocate, or choose among alternative solutions.

The order of information in an executive summary does not have to follow the exact order of the report itself. In fact, some executive summaries start with recommendations. Find out your boss's preference.

▮ Evaluative Summaries

To write an evaluative summary, also called a **critique,** follow all the guidelines on pages 413–417. As with executive summaries, you will be expected to provide a commentary on the material, that is, give your opinion.

Your employers may often ask you to summarize and assess what you have read. For example, you may be asked to condense and judge the merits of a report, paying special attention to whether its recommendations should be followed, modified, or ignored. Your company or agency may also ask you to write short evaluative summaries of job applications, sales proposals, or conferences.

A Report on Providing Better Training at Techtron Sites

Management has commissioned this report to investigate ways to prepare for the OSHA audits scheduled between February and June 2009, at our seven regional Techtron plants. Most directly, this report focuses on our ability to complete Phase One of ISO 14001 certification.

Starts with purpose of report

Currently, the Techtron safety training programs are inadequate; they are neither comprehensive nor up-to-date. We lack necessary software to instruct employees about the EPA and OSHA regulations and requirements that apply to hazardous materials or procedures used in our company. Consequently, safety violations have occurred with lockouts, confined spaces, fall protection, and the "Right to Know Law" concerning labeling of chemicals.

Identifies problem and why it is important

Exploring better ways to conduct our training sessions, we purchased a copy of the software program **EPA/OSHA Trainer**, regarded as the best on the market (available from EDI @ $800 per copy). The **Trainer** offers effective guidelines on developing safety meetings and giving demonstrations. It also includes instructions, written in clear, nontechnical language, on how to identify, collect, and document hazardous materials. Additionally, the **Trainer** supplies the full text of EPA/OSHA regulations, with updates issued quarterly.

Explains the solution tested

To test the effectiveness of the **Trainer** software, we scheduled an internal audit at our Hendersonville site last month. After progressing through the **Trainer** module, a core group of employees interviewed by management successfully completed all required regulatory training. Subsequently, employees who had undergone such training were able to instruct and monitor the performance of other employees in the program.

Verifies effectiveness of solution

To ensure the safety of our employees and to compete in a global marketplace, Techtron must pass the OSHA 14001 certification. Purchasing seven additional copies of the **EPA/OSHA Trainer** software (7 @ $800 = $5,600) is a wise and necessary investment.

Ends by stressing action to be taken

Guidelines for Writing a Successful Evaluative Summary

To write a careful evaluative summary, follow these guidelines:

- Keep the summary short—5 to 10 percent of the length of the original.
- Blend your evaluations with your summary; do not save your evaluations for the end of the summary.

- Place each evaluation near the summarized points to which it applies so readers will see your remarks in context.
- Include a pertinent quotation from the original to emphasize your recommendation.
- Comment on both the content and the style of the original.

Evaluating the Content

Answer these questions on content for your readers:

1. **How carefully and completely is the subject researched?** Is the material accurate and up-to-date? Are important details missing? Exactly what has the author left out? Where could the reader find the missing information? If the material is inaccurate, will the whole work be affected or just part of it?

2. **Is the writer or speaker objective?** Are conclusions supported by evidence? Is the writer or speaker following a particular theory, program, or school of thought? Is that fact made clear in the source? Has the author or speaker emphasized one point at the expense of others? What are the writer's qualifications and background?

3. **Does the work achieve its goal?** Is the topic too large to be adequately discussed in a single talk, article, or report? Is the work sketchy? Are there digressions, tangents, or irrelevant materials? Do the recommendations make sense?

4. **Is the material relevant to your audience?** How would the audience use it? Is the entire work relevant or just part of it? Why? Would the work be useful for all employees of your company or only for those working in certain areas? Why? What answers offered by the work would help to solve a specific problem you or others have encountered on the job?

Evaluating the Style

Answer these questions on style for the readers of your evaluative summary:

1. **Is the material readable?** Is it well written and easy to follow? Does it contain helpful headings, careful summaries, and appropriate examples?

2. **What kind of vocabulary does the writer or speaker use?** Are there many technical terms or jargon? Is it written for the layperson? Is the language precise or vague? Would readers have to skip certain sections that are too complicated?

3. **What visuals are included?** Charts? Graphs? Photographs? How are they used? Are they used effectively? Are there too many or too few?

An evaluative summary of an article.   **Figure 9.7**

Abbasi, Sami M., Kenneth W. Hollman, and Robert Hayes. "Bad Bosses and How Not to Be One." *Information Management Journal* 42.1 (2008): 52–56.

According to this practical and convincingly researched article, the way employees are managed determines a company's success. The authors helpfully begin by describing the new twenty-first-century workplace where power has shifted from a top-down authoritative management style to one respecting employees as "knowledge workers" whose professional contributions are essential in an Internet culture. The article then turns to a classification of six types of difficult bosses, ranging from incompetents, crooks, and bullies to dodgers, know-it-alls, and "walking policy manual[s]" who stick to a policy, however dated or contradictory. These bad bosses use intimidation, manipulation, blame, conflict, and cover-ups to exert or protect their power. Effective bosses, on the other hand, remove fear from the workplace, build trust, encourage feedback, and act as advocates for their employees with upper management. Although aimed at information managers, the guidelines in this readable article apply to anyone who wants to be a good—or better—boss.

Figures 9.7, 9.8, and 9.9 contain evaluative summaries. Note how the writers' assessments are woven into the condensed versions of the originals. Figure 9.7 is a student's opinion of an article summarized for a class in information management. Figure 9.8 is an evaluative summary in memo format collaboratively written by two employees who have just returned from a seminar. They have divided their labor, one writing the opening paragraph and the summary of "Techniques of Health Assessment" and the other doing the summaries of "Assessment of the Heart and Lungs" and "Assessment of the Abdomen." Together they drafted and revised the "Recommendations" and prepared the final copy of the memo.

Another kind of evaluative summary—a book review—is shown in Figure 9.9. Many journals and websites print book reviews to inform their professional audiences about the most recent studies in their fields. Reviews condense and assess books, reports, government studies, websites, films, and other materials. The short review in Figure 9.9 comments briefly on usefulness to the intended audience and on style, provides clarifying information, and explains how the book is developed. The Web contains numerous sites that run book reviews, including Amazon (*http://www.amazon.com*), Barnes and Noble (*http://www.barnesandnoble.com*), Books on Line (*http://www.booksonline.co.uk*), Business Book Review.com (*http://www.businessbookreview.com*), and eBookReviews.net (*http://www .ebookreviews.net*).

Figure 9.8   A collaboratively written evaluative summary of a seminar.

SABINE COUNTY HOSPITAL
7200 Medical Blvd.
Sabine, TX 77231-0011
512-555-6734
www.sabine.org

TO: Mohammed Lau, M.S.N.   SUBJECT: Evaluation of Physical
        Director of Nurses   Assessment Seminar

FROM: Elena Roja, R.N.   DATE: September 13, 2009
        Lee Schoppe, R.N.

Gives overall structure of seminar

On September 7, Doris Fujimoto, R.N., and Rick Ponce, R.N., both on the staff of Houston Presbyterian Hospital, conducted a practical and beneficial seminar on physical assessment. The one-day seminar was divided into three units: (1) **Techniques of Health Assessment**, (2) **Assessment of the Heart and Lungs**, and (3) **Assessment of the Abdomen**.

Techniques of Health Assessment

Describes and evaluates each part of seminar

Four procedures used in physical assessment—inspection, percussion, palpation, auscultation—were defined and demonstrated. Return demonstrations, used throughout the seminar, meant we did not have to wait until we went back to work to practice our skills. The instructors stressed the proper use of the stethoscope and the ways of taking a patient's medical history. We were also asked to take the medical history of the person next to us.

Assessment of the Heart and Lungs

Explains why one part was unsuccessful

After we inspected the chest externally, we covered the proper placement of hands for percussion and palpation and the significance of various breath sounds. The instructors helped us find areas of the lung and identify heart sounds. However, the film on examining the heart and lungs was ineffective because it included too lengthy information for many seminar participants.

Continued

2.

Assessment of the Abdomen
The instructors warned that the order of examination of the abdomen differs from that of the chest cavity. Auscultation, not percussion, follows inspection so that bowel sounds are not activated. The instructors then clearly identified how to detect bowel sounds and how to locate the abdomen and palpate organs.

Recommendations
We strongly recommend a seminar like this for all nurses whose expanding role in the health care system requires more physical assessments. Although the seminar covered a wealth of information, the instructors admitted that they discussed only basics. In the future, however, it would be better to offer follow-up seminars on specific body systems (chest cavity, abdomen, central nervous system) instead of combining topics because of the amount of information involved and the time required for demonstrations.

Ends with endorsement but offers suggestions for improving seminar

A book review includes the most important and useful—to a key audience—information about a book or report. Reviews are also important for the kinds of information that they do *not* include—details and irrelevant (for the audience) information that would only clog a summary. For example, the review in Figure 9.9 indicates that the book is an excellent guide to team building, but does not go into detailed descriptions of the author's recommended tips, strategies, and instruments.

Abstracts

In addition to writing summaries, your employer may ask you to write abstracts. Abstracts are found in several key documents in the world of work and are the staple of the Internet.

Differences Between a Summary and an Abstract

The terms *summary* and *abstract* are often used interchangeably, resulting in some confusion. That problem arises because there are two distinct types of abstracts: *descriptive abstracts* and *informative abstracts*. An informative abstract is another name for a summary; the descriptive abstract is not. Why? An informative abstract (or summary) gives readers conclusions and indicates the results or causes. Look at the summary in Figure 9.4. It explains why virtual reality should be included in law

Figure 9.9   A book review.

Book Review

Team Troubleshooting: How to Find and Fix Team Problems
By ROBERT W. BARNER
Davies-Black Publishing, Palo Alto, Calif.
Representation: Anita Halton Associates, (949) 494-8564, 326 pages, $32.95

Overall Recommendation

DON'T BOTHER	BORROW	BUY

Engaging: 4   Innovative: 4   Usefulness: 5   Visual Aids: 4

Team building with at twist: While most books on the subject limit themselves to talk about getting a new team up and running or fixing a team after it breaks, Robert W. Barner deals with both and then goes a step farther. *Team Troubleshooting* promotes the concept of maintaining a healthy team through anticipating problems, performing regular team tune-ups and taking proactive measures to head off trouble before it begins. High points: strategies to foresee and develop plans for dealing with change, rather than becoming a victim of it; how-to tips to extend beyond the team itself to the problems of creating, maintaining and mending external relationships; ideas to cope with the fact that the team is not an isolated, independent entity, but is part of a bigger system and may be at the mercy of management and other forces.

The unusually user-friendly format makes the book truly serve as a guide.

Useful organization, clear chapter headings and cross-referencing help the reader find information quickly. Barner also does more than just name problems and offer quick fixes—he helps to assure accurate problem diagnosis by clearly describing both symptoms and underlying causes. Dozens of new instruments—not just the same tired old quizzes that appear in so many other teamwork books—lead both team leaders and members (and trainers or consultants involved in the "team" business) through exercises. They include scripting out scenario forecasts, identifying early warning signs of trouble, performing stakeholder analyses, and mapping relationships. It's been a long time since anything new has been said about teams—Barner's book deserves a space on the shelf of anyone truly interested in making a team work.

—*Jane Bozarth*

enforcement training: because virtual reality gives officers field-translatable training. Informative abstracts are found at the beginning of long reports. Descriptive abstracts, on the other hand, do not give conclusions.

All abstracts share two characteristics: the writer never uses "I" and avoids footnotes.

Writing the Informative Abstract

An informative abstract is not as long as an executive summary, which gives more supporting details. As a part of your course work or your job you will probably have to write informative abstracts for long reports (see Chapter 15).

One way to approach writing the abstract of a report is to think of it as a table of contents in sentence form. The table of contents is, in effect, the final outline; it is easily fleshed out into an abstract, as Figure 9.10 shows. On the left is a table of contents, and on the right is the abstract written from that outline.

This system works only if your table of contents is neither too detailed nor too skimpy. Starting off with a good outline of an article or a report provides the best beginning for your abstract. Make sure your sentences are complete and grammatical. Do not omit verbs, conjunctions, and articles. Proper subordination is essential. You should expect to condense a whole paragraph of the original to a sentence, an individual sentence to a phrase, and a phrase to a single word.

Abstract written from a table of contents.   **Figure 9.10**

Table of Contents	Abstract

Need for Genetic Counseling
    Definition of Genetic Counseling
    Statistics on Genetic Counseling
	Genetic counseling is a service for people with a history of hereditary disease. One in 17 births contains some defect; one-fourth of the patients in hospitals are victims of genetic diseases (including diabetes, mental retardation, and anemia). One of every 200 children born has chromosome abnormalities.

Purpose of Genetic Counseling	Genetic counseling offers advice to parents who may give birth to children with genetic diseases and assistance to those with children already afflicted.

The Counseling Process
    Evaluating the Needs of the
        Counselees
    Taking a Family History
    Estimating the Risks
    Counseling the Family
	The first step in counseling is to evaluate the needs of the parents. A family history is prepared and risks of future children being afflicted are evaluated. The life expectancy and possible methods of treatment of any afflicted child also can be determined. Alternatives are presented.

Determination of a Genetic Disorder
    Amniocentesis
    Karyotyping
    Fluorescent Banding
    Staining
	Four prenatal tests are used to determine if a genetic disorder is present: amniocentesis, karyotyping, fluorescent banding, and staining.

Advantages of Genetic Screening
    Lower Cost
    Increased Availability
	The development of these four relatively simple methods has lowered the cost of genetic counseling and increased its availability.

Reprinted by permission of Professor Mary Scotto.

Writing the Descriptive Abstract

A descriptive abstract is short, usually only a few sentences; it does not go into any detail or give conclusions. A descriptive abstract provides information on what topics a work discusses but not how or why they are discussed. Busy readers rely on a descriptive abstract to decide whether they want or need to consult the work itself. Here is a descriptive abstract of the article summarized in Figure 9.4.

> Virtual reality can be used to teach law enforcement officers firearms training, SWAT team assaults, incident re-creation, and crime location processing. This training technology will be of interest to law enforcement administrators.

Figure 9.11 reproduces two descriptive abstracts from the reference work *Library & Information Science Abstracts* as well as two signed abstracts from the *Journal of Interactive Marketing*, a publication that includes abstracts as a way to help readers learn about research in this specialized discipline.

■ Writing Successful News Releases

Visit www
.cengage.com/
english/kolin for
an online exercise,
"Comparing and
Contrasting Two
Different Online
News Releases."

A **news release**, sometimes called a **press release** or **media release**, is another type of on-the-job document that requires you to summarize key information for a variety of readers. Basically, a news release is an announcement (usually one page or a single screen on the Web) about your company's or agency's specific product, services, or personnel. It should be crisp and highlight only the most important and relevant facts clearly and straightforwardly, as the other summaries you have studied in this chapter do.

Such releases are frequently posted on a company's or agency's website (as well as sent in the mail or as a fax). They inform readers—the news media, potential and current customers, other agencies—of newsworthy events that promote the professional accomplishments of a company. Because your company's image can be enhanced or tarnished by the release you write, double-check all your facts, particularly names, dates, places, and any warranty as sales conditions. The best releases project a professional, customer-centered, and quality-focused corporate or agency image.

Figures 9.12 and 9.13 contain sample news releases. The release shown in Figure 9.12 was distributed over the Web whereas the release in Figure 9.13 was distributed in hard-copy form. Both figures label the various parts of a news release, which are discussed on pages 423–427.

Subjects Appropriate for News Releases

News releases should be written only about newsworthy subjects, such as those listed on page 422.

Descriptive abstracts of journal articles.   Figure 9.11

2002-01577

Mann, Charles C **Electronic paper turns the page.** *Technology Review* **104**(2): 42–48 (March 1, 2001) (ISSN: 1099-274X) In English.
Journal URL: http://www.techreview.com

Reports that the key to enabling electronic books to revolutionize the publishing industry is the development of electronic paper. Draws a parallel between the general availability of paper and the development of the first moveable-type printing press, noting that the technology to create the printing press existed for 100 years before the widespread use of ordinary paper provided the means to make the press practical. Notes that several companies are engaged in research and development of electronic paper, adding that at least one working prototype already exists. Indicates that the recent discovery of electrical conducting properties in plastic will enable researchers to overcome the issues of flexibility and cost that have slowed development in the past.

28680312

Murphy, Jeannette **Globalization: Implications for health information professionals.** *Health Information & Libraries Journal* **25**(1): 62–68 (March 2008) (ISSN: 1471-1834) (DOI:10.1111/j.1471-1842.2007.00761.x)

The article discusses how globalization processes may affect health information personnel. The author notes how globalization has led to a global economy and global culture through the transmission of services and information and the creation of international organizations and polices. She suggests that international organizations such as the World Bank and the World Health Organization will affect international health governance through the influence of health librarians. Information technology in fact may improve health education and communication of health information. Globalization has also led to outsourcing of health care services such as information processing. Medical tourism and physician mobility have increased.

Strategic and Ethical Considerations in Managing Digital Privacy

Ravi Sarathy and Christopher J. Robinson (August 2003), *Journal of Business Ethics,* 46(2), pp. 111–126.

Information about customers and prospects is readily available through a variety of digital sources. The questions a marketer must answer is how much of this available data should be used for commercial purposes and how much should remain privileged and off limits. In this paper the authors develop a model of the factors influencing privacy strategy. This model incorporates external, ethical, and firm-specific factors that impact customer privacy protection strategy formulation. The model is then applied to various scenarios to determine the firm's most likely customer privacy strategy. International implications of the model are also discussed.

Scovotti. (8, 13)

The Professional Service Encounter in the Age of the Internet: An Exploratory Study

Gillian Hogg, Angus Laing, and Dan Winkelman (2003), *The Journal of Services Marketing,* 17(5), pp. 476–495.

The Internet, by providing access to an unprecedented amount of healthcare-related information, is changing the balance of power in the relationship between healthcare consumers and professionals. Patients play a more active role in the relationship, interacting with healthcare professionals and other consumers to understand their illnesses. This situation changes the nature of the doctor/patient relationship—where the doctor becomes only one of the *advisors* in the service encounter. The implications of this research extend to other types of service encounters, where consumers may be engaging in virtual, parallel service encounters.

1. **New products, services, or publications**

   - a new or improved model, line of equipment, or website
   - new or expanded technical or customer-friendly services
   - the entrance of your company into a different market
   - the application of the latest technology in creating a product

2. **New policies and/or procedures**

   - acquisition of new technologies to lower production costs
   - changes in production to improve safety, delivery, accuracy
   - cooperation and collaboration between two agencies or branches

3. **Personnel changes**

   - appointments
   - promotions or recognition for winning an award
   - hiring additional staff
   - retirements

4. **New construction/developments**

   - plant or office openings
   - new satellite, branch, or overseas offices
   - expansion of an existing site

5. **Financial and business news**

   - company reorganizations—mergers, acquisitions
   - stock reports
   - quarterly earnings and dividends
   - sales figures
   - community service programs

6. **Special events**

   - training seminars
   - demonstrations of new equipment
   - charity benefits
   - visits of national speakers
   - dedications

News Releases About Bad News

While the topics above focus on a company's or organization's achievements—the positive contributions a firm or an agency makes to its consumers and its community—not all the news you may be asked to announce is pleasant. At times you may have to report on events that are difficult—product recalls, work stoppages or strikes, downsizing, plant closures, limited availability or unavailability of parts, fires, computer viruses, alerts, higher prices, declining enrollments, or canceled events.

Even when you have to write releases about such unpleasant events, portray your company honestly in the most professional and conscientious light. Regardless of the news, be accurate, ethical, honest and straightforward, and available.

Topics That Do Not Warrant a News Release

Not every event or change at your company or agency will warrant a news release. Releases about events that have already happened are old, tired news. Keep in mind as well that things in the works may as yet be too new for you to announce. Below are some topics that hold little or no interest and about which you would *not* write a release.

1. **Well-known products or services.** Don't repeat the obvious.

2. **Products or services still in the planning stages.** Not only would such news be premature, but releasing it ahead of time might meet with your boss's severe disapproval, not to mention your competition's delight.

3. **Controversial events or company problems.** Refrain from writing a news release on these subjects unless your boss instructs you to do so.

4. **A history of your department, division, or company.** Save remarks about your firm's history for the annual report to stake holders.

5. **An obviously padded tribute to your boss or to your company or to your customers.**

Organization of a News Release

The following sections contain guidelines for organizing and writing the different parts of your news release. Note how these various parts flow together in the news releases in Figures 9.12 and 9.13.

The cardinal rule in writing a news release is to put the most important piece of news first. Don't bury it in the middle or wait until the end. Everything in a news release should be arranged in descending order of importance so that your first paragraph contains only the most significant fact and ideas. Think of your release as an inverted pyramid with the top summarizing only the most crucial points.

The Three Parts of a News Release

The three components of a news release are the slug (headline), lead, and body, as described below.

The Slug, or Headline. One of the most crucial parts of your release is the headline, or **slug**. It should announce a specific subject for readers and draw them into it. Write a slug that entices or grabs your readers, but also one that informs them quickly and clearly. Note how the slugs in Figures 9.12 and 9.13 point readers to the main point in the first sentence and use key words to help readers track the subject.

Figure 9.12   A news release from the Web.

CDC Home | Search | Health Topics A-Z

OFFICE OF COMMUNICATION   **MEDIA**
1000 Clifton Rd., MS D25, Atlanta, GA 30333 - Ph. 404-639-3286 FAX: 404-639-7394   **RELATIONS**

March 18, 2008   (404) 639-3286
Contact: CDC, Jennifer Marcone   http://bookstore.phf.org

Local Content

Email Us
On-Line
Publications
Image LIbrary
Press Kit
Press Releases
Daily Updates

Quick Jumps

Health Statistics
Health Topics
A-Z
Global Health
Odyssey

Press Release

CDC Releases New Interactive Website for
Health Information for International Travel **(The Yellow Book)**

ATLANTA–The Centers for Disease Control and Prevention (CDC) unveiled a re-designed, interactive website of one of CDC's most widely disseminated publications, *Health Information for International Travel,* commonly dubbed "The Yellow Book." The Yellow Book is considered by many health care providers, travel professionals, airlines, cruise lines, and humanitarian institutions to be the gold standard for health recommendations for international travel.

The Yellow Book and companion website **www.cdc.gov/travel/yb** are published biennially by CDC's Division of Global Migration and Quarantine as a reference for those who advise international travelers of health risks.

The key to the new, interactive website is the use of drop-down menus for all of the major subjects in the Yellow Book, including vaccination information, yellow fever requirements, malaria information, geographic distribution, and health hints. Users can obtain customized reports for individual travel plans and locate particular subjects or destinations in the text without having to search through unrelated topics.

"International travelers will find the interactive Yellow Book website to be very helpful," said CDC Director Jeffrey P. Koplan, MD, MPH. "The new features make it easier to find information about preventive measures travelers can take to protect their health."

The handy, interactive design of the Yellow Book website incorporated many comments and suggestions from the public and health care providers seeking more customized information in a user-friendly format. CDC encourages users to submit comments about the new website through the "comments section" on the site's homepage.

30

http://bookstore.phf.org

Slug empha-sizes impor-tance of topic

Lead answers who, when, how, what, and why

Body empha-sizes benefits for audience

Quotation from expert endorses new design and features

Conclusion invites reader interaction

Number 30 signals end of release

A news release circulated in hard copy.   Figure 9.13

NEWS RELEASE

United States Department of Commerce • 1401 Constitution Avenue, NW • Washington, DC 20230 • Web: http://www.commerce.gov

Release No. 0150.01

Cheryl Mendonsa
Cheryl.Mendonsa@technology.gov
202-482-8321

Marjorie Weisskohl
Marjorie.Weisskohl@technology.gov
202-482-0149

Contact information

U.S. Department of Commerce Announces Website Focused on Resources for Tech-Based Economic Development (TBED)

Slug

Under Secretary of Commerce Phillip J. Bond today announced a new resource that will help communities working to encourage the growth of technology companies and build tech-based economies. A new website, the TBED Resource Center **(http://www.tbedresourcecenter.org),** gives users the chance to learn from others' experiences and to benefit from the latest research on creating a tech-based economy.

Lead, or hook

The TBED Resource Center categorizes and provides links to more than 1,300 research reports, strategic plans, best practices and impact analyses from state and federal government, university researchers, and foundations. The website is a result of a cooperative project of the U.S. Department of Commerce's (DOC) Office of Technology Policy and State Science and Technology Institute, based in Westerville, Ohio.

Body

"This website will be an invaluable tool to anyone involved in economic development," said Under Secretary Bond. "Access to information on what has worked in various communities will help policymakers and business people make efficient use of valuable time and resources, and provide productive ideas and contacts that others might find useful." Bond made the announcement via live telecast hosted by Commerce's Economic Development Administration and the National Association of Regional Councils.

Quote from authority

The user-friendly search options allow for easy navigation with abundant topics spanning a wide field of interests. Users can search for reports based upon geography, topic, type, or keyword with the option of selecting multiple fields. Numerous international reports also are included, offering a globally diverse perspective in tech-based economic development.

Emphasizes benefits

Appropriate positive tone

Reports fall into one of the 35 topics, such as brain drain, education, entrepreneurship, innovation, research and development, and work force.

To keep current with TBED trends, the website is continually updated as new reports are released.

Signals end of release

#

The Lead. The first (and most significant) sentence of a news release is called the **lead**. It introduces and aptly summarizes your topic, sets the tone, and continues to keep the readers' attention. For that reason it is also called the **hook**. See how the leads in Figures 9.12 and 9.13 grab readers' attention. The best leads easily answer the basic questions **Who? What? When? Where? Why?**—the **five W's**—and **How?** (or **How much?**). Not every lead will answer all these questions, but the more of them you do answer, the better your chances will be of capturing your audience's attention. Here are some effective leads that answer these crucial questions:

> *who*   *when*
> Massey Labs announced today that the FDA has approved the marketing
>
> *what*   *why*
> of its vaccine—Viobal—to retard recurrent lesions of herpes simplex.

> *who*   *what*
> Maryville Engineering, Inc., has been awarded a contract by Aerodynamics, Inc.,
>
> *why*
> to develop an acoustical system to measure and monitor stress levels at the
>
> *where*   *when*
> Knoxville aircraft plant, district manager Carmelita Stinn, P.E., announced today.

Misleading leads that deviate substantially from the guidelines just discussed annoy readers and risk losing their interest. Resist the temptation to start off with a question or with folksy humor.

Feeble Humor:	Slap, slap, scratch, scratch, itch. This is how millions of people will handle their mosquito problems this summer.
Boastful Start:	If you thought 2009 was great, you ain't seen nothin' yet!
Undirected Question:	Does the thought of hypothermia send chills up your spine?
Exaggeration:	I bet you wonder how you ever did without the new Fourier scanner.

The Body. What kinds of information should you give readers in the second and subsequent paragraphs of your release? If the five W's are answered in your lead, the following paragraphs fill in only the most necessary supporting details. Regard your lead as a summary of a summary. The body of your release then amplifies the *Why? How?*, and may also get into the *So what?*

But avoid filling your news release with unnecessary technical details, such as scientific formulas or intricate speculations. Instead, relate your product or service to your targeted audience's needs by emphasizing the benefits to them without loading your news release with hype.

Use quotations selectively and to clarify and highlight, not to apple-polish. Use a quotation only to report vital facts or to cite an authority. Do not turn your release into an interview with your employer or favorite customer.

Style and Tone of a News Release

The style and tone of your news release need to be objective and take into account your audience's background. Keep your style simple and to the point. Here are some suggestions that will help you:

1. **Write concisely.** Use easy-to-read paragraphs of three to five sentences, and keep your sentences short (under twenty words is ideal). Avoid long, convoluted sentences and fillers that start off with such phrases as "It has been noted that. . .".

2. **Emphasize benefits to the reader.** Use graphic, understandable language that applies to the reader's life. Avoid unfamiliar jargon—it's deadly. A news release summarizing the benefits of a hospital counseling center lost potential seminar participants by describing it in jargon: "Milford Hospital is pleased to offer an adventure-intensive counseling workshop that incorporates trust sequencing; high element activities; and quantifiable decision modules."

3. **Keep your tone upbeat, easygoing, and direct; stress the human side.** But stay away from adjectives dripping with a pushy sell—"incomparable," "fantastic," "incredible." See Figures 9.12 and 9.13 with their emphasis on benefits to travelers and communities interested in technology-based economic development, respectively.

As with other types of summaries described in this chapter, news releases give readers essential, useful information quickly and concisely. They provide the "big picture." And by doing so, they capture the reader's interest and emphasize the goals and contributions of the agency at the same time.

✓ Revision Checklist

☐ Read and reread the original thoroughly to gain a clear understanding of the purpose and scope of the work.

☐ Underlined key transitional words, main points, significant findings, applications, solutions, conclusions, recommendations.

☐ Separated main points clearly from (a) minor ones, (b) background information, (c) illustrations, and (d) inconclusive findings.

Continued

(Continued)

☐ Excluded examples, explanations, and statistics from the summary or abstract.
☐ Deleted information not useful to audience—information too technical or irrelevant.
☐ Changed language of original to my own words so I am not guilty of plagiarism.
☐ Made sure that emphasis of summary matches emphasis in original.
☐ Determined that sequence of information in summary follows sequence of original.
☐ Added necessary connective words that accurately convey relationships between main points in original.
☐ Edited to eliminate wordiness and repetition from summary.
☐ Cited source of original correctly and completely.
☐ Avoided phrases that draw attention to the fact that I am writing a summary or abstract.
☐ Summarized material objectively without adding commentary (for informative summary).
☐ Commented on both content and style (in evaluative summary).
☐ Interspersed evaluative commentary throughout summary so that assessments appear near relevant points.
☐ Included a direct quotation to illustrate or reinforce my recommendation (for evaluative summary and news releases).
☐ Ensured that descriptive abstract is short and to the point and does not offer a judgment.
☐ Prepared news releases that project a professional image of my employer.
☐ Summarized only essential information in my news release.
☐ Arranged information from top down, with the most important first.
☐ Wrote clear, crisp slug and lead and a concise body.
☐ Sent news release on newsworthy topic to appropriate publications, organizations, and audiences.

Exercises

1. Summarize a chapter of a textbook you are now using for a course in your major field. Provide an accurate bibliographic reference for that chapter (author of the textbook, title of the chapter, title of the book, place of publication, publisher's name, date of publication, and page numbers of the chapter).

2. Summarize a lecture you heard recently. Limit your summary to one page. Identify in a bibliographic citation the speaker's name, date, and place of delivery.

3. Listen to a television network evening newscast and to a later news update on the same station. Select one major story covered on the evening news and indicate which details from it were omitted in the news update.

4. Write a summary of the marketing report in Chapter 8 (pp. 375–387) or the business report on non-native speakers of English in the work force (pp. 653–667).

5. Bring to class an article from *Reader's Digest* and the original material it condensed, usually an article in a journal or magazine published six months to a year earlier. In a paragraph or two indicate what the *Digest* article omits from the original. Also point out how the condensed version is written so that the omitted material is not missed and how the condensation does not misrepresent the main points of the article.

6. Assume that you are applying for a job and that the personnel manager asks you to summarize your qualifications for the job in two or three paragraphs. Write those paragraphs and indicate how your background and interests make you suited for the specific job. Mention the job by title at the beginning of your first paragraph.

7. Write a summary of one of the following articles.
   a. "Microwaves," in Chapter 1 on pages 41–42.
   b. "Protecting Personal Information," below.

8. Write a descriptive abstract of the article you selected in Exercise 7.

Protecting Personal Information: A Guide for Businesses
Small Business Administration

Companies keep sensitive personal information in their files—names, social security numbers, credit card or other account data—that identifies customers or employees. This information often is necessary to fill orders, meet payroll, or perform other necessary business functions. However, if sensitive data falls into the wrong hands, it can lead to fraud, identity theft, or similar troubles. Given the high cost of a security breach—losing your customers' trust and perhaps even defending yourself against a lawsuit—it is just plain good business to safeguard personal information.

Protect the Information That You Keep
What's the best way to protect the sensitive, personal information your company needs to keep? It depends on the kind of information and how it's stored. The most effective data security plans deal with four key elements: physical security, electronic security, password management, and employee training, discussed below.

Physical Security
Many data compromises happen the old-fashioned way—through lost or stolen paper documents. Often, the best defense is a locked door or an alert employee. Here are some helpful ways to protect information from falling into the wrong hands:

First of all, store paper documents or files, as well as CDs, floppy disks, Zip drives, tapes, and backups containing personally identifiable information in a locked room or in a locked file cabinet. Limit access to employees with a legitimate business need. Control who has a key, and the number of keys. Then require that files containing personally identifiable information be kept in locked file cabinets except when an employee is working on the file. Remind employees not to leave sensitive papers out on their desks when they are away from their workstations, to put files away, log off their computers, and lock their file cabinets and office doors at the end of the day.

Next, implement appropriate access controls for your building. Tell employees what to do and whom to call if they see an unfamiliar person in the office or elsewhere on the premises. If you maintain off-site storage facilities, limit employee access to those with a legitimate business need. Know if and when someone accesses the storage site. For example, if you ship sensitive information using outside carriers or contractors, encrypt the information and keep an inventory of the information being shipped. Also, use an overnight shipping service that will allow you to track the delivery of your information.

Electronic Security

Computer security isn't just the realm of your IT staff. Make it your business to understand the vulnerabilities of your computer system, and follow the advice of experts in the field. To do this, identify the computers or servers where sensitive personal information is stored. Make sure, too, that you identify all connections to the computers where you store sensitive information. These may include the Internet, electronic cash registers, computers at your branch offices, computers used by service providers to support your network, and wireless devices like inventory scanners or cell phones.

Moreover, encrypt all sensitive information that you send to third parties over public networks (like the Internet), and consider encrypting sensitive information stored on your computer network or on disks or portable storage devices used by your employees. It is a smart idea to encrypt e-mail transmissions as well—within your company if they contain personally identifying information. In addition, regularly run up-to-date anti-virus and anti-spyware programs on individual computers and on servers on your network. Check expert websites (such as *www.sans.org*) and your software vendors' websites regularly for alerts about new threats and to learn about implementing new policies for installing vendor-approved patches to correct problems.

Most important of all, scan the computers on your network to identify and profile the operating system and open network services. If you find services that you don't need, disable them to prevent hacking or other potential security problems. For example, if e-mail service or an Internet connection is not necessary on a certain computer, close the ports to those services on that computer to prevent unauthorized access to that machine. When you receive or transmit credit card information or other sensitive financial data, use Secure Sockets Layer (SSL) or another secure connection that protects the information in transit.

Pay particular attention to the security of your Web applications—the software used to give information to visitors to your website and to retrieve information from them. Web applications may be particularly vulnerable to a variety of hacker attacks. In one variation called an "injection attack," a hacker inserts malicious commands into what looks like a legitimate request for information. Once in your system, hackers transfer sensitive information from your network to their computers. Be on guard against these attacks by more closely scrutinizing requests.

Password Management

Also essential to any business's security is password protection. You can control access to sensitive information by requiring your employees to use "strong" passwords. Technical security experts advise that the longer the password, the better. Because simple passwords—like common dictionary words—can be easily guessed, insist that your employees choose passwords with a mix of letters, numbers, and characters. Require an employee's user name and password to be different, and insist on frequent changes in passwords. Lock out users who don't enter the correct password within a designated number of log-on attempts. Password-activated screen savers can lock employee computers after a period of inactivity.

Another wise move is to warn employees about possible calls from identity thieves attempting to deceive them into giving out their passwords by impersonating members of your IT staff. Let employees know that calls like this are always fraudulent, and that no one should be asking them to reveal their passwords. Set up clear and effective procedures through which employees can communicate with IT staff.

When installing new software, make sure your company immediately switches vendor-supplied default passwords to a more secure strong password and cautions employees against transmitting sensitive personally identifying data—social security numbers, passwords, account information—via e-mail. Unencrypted e-mail is not a secure way to transmit any information.

Employee Training

Your data security plan may look great on paper, but it's only as strong as the employees who implement it. Take time to explain the rules to your staff, and train them to spot security vulnerabilities. Periodic training emphasizes the importance you place on data security practices. A well-trained work force is a company's best defense against identity theft and data breaches.

You can also create a "culture of security" by implementing a regular schedule of employee training. Update employees as you find out about new risks and vulnerabilities. Make sure training extends to employees at satellite offices and to temporary help and seasonal workers as well. If employees don't attend, consider blocking their access to the network. If you train employees to recognize security threats, you can cut down on the risks such thefts pose. Tell your employees how to report suspicious activity and publicly reward employees who alert you to vulnerabilities.

Finally, ask every new employee to sign an agreement to follow your company's confidentiality and security standards for handling sensitive data. Make sure these most recent members of your work force understand that abiding by your company's data security plan will be an essential part of their duties. And regularly remind all employees of your company's policy—and any legal requirement—to keep customer information secure and confidential.

Source: Federal Trade Commission, *Protecting Personal Information: A Guide for Businesses*. Washington, DC: GPO, 2008.

9. Below are two sloppily written news releases that are poorly organized and, in a few instances, unethical. Reorganize and rewrite them according to the guidelines presented in this chapter.

a. Friday   Alan Bowerstock

Metropolitan State University is a four-year urban institution of higher education offering majors in many fields. Located two miles west of Taylorsville Tech Park, MSU currently boasts more than 7,000 undergraduate students, more than our rival, Central Tech.

Among the many student services currently available at MSU is the Division of Career Placement; this division is located in the Student Services Building, Room 301, just across the hall from the Department for Greek Life.

In the last year, the Division has assisted more than 2,000 MSU students to find part-time jobs. Full-time jobs, too.

The goal of the Division is to help students earn money for their college expenses. The Division also wants to assist local businesses in contacting MSU qualified undergraduates.

The MSU family is well represented on the homepage. Every department from the University has been encouraged to report on only its most favorable activities. The Division of Career Placement is also on the homepage.

Counselors at the Division can help MSU students prepare a four- or five-line ad about their qualifications. The Division will run these ads in their homepage. The university will also run ads looking for job candidates from the state employment agency and the greater Taylorsville area.

b. For General Information   Frank Day

J.T. Bushart, CEO of Bonnetti and Blount Construction for the last three years, asserted today that the company is devoted to progress and change. Bushart came to B&B Engineering after several years working for Capitol City Engineering, a less progressive firm.

```
Bonnetti and Blount is a leading firm of contractors
and has worked for both national and international cor-
porations.

The firm specializes in construction projects that
require special expertise because of their challenges
in difficult terrains.

B&B has just been awarded a contract to work on the 10
million dollar renovation of two major Fairfax dams.
The firm anticipates hiring more than 200 new workers.
These new employees will work on the dams that present
dangerous conditions to residents.

When completed, the two new dams will further assist
residents of Fairfax and Hamilton Counties receive all
the necessary irrigation and hydroelectric energy they
need.

Engineering sketches and blueprints are in the works.
```

10. Write an appropriate news release on one of the following newsworthy topics to be included in your company's website.
   a. hiring a new webmaster
   b. premiering a new product or service
   c. acquiring a smaller firm whose products and services are very different from those of your company
   d. providing an environmentally sensitive service that enhances life in the community in which your employer's headquarters is located
   e. promoting an employee who has been with your company for at least five years
   f. offering highly competitive warranties on a new line of products

11. Write a news release on one of the following unpleasant topics while still projecting a positive image of your company.
   a. inconveniences because of recent construction
   b. a temporary power outage in a neighborhood or city
   c. a company/organization server is down
   d. a change in the hours of operation of a store or plant
   e. an increase in insurance premiums
   f. a reduction in number of times trash is picked up per week
   g. an order page is down on a frequently visited e-business site because of a computer virus
   h. a sports injury has benched a star player on a local team for the next month

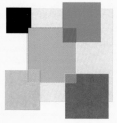

PART IV

Preparing Documents and Visuals

10 Designing Clear Visuals

11 Designing Successful Documents and Websites

12 Writing Instructions and Procedures

13 Writing Winning Proposals

14 Writing Effective Short Reports

15 Writing Careful Long Reports

16 Making Successful Presentations at Work

Designing Clear Visuals

Visuals are essential in the world of work. They are vital to the success of newsletters, reports, proposals, instructions, PowerPoint presentations, blogs, and, most certainly, websites. Even your company's logo reveals a great deal about its image and mission. The reason visuals are so important is that they help readers understand your work. Experts estimate that as much as 80 percent of our learning comes through our sense of sight.

Visuals are also crucial to the success of your written work. In fact, visuals work in conjunction with your writing to inform, illustrate, and persuade. The tables, graphs, and charts you use to simplify financial and other statistical information in a report, or the photos and other graphics you use have a major impact on how readers will judge your work. A poorly designed visual—one that is incorrect, incomplete, unclear, or unethical—signals to readers that your research and writing may also be flawed.

Visual Thinking in the Workplace

Thanks in part to the Internet, that "virtual library," today's global culture highly prizes visual thinking, or presenting/seeing information in terms of its spatial and organizational appearance. Visuals and other types of graphics not only capture a reader's attention but also help him or her better understand your message—whether it is giving the big picture or showing how numbers can be crunched. In addition, visuals give readers a lot of information quickly and persuasively. Workplace audiences are accustomed to reading visuals and will expect you to be skilled at presenting such graphics in your work to assist them in understanding your message. In a Web-based, global culture, icons and other visuals help readers access and navigate the Internet.

Chapter 10 surveys the kinds of visuals you will encounter most frequently in the world of work. It shows you how to read, locate, create, and write about them—and it also describes the types of visuals and visual configurations you can create and copy with graphics software packages. Finally, this chapter discusses the

Visit www .cengage.com/ english/kolin for this chapter's online exercises, ACE quizzes, and Web links.

ethical ways to use visuals and how to select the most appropriate ones for global audiences. But, as you have already seen, a discussion of visuals in this book is not confined just to this chapter. Visuals are discussed as part of preparing a variety of business documents covered in other chapters—from letters and instructions to proposals, reports, and websites.

The Purpose of Visuals

Here are several reasons why you should use visuals. Each point is graphically reinforced in Figure 10.1.

1. **Visuals arouse readers' immediate interest.** They catch the reader's eye quickly by setting important information apart and by giving relief from sentences and paragraphs. Note the eye-catching quality of the visual in Figure 10.1.

2. **Visuals increase readers' understanding by simplifying concepts.** Visuals are especially important if you have to explain a technical process to a nonspecialist audience. They can make a complex set of numbers easier to comprehend, helping readers see percentages, trends, comparisons, and contrasts. Figure 10.1, for example, shows at a glance the growth of e-commerce.

Figure 10.1   A line-and-bar chart depicting the growth of e-commerce compared to traditional businesses.

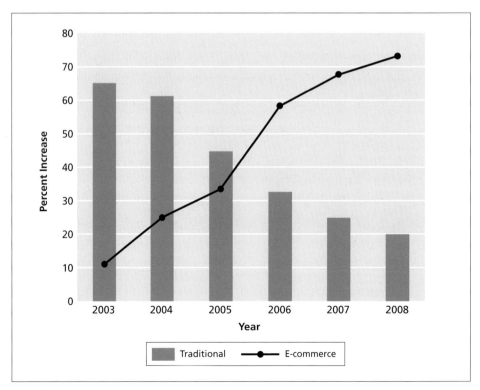

3. **Visuals are especially important for non-native speakers of English and multicultural audiences.** Given the growing international audience for many business documents, visuals, if prepared carefully, can make your communication with them easier and clearer. Study the visuals in Terri Smith Ruckel's report in Figure 15.3.

4. **Visuals emphasize key relationships.** Through their arrangement and form, visuals quickly show contrasts, similarities, growth rates, and downward and upward movements, as well as fluctuations in time, money, and space.

5. **Visuals condense and summarize a large quantity of information into a relatively small space.** A visual allows you to streamline your message by saving words. It can record a great deal of statistical data in far less space than it would take to describe those facts in words alone. Note how in Figure 10.1 the growth rates of two different types of businesses are concisely expressed and documented.

6. **Visuals are highly persuasive.** Visuals have sales appeal. They can convince readers to buy your product or service or to accept your point of view and reject your competitor's. A visual can graphically display, explain, and reinforce the benefits and opportunities of the plan you are advocating.

Types of Visuals and Their Functions

Table 10.1 shows some of the most frequent types of visuals found in workplace writing and why and how you can use them. Each of these visuals is discussed in this chapter.

Choosing Effective Visuals

Select your visuals carefully. You'll find that many of the visuals you use will originate from the Internet or from a graphics program on your computer. But you can also scan hard copy visuals and upload digital photographs (see the Tech Notes on generating graphics in this chapter). The following suggestions will help you to choose effective visuals.

1. **Include visuals only when they are relevant for your purpose and audience.** Never include a visual simply as a decoration. A short report on fire drills, for example, does not need a picture of a fire station. Similarly, avoid any visual that is too technical for your readers or that includes more details than you need to show.

2. **Use visuals in conjunction with—not as a substitute for—written work.** Visuals do not take the place of words. In fact, you may need to explain information contained in a visual (see pp. 447–448). A set of illustrations or group of tables alone may not satisfy readers looking for a summary or a recommendation. Note how the visual in conjunction with the description of a magnetic resonance imager (MRI) in

TABLE 10.1   Types of Workplace Visuals

Table		■ Classifies data (figures/facts) into easy-to-read categories using rows and columns ■ Gives statistical data or verbal descriptions clearly and concisely ■ Allows readers to see comparisons and contrasts more quickly than just embedding numbers in the text
Line graph	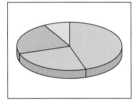	■ Transforms numbers into a picture (curves, shapes, patterns) that represents movement in time or space ■ Displays data that change often—costs, sales, rates, employment, production, temperature, etc. ■ Forecasts trends in terms of variables
Circle or pie chart	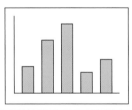	■ Reveals the parts/percentages (budgets, shares, time allotments, etc.) that make up the whole ■ Shows the proportion of each part to another; relates each part to a whole ■ Easily understood by audiences worldwide
Bar chart	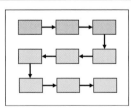	■ Uses vertical or horizontal bars to measure different data in space or time ■ Represents comparisons/changes through the comparative length of bars ■ Can be segmented (divided) to show multiple percentages within a single bar ■ Less technical than a graph
Flow chart	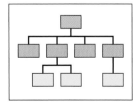	■ Reveals stages in an activity or process ■ Blueprints the actual steps to take, first to last, to complete a process/job ■ Identifies places where retracing or skipping steps is required
Organiza-tional chart		■ Shows structure of an organization from chief executive to divisions and departments to employees ■ Provides quick view of areas of authority and responsibility ■ Helps in routing material to appropriate readers

TABLE 10.1 (Continued)

Pictograph	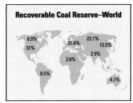	■ Represents statistical data by means of pictures varying in size, numbers, or color ■ Easy for a global audience to understand if the symbols/icons (pictograms) are chosen carefully and appropriately ■ Must supply a key stating how much each symbol represents
Map		■ Details specific geographic features (elevation, lakes/rivers, forests, etc.) ■ Identifies roads, businesses (every company location, TV station), etc. ■ Displays census data (population density, number of households with children)
Cutaway drawing	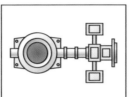	■ Reveals the interior view of an object by removing the exterior view ■ Helps explain how something works ■ Uncovers an internal mechanism that a photo or simple drawing cannot
Exploded drawing	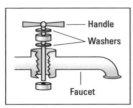	■ Shows an object with parts separated to indicate the relationship of the parts to one another ■ Illustrates how parts fit together ■ Instructs how a piece of equipment is constructed and should be assembled
Photograph		■ Demonstrates the actual appearance of something ■ Pinpoints color accurately—no guesswork ■ Cannot reveal interior parts
Clip art		■ Provides ready-made icons, images, symbols, and pictures ■ Available online or in print form ■ Must be appropriate for audience, especially international readers ■ Should not look unprofessional or cartoonish

Figure 10.2   A visual used in conjunction with written work.

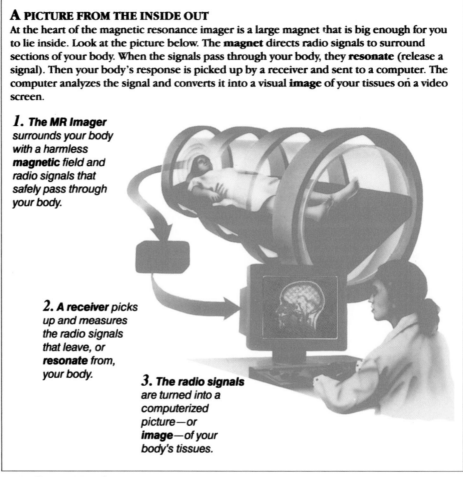

A PICTURE FROM THE INSIDE OUT

At the heart of the magnetic resonance imager is a large magnet that is big enough for you to lie inside. Look at the picture below. The **magnet** directs radio signals to surround sections of your body. When the signals pass through your body, they **resonate** (release a signal). Then your body's response is picked up by a receiver and sent to a computer. The computer analyzes the signal and converts it into a visual **image** of your tissues on a video screen.

1. **The MR Imager** *surrounds your body with a harmless* **magnetic** *field and radio signals that safely pass through your body.*

2. **A receiver** *picks up and measures the radio signals that leave, or* **resonate** *from, your body.*

3. **The radio signals** *are turned into a computerized picture—or* **image**—*of your body's tissues.*

Reprinted by permission of Krames Communications.

Figure 10.2 makes the procedure easier to understand than if the writer had used only words or only a visual. This visual and verbal description are appropriately included in a brochure teaching patients about MRI procedures.

3. Experiment with several visuals. Evaluate a variety of options before you select a particular visual. For instance, software such as PowerPoint (see pp. 676–678) allows you to represent statistical data in several ways. Preview a few different versions of a visual or even different types of visuals to determine which one would be best.

4. Always use easy-to-read and relevant visuals. If readers have trouble understanding its function and arrangement, your visual probably is not appropriate and you need to change or revise it.

5. Be prepared to revise and edit your visuals. Just as you draft, revise, and edit your written work to meet your audience's needs, create several versions of your

An ineffective visual: Too much information is crowded into one graphic.   Figure 10.3

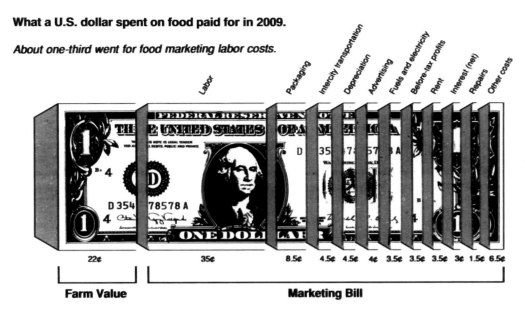

visual to get it right. Expect to change shapes or proportions; experiment with different colors, shadings, labels, and various sizes. Replace any visuals that may distract the reader or contradict your message. Also replace or revise any visuals that are unethically distorted or exaggerated. (See "Using Visuals Ethically" on pages 472–478.)

6. Consider how your visuals will look on the page. Visuals should add to the overall appearance of your work, not detract from it. Don't cram visuals onto a page or allow them to spill over your text or margins. Many programs help you move your visuals directly into your word processor document so you can insert them properly. Go to pages 496–501 for advice on effective page layouts.

Ineffective Visuals: What *Not* to Do

Here are some guidelines on what to avoid with visuals:

- Avoid visuals that include more details than your audience needs.
- Never use a visual that distracts from your work (for example, one that is too small, too large, does not use the right type of shading).
- Never use a visual that presents information that contradicts your work.
- Never distort a visual for emphasis or decoration.
- Be careful that you don't omit anything when you reproduce an existing visual.
- Never use visuals that discriminate or stereotype (for example, avoid pictures of a work force that excludes female employees).
- Don't use a visual that looks fuzzy, dotted, or streaked.

See how Figure 10.3 violates many of these rules. It divides the U.S. dollar bill into too many slices. Confronted with so many different wedges the reader would have trouble identifying, separating, comparing, and understanding the costs.

Tech Note

Scanners and Scanning

Scanners allow you to produce a digital copy of an image or a document to store and use on a computer. Using technology similar to that of a photocopier, a scanner captures an image and creates a digital reproduction that can then be altered or enhanced and printed, used as a visual in a document, or stored as a file in your computer. Scanners provide an efficient means of incorporating into your documents those visuals unavailable in digital form, such as older photographs, sketched diagrams, or pictures from books. For example, if you need to attach a diagram of a product to a document, but have only a hard copy, you can scan the image and use it within your document rather than attaching a hard copy. Scanning has begun to replace facsimiles in the business world to save time and money by transferring documents more efficiently. Unlike a fax, scanning allows you to attach the document to an e-mail message and send it directly. Scanning documents also preserves color images that may be distorted in faxing.

Using high-resolution digital imaging, with which scanners are equipped, you may produce high-quality replications of photographs. Using optical character recognition (OCR) software, scanners can reproduce text on a word processing or spreadsheet program. This is particularly useful if you need to edit a document that is available only in hard copy, thus ensuring a virtually error-free transition of text from hard copy onto your computer.

Scanning helps companies to maintain an accurate, comprehensive, and searchable filing system that saves time, space, and money. Scanners contribute to a paperless office by capturing, storing, and cataloging documents, particularly useful for legal records and tax documents, which have to be stored for several years. Scanning these documents allows for quick retrieval and easy reproduction, and reduces the risk of losing paper copies.

Lori Brister
Anna Gibson

Inserting and Writing About Visuals: Some Guidelines

Using a visual requires more of you as a writer than simply inserting it into your written work. You need to *use visuals in conjunction with what you write*. The following guidelines will help you to (1) identify, (2) cite, (3) insert, (4) introduce, and (5) interpret visuals for your readers.

Identify Your Visuals

Give each visual a number and caption (title) that indicates the subject or explains what the visual illustrates. An unidentified visual is meaningless. A caption helps your audience to interpret your visual—to see it with your purpose in mind. Tell your readers what you want them to look for.

- Use a different typeface (bold) and size in your caption than what you use in the visual itself.
- Include key words about the function and the subject of your visual in a caption.
- Make sure any terms you cite in a caption are consistent with the units of measurement and the scope (years, months, seasons) of your visual.

Tables and figures should be numbered separately throughout the text—Table 1 or Figure 3.5, for example. (In the latter case, Figure 3.5 is the fifth figure to appear in Chapter 3.) Below are some examples of figure and table numbers and titles.

- Table 2. Paul Jordan's work schedule, January 15–23
- Figure 4.6. The proper way to apply for a small business loan
- Figure 12. Income Estimation Figures for North Point Technologies

Cite the Source for Visuals

If you use a visual that is not your own work, give credit to your source (newspaper, magazine, textbook, company, federal agency, individual, or website). If your paper or report is intended for publication, you must obtain permission to reproduce copyrighted visuals from the copyright holder.

Insert Your Visuals Appropriately

Since many of the images you use will come from outside sources, especially the Web, you have to incorporate them clearly and in appropriate places in your written work. See the Tech Note "Importing Visuals from the Web" on the following page.

Here are some guidelines to help you incorporate the visuals in the most appropriate places for your readers.

- Never introduce a visual *before* a discussion of it; readers will wonder why it is there. Use a sentence or two to introduce your visuals.
- Always mention in the text of your paper or report that you are including a visual. Tell readers where it is found—"below," "on the following page," "to the right," "at the bottom of page 3."
- Place visuals as close as possible to the first mention of them in the text. Try not to put a visual more than one page after the discussion of it. Never wait two or three pages to present it. By inserting a visual near the beginning of your discussion, you help readers better understand your explanation.
- Use an appropriate size for your visual. Don't make it too large or too small. Gauge size by how much data you are presenting and where the visual best fits into your discussion of it.
- If the visual is small enough, insert it directly in the text rather than on a separate page. If your visual occupies an entire page, place that page containing

your visual on the facing page or immediately after the page on which the first reference to it appears.

- Center your visual and, if necessary, box it. Leave at least 1 inch of white space around it. Squeezing visuals toward the left or right margins looks unprofessional.
- Never collect all your visuals and put them in an appendix. Readers need to see them at those points in your discussion where they are most pertinent.

Tech Note

Importing Visuals from the Web

To incorporate visuals from the Internet in your report you will need to save the image to your hard disk or copy it into your computer's memory. Saving an image to your hard disk should only be done if you wish to store the image for future use, since otherwise it will clutter your hard drive. If you wish to use an image only once, copy it to your computer's temporary memory and paste it into the document.

When importing images from online sources, keep the following guidelines in mind:

- Be careful not to resize your image so that it becomes distorted. Images are limited to a certain size by the number of pixels (or dots) per inch (dpi). If you use your word processor's resize tool to make the image too large, it will become fuzzy and look unprofessional.
- When you copy images from the Web, your computer can store only one image at a time in its temporary memory. Make sure you paste the image before you copy another, or it will be lost.
- Many websites will not allow you to copy their images. When you right-click on an image, a dialogue box will appear informing you that the image is protected and cannot be copied or saved. In order to obtain this image, you will need to e-mail the site's webmaster and request permission to use the visual.
- If you save an image to your hard drive for use in a document, give it an easy-to-remember file name, since your computer will automatically take the code word (often a combination of numbers and letters, as in a serial number) as the file name for the image. Save the file as a name you can easily recall.
- Be sure to format your visual so that the text wraps around it and does not become obscured by the image.

Lori Brister
Anna Gibson

Introduce Your Visuals

Refer to each visual by its number and, if necessary, mention the title as well. In introducing the visual, though, do not just insert a reference to it, such as "See Figure 3.4" or "Look at Table 1." Help readers to understand the relationships in your visual. Here are two ways of writing a lead-in sentence for a visual.

> Poor: Our store saw a dramatic rise in the shipment of electric ranges over the five-year period as opposed to the less impressive increase in washing machines. (See Figure 3.)

This sentence does not tie the visual (Figure 3) into the sentence where it belongs. The visual just trails insignificantly behind.

> Better: As Figure 3 shows, our store saw a dramatic rise in the shipment of electric ranges over the five-year period as opposed to the less impressive increase in washing machines.

Mentioning the visual in Figure 3 alerts readers to its presence and function in your work and helps them to more easily understand your message.

Interpret Your Visuals

Give readers help in understanding your visual and in knowing what to look for. Let them know what is most significant about the visual. Do not expect the visual to explain itself. Inform readers what the numbers or images in your visual mean. What types of conclusions can you and your readers logically make after seeing the visual?

In a report on the benefits of vanpooling, the writer supplied the following visual, a table:

TABLE 1.   Travel Time (in minutes): Automobile Versus Vanpool

Private Automobile	Vanpool
25	32.5
30	39.0
35	45.5
40	52.0
45	58.5
50	65.0
55	71.5
60	78.0

Source: U.S. Department of Transportation. *Increased Transportation Efficiency Through Ridesharing: The Brokerage Approach* (Washington, D.C., DOT-OS—40096): 45.

To interpret the table, the writer called attention to it in the context of the report on transportation efficiency.

> Although, as Table 1 above suggests, the travel time in a vanpool may be as much as 30 percent longer than in a private automobile (to allow for pickups), the total trip time for the vanpool user can be about the same as with a private automobile because vanpools eliminate the need to search for parking spaces and to walk to the employment site entrance.[1]

■ Two Categories of Visuals: Tables and Figures

Visuals can be divided into two categories—tables and figures. A **table** arranges information—numbers and/or words—in parallel columns or rows for easy comparison of data. Anything that is not a table is considered a figure. **Figures** include graphs, circle charts, bar charts, organizational charts, flow charts, pictographs, maps, photographs, and drawings. Expect to use both in your work.

■ Tables

Visit www
.cengage.com/
english/kolin for
an online exercise,
"Creating Tables
and Charts
Using Your
Word-Processing
Program."

Tables are parallel columns or rows of information organized and arranged into categories to show changes in time, distance, cost, employment, or some other distinguishable or quantifiable variable in a compact space. Electronic spreadsheet programs heavily rely on tables to convey information, as in Table 10.2. Tables also summarize material for easy recall—causes of wars, provisions of a law, or differences between a common cold, flu, and pneumonia, as in Table 10.3. Observe how Table 10.4 (p. 451) also easily condenses much information and arranges it in quickly identifiable categories.

Parts of a Table

To use a table properly, you need to know the parts that constitute it. Refer to Table 10.4, which labels these parts, as you read the following:

- The main **column** is "Amount Needed to Satisfy Minimum Daily Requirement," and the **subcolumns** are the protein sources for which the table gives data.
- The **stub** refers to the first vertical column on the left-hand side. The stub column heading is "Source." The stub lists the foods for which information is broken down in the subcolumns.
- A **rule** (or line) across the top of the table separates the headings from the body of the table.

---

[1]James A. Devine, "Vanpooling: A New Economic Tool," *AIDC Journal*.

TABLE 10.2   A Spreadsheet Table

	A	B	C	D	F	G	H
1	Acct#	Account Description	YTD	April	March	February	Janua
2							
3	**Employee Costs**						
4	110	Payroll	$171,465	$29,750	$28,547	$27,540	
5	120	IRS/FICA/Wk comp/State/SDI	$46,295	$8,032	$7,707	$7,435	
6	130	Commissions	$56,436	$9,520	$8,875	$9,825	
7	140	Retirement Plan	$20,575	$3,570	$3,425	$3,304	
8	150	Insurance	$9,000	$1,500	$1,500	$1,500	
9							
11	**Subcontractors & Services**						
12	201	Telecommunication Services	$2,616	$436	$436	$436	
13	202	Design Consultants	$875	$500	$375	$0	
14	203	Photo/Video Services	$535	$45	$325	$45	
15	250	Graphic Services	$2,957	$765	$95	$375	
16	251	Photo/Stats	$1,612	$568	$755	$0	
17	252	Typesetting	$1,453	$388	$325	$195	
18	253	Printing Services	$6,186	$951	$849	$325	
19	254	Legal & Accounting					
20							
21	**Supplies and Materials**						
22	301	Office Supplies	$875	$500	$732	$433	
23	302	Office Postage	$535	$45	$255	$325	
24	303	Office Equipment & Furniture	$2,957	$765	$78	$21	
25	304	Miscellaneous Supplies	$1,612	$568	$49	$36	
26							
27	**Facilities Overhead**						
28	405	Plant	$1,612	$568	$1,700	$1,700	

Sheet1 / Sheet2 / Sheet3 /

Guidelines for Using Tables

When you include a table in your work, follow these guidelines.

- Number the tables according to the order in which they are discussed (Table 1, Table 2, Table 3).
- Include the table on the same page, where it is most appropriate. It is hard for readers to follow a table spread across different pages.
- Give each table a concise and descriptive title to show exactly what is being represented or compared.
- Use words in the stub (a list of items about which information is given), but put numbers under column headings. The Source column in Table 10.4 is the stub.
- Supply footnotes (often indicated by small raised letters: [a], [b]) if something in the table needs to be qualified, for example, the number of cups of milk in Table 10.4. Then put that information below the table.
- List items in alphabetical, chronological, or other logical order.

TABLE 10.3   Table Showing Differences Between a Common Cold, Influenza, and Pneumonia

Symptoms	Cold	Influenza	Pneumonia
Fever	Rare	Characteristic high (100.4–104°F); sudden onset, lasts 3 to 4 days	May or may not be high
Headache	Occasional	Prominent	Occasional
General aches and pains	Slight	Usual; often quite severe	Occasionally quite severe
Fatigue and weakness	Quite mild	Extreme; can last up to a month	May occur depending on type
Exhaustion	Never	May occur early and prominent	May occur depending on type
Runny, stuffy nose	Common	Sometimes	Not characteristic
Sneezing	Usual	Sometimes	Not characteristic
Sore throat	Common	Sometimes	Not characteristic
Chest discomfort, cough	Mild to moderate; hacking cough	Can become severe	Frequent and may be severe
Complications	Sinus and ear infections	Bronchitis, pneumonia; can be life-threatening	Widespread infections of other organs; can be life-threatening, especially in elderly and debilitated persons

Source: Jacquelyn G. Black, *Microbiology: Principles and Explorations*. Upper Saddle River, NJ: Prentice-Hall, 1996.

- Arrange the data you want to compare vertically, not horizontally; it is easier to read down than across a series of rows.
- Place tables at the top (preferable) or bottom of the page and center them on the page rather than placing them up against the right or left margin.
- Don't use more than five or six columns; tables wider than that are more difficult for readers to use.
- Round off numbers in your columns to the nearest whole number to assist readers in following and retaining information.
- Always give credit to the source (the supplier of the statistical information) on which your table is based.

TABLE 10.4   Parts of a Table

Table number

TABLE 1 Efficiency of Some Protein Sources in Meeting an Adult's Minimum Daily Requirements				
Source	**Percent of Protein**	**Percent of Amino Acids**	**Amount Needed to Satisfy Minimum Daily Requirement**	
			(grams)	*(ounces)*
Cheese[a]	27	70	227	7.2
Corn	10	50	860	30.0
Eggs	11	97	403	14.1
Fish[a]	22	80	244	8.5
Kidney beans	23	40	468	16.4
Meat[a]	25	68	253	8.8
Milk	4	82	1,311	45.9[b]
Soybeans	34	60	210	7.3

Title
Rule
Column heading
Subheading
Stub
Origin of data
Footnotes

Source: From Starr/Taggart, *Biology: The Unity and Diversity of Life*, 4th ed. Copyright © 1987 Brooks/Cole, a part of Cengage Learning, Inc. Reprinted with permission. www.cengage.com/permissions.

[a] = Average value
[b] = Equivalent of 6 cups

Figures

As we saw, any visual that is not a table is classified as a figure. The types of figures we examine next are

- graphs
- circle, or pie, charts
- bar charts
- organizational charts
- flow charts
- pictographs
- maps
- photographs
- drawings

Graphs

Graphs transform numbers into pictures. They take statistical data presented in tables and put them into rising and falling lines, steep or gentle curves. The three types of graphs are (1) simple line graphs, (2) multiple-line graphs, and (3) area graphs.

Functions of Graphs

Graphs vividly portray information that changes, such as

- sales
- costs
- trends
- distributions
- employment
- energy levels
- temperatures

Graphs not only describe past and current situations but also forecast trends.

Tech Note

Computer-Generated Graphs and Charts

You can easily generate charts, graphs, or tables by using the templates available in your word processing software, such as Microsoft Word, and by selecting Insert Chart or Insert Table. In fact, many of the visuals in this book, especially those in the long reports in Chapters 8 and 15, were created using such software. Graph and chart templates allow you to insert your raw numerical data into the appropriate blanks and then add titles, labels for axes, and color-coded keys. Such software not only helps construct different visuals, including pie charts and flow charts, but it also allows writers to show visuals from different perspectives and emphases, such as shading, 3-D, exploded view, and many others. Spreadsheet programs, such as Microsoft Excel, also provide easy-to-use templates for creating graphs and charts using the numerical data you have entered into your spreadsheet. When using these software programs, avoid the temptation to make visuals too showy. Although dazzling graphics may look interesting, you don't want them to take over the report. Computer-generated graphs and charts should work with the writer's words, not overshadow them.

Graphs Versus Tables

Because graphs actually show change, they are more dramatic than tables. You will make the reader's job easier by using a graph rather than a table. Many financial websites and print publications—the *Wall Street Journal* and *USA Today*, for example—open with a graph for the benefit of busy readers who want a great deal of financial information summarized quickly.

Simple Line Graphs

Basically, a simple graph consists of two sides—a **vertical** or **y-axis** and a **horizontal** or **x-axis**—that intersect to form a right angle, as in Figure 10.4. The space between the two axes contains the picture made by the graph—the amount of snowfall in Springfield between November 2009 and April 2010. The vertical line represents the **dependent variable** (the snowfall in inches), the horizontal line, the **independent variable** (time in months). The dependent variable is influenced most directly by the independent variable, which almost always is expressed in terms of time or distance. The vertical axis is read from bottom to top; the horizontal axis from left to right.

Multiple-Line Graphs

The graph in Figure 10.4 contains only one line per category. But a graph can have multiple lines to show how a number of dependent variables (conditions, products) compare with one another.

The six-month sales figures for three salespeople can be seen in the graph in Figure 10.5. The graph contains a separate line for each of the three salespersons. At a glance readers can see how the three compare and how many dollars each salesperson

Figure 10.4

A simple line graph showing the amount of snowfall in Springfield from November 2009 to April 2010.

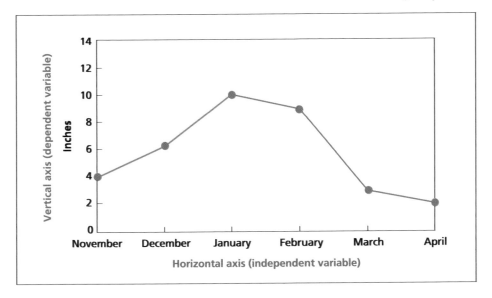

Figure 10.5

A multiple-line graph showing sales figures for the first six months of 2009 for three salespeople.

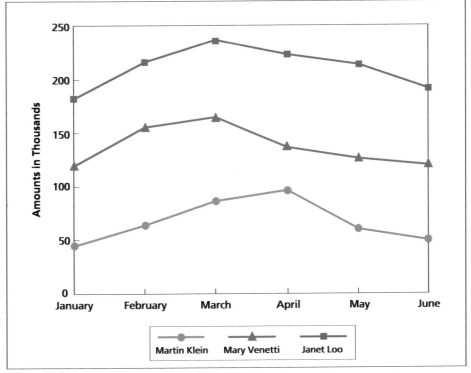

Uses different colors and symbols to represent each salesperson

Legend explains what different symbols and colors represent

generated per month. Note how the line representing each person is clearly differentiated from the others by symbols and colors. Each line is clearly tied to a **legend** (an explanatory key below the graph) specifying the three salespersons.

Area graphs

Figure 10.6 is another graphic representation of the information in Figure 10.5. The graph in Figure 10.6, known as an **area,** or **multiband, graph,** shows relationships among the three salespeople without providing the exact mathematical documentation of Figure 10.5. In Figure 10.6 the spaces between the different curves are filled in, with colors keyed in the legend to identify the three salespersons.

Guidelines on Creating Graphs

1. Use no more than three lines in a multiple-line graph, so readers can interpret the graph more easily. If the lines run close together, use a legend to identify individual lines.
2. Label each line to identify what it represents for readers. Include a key, or legend, as in Figure 10.5.
3. Keep each line distinct in a multiple-line graph by using different colors, dots or dashes, or symbols. Note the different symbols in Figure 10.5.
4. Make sure you plot enough points to show a reasonable and ethical range of the data. Using only three or four points may distort the evidence.

Figure 10.6   An area, or multiband, graph of Figure 10.5.

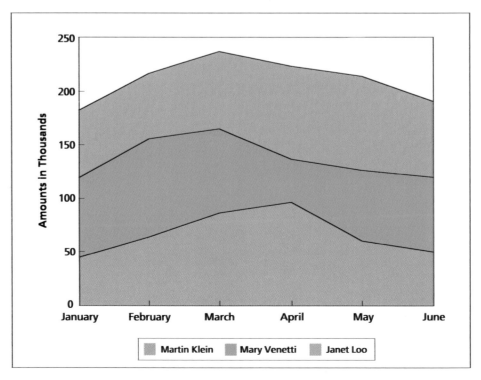

5. Keep the scale consistent and realistic. If you start with hours, do not switch to days or vice versa. If you are recording annual rates or accounts, do not skip a year or two in order to save time or be more concise.

6. For some graphs there is no need to begin with a zero. This is called a **suppressed zero graph**, which automatically begins with a larger number when it would be impossible to start with zero. For others, you may not have to include numbers beyond seven or eight. Your subject and purpose will determine the range of data you need to show.

Charts

Although charts and graphs may seem similar, there is a big difference between them. Graphs are usually more complex and plotted according to specific mathematical coordinates. Charts, on the other hand, do not display exact and complex mathematical data. Instead, they present an overall picture of how individual pieces of data (from a graph or table) fall into place to express relationships.

Among the most frequently used charts are (1) circle, or pie, charts, (2) bar charts, (3) organizational charts, and (4) flow charts.

Circle or Pie Charts

Circle charts are also known as **pie charts**, a name that descriptively points to their construction and interpretation. Tables are more technical and detailed than circle charts. Figure 10.7 shows an example of a pie chart used in a government document. A table or graph with a more detailed breakdown of, say, a city's budget would be much more appropriate for a technical audience (auditors, budget and city planners).

The full circle, or pie, represents the whole amount (100 percent or 360 degrees) of something: the entire budget of a company or a family, a population group, an area of land, the resources of an organization or institution. Each slice or wedge represents a percentage or portion of the whole.

A circle chart effectively allows readers to see two things at once: the relationship of the parts to one another and the relationship of the parts to the whole.

A three-dimensional circle chart showing the breakdown by department of the   Figure 10.7
proposed Midtown city budget for 2009.

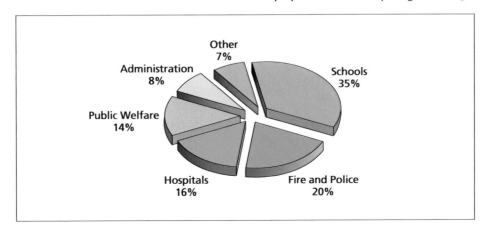

Preparing a Circle Chart
Follow these seven rules to create and present your circle chart:

1. **Keep your circle chart simple.** Don't try to illustrate technical statistical data in a pie chart. Pie charts are primarily used for general audiences.

2. **Do not divide a circle, or pie, into too few or too many slices.** If you have only three wedges, use another visual to display them (a bar chart, for example, discussed below). If you have more than seven or eight wedges, you will divide the pie too narrowly, and overcrowding will destroy the dramatic effect. Instead, combine several slices of small percentages (2 percent, 3 percent, 4 percent) into one slice labeled "Other," "Miscellaneous," or "Related Items."

3. **Make sure the individual slices total 100 percent, or 360 degrees.** For example, if you are constructing a circle chart to represent a family's budget, the breakdown percentages might be as follows:

Category	Percentage	Angle of slice
Housing	25%	90.0°
Food	22%	79.2°
Energy	20%	72.0°
Clothes	13%	46.8°
Health Care	12%	43.2°
Miscellaneous	8%	28.8°
Total	100%	360°

4. **Put the largest slice first, at the 12 o'clock position, then move clockwise with proportionately smaller slices.** Schools occupy the largest slice in Figure 10.7 because they receive the biggest share of taxes.

5. **Label each slice of the pie horizontally.** Key in the identifying term or quantity inside, but make sure the label is big enough to read. Do not put in a label upside down or slide it in vertically. If the individual slice of the pie is small, draw a connecting line from the slice to a label positioned outside the pie.

6. **Shade, color, or cross-hatch slices of the pie to further separate and distinguish the parts.** Figure 10.7 effectively uses color. But be careful not to obscure labels and percentages; also make certain that adjacent slices can be distinguished readily from each other. Do not use the same color or similar colors for two slices.

7. **Give percentages for each slice to further assist readers,** as in Figure 10.7.

Bar Charts

A bar chart consists of a series of vertical or horizontal bars that indicate comparisons of statistical data. For instance, in Figure 10.8 vertical bars depict increases in numbers of working mothers. Figure 10.9 uses horizontal bars to depict the nation's top 20 metropolitan areas, based on building permits. The length of the bars is determined according to a scale that your computer software can easily compute.

Bar Charts, Graphs, and Tables—Which Should You Use?
When should you use a bar chart rather than a table or a line graph? Your audience will help you decide. If you are asked to present statistics on costs for the company

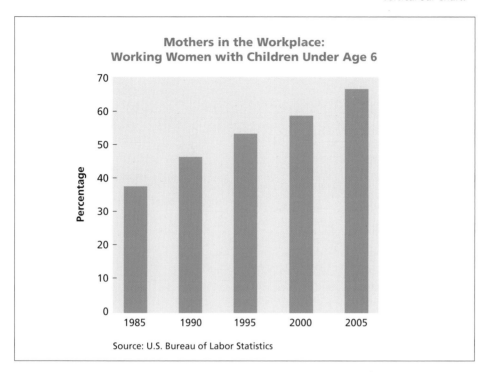

Mothers in the Workplace:
Working Women with Children Under Age 6

Source: U.S. Bureau of Labor Statistics

accountant, use a table. Because a bar chart is limited to a few columns, however, it cannot convey as much information as a table or graph. However, if you are presenting the same information to a group of stockholders or to a diverse group of employees, a bar chart may be more relevant and persuasive.

Types of Bar Charts
There are three types of bar charts.

   1. Simple bar charts. Figure 10.8 is the most basic form of bar chart. Each bar represents the percentage of working women with children under the age of 6, and the height of the bar corresponds to the number of percentage points for a given year.

   2. Multiple-bar charts. Figure 10.10 shows a variation of the simple, vertical bar chart. Four different colored bars are used for each year to represent the amount of money spent on different types of advertising in 2006, 2007, 2008, and 2009. A legend at the top of the chart explains what each bar stands for. But avoid using more than four bars in a group for any one year. More bars will make your chart crowded and difficult to read.

   3. Segmented, or cross-hatched (divided), bar charts. To show the different components that constitute a measured whole, use a segmented bar chart like the one in Figure 10.11. A single segmented bar lists the travel expenses of Weemco, a small firm, in October 2009. The entire bar equals the travel total—$137,000—which was spent in four areas: airfare, ground transportation, lodging, and meals. Each of these expenses is represented by a different type of shading on the single column. A group of

Figure 10.9   Horizontal bar chart.

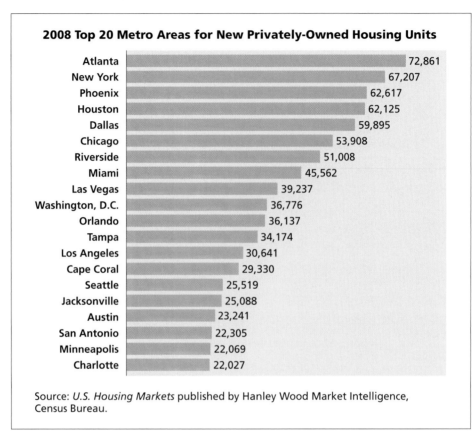

2008 Top 20 Metro Areas for New Privately-Owned Housing Units

Metro Area	Units
Atlanta	72,861
New York	67,207
Phoenix	62,617
Houston	62,125
Dallas	59,895
Chicago	53,908
Riverside	51,008
Miami	45,562
Las Vegas	39,237
Washington, D.C.	36,776
Orlando	36,137
Tampa	34,174
Los Angeles	30,641
Cape Coral	29,330
Seattle	25,519
Jacksonville	25,088
Austin	23,241
San Antonio	22,305
Minneapolis	22,069
Charlotte	22,027

Source: *U.S. Housing Markets* published by Hanley Wood Market Intelligence, Census Bureau.

segmented bars can be used to show multiple comparisons among many categories, as in Figure 10.12, which depicts energy consumption levels and types in five states.

Organizational Charts

An **organizational chart** pictures the chain of command in a company or agency, with the lines of authority stretching down from the chief executive, manager, or administrator to assistant manager, department heads, or supervisors to the work force of employees. Figure 10.13 (page 461) shows a hospital's organizational chart for its nursing services.

Organizational charts have these functions:

- to inform employees and customers about the makeup of a company
- to depict the various offices, departments, and units
- to show where people work in relationship to one another in a business
- to coordinate employee efforts in routing information to appropriate departments

Flow Charts

A **flow chart** displays the stages in which something is manufactured, is accomplished, develops, or operates. Flow charts are highly effective in showing the steps in following a procedure. They can also be used to plan the day's or week's activities.

A multiple-bar chart showing advertising expenditures by major media, 2006–2009.   Figure 10.10

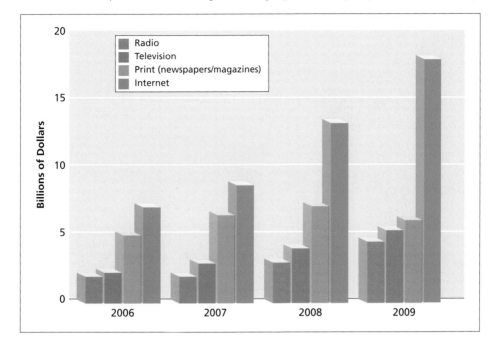

A segmented bar chart representing total travel expenditures for Weemco Communications for October 2009.   Figure 10.11

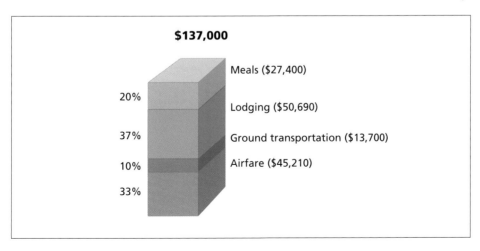

A flow chart tells a story with arrows, boxes, and sometimes pictures. Boxes are connected by arrows to visualize the stages of a process. The presence and direction of the arrows tell the reader the order and movement of events involved in the process.

Figure 10.12   A multiple-bar, segmented bar chart showing energy consumption by sector in five states that consumed the most energy in a given year.

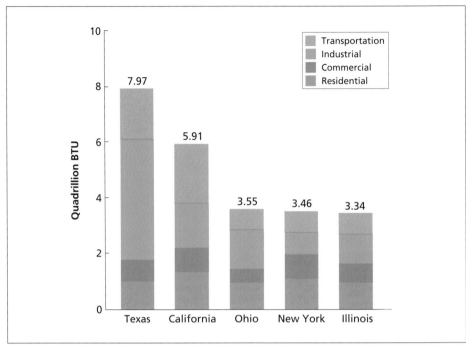

Source: U.S. Energy Information Administration.

Flow charts often proceed from left to right and back again, as in the flow chart below, showing the steps to be taken before graduation.

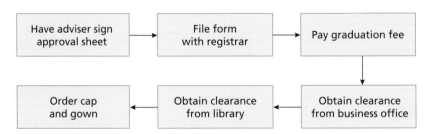

Flow charts can also be constructed to read from top to bottom. Programming instructions frequently are written that way. See, for example, Figure 10.14 (p. 462) which uses a programming flow chart to show the steps a student must follow in writing a research paper.

A flow chart should clarify, not complicate, a process. Do not omit any important stages, but at the same time do not introduce unnecessary or unduly detailed information. Show at least three or four stages and make sure the various stages appear in the correct sequence.

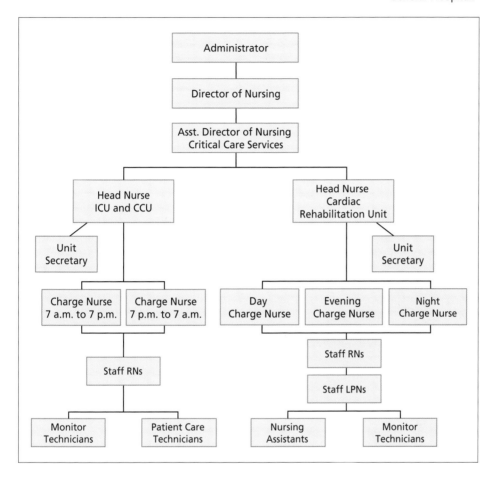

Pictographs

Similar to a bar chart, a **pictograph** uses picture symbols (called **pictograms**) to represent differences in statistical data, as in Figure 10.15 (p. 463). A pictograph repeats the same symbol or icon to depict the quantity of items being measured. Each symbol stands for a specific number, quantity, or value. Pictographs are visually appealing and dramatic and are far more appropriate for a nontechnical than a technical audience.

Guidelines on Creating Pictographs

When you create a pictograph, follow these four guidelines:

1. Choose an appropriate, easily identifiable symbol for the topic.
2. Indicate the precise quantities each icon represents by placing numbers after or at the top of the visual.
3. Increase the number of symbols rather than their sizes because differences in sizes are often difficult to construct or interpret.

Figure 10.14   A programming flow chart showing steps in writing a research report.

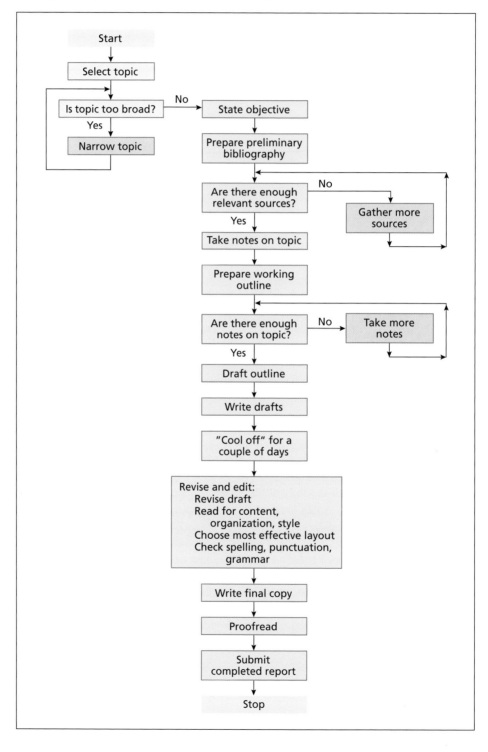

A pictograph showing the growth of one state's retirement assets.   Figure 10.15

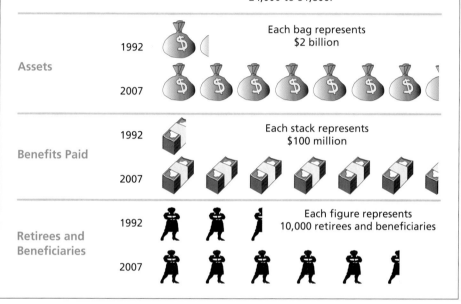

How does the Public Employees' Retirement System (PERS) compare today with 15 years ago?

During that time, PERS experienced spectacular growth, with both assets and benefits paid increasing more than sixfold.

- The market value of assets increased from $2.3 billion to $14.9 billion.
- PERS paid benefits of $94.3 million in FY '92. The total reached $601.1 million during FY '07.
- Retirees and beneficiaries drawing benefits more than doubled, from 24,000 to 51,800.

Assets

1992

2007

Each bag represents $2 billion

Benefits Paid

1992

2007

Each stack represents $100 million

Retirees and Beneficiaries

1992

2007

Each figure represents 10,000 retirees and beneficiaries

Source: PERS Member Newsletter, August 1999. Official Publication of the Public Employees' Retirement System of Mississippi. By permission of Public Employees' Retirement System.

4. Avoid crowding too much information into a pictograph, such as the one in Figure 10.3, in which a U.S. dollar bill is divided into too many sections.

Maps

The maps you use on the job may range from highly sophisticated and detailed geographic tools to simple sketches such as the map in Figure 10.16, which shows the location of a town's water filter plants and pumping stations. This is a **large-scale map** that displays social, economic, or physical data for a small area. The map in Figure 10.17, on the other hand, gives a state-by-state account of the number of U.S. representatives and the change in those numbers over a ten-year period.

You may have to construct your own map, like the one in Figure 10.16, or scan one in a published source or one on the Internet. If you scan a map, be sure to obtain permission to use it from the copyright holder.

www

Visit www .cengage.com/ english/kolin for an online exercise, "Assessing Online Maps, Photographs, Drawings, and Clip Art."

Figure 10.16   A map showing the location of Smithville Water Department's filter plants and pumping stations.

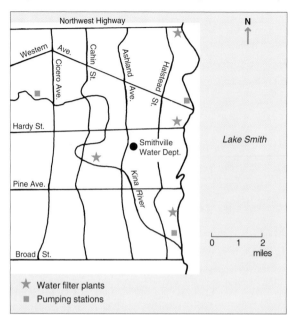

Guidelines for Creating a Map.   Follow these steps when you create a map:
1. Always acknowledge your source if you did not construct the map yourself.
2. Use dots, lines, colors, symbols, and shading to indicate features. Markings should be clear and distinct.
3. If necessary, include a legend, or map key, explaining dotted lines, colors, shading, and symbols as in Figure 10.16.
4. Exclude features (rivers, elevations, county seats) that do not directly relate to your topic. For example, a map showing the crops grown in two adjacent counties need not show all the roads and highways in those counties. A map showing the presence of strip mining or features such as hills or valleys needs to indicate elevation, but a map depicting population or religious affiliation need not include topographical (physical) detail.
5. Indicate direction. Conventionally, maps show north, often by including an arrow and the letter *N*: *N* ↑.

Photographs

Correctly taken or scanned, photographs are an extremely helpful addition to job-related writing. A photograph's chief virtues are realism and clarity. Among its many advantages, a photo can

- show what an object looks like (Figure 10.18)
- demonstrate how to perform a certain procedure (Figure 10.19)
- compare relative sizes and shapes of objects (Figures 10.18 and 10.19)
- compare and contrast scenes or procedures (Figure 10.20)

Apportionment of the U.S. House of Representatives for the 108th Congress.   Figure 10.17

Tech Note

Digital Photography

Digital cameras and camera phones allow you to easily supply professional-looking, customized photos with your written work. Digitalized photography offers the following benefits in the global marketplace:

■ Shows conditions or sites right away so you can include a photo in an incident or credit report—you do not need to wait for film to be developed

■ Gives you more than one opportunity to take a picture—helps you select, highlight, and edit

■ Allows you to send photographs easily and quickly over the Internet, store them on your computer or on a CD, and upload them to a website

Continued

Continued

■ Gives you the opportunity to edit photographs for color, sharpness, contrast, brightness, size, and red-eye—you can also crop out unnecessary details

To achieve these goals, be careful when using a digital camera or camera phone that it is set to the correct photo-quality level. The poorer the quality, the more pictures your camera will hold. However, if you will be using photographs for a newsletter or company magazine or if you wish to print photographic-quality images, make sure your camera is set on a higher quality level.

Figure 10.18   A photo showing what a piece of equipment looks like.

Source: Corbis Royalty Free (*http://www.corbis.com*).

A photo showing how to perform a procedure and comparing relative sizes and shapes of objects. **Figure 10.19**

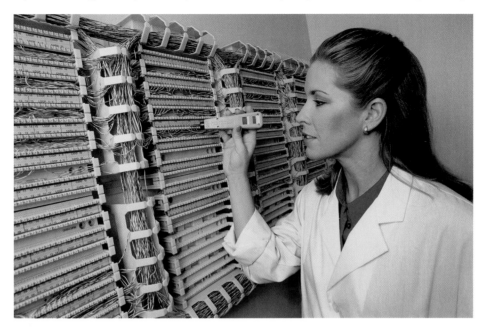

Source: Corbis Royalty Free (*http://www.corbis.com*).

Guidelines for Taking Photographs

Whether you use a traditional or digital camera, observe the following guidelines:

1. **Take the photo from the most appropriate distance.** Decide how much foreground and background information your audience needs. For example, if you are photographing a three-story office building, your picture may misleadingly show only one or two stories if you are standing too close to the building.

2. **Select the correct angle.** Take the photo from an angle that makes sense, usually straight on so that the object or person photographed can be viewed in full.

3. **Include only the details that are necessary and relevant for your purpose.** Crop (edit out) any unnecessary details from the photograph.

4. **Provide a sense of scale.** So that readers understand the size of the object, include in the photograph a person (if the object is very large), a hand (if the object is small), or a ruler.

5. **Make sure you consider lighting and resolution.** Take your photograph in appropriate light, so that the subject of your photograph is clear and crisp. If you are using a digital camera, take a high-resolution photograph for maximum clarity.

6. **Always ask for permission before you take a photograph of a person, place, or thing, unless you are photographing a public park or building.** Do not take photographs of copyrighted materials, for instance, a work of art, a page from a copyrighted book, or a movie or television screen. Also obtain permission from your boss if you are photographing your own company's equipment, sites, or designs.

Figure 10.20   Photos showing comparison/contrast of using gasoline versus electricity to power an automobile.

© moodboard/Super Value Royalty-Free/Corbis

© Dean Siracusa/Corbis

To get a graphic sense of the effects of taking a photo the right and wrong way, study the photographs in Figures 10.21 and 10.22. A clear and useful picture of a hydraulic truck (often called a "cherry picker") used to cut high branches can be seen in Figure 10.21. The photographer rightly placed the truck in the foreground, but included enough background information to indicate the truck's function. The worker in the bucket helps to show the truck in operation.

In Figure 10.22 everything merges because the shot was taken from the wrong angle. The reader has no sense of the parts of the truck, their size, or their function.

Drawings

Drawings can show where an object is located, how a tool or machine is put together, or what signals are given or steps taken in a particular situation. A drawing can be simple, like the one in Figure 10.23, which shows readers exactly where to place smoke detectors in a house.

A more detailed drawing can reveal the interior of an object. Such sketches are called **cutaway drawings** because they show internal parts normally concealed from view. Figure 10.24 is a cutaway drawing of an extended-range electric vehicle, the GM Volt. The drawing shows the passenger interior and the battery-powered engine, which allows the car to drive up to forty miles without the simultaneous use of gasoline.

An effective photograph—truck in foreground, enough background information, and a worker to show the size and function of the truck.   Figure 10.21

Photograph by David Longmire.

Another kind of sketch is an **exploded drawing**, which blows the entire object (a notebook) up and apart as in Figure 10.25 (p. 473) to show how the individual parts are arranged. An exploded drawing comes with most computers and uses **callouts**, or labels, to identify the components. The labels are often attached to the drawing with arrows or lines.

Guidelines for Using Drawings

1. Keep your drawing simple. Include only as much detail as your reader will need to understand what to do, be it to assemble or to operate a mechanism.
2. Clearly label all parts so that your reader can identify and separate them.
3. Decide on the most appropriate view of the object to illustrate—aerial, frontal, lateral, reverse, exterior, interior—and indicate in the title which view it is.
4. Keep the parts of the drawing proportionate unless you are purposely enlarging one section.

Clip Art

Clip art refers to ready-to-use electronic images. These small cartoon-style representations and business-related photographs, such as the ones shown in Figure 10.26 (p. 473), depict almost anything you can think of and pertain to almost every field—architecture, information science, criminal justice, international marketing, sports,

Figure 10.22   A poor photograph—taken from the wrong angle so that everything merges and becomes confusing.

Photograph by David Longmire.

A simple drawing showing where to place smoke detectors in a house. Figure 10.23

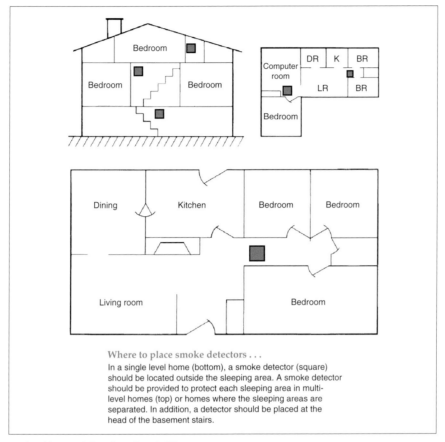

Where to place smoke detectors . . .
In a single level home (bottom), a smoke detector (square) should be located outside the sleeping area. A smoke detector should be provided to protect each sleeping area in multi-level homes (top) or homes where the sleeping areas are separated. In addition, a detector should be placed at the head of the basement stairs.

Reprinted by permission of *Southern Building*.

government, and health care. Figure 10.27 (p. 474) uses appropriate clip art in instructions to stockholders on how they can record their vote for a change in company policy.

Many graphics packages come with comprehensive clip art libraries grouped under such useful headings as energy, government, leisure, health, money, the outdoors, food, technology, transportation. Microsoft Office includes a clip art library with links to a larger online database of free clip art. There are also free clip art and photo-illustration databases such as Yahoo! Picture Gallery at *http://gallery.yahoo.com*. You can use clip art for a variety of projects.

When you use clip art, follow these guidelines:

1. **Use simple, easy-to-understand images.** Select an image that conveys your idea quickly and directly. Avoid using an image of an unfamiliar object or of a drawing or silhouette that might confuse your audience.
2. **Use clip art functionally.** Do not insert clip art as decorations. Using too many will make your work look unprofessional. Each piece of clip art should contribute significantly to, not compete with, your message.

Figure 10.24   Cutaway drawing of a hybrid car.

Source: General Motors Corporation. Used with permission, GM Media Archives.

3. **Make sure the clip art is relevant for your audience and your message.** A clip art airplane does not belong in a technical report on fuel capacity or jet engine design.
4. **Make sure your clip art is professional.** Some clip art is humorous, even silly, which may not be appropriate for a professional business report or proposal.

Using Visuals Ethically

Make sure your visuals, whether you create or import them, are ethical. Like your words, your visuals should represent you, your company, and the data truthfully. Ethical visuals convey and interpret statistical information and other types of data, products and equipment, locations, and even individuals without misinterpretation. Your visuals should neither distort events and data nor mislead readers. Ethical visuals should be:

- accurate
- honest, fair
- complete
- appropriate
- easy to read
- clearly labeled
- uncluttered
- consistent with conventions

Exploded drawing of a notebook.   Figure 10.25

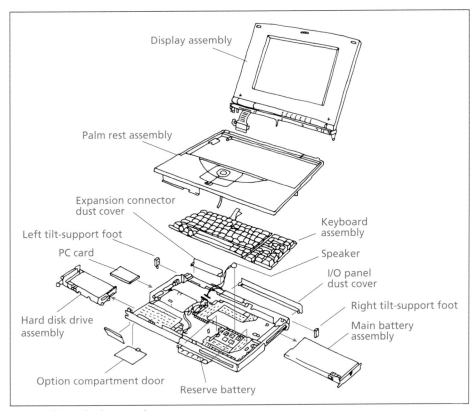

Display assembly

Palm rest assembly

Expansion connector
dust cover

Left tilt-support foot

PC card

Keyboard
assembly

Speaker

I/O panel
dust cover

Right tilt-support foot

Main battery
assembly

Hard disk drive
assembly

Option compartment door   Reserve battery

Examples of clip art.   Figure 10.26

Figure 10.27   Clip art in a business document

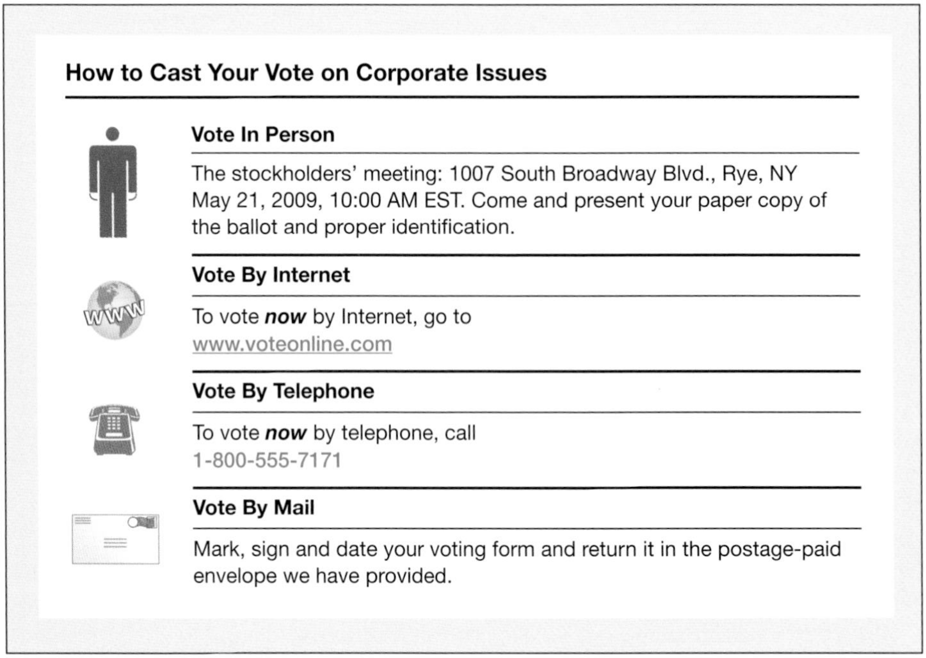

How to Cast Your Vote on Corporate Issues

Vote In Person

The stockholders' meeting: 1007 South Broadway Blvd., Rye, NY May 21, 2009, 10:00 AM EST. Come and present your paper copy of the ballot and proper identification.

Vote By Internet

To vote *now* by Internet, go to www.voteonline.com

Vote By Telephone

To vote *now* by telephone, call 1-800-555-7171

Vote By Mail

Mark, sign and date your voting form and return it in the postage-paid envelope we have provided.

Guidelines for Using Visuals Ethically

To ensure that your visuals are ethical, honest, accurate, and easy to read, avoid the following unethical practices no matter what type of visual you use.

Photos

- Do not distort a photo by omitting key details or by misrepresenting dimensions, angles, sizes, or surroundings or by superimposing one image over another.
- Do not take a photo of your most expensive, top-of-the-line product/model but then place the cost of your lowest-priced product/model under it.
- Do not misrepresent location. For example, taking a photo in a "doctored" or off-site location, studio, or lab and then claim it as an "actual" location shot.
- Never take a photo of an individual for business purposes (e.g., putting it in a newsletter, ad, on the Web) without his or her permission.
- Don't counterfeit or subtly alter a company's logo to sell, distribute, or promote an imitation as the real thing.

Graphs

- Don't distort a graph by plotting it in misleading or unequal intervals. For example, omitting certain years or dates to hide a decline in profits (contrast Figure 10.28, and its misleading interpretation, with the ethical revision in Figure 10.29).

Unethical visual and misleading interpretation.   Figure 10.28

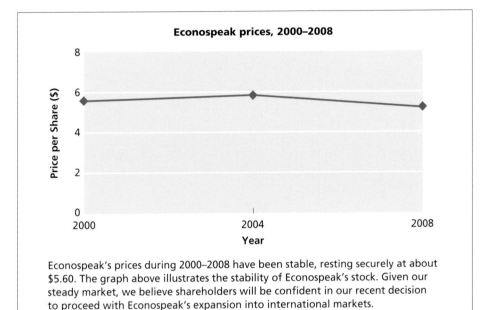

Econospeak's prices during 2000–2008 have been stable, resting securely at about $5.60. The graph above illustrates the stability of Econospeak's stock. Given our steady market, we believe shareholders will be confident in our recent decision to proceed with Econospeak's expansion into international markets.

Ethical revision of Figure 10.28   Figure 10.29

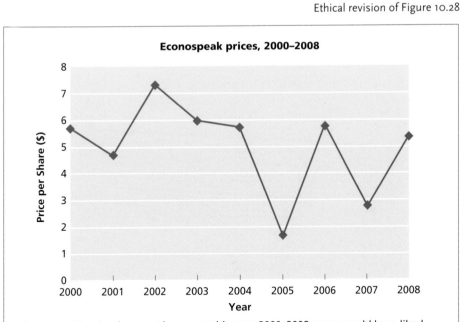

Econospeak's prices have not been as stable over 2000–2008 as we would have liked. The graph above illustrates the difficulties the company has faced in the market in the past decade, resulting in shifts by shareholders. We believe, however, that Econospeak's expansion into the global market will be possible by 2009.

- Include information in correct chronological sequence along the horizontal axis.
- Don't switch the type of information usually given along the vertical axis with the horizontal axis.
- Don't project growth or increases without having reliable and valid reasons.
- Don't misrepresent data or trends by making increments along the vertical axis too limited, leaving a much smaller (and incomplete) area to represent. When data are plotted wrongly this way, readers are unethically led to misinterpret the numbers—to read that there was little loss in revenue, or no change in sales, for example. For instance, if the horizontal axis begins at $5 and advances to $6 a share, you leave only an intentionally small and misleading area to measure. If stocks fell below $5 a share, your graph would unethically not represent those declines.

Bar Charts

- Don't use color or shading to mislead or distort—for example, shading one bar to make it more prominent than the others.
- Make sure the height and width of each bar truthfully represents the data it purports to. That is, don't make one of the bars larger to maximize the profits, products, or sales in any one year.
- Show bars for every year (or other sales period) covered. Note how the unethical bar chart (and accompanying text) in Figure 10.30 violates the rule but the chart in Figure 10.31 faithfully represents the data.

Pie Charts

- Don't use 3-D to distort the thickness and unjustified emphasis of one slice of the circle to misleadingly deemphasize others.
- Don't conceal negative information (losses, expenses, etc.) by silently including the information in another category or slice or lumping it into a category marked "other" or "miscellaneous."
- Make sure percentages match the number and size of the wedges or slices of the pie chart. See Figures 10.32 and 10.33 on page 478. Note how a larger expense for Guest Speakers (35% of budget) is unethically misrepresented in Figure 10.31 by using a smaller-sized wedge, while the expenses for Venue Rental are actually less than for Guest Speaker expenses but drawn larger to misrepresent costs.

Drawings

- Avoid any clutter that hides features.
- Label all parts correctly.
- Do not omit or shadow any necessary parts.
- Draw an object accurately. Tell readers if your drawing is the actual size of the object or equipment or if it is drawn to scale. Provide a scale.

Unethical visual.   Figure 10.30

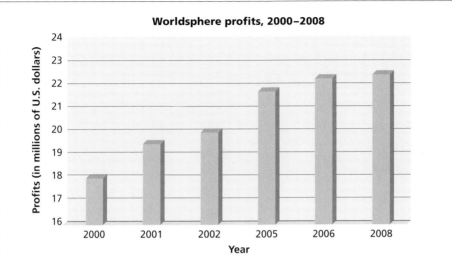

Profits at Worldsphere have seen a healthy increase over recent years. The above bar chart demonstrates the steady increase in profits, which have risen $4.5 million since 2000. Given the profit history of Worldsphere, our investors can be confident of future profits and the security of their stock with our company.

Ethical revision of Figure 10.30.   Figure 10.31

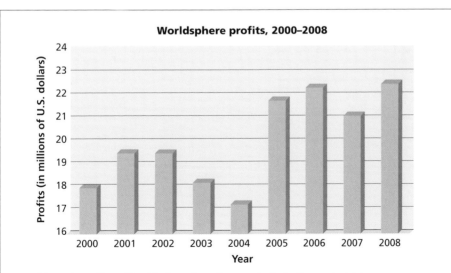

Although profits at Worldsphere have been volatile in the past eight years, we are now at our highest profit margin yet. The bar chart above shows the effects of market difficulties for the period 2002–2004, when the industry suffered major cutbacks. However, Worldsphere achieved a successful turnaround in 2005, with profits regaining strength due to revised marketing.

Figure 10.32   Unethical visual of pie chart.

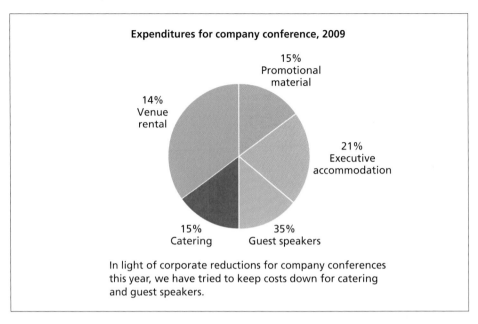

Figure 10.33   Ethical revision of Figure 10.32.

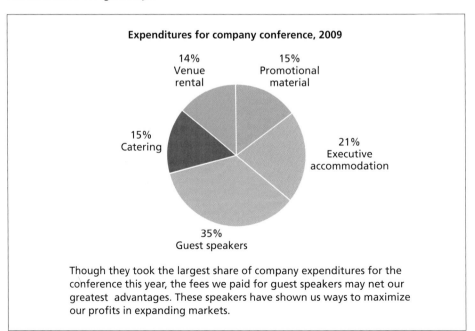

Using Appropriate Visuals for International Audiences

Whether you are writing for an expanding international business community in South America, China, or the countries that formerly made up the Soviet Union or for multicultural readers in the United States, you will have to prepare numerous documents that require visuals. These can range from instructions containing warning and caution statements and computer manuals explaining technical details to sales documents, reports, and online presentations.

Choose your visuals for international readers just as carefully as you do your words. Visuals should be designed to help international audiences understand your message clearly and without bias. Visuals and other graphics must be consistent with your message, not detract from or distort it.

Visuals Do Not Always Translate from One Culture to Another

Keep in mind that visuals do not automatically transfer from one culture to another. They are not boilerplate images that have the same meaning for all readers across the globe. Like words, visuals—photos, clip art, designs, colors, and other graphic devices—may have one meaning or use in the United States and a radically different one in other countries around the globe.

Of course, there are some exceptions. Groups such as the American Institute of Graphic Arts (AIGA) and the International Organization for Standardization (ISO) have promoted the worldwide use of informational and safety icons. In the 1970s, the AIGA, in conjunction with the U.S. Department of Transportation, developed a system of 50 passenger and pedestrian icons to be used in major transportation centers and at international events, and this system is now accepted in most countries. The ISO, with over 160 countries in its membership, has developed more than 16,000 agreed-on standards for product quality, safety, efficiency, and environmental responsibility. Chief among these standards are international icons to communicate warnings and cautions. Figure 10.34 shows some of these internationally recognized icons.

To avoid confusing or offending an international audience, investigate the meaning and cultural significance of a visual before you include it. Consult a native

Figure 10.34   Internationally recognized icons.

speaker from your audience's country to see if your visuals are culturally accept-able; then follow up with your collaborative team and any graphic artists assigned to the project to assure every visual is acceptable.

Guidelines for Using Visuals for International Audiences

To communicate appropriately and respectfully with your international readers through visuals and other graphics, follow these guidelines:

1. **Do not use any images that ethnically or racially stereotype your readers.** A large clothing manufacturer was criticized by the Asian American community when it released T-shirts depicting two men with slanted eyes and rice paddy hats as laundromat owners. Similarly, depicting Native Americans through clip art im-ages of red-faced chiefs is insulting. Rather than using an ethnic or racial pictogram, use neutral stick figures or clip art.

2. **Be respectful of religious symbols and images.** A cross to a U.S. audience rep-resents a church, first aid, or a hospital, and, when the cross is a large red one on a white background, the Red Cross. But the symbol for that humanitarian agency in Turkey and in other Muslim countries is the crescent. Portraying a smiling Buddha to sell products is considered disrespectful to residents in Southeast Asia.

3. **Avoid using culturally insensitive or objectionable photographs.** What is ac-ceptable in photographs in the United States may be taboo elsewhere in the world. Portraying men and women eating together at a business conference is unaccept-able in Saudi Arabia; a photograph showing men and women swimming together in a hotel pool would go against local customs in China. Moreover, depicting individ-uals seated with one leg crossed over the other is regarded as a disrespectful gesture in many countries in the world. Even showing one body part—a leg or forearm—is offensive in Muslim countries.

4. **Avoid any icons or clip art that international readers would misunderstand.** Symbolism varies extensively from one country to another. In the United States, an owl can stand for wisdom, thrift, and memory, while in Japan, Romania, and some African countries it is a symbol for death. A blindfolded woman balancing a set of scales, the symbol for justice in the United States, is meaningless in this context for readers in Africa or East Asia. A computer programmer who used the symbol of a highway sign for Route 66 to convey an "IP router" failed to realize that this icon would confuse international audiences and perhaps some Americans as well. A soft-ware program showing the icon of a mailbox (below) for the designation "You Have Mail" confused readers in other countries who thought the icon represented a birdhouse. A better alternative would be an icon of an envelope.

TABLE 10.5   Different Cultural Meanings of Various Gestures

Gesture	Meaning in the United States	Meaning in Other Countries
OK sign Index finger joined to thumb in a circle	All right; agreement	Sexual insult in Brazil, Germany, Russia; sign for zero, worthlessness in France
Thumbs-up	A winning gesture; good job; approval	Offensive gesture in Muslim countries
Waving or holding out open palm	Stop	Obscene in Greece— equivalent to throwing garbage at someone
Pointing with the index finger	This way; pay attention; turn the page	Rude, insulting in Japan and Venezuela; in Saudi Arabia used only for animals, not people
Nodding head up and down	Agreement, saying yes	Greek version of saying "no"; in China means "I understand," **not** "I agree"
Motioning with index finger	Signal to "come over here"	Insulting in China—instead, extend arm and wave to ask "come over"

5. Be cautious about importing images/photos with hand gestures in your work, especially in manuals or other instructional materials. Many gestures are culture-specific; they do not necessarily mean the same thing in other countries that they do in the United States. Table 10.5 lists cultural differences around the globe for some common gestures.

6. Don't offend international readers by using colors that are culturally inappropriate. A color may have a far different meaning for one group than it does for another. For instance, emphasizing the color red in a document aimed at Native Americans or the color yellow in a document for Asians or Asian Americans is likely to be seen as offensive. So, too, overusing red in a document for Russian or Eastern European readers may be interpreted as a deliberate reference to Communism.

Always do some research (on the Internet or by consulting with a representative from the audience you want to reach) to find out if the color scheme you have chosen for your document is appropriate for your target audience. For example, purple is the color of death and mourning in Thailand, as is white in China. Although orange is the symbolic color of Northern Ireland, you should avoid it when writing to readers whose culture does not value that color.

7. Be careful using directional signs and shapes. While road signs tend to be fairly recognizable throughout the world—for example, the octagon is the shape for the stop sign—there are country-specific signs and code books. For instance, a pennant on American highways, exclusively signaling a No-passing zone, may not have the same meaning in Nigeria or India. The symbol below points to a railroad crossing for American readers but would baffle an audience in the Czech Republic.

8. Avoid confusing an international audience with punctuation and other writing symbols used in the United States. Punctuation use and symbols vary widely around the globe. Not all cultures use a **?** to end a sentence asking a question, to represent the **Help** function in a computer program, or for an **FAQ** link on a website. For instance, elipses (**. . .**) and slashes **/** (and/or) may not be a part of the language your international audience reads and writes. Similarly, be careful about using the following graphic symbols, familiar to writers and speakers of U.S. English, but not necessarily to an international audience:

#	pound	&	ampersand (and)
©	copyright	*	asterisk
™	trademark	%	percentage

Include a glossary or a key to these symbols or any other graphics your reader may not understand or revise your sentences to avoid these symbols.

Conclusion

This chapter has introduced you to the types of visuals you can expect to use frequently in the world of work—and how to construct or import, label, insert, and introduce them in your writing. Following the guidelines here will make your documents businesslike, more persuasive, and easier for your readers to follow. Equally important, whenever you use a visual, always keep in mind the importance of following the same ethical standards that you follow for your written work. You also need to be respectful of your audience's culture and the context in which your visual is found.

Revision Checklist

☐ Located all places where a visual would help readers better understand my message.
☐ Used appropriate Internet images as a virtual library of visuals.
☐ Chose most effective visual (table, chart, graph, drawing, photograph) to represent information the audience needs.
☐ Experimented with different graphics software programs.
☐ Drafted and edited visual until it meets readers' needs.
☐ Verified statistical data that visual portrays.
☐ Made sure that visual does not simply repeat information in my written work.
☐ Selected right amount of technical detail to include in visual.
☐ Ensured that every visual is attractive, clear, complete, and relevant.
☐ Gave each visual a number, a title, and, where necessary, a legend and/or a callout.
☐ Numbered tables and figures consecutively.
☐ Inserted visual near the written description or commentary to which it pertains.
☐ Introduced and interpreted each visual in appropriate place in report or paper.
☐ Acknowledged sources for any copyrighted visuals and gave credit to individuals whose statistical data are the basis of a visual.
☐ Used and interpreted visuals ethically.
☐ Used visuals that respect the cultural traditions of international readers.

Exercises

1. Bring to class three or four website homepages that use especially effective visuals. In a short memo or e-mail (three or four paragraphs) to your instructor, indicate why and how each visual is appropriate for and convincing to a particular audience. What would each homepage look like without its visual?

2. Record the highest temperature reached in your town for the next five days. Then collect data on the highest temperature reached in three of the following cities—Boston, Chicago, Dallas, Denver, Los Angeles, Miami, New Orleans, New York, Philadelphia, Phoenix, Salt Lake City, San Francisco, Seattle—over the same five days. (You can get this information from a printed newspaper or on the Internet.) Prepare a table showing the differences for the five-day period.

3. Go to a supermarket and get the prices of four different brands of the same product (hair spray, aspirin, a soft drink, a box of cereal). Present your findings in the form of a table.

4. One government agency supplied the following statistics on the world production of oranges (including tangerines) in thousands of metric tons for the following countries during the years 2005–2008: Brazil, 2,005, 2,132, 2,760, 2,872; Israel, 909, 1,076, 1,148, 1,221; Italy, 1,669, 1,599, 1,766, 1,604; Japan, 2,424, 2,994, 2,885, 4,070; Mexico, 937, 1,405, 1,114, 1,270; Spain, 2,135, 2,005, 2,179, 2,642; and the United States, 7,658, 7,875, 7,889, 9,245. Prepare a table with that information and then write a paragraph in which you introduce and refer to the table and draw conclusions from it.

5. Keep a record for one week of the number of miles you walk, ride, or drive each day. Then prepare a line graph depicting that information.

6. Prepare a table to show the following statistical data: According to a 2000 census, the town of Ardmore had a population of 34,567. By a 2005 census the town's population had decreased by 4,500. In the 2000 census the town of Morrison had a population of 23,809, but by the 2005 census the population had increased by 3,689. The 2005 census figure for the town of Berkesville was 25,675, which was an increase of 2,768 from the 2000 census.

7. Prepare a line graph for the information in Exercise 6.

8. Prepare a bar graph for the information in Exercise 6.

9. Write a paragraph introducing and interpreting the following table.

Year	Soft Drink Companies	Bottling Plants	Per Capita Consumption (gallons)
1940	750	750	10.3
1945	578	611	12.5
1950	457	466	18.6
1955	380	407	17.2
1960	231	292	15.9
1970	171	229	15.4
1975	118	197	16.0
1980	92	154	18.7
1985	54	102	21.1
1990	43	88	23.1
1995	45	82	25.3
2000	37	78	27.6
2005	34	72	30.1

10. According to a municipal study in 2007 the distribution of all companies classified in each enterprise industry in that city was as follows: minerals, 0.4%; selected services, 33.3%; retail trade, 36.7%; wholesale trade, 6.5%; manufacturing, 5.3%; and construction, 17.8%. Make a circle chart to represent the distribution

and write a one- or two-paragraph interpretation to accompany (and explain the significance of) your visual.

11. Construct a segmented bar chart to represent the kinds and numbers of courses you took in a two-semester period or during your last year in high school.

12. Prepare a bar chart for the different brands of one of the products in Exercise 3. Write a paragraph introducing your illustration.

13. Find a pictograph in a math or business textbook, in a magazine (try *Business Week*, *Newsweek*, or *U.S. News & World Report*) online, or on a website from the Department of Labor Statistics. Make a bar graph from the information contained in the pictograph, and then write a paragraph introducing the bar graph and drawing conclusions from it.

14. Construct an organizational chart for a business or an agency you worked for recently. Include part-time and full-time employees, but indicate their titles or functions with different kinds of shapes or lines. Then attach the chart to a brief e-mail to your employer explaining why this kind of organizational chart should be distributed to all employees. Focus on the types of problems that could be avoided if employees had access to such a chart.

15. Prepare a flow chart for one of the following activities:
    a. jumping a "dead" car battery
    b. giving an injection
    c. making a reservation online
    d. using an iPhone
    e. checking your credit online
    f. putting out an electrical fire
    g. joining a chat group on the Internet
    h. preparing a visual using a graphics software package
    i. changing your e-mail account password
    j. putting your blog online

16. Draw an interior view of a piece of equipment you use in your major, and then identify the relevant parts using callouts.

17. Prepare a drawing of one of the following tools and include appropriate callouts with your visual.
    a. PDA
    b. iPod
    c. pliers
    d. stethoscope
    e. swivel chair
    f. DVD player
    g. ballpoint pen
    h. soldering iron
    i. high definition TV
    j. pair of eyeglasses

18. Prepare appropriate visuals to illustrate the data listed below. In a paragraph immediately after the visual explain why the type of visual you selected is appropriate for this information.
    a. Life expectancy is increasing in the United States. This growth can be dramatically measured by comparing the number of teenagers with the number of older adults (over age 65) in the United States during the past few years and then projecting those figures. In 1970 there were approximately 28 million teenagers and 20 million older adults. By 1980 the number of teenagers climbed to 30 million and the number of older adults increased to 25 million. In 1990 there were 27 million teenagers and 31 million older adults. In 2000 the number of teenagers had leveled off to 23 million, but the number of older adults soared to more than 36 million.
    b. Researchers estimate that for every adult in the United States 3,985 cigarettes were purchased in 1985; 4,100 in 1990; 3,875 in 1995; 3,490 in 2000; and 2,910 in 2005.

19. Find a photograph that contains some irrelevant clutter. Write a letter to the marketing department of the company for which you work that wants to use the photograph. Tell the department what to delete and why.

20. Below are four examples of poorly prepared visuals with brief explanations of how they were intended to be used. Redo one of the visuals to make it easier to read and to organize information. Supply a paragraph to accompany your new visual.
    a. To accompany a report on problems that pilots have encountered with a particular model of jet engine.

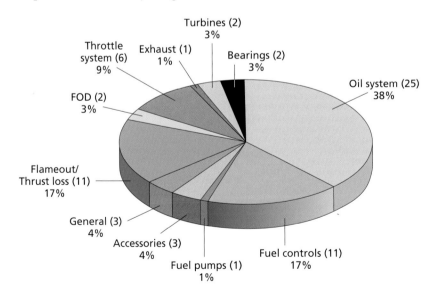

b. To show that a hiking trail is compatible with wheelchair access laws.

c. To show the percentage of white oak acorns in each quality category, by year, that were collected from traps from 2006 to 2010 in national forests.

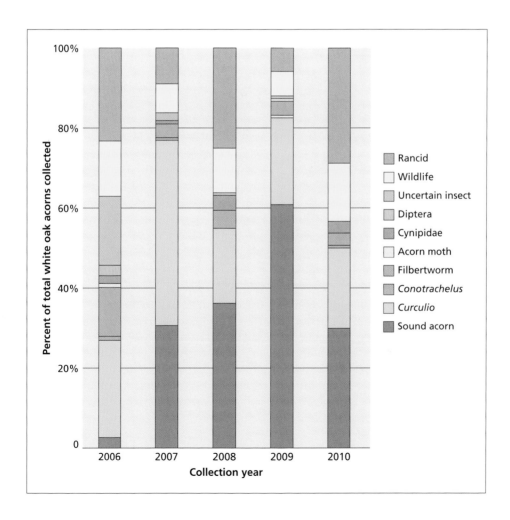

d. To accompany a report on where a business invested its funds.

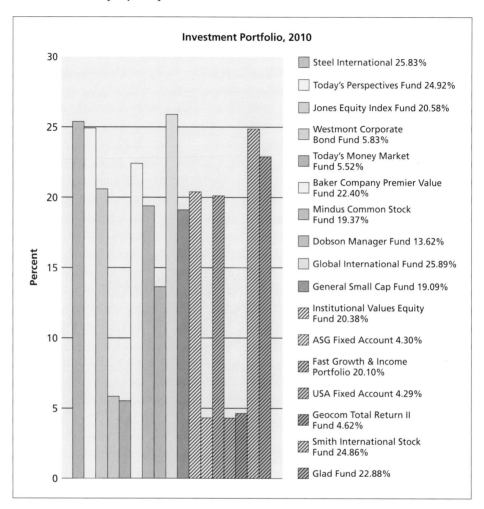

21. You work for a large manufacturer of industrial heat pumps and have been asked to help write a section of a report on the increased business your firm has been doing overseas. Based on the sales figures below for the years 2004, 2005, and 2006 (listed in that order) for each of the following countries, prepare two different yet complementary visuals. Also, supply a one-page description and interpretation of the statistics represented in your visuals. You may work collaboratively with one or more students in your class to prepare the visuals as well as to write the section of the report on international sales.

Argentina, 45, 53, 34; Australia, 78, 90, 115; Bolivia, 23, 43, 52; Brazil, 29, 34, 35; Canada, 116, 234, 256; Denmark, 65, 54, 87; England, 256, 345, 476; France, 198, 167, 345; Germany, 234, 398, 429; Holland, 65, 80, 89; Italy, 49, 52, 97; Japan, 67, 43, 29; New Zealand, 12, 69, 114; Norway, 33, 92, 104; Switzerland, 164, 266, 306; Sweden, 145, 217, 266; People's Republic of China, 7, 100, 296; Korea, 55, 43, 28.

In your written report, take into account trends, shifts in sales, and possible consequences for further marketing and conclude with a specific recommendation to your employer.

22. Supply an appropriate example of clip art to accompany each of the numbered guidelines in the "Top 10 Boating Safety Tips" list below. Explain your choices in a short report to your instructor.

Top 10 Boating Safety Tips

1. **Wear Your Lifejacket or PFD.** An approved Personal Floatation Device (PFD) is required by law for each person on board. Remember: It won't work if you don't wear it.
2. **Boating and Booze Don't Mix.** Alcohol impairs your ability to make good, quick decisions. This is critical when operating a fast and easily maneuverable personal watercraft.
3. **Know Your Craft.** Study the manufacturer's manual and practice handling your craft under experienced supervision and in open water well away from other boaters.
4. **Take a Boating Safety Course.** Learn the common boating rules, regulations, and safe practices.
5. **Look Out.** Ride defensively. Collisions with other boats or stationary objects like rafts or docks are the number one cause of personal watercraft injuries.
6. **Watch the Weather.** Check the weather forecast before starting out. Be alert for the wave, wind, and cloud changes that signal the approach of bad weather.
7. **Be Prepared for Cold Water.** Cold water robs body heat 25 times faster than does cold air of the same temperature. If you fall off your craft into cold water, immediately reboard.
8. **Know the Area.** Do not assume the water is clear of obstructions. Rocks, shoals, sandbars, and submerged pilings can seriously damage the craft or those on board. Check marine charts and stay in marked channels.
9. **Carry Safety Equipment.** Besides approved PFDs and a sound-signaling device (like a whistle), it is prudent to carry a tow rope and, when operating on a large body of water, some small type-B flares in a watertight container.
10. **Don't Ride at Night.** Most personal watercraft do not have the lights that the law requires for night riding; therefore, it is unsafe to be on one after dark.

23. Explain why the following would be inappropriate in communicating with an international audience and how you would revise the document containing such visuals/graphics symbols:
    a. clip art showing a string tied around an index finger
    b. picture of a man with a sombrero on a website for Pronto Check Cashing Company
    c. clip art/drawing of a light bulb and logo "Smart Ideas" for a CPA firm

d. clip art showing someone crossing the middle finger over the index finger (wishing sign)
e. drawing of a cupid figure for a caterer
f. a sales brochure showing a white and blue flag for a French audience
g. a poster showing a woman's track and field team used to advertise a brand of footware to an Arabic-speaking audience
h. a satisfied customer making the gesture of OK in an ad aimed at a Japanese audience
i. an image of a rabbit on a website for an automobile manufacturer to stress how fast its cars are
j. a photograph of a roll of Scotch tape to show international readers that your company can solve problems quickly

CHAPTER 11

Designing Successful Documents and Websites

Visit www
.cengage.com/
english/kolin for
this chapter's
online exercises,
ACE quizzes, and
Web links.

The success of your document (letter, proposal, report) or website depends as much on how it looks as what it says. In designing print documents and websites, you need to project a positive, professional image of yourself, your company, your product. A report filled with nothing but thick paragraphs of type, with no visual clues to break them up or to make information stand out, is sure to intimidate readers and turn them away. They will conclude that your work is too complex and not worth their effort or time. And if there is one thing that the world of work especially dislikes it is some-one who cannot get to the main point—the bottom line—quickly.

Websites and documents need to look user-friendly, signaling to your audience that your message is

- easy to read
- easy to follow and understand
- easy to recall

Don't bog your audience down in unbroken long paragraphs or crowded screens of information. Instead, separate material into smaller units that are visually appealing. You can help readers find key points at a glance through **chunking** (writing smaller paragraphs) and using lists, boldface type, and bullets. That way they can find your ideas easily the first time through or on a second reading if they have to double back to check or verify a point.

This chapter will show you how to design professional-looking, reader-friendly documents and websites that are a credit to you and your company.

Organizing Information Visually

Take a quick look at Figures 11.2 (p. 497) and 11.3 (pp. 498–500). The same infor-mation is contained in each figure. Which appeals more to you? Which do you think would be easier to read? Which is better designed? As the two figures show, design or layout plays a crucial role in an audience's overall acceptance of your work. Infor-mation is organized (and perceived) visually as well as verbally. A document or website that looks logical and easy to read and that is clearly organized with the audience in mind will increase your credibility. If you offer your audience densely packed material, with few visual road signs to help them navigate through it, they will conclude that you are poorly organized and that your thinking is not very clear or direct.

Your company, too, will win or lose points because of your design choices. A visually appealing website or document will enhance a company's reputation and improve its sales. A poorly designed one will not. Consumers will think your firm is inflexible, hard to do business with, and uncaring about specific problems customers may have over a policy or a set of instructions. Your company can easily be perceived as not respecting the individuals it does business with or for being old-fashioned, hard-nosed, or just plain unprofessional.

Characteristics of Effective Design

As you read this chapter and apply its principles to your own work, make sure your documents and websites offer your audience the qualities of effective design on the left and avoid those on the right:

Effective	Ineffective
■ visual appeal	■ crowded
■ logical organization	■ disorganized
■ clarity	■ hard to follow
■ accessibility	■ difficult to read
■ variety	■ boring, repetitious
■ relevance	■ inconsistent

Figure 11.3 displays the positive design characteristics on the left while Figure 11.2 displays the unfavorable ones on the right, study the annotations to these figures to see how the errors in Figure 11.2 are corrected in Figure 11.3. By modeling your documents on the layout and design of Figure 11.3 you can guarantee that your message will be clearly and well received.

Tech Note

Desktop Publishing Programs

Word processing programs such as Microsoft Word and Corel WordPerfect have evolved into more and more powerful desktop publishing programs. In addition, files created in word processing programs can be imported into sophisticated desktop publishing programs such as Adobe InDesign, Apple Pages, Microsoft Publisher, and QuarkXPress, as can graphics created or edited in programs such as Adobe Photoshop, PaintShop Pro, Adobe Illustrator, CorelDRAW, and Microsoft PhotoDraw.

Desktop Publishing

Desktop publishing programs, sometimes referred to as **page layout software**, provide an inexpensive alternative to a professional print shop. Because desktop publishing software permits users to design page layouts, include visuals, and produce high-quality final copies, you can create printed documents right in your own home or office.

With desktop publishing you can

- select from templates specifically designed for reports, newsletter, brochures, and other formats
- design your own templates if you often create similar documents
- take advantage of numerous font choices
- delete, insert, and move entire blocks of text
- integrate various changes in typeface—such as bold, italics, and underlining
- vary type sizes
- justify margins
- change line spacing
- center words, titles, or lines of text
- break and number pages
- arrange text in multiple columns
- insert sidebars, tables, footnotes, headers, footers, and sidebar quotes
- import graphics, such as drawings, photographs, icons, and logos, from your hard drive, a CD, or the Web
- insert graphics available within your software program, such as clip art (usually located in a separate clip art file—see p. 495), special symbols and shapes (typically located in a Symbols menu), illustrations you have created using the drawing tools and charts/graphs of your software program.

Type

There are many different styles of type (see pp. 501–505). All computers come equipped with software that contains a large number of typefaces (anywhere from 100 to 1,000). For example, WordPerfect 12 features more than 1,000 high-quality fonts. Different fonts are also available over the Internet at such sites as *http://1001freefonts.com* and *http://www.fonts.com/*.

Templates

Desktop publishing software is equipped with predesigned **templates**. Templates are patterns and predesigned text columns, like those shown in Figure 11.1, that offer numerous page layout formats for reports, newsletters, brochures, and other marketing and communications documents. A template for a report, for example, would contain all the headings and divisions you need. You can also create and save your own template for an original format you use frequently.

Figure 11.1   Examples of templates.

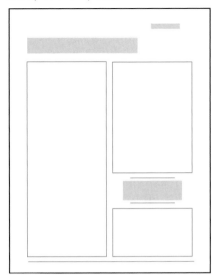 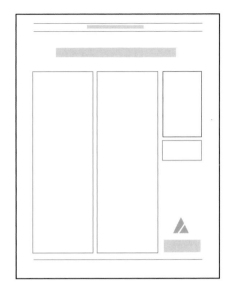

Graphics

A graphics program supplies shapes and lines that can be manipulated (skewed, enlarged) and offers options for sophisticated use of color and shading. With a drawing program you can create diagrams, charts, and illustrations that can be saved as electronic files and imported into word processing or desktop publishing documents.

You can move and place graphics anywhere in a document. Graphic design programs provide customized graphs, charts, tables, and expanded font (design) sizes and shapes.

Here are some types of graphics available to you.

1. Drawing tools. Desktop publishing programs contain tools that allow you to create a variety of shapes, rules, borders, and arrows. Having drawn a box, circle, triangle, or whatever, you can use other tools to fill in or alter the appearance of the shape or to add words to it; then you can rearrange all the elements into a graphic that communicates quickly and effectively.

2. Icons. Icons are symbols or visual representations of concepts or actions. The skull and crossbones on a container of poison is an icon that warns of danger. Many highway signs are icons that tell us quickly what to expect ahead: an S-curve, merging traffic, a hospital. *Graphic icons* are simply pictures that communicate directly. No matter what language we speak—and even if we cannot read—icons tell us at a glance which restroom to use or how to fasten the seat belt in an airplane. A nation's flag is an icon, and so are many religious symbols and most company logos.

Computer software often relies on icons to help us perform common actions without having to remember keyboard commands. Arrows in the scroll bars move us easily around a window; a file folder helps us group related documents; and a trash can holds files to be deleted later. Icons in the menu bar make it easy to print a file, open an address book, search for specific text, cut copy, or dial the phone. Examples of public service information icons are shown below:

3. Clip art. Clip art is a library of simple drawings, often classified by themes, that can be imported (copied) from floppy disk, CD, or the Internet to your document. Some clip art software packages offer as many as 500,000 images, arranged into such diverse categories as animals, computers, holidays, famous people, food, and various technologies. Clip art is widely used to make business documents attractive and appealing. If the graphic is symbolic or represents something unique about the business, it can also be the basis for a company logo. Review pages 470–472.

4. Stock photos/stock art. Assembled photos and art on a CD or the Internet can be imported, sometimes with a permissions fee, sometimes free with the purchase of the CD.

Before Choosing a Design

Before you begin designing a document, you need to know exactly what you are designing, for whom, and how. Think carefully about how your work will look on the page, what you want to achieve, and what your resources are. Always get your plans approved by your boss or a print publications committee. Here are questions to ask:

1. **Who is your target audience?** Will they be local residents or global consumers? What do they have in common—gender, locality, educational level? How you reach and appeal to this group must be taken into consideration before you decide on a "look and feel" for your document. Print size, style, and so forth should be selected based on the audience's characteristics and needs.

2. **What is the purpose of your document?** Are you going to have to motivate the reader to look at it? Use an eye-catching cover page, attractive colors, and short paragraphs. Or is your document required reading for co-workers and therefore subject to "house style"? Are you designing a newsletter for a professional group that needs to comply with a set of guidelines about how it should be laid out?

3. **How will your document be reproduced?** Will it be set by a professional printer or is it an internal document that can be scanned and distributed? This

will affect the number and type of colors you use, the type of paper you use, and the size of the paper.

4. **How will it be distributed?** Will it fit in a standard size envelope? Will it need to be folded more than once? Do you need larger envelopes or a smaller page size? Will the document be read by more than one reader or more than once? If so, the paper will need to be more sturdy.

5. **How much will it cost?** Financial decisions are crucial in making design choices. Printing in color can be much more expensive than standard black-and-white printing. If you use color, try to stick to one or two colors throughout to keep the costs down. Talk to your employer about the costs of reproduction before you decide on anything.

The ABCs of Print Document Design

Visit www
.cengage.com/
english/kolin for
an online exer-
cise, "Designing
a Document for
an International
Audience Using
Online Visuals."

The basic features of printed document design are

- page layout
- typography, or type design
- graphics
- color

Every page of your document should integrate those four elements. The proper arrangement and balance of page layout elements, type, graphics, and color involve the same level of preparation that you would spend on your research, drafting, revising, and editing. Your document should cooperate with—not detract from—your message. Just as you research your information, you have to research and experiment in order to adopt the most effective design for your document.

Page Layout

Each of your pages needs to coordinate space and text pleasingly. Too much or too little of one or the other can jeopardize the reader's acceptance of your message. To make sure you design an effective and attractive page layout, pay attention to the following elements:

1. **White space.** White space, which refers to the open areas on the page that are free of text and visuals, can help you increase the impact of your message. If you pack too much information or too many design elements into a document, you will distract the reader from the message you want to convey. Skimping on white space can frustrate a reader eager to locate parts of a document quickly and to process them easily. White space can entice, comfort, and appeal to the reader's "psychology of space" by

- attracting the reader's attention
- assuring the reader that information is presented logically
- announcing that information is easy to follow
- assisting the reader to organize information visually
- allowing the reader to forecast and highlight important information

Compare Figures 11.2 and 11.3 again. Which document was designed by someone who understands the importance of white space?

A poorly designed document.   Figure 11.2

No title

The results for the recent cholesterol screening at our company's Health Fair were distributed to each employee last week. Many employees wanted to know more information about cholesterol in general, the different types of cholesterol, what the results mean, and the foods that are high or low in cholesterol.

Single-spacing makes document difficult to read

We hope the information provided below will help employees better answer their questions concerning cholesterol and our cholesterol screening program.

High cholesterol, along with high blood pressure and obesity, is one of the primary risk factors that may contribute to the development of coronary heart disease and may eventually lead to a heart attack or stroke. Cholesterol is a fatty, sticky substance found in the bloodstream. Excessive amounts of the bad type of cholesterol can deposit on the walls of the heart arteries. *This deposit is called plaque and over a long period of time plaque can narrow or even block the blood flow through the arteries.*

Lack of headings makes it hard for readers to organize material

Total cholesterol is divided into three parts—LDL (low-density lipoprotein), or bad cholesterol; HDL (high-density lipoprotein), or good cholesterol; and VLDL (very low-density lipoprotein), a much smaller component of cholesterol you don't have to worry about. Bad (LDL) cholesterol forms on the walls of your arteries and can cause a lot of damage. Good cholesterol, on the other hand, functions like a sponge, mopping up cholesterol and carrying it out of the bloodstream.

You should have received *three cholesterol numbers.* One is for your HDL, or good cholesterol, and the other is for your LDL, or bad cholesterol, reading. These two numbers are added to give you the third, or composite, level of your total cholesterol. You are doing fine.

Frequent font changes are confusing for readers

As you can see, a total cholesterol reading of below 200 is considered safe. Continue what you have been doing. If your reading falls in the moderate risk range of 200–239, you need to modify your diet, get more exercise, and have your cholesterol checked again in six months. *If your reading is above 240, see your doctor.* You may need to take cholesterol-lowering medication, if your doctor prescribes it. Reducing your total cholesterol by even as little as 25% can decrease your risk of a heart attack by 50%.

Uneven presentation of numbers

The Surgeon General recommends that your LDL, or bad, cholesterol should be below 130; and your HDL, or good, cholesterol needs to be at least above 36. Ideally, the ratio between the two numbers should not be greater than 5 to 1. That is, your HDL should be at least 20% of your LDL. The higher your HDL is, the better, of course. So even if you have a high LDL reading, if your HDL is correspondingly high you will be at less risk.

Lack of margins makes document look dense and complex

One of the easiest ways to decrease your cholesterol is to modify your diet. Cholesterol is found in foods that are high in saturated fat. *Saturated fat comes from animal sources and also from certain vegetable sources.* Foods high in bad cholesterol that you should restrict, or avoid, include whole milk, red meat, eggs, cheese, butter, shrimp, oils such as palm and coconut, and avocados. Generally, food groups low in cholesterol include fruits, vegetables, and whole grains (wheat breads, oatmeal, and certain cereals), lean meats (fish, chicken), and beans.

The goal of our cholesterol screening is to help each employee lower his or her cholesterol level and eventually reduce the risk of heart disease. **Besides the advice given above,** you can do the following: get regular aerobic exercise—bicycling, brisk walking, swimming, rowing—for at least 30 minutes 3–4 times a week. But get your doctor's approval first. *Eat foods low in cholesterol* but high in dietary fiber (beans, oatmeal, brown rice). Maintain a healthy weight for your frame to lower your body fat. Minimize stress, which can increase cholesterol. Learn relaxation techniques.

Inconsistent use of italics and boldfacing

Figure 11.3   An effectively designed document with the same text as Figure 11.2.

Title clearly set apart from text by caps and boldfacing

Text is double-spaced with generous margins, making it more readable

Text does not look cluttered

Headings divide material into easy-to-follow units for readers

Consistent use of one font

Only key terms are boldfaced

Types of cholesterol are labeled with numbers

Cholesterol Screening

The results for the recent cholesterol screening at our company's Health Fair were distributed to each employee last week. Many employees wanted to know more information about cholesterol in general, the different types of cholesterol, what the results mean, and the foods that are high or low in cholesterol. We hope the information provided below will help employees better answer their questions concerning cholesterol and our cholesterol screening program.

Determining Risk Factors

High cholesterol, along with high blood pressure and obesity, is one of the primary risk factors that may contribute to the development of coronary heart disease and may eventually lead to a heart attack or stroke. Cholesterol is a fatty, sticky substance found in the bloodstream. Excessive amounts of the bad type of cholesterol can deposit on the walls of the heart arteries. This deposit is called **plaque** and over a long period of time plaque can narrow or even block the blood flow through the arteries.

Separating Types of Cholesterol

Total cholesterol is divided into three parts: (1) **LDL** (low-density lipoprotein), or bad cholesterol; (2) **HDL** (high-density lipoprotein), or good cholesterol; and (3) **VLDL** (very low-density lipoprotein), a much smaller component of cholesterol you don't have to worry about. Bad (LDL) cholesterol forms on the walls of your arteries and can cause a lot of damage. Good cholesterol, on the other hand, functions like a sponge, mopping up cholesterol and carrying it out of the bloodstream.

(Continued)   Figure 11.3

Understanding Your Cholesterol Results

You should have received three cholesterol numbers. One is for your **HDL** (or good cholesterol) and the other is for your **LDL** (or bad cholesterol) reading. These two numbers are added to give you the third, or composite, level of your total cholesterol.

Cholesterol levels can be classified as follows:

Minimal Risk	Moderate Risk	High Risk
below 200	200–239	above 240

Concise paragraph provides clear opening for new section

Emphasizes range of risk categories by setting them apart and shading

As you can see, a total cholesterol reading of below 200 is considered safe. You are doing fine. Continue what you have been doing. If your reading falls in the moderate risk range of 200–239, you need to modify your diet, get more exercise, and have your cholesterol checked again in six months. If your reading is above 240, see your doctor. You may need to take cholesterol-lowering medication, if your doctor prescribes it. Reducing your total cholesterol by even as little as 25% can decrease your risk of a heart attack by 50%.

Includes more space between sections

Relationship Between Bad and Good Cholesterol

The Surgeon General recommends that your **LDL**, or bad, cholesterol should be below 130. And your **HDL**, or good, cholesterol needs to be at least above 36. Ideally, the ratio between the two numbers should not be greater than 5 to 1. That is, your **HDL** should be at least 20% of your **LDL**. The higher your **HDL** is, the better, of course. So even if you have a high **LDL** reading, if your **HDL** is correspondingly high you will be at less risk.

Paragraphs are neither too long nor too short

Recognizing Food Sources of Cholesterol

One of the easiest ways to decrease your cholesterol is to modify your diet. Cholesterol is found in foods that are high in saturated fat. Saturated fat comes

2

Page is numbered

Continued

Figure 11.3   (Continued)

A double spaced numbered list helps readers identify foods with bad cholesterol

from animal sources and also from certain vegetable sources. Foods high in bad cholesterol that you should restrict include:

1. whole milk
2. red meat
3. eggs
4. cheese
5. butter
6. shrimp
7. oils such as palm and coconut
8. avocados

Easy-to-follow examples of foods low in cholesterol

Generally, food groups low in cholesterol include fruits, vegetables, and whole grains (wheat breads, oatmeal, and certain cereals), lean meats (fish, chicken), and beans.

Realizing It Is Up to You

Heading signals conclusion

The goals of our cholesterol screening program are to help each employee lower his or her cholesterol level and eventually reduce the risk of heart disease. Besides the advice given above, you can do the following:

Bulleted list serves as both summary and plan for future action

- Get regular aerobic exercise—bicycling, brisk walking, swimming, rowing—for at least 30 minutes 3–4 times a week. But get your doctor's approval first.
- Eat foods low in cholesterol but high in dietary fiber (beans, oatmeal, and brown rice).
- Maintain a healthy weight for your frame to lower your body fat.
- Minimize stress, which can increase cholesterol. Learn relaxation techniques.

3

Source: Thanks to Sgt. Mannie E. Hall of the U.S. Army for creating this document.

2. Margins. Use wide margins, usually 1 to ½ inches, to "frame" your document with white space surrounding text and visuals. Most documents easily accommodate equal margins on all four sides, a technique that creates balance and prevents the document from looking cluttered or overcrowded. If your document requires binding, you may have to leave a wider left margin (2 inches).

3. Line length. Most readers find a text line of 10 to 14 words, or 50 to 70 characters (depending on the type size you choose), comfortable and enjoyable reading. Line length, or "line measuring," depends, of course, on font type and size, number of columns, width of margins, and space between words. Never exceed your margin settings. Excessively long lines that jump into the margins signal that your work is difficult to read. In the example below, note how the extra-long lines unsettle your reading and tax your eye movement; they signal rough going.

In order to succeed in the world of business, workers must learn to brush up on their networking skills. The network process has many benefits that you need to be aware of. These benefits range from finding a better job to accomplishing your job more easily and efficiently. Through networking you are able to expand the number of contacts who can help you. Networking means sharing news and opportunities. The Internet is the key to successful networking.

Conversely, do not print a document with too short or extremely uneven lines.

> In order to succeed in the world of
> business, workers must learn
> to brush up on their networking skills. The
> network process has many
> benefits you need to be
> aware of.

Readers will suspect your ideas are incomplete, superficial, or even simple-minded.

4. Columns. Document text usually is organized in either single-column or multi-column formats. Memos, letters, and reports are usually formatted without columns, whereas documents that intersperse text and visuals (such as newsletters and magazines) work better in multicolumn formats.

Typography

Typography consists of typeface, type size, type styles, justification, heads and subheads, lists, and captions.

Typeface

Readability of your text is crucial. Select a typeface, therefore, that ensures your text is

- legible
- attractive
- functional
- appropriate for your message
- complementary with accompanying graphics

The most familiar typefaces are the following four.

Times Roman   Palatino

Helvetica   **Arial**

Avoid using a typeface that looks like script, and don't mix and switch typefaces. The result makes your work look amateurish and disorganized.

Type Size

Type size options are almost unlimited, depending again on your software package and printer capabilities. Type size is measured in units called **points**, 72 points to the inch: the bigger the point size, the larger the type. Most business and educational documents use from 10-point to 12-point type, although the range today varies from 6 to 72 points (and beyond). Newspaper classified ads are in small 6-point to 8-point type, while headlines are set in much larger, 30-point to 36-point type. The text in this book is set in 10-point type. Never print your letter or report in 6- or 8-point newspaper classified ad type or in a size larger than 12-point type. Here are some suggestions:

Times 8 point = footnotes/endnotes

Helvetica 12 point = letters, reports

Palatino 14 point = headings

Arial 22 point = title of your report

Type Styles

Also known as **attributes**, these include boldface, italics, shadow, underlining, small caps, and shading.

Boldface
Italics
Shadow
<u>Underlining</u>
Small Caps
Shading

Avoid overusing boldface and italics. Use them only when necessary and not for decoration. Do *not* underline text unless absolutely necessary. Not only will too many special effects make your document harder to read, but also you will lose the dramatic impact these styles have to distinguish and emphasize key points that do deserve to be highlighted.

Justification

Sometimes referred to as alignment, justification consists of left, right, full, and center options. Left-justified (also called **unjustified** or **ragged right**) is the preferred method because it allows space between words in lines of text to remain constant and it is easier to read. In fully justified alignment (both left *and* right) the word spacing varies from line to line. Left justification gives a document a less formal look than full justification. Narrow columns of text should be set left-justified to avoid awkward gaps between words and excessive hyphenation.

Our new website offers consumers a mall on the Internet. It gives shoppers access to our products and services and makes buying easy and fun. Our new website offers consumers a mall on the Internet. It gives shoppers access to our

Our new website offers consumers a mall on the Internet. It gives shoppers access to our products and services and makes buying easy and fun. Our new website offers consumers a mall on the Internet. It gives shoppers access to our

Left-justified text

Right-justified text

> Our new website offers consumers a mall on the Internet. It gives shoppers access to our products and services and makes buying easy and fun. Our new website offers consumers a mall on the Internet. It gives shoppers access to our products and

Full-justified text

> Our new website offers consumers a mall on the Internet. It gives shoppers access to our products and services and makes buying easy and fun. Our new website offers consumers a mall on the Internet. It gives shoppers access to our

Centered text

Heads and Subheads

Heads and subheads are brief descriptive words or phrases that signal starting points and major divisions in your document. They provide helpful road signs for readers charting their course through a document, as in Figure 11.3. Without heads and subheads your work will look unorganized and cluttered. Heads and subheads should also parallel each other grammatically. Note how the headings below, from a poorly organized proposal from the Acme Company, are not parallel.

- What Is the Problem?
- Describing What Acme Can Do to Solve the Problem
- It's a Matter of Time . . .
- Fees Acme Will Charge
- When You Need to Pay
- Finding Out Who's Who

Revised, the heads are parallel and easier for a reader to understand and follow.

- A Brief History of the Problem
- A Description of Acme Solutions
- A Timetable Acme Can Follow
- A Breakdown of Acme's Fees
- A Payment Plan
- A Listing of Acme's Staff

Heads and subheads immediately attract attention and quickly inform readers about the function, scope, purpose, or contents of the document or section. They help readers prioritize information, too, by emphasizing the main points they need to look for and remember. Additionally, heads and subheads introduce helpful white space to separate text and organize your document. The space around a heading is like an oasis for the reader, signaling both a rest and a new beginning.

Subheads allow the reader to skim or review a document in an outline type of format. Headings and subheadings follow an established order or hierarchy.

In designing a document with heads and subheads, follow these guidelines:

- Use larger type size for heads and subheads than for text; major heads should be larger than subheads. If your text is in 10-point type, your heads may be in 16-point type and your subheads in 12- or 14-point type.

> # Sixteen-Point Head
> ## Subhead in 12 Point
> Use larger type for heads than you do for text;
> major headings should be larger than subheads.
> If your text is in 10-point type, your heads may
> be in 14-point type and your subheads in 12 or 14.

- Further differentiate heads and subheads by using all capital letters, initial capital letters (the first letter of each word capitalized), boldface, or italics.
- Establish a horizontal position for a head, such as centered or aligned left, and keep it consistent throughout the document.
- Do not overuse heads (e.g., inserting them after each paragraph); having too many heads is as bad as having too few.
- Allow additional space, or leading, above a head to set it off from the preceding section. You may also want to allow additional space below a head.
- If you are using a color printer, consider using a second color for major headings.

Lists

Placing items in a list helps readers by dividing, organizing, and ranking information. Lists emphasize important points and contribute to an easy-to-read page design. Lists can be numbered, lettered, or bulleted. Take a look at the memos, proposals, and reports in Chapters 3, 13, 14, and 15 that effectively use lists.

Captions

Used to accompany, explain, highlight, or reference pictures or other graphics (like charts and graphs), captions are titles that help a reader identify a visual and quickly explain the nature of the picture or other graphic. (Chapter 10 discusses using captions with visuals in a document.)

Graphics

Like other visuals, graphics work in conjunction with your words. A document without visuals or graphics may look boring, confusing, or unattractive. You will have to judge how and when a graphic can improve your message and convince your reader. The list below identifies some common graphics included in desktop publishing software packages.

1. **Clip art.** See page 495 and also pages 470–472 in Chapter 10.
2. **Boxes.** These lines isolate or highlight text or visuals. Tables 10.2, 10.3, and 10.4 in Chapter 10 use boxes.

3. **Rules.** These lines are classified as either vertical or horizontal. Vertical rules are used to separate columns of text, while horizontal rules separate sections introduced by subheads.

4. **Letterheads and logos.** A company's corporate image is represented and symbolized by its letterhead, usually consisting of a graphic, or icon, integrated with the company name. Often the company's street address, e-mail address, and website are included in the letterhead. Company letterhead conveys the firm's message and expresses its character. Typically, logos and sometimes the entire letterhead are imported as graphics files. A letterhead/logo should creatively set one company apart from others. See Figures 6.3 and 6.8.

Using Color

Nothing attracts attention and communicates more quickly and more powerfully than color. Color can set moods and create impressions. It may be one of the most important tools to increase sales, gain faster project approval, or provide a major advantage over competitors. Color helps to sell ideas 85 percent more effectively than black-and-white communications.

Color can be used in many ways to organize written information and therefore enhance readability. Color visually breaks up long segments of text and can tie important ideas together. For example, using colored boxes to identify special notes in the text helps readers locate important information quickly. Use color for borders and graphic accents, headings, titles, key words, Internet addresses, sidebars, rules, and boxes that link related facts, figures, or information.

Guidelines on Using Color Effectively

Here are some guidelines to follow when you use color in your documents:

- Match the most appropriate color with the object you want to display. Keep in mind that an object's perceived size is affected by color. Light colors make objects look larger; dark colors make objects appear smaller.
- Estimate how the color will look on the page—colors look different on the screen than they do on a sheet of paper. Print a sample page.
- Make sure text colors contrast sharply with background colors.
- Use no more than two or three colors on a page unless there are photographs, illustrations, or graphics.
- For graphs and charts, place similar colors close to one another to maximize differences.
- Too many bright colors overwhelm the eye, so use them sparingly to call attention to important elements.
- Select "cool" colors, such as blue, turquoise, purple, and magenta, for backgrounds. However, avoid light blue text, which is hard to read against a dark background.
- Use colors that respect an international reader's cultural heritage. See page 481.

Poor Document Design: What *Not* to Do

Up to this point we have introduced the various elements of effective document design. Now, in contrast, we'll look at what you need to avoid. By knowing what looks bad from a reader's point of view, you will be much better able to design a document that works well visually for your reader.

Figure 11.2, a poorly designed document, illustrates many of the following common mistakes in document design. But note how Figure 11.3, which contains the same information as Figure 11.2, incorporates many of the effective techniques just described. Avoid the following errors when you design your document.

1. Insufficient white space. Narrow margins and limited space between headings and text are classic mistakes. A lack of white space between paragraphs or heads or around the borders of a document tells a reader, "Roll up your sleeves: This will be tough, unenjoyable reading."

2. Inappropriate line length. Excessively long lines are hard to read and signal to your audience that your work is highly complex and unrewarding.

3. Overuse of visuals. Too many visuals (boxes, rules, and clip art images) can create confusion and crowd your pages. Establish a balance between text and visuals. Never use visuals just for decorations or to fill space. Review the guidelines on using visuals (see p. 443).

4. Mixing typefaces. The large number of fonts available in desktop publishing packages makes it easy to use too many of them. Be aware of this temptation, as the result often looks amateurish and disorganized. Don't use more than two type families on a page.

5. Using hard-to-read fonts for body copy. Some typefaces, such as Avant Garde, may look great in headlines, but are difficult to read in the text of your document.

6. Using small type and tight leading. Most typefaces work best for body copy when sized between 9 and 11 points. Leave plenty of line space (leading) between each line, too. A good rule of thumb is to set leading two or three points larger than the body text size (e.g., 12-point leading with 10-point type).

7. Too few or no heads and subheads. Without headings as useful guideposts your document will seem unorganized and illogical. As we saw in Figure 11.3, heads and subheads are typographical markers that signal starting points and major divisions; they thus provide helpful landmarks for readers charting their course through your document.

8. Excessive spacing. Too much space distorts the document and its message. Leaving three or four spaces between consecutive lines of a text signals to readers that your ideas may be lightweight and not well researched. Also, including too much space around a visual or around the borders of your text can call into question the overall professional status of your work.

To avoid such errors, follow these tips:

- Use only one space after a period, not the traditional two spaces.
- Don't indent the first line of text after a head or subhead.

- Avoid full justification on narrow columns.
- Eliminate excessive spacing in lists between the bullets, numbers, or symbols and the actual text entries.

9. Misusing capitals, boldface, or italics. Printing an entire document with capitals (large or small) makes it hard to read. (But printing heads or subheads in capitals will differentiate them from your text.) Similarly, avoid overusing boldface or italics. Do not underline text unless absolutely necessary. Not only will too many special effects make your work difficult to follow, but you will also lose the dramatic impact those attributes have.

Writing for and Designing Websites*

In the new global village, you will need to navigate as well as write for websites. Such knowledge is essential to communicate effectively with colleagues, your boss, and customers around the world.

Designing a website is a complex, highly technical process. Your employer will not expect you to have the technical proficiency of a webmaster, and you won't be asked to construct a company website on your own. Instead, you will likely be part of a team of iTech, graphics, and marketing specialists responsible for your firm's Web presence. However, your employer will expect you to keep up with the latest visuals and features of Web design. This part of Chapter 11 stresses principles and guidelines you will need to

- help readers navigate your website easily
- ensure that your site is current
- make your site user friendly (for global readers as well as English-speaking audiences)
- create a visually attractive site
- keep your design and content ethical

The Organization of a Website—The Basics

The Web connects sets of information—pages—at millions of sites around the world. Some pages are quite long, while others take up just one computer screen. Websites can be personal (as we saw in Chapter 7), corporate, informational, and so forth. Regardless of where it originates, a website has to arouse a reader's interest; show the appeal of a product, service, or topic; demonstrate the application of the product, service, or topic; and include a call to action.

A website can consist of one screen or many screens. In fact, a website for a government agency such as E-Government, the U.S. Office of Management and Budget (*www.whitehouse.gov/omb/egov/*), can translate into hundreds of pages of

---

*Some of the material in this section comes from Michael Tracey, of Bay St. Louis, Mississippi, who designs websites and builds and upgrades computer hardware.

printed text. The first page of a website—the homepage—is the thread that connects subsequent pages into a seamless presentation. To design and write an effective homepage, you should always provide a menu that allows easy navigation to the rest of the site. Also, each page should include a menu button that links directly back to the homepage, as illustrated below.

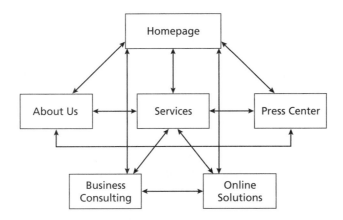

The homepage is often but not always where visitors to a website will go first; if it catches their attention, they are likely to click onto your other pages. As you read the rest of this section, look at the sample homepages in Figures 11.5 through 11.10. Various programs are available to create a website, including Microsoft Expression Web, Adobe Dreamweaver, and others.

In the following sections of this chapter, we'll discuss the differences between a website and a printed document, identify various parts or elements of a website, and help you to write for and contribute to the design of your own or your employer's website.

Web Versus Paper Pages

Writing and designing a website requires the same emphasis on planning, drafting, formatting, and testing that traditional business documents printed on paper do. Since webpages include many of the same elements, the four keys to effective writing (from Chapter 1, pp. 8–20) apply to online content as much as they do to business writing in print:

1. Identify your audience.
2. Establish your purpose.
3. Formulate your message.
4. Select your style and tone.

Yet, clearly, a website differs from a paper memo, letter, or report. Webpages include images, sometimes worked into Flash graphics animations or Java applets that fade one image into another or create blinking, movable elements. In addition, websites are more colorful and their content is organized differently from a print source. And they are not fixed on the screen. They can change, be edited, and be available to a far greater audience far more quickly, than any print source ever conceivably could be. Every website is really a work in progress. But in order to know how to write for a website you need to know how it differs from printed documents.

Web Versus Print Readers

Webpages are read differently from the way print documents are. People do not generally read a website word for word hunting for information, carefully studying each sentence. They want to find information at a glance. On the average, Web readers will spend 10 to 20 seconds scanning a page. They will not waste their time scrolling in the hopes of finally finding what they need. They'll just click to another site. For that reason, Web articles, news stories, and features are shorter, more condensed, and more strategically arranged to give only essential information than those in print publications. Webpages are designed to be read quickly. Some experts advise using half the word count of a paper document for a website.

Keep in mind that text on the Web needs to be shorter because of the size of the screen. Text displayed on a screen will be harder to read than on a piece of paper. Lots of text, along with clip art and images, on a website squeezed onto a screen, coupled with the problem of glare on the computer screen, can lead to eyestrain and fatigue for a reader more quickly than in a print source.

Moreover, a Web audience will not necessarily read your entire website. In fact, they may not have even begun their search at your site. So you can't assume your Web audience, like one for a printed source, will read from the start to the finish of your site. Don't expect to write to build up a climax or to postpone key facts. Web readers may navigate through your pages in a predetermined order or read all of its parts. They can start with a home- or later page and then go to any other page on that site. This flexibility on the part of Web readers makes the organization of each page—and your overall site—crucial factors to consider when you write for, edit, or design a website.

Finally, Web readers will expect navigational cues that readers of a print source do not. A Web audience will be looking for visual markers such as highlighted keywords to click on or bulleted lists, different colored text, commands such as *search*, *click here*, *go to*, *go back*, *contact us*, arrows and crosses, or addresses of other sites to visit. Note how these cues are incorporated into the websites in Figures 11.4 through 11.7.

Case Study: Converting a Print Document into a Website

To illustrate how a print document differs from an online one, let's consider how the article "Virtual Reality: The Future of Law Enforcement," which appeared in Chapter 9, can be recreated for a Web audience. The print version had fourteen sections presented in the following order:

1. Introduction
2. Traditional Training Limitations

3. Training with Virtual Reality
4. What Is Virtual Reality?
5. How Does Virtual Reality Work?
6. Uses for Virtual Reality
7. Law Enforcement Training
8. Pursuit Driving
9. Firearms Training
10. High-Risk Incident Management
11. Incident Re-creation
12. Crime Scene Processing
13. Is Virtual Reality Virtually Perfect?
14. Physical Limitations and Effects

Had that article been published as a website, the author might have presented the entire table of contents on a homepage, as in Figure 11.4, allowing readers to view the sections they wanted in the order they wanted. Instead of turning pages, they would click menu links.

A re-creation of the homepage for a website based on the virtual reality article.   **Figure 11.4**

VIRTUAL REALITY
- What Is Virtual Reality?
- How Does Virtual Reality Work?
- Uses for Virtual Reality
- Is Virtual Reality Virtually Perfect?
- Physical Limitations and Effects

Bulleted hyperlinks allow for easy navigation

Uses appropriate visual to catch reader attention

LAW ENFORCEMENT
- Law Enforcement Training
- Pursuit Driving
- Firearms Training
- High-Risk Incident Management
- Incident Re-Creation
- Crime Scene Processing

Short chunk of text designed to guide reading

THE FUTURE OF LAW ENFORCEMENT

Introduction
Traditional Training Limitations
Training with Virtual Reality
About this Website
Contact Information

A late night police pursuit of a suspected drunk driver winds through abandoned city streets. The short vehicle chase ends in a warehouse district where the suspect abandons his vehicle and continues on foot. The rookie patrol officer exits his vehicle and gives chase. . . **[More]**

Guidelines for Designing and Writing a Successful Homepage

A homepage has two main functions. First, it has to catch visitors' attention. If it fails to do that, everything else is a waste of time; your visitors will click their mouse and be gone. Second, it has is to sell your company's product or service or to introduce your organization. That may seem backward, but unless you attract readers to your website, you cannot promote your product or service. Successful homepages accomplish both of these objectives clearly and effectively. For example, students applying for financial aid can find it a daunting process, but the design of the FinAid homepage, Figure 11.5, keeps things simple and upbeat. Even the image

Figure 11.5   Sample homepage from FinAid.

Left column headers use appropriate icons

Uses concise and clear language

Page is carefully organized, visually balanced, and uncluttered

Links arranged by audience needs

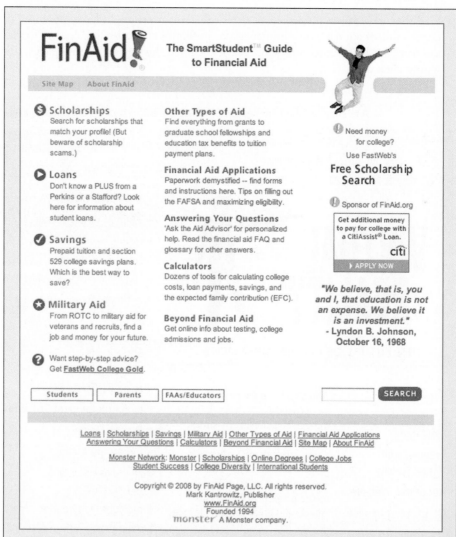

Examples of navigational links.   Figure 11.6

Site translated for international readers

Uses appropriate icons

Site is easy to navigate because of images and menus

Solicits feedback

of the student "jumping for joy" contributes to the page's user-friendliness. Similarly, the EPA homepage on global warming in Figure 11.6 uses environmentally appropriate images to emphasize its purpose.

Figure 11.7   Homepage of the International Chamber of Commerce.

Reprinted courtesy of the International Chamber of Commerce.

Designing a website requires you to follow many of the rules you read about earlier in this chapter for printed pages. (See pp. 496–508.) In addition, you should adhere to the following eight guidelines to create an effectively designed and written website that will capture your audience's attention.

1. **Make your site easy to find.**

   - Obtain a registered domain name for your site. Choose a professional domain name that clearly and quickly tells readers what you do. Avoid cute spellings and fanciful names, e.g., happihouse.com for toledohouse-painters.com. If your Internet provider goes out of business, you can easily transfer your site to another server without changing the name and, in the process, losing customers. A good place to start is the InterNIC website (*http://www.internic.net*), which provides updated information on Internet domain name registrations.
   - Submit your website to the various search engines and directories to make sure a broad audience knows about your site. See page 342 for some of the most notable search engines. You may also want to take advantage of website submission services offered by companies such as Microsoft Small Business Services (*http://www.bcentral.com*) and Overture.com (*http://www.overture.com/d/home/*). For a subscription fee, these companies will promote your site to various search engines on an ongoing basis.
   - Make your site search-engine ready (see the Tech Note "Writing for Search Engines" on p. 516).

2. **Make your site easy to download and navigate.**

   - First make your site simple and easy to download. If readers have trouble even getting onto your site because of too many sound, video, and graphics elements, they may well skip it. Help your visitors find their way easily through your site with logical and effective navigation tools such as:

in-text hyperlinks	search engines
navigation bars	button links
indexes and menus	rollover icons
site maps and tables of contents	previous, next, and back links

     Note how, in Figure 11.6, the EPA provides multiple navigation aids: menu buttons that are labeled and illustrated with images and that elaborate on the menu on the left-hand side.
   - Test every link to make sure it is current, that it works, and that it connects to related sites.
   - Don't overload your page with images so your pages look crowded. Note how Figure 11.5 uses a single image, four smaller icons, and boxes at the bottom right. The carefully organized photos, images, and icons displayed in Figure 11.7 also make the site easy to navigate.
   - Make sure all your pages are linked and that all the links work. A website that does not identify one of its links or that contains a broken link will drive readers/customers away.

Tech Note

Writing for Search Engines

Search engines do not "see" webpages the way visitors do. They do not see designed pages or colorful graphics. Instead, they see only the HTML (hypertext markup language) source code, which is the "skeleton" of a webpage.

Because search engines see only the source code of a page, you need to include the most important keywords related to your webpage in the most likely places for a search engine to find them—near the top of your webpage in a section called the "head." Since search engines read webpages from top to bottom, the keyword-containing text should appear in the "head" section and at the top of your page.

This holds true whether you are creating an electronic résumé (see pp. 266–273) or a company website. If your firm manufactures retractable awnings, you need to include the keywords *retractable* and *awnings* in the "head" tags and in the body text, thus proving definitively to the search engines that your page is indeed about "retractable awnings." Then, when a user visits a search engine to search for that keyword phrase, the likelihood of your site appearing toward the top of the rankings increases.

You will want to include your keyword phrase in the title, description, and keyword tags in the head section, and you want the title and description tags to be "selling" tools for your company. For instance, you might say "Discount Retractable Awnings" in your title tag to emphasize customer savings.

In the competitive online business world, the most important goal is to get your company's website in front of your customers. Knowing how search engines work can ensure that customers find your firm online.

Source: Adapted from "Knowing How to Build Web Sites Can Insure More Net Traffic," by Robin Noble, August 2002. Reprinted by permission of Robin Noble.

3. **Make your site informative.**

   - Provide all essential information on your homepage, including your company's name, address, e-mail, and phone numbers. The more helpful your site is, the more likely it will draw repeat visitors.
   - Tell the visitor what products or services you offer. For example, the U.S. Postal Service promotes its many services on its homepage.
   - Indicate what type of information may be obtained through the website, including links to your sales force, customer service, and technical support.
   - Offer readers different types of interaction—FAQs, bulletin boards, animated product demonstrations, and free e-mail subscriptions.

4. Make your site easy to read for both native and non-native English speakers.

- Put the most important point first in the topic sentence.
- Get to the point right away.
- Write short paragraphs (chunks of content)—no more than three to four sentences long.
- Provide headings with keywords and bulleted lists to help readers locate information quickly.
- Use plain, concise English—concrete nouns, action verbs. And keep sentences short—use only active voice.
- Use plenty of white space but not after each sentence. Single-sentence paragraphs are harder to process.
- Include scannable terms and hyperlinks; always highlight them to make them stand out.
- Select fonts that are easy to read (review pp. 501–505 on typography). Use larger fonts than you would for a print document.

5. Keep your site updated.

- New information is vital for product and service company sites. Otherwise, readers have no reason to make return visits. Feature insider news updates about your business or preproduction information on products or services. Build in hyperlinks to product reviews, conferences, awards, and so on.
- Revise the design of your site if your company offers a new product or service or a new promotional element on its homepage. Clearly, your site does not need a major design overhaul every week, but a new page alerts customers to new products and services.
- Indicate when the site was last updated, so readers will know your information is kept current.
- Periodically check your external links to make sure they work.

6. Use images and icons effectively.

- Arrange images and photos so they do not interfere with text or layout. Proportion is important for achieving a balance between different page elements. Note how the sizes of the images in Figure 11.7 are similar.
- Choose appropriate icons or images to illustrate menus and page sections. As with any textual document, images should serve a purpose.
- Be conservative in using animations or anything that might be viewed as a gimmick, because they may distract readers from other content on the page.
- Keep images proportional so that they are not too big or small for the page.

7. Encourage visitor interaction by soliciting feedback.

- Ask readers to e-mail you about your product, service, or website. And make sure a procedure for handling that feedback is developed within your organization.
- Include a feedback form or survey with specifically targeted questions (see pp. 322–323), including multiple-choice pull-down menus and comment boxes, on your website to encourage visitors to leave useful comments.

- Conduct a usability test. Ask readers what you could do better or what they like about your website.

8. **Make sure your website is ethical.**

   - Do not plagiarize from another Web (or print) source. If you include any information from another site, including quotations, visuals, or statistics. Get permission and acknowledge the source on your site.
   - Avoid sexist, racist, and other biased forms of language. Moreover do not offend an international audience by using terms or names that are insulting, stereotypical, or condescending. See pages 6–8.
   - Never make false or exaggerated claims. Be honest and accurate. Earn your readers' trust.
   - Provide accurate and recent statistics and other data.

Creating Storyboards for Websites and Other Documents

Visit www
.cengage.com/
english/kolin for
an online exer-
cise, "Creating a
Storyboard for
an Existing
Website."

To help you plan the design of a website or a print document such as a brochure, newsletter, or even a handbook, you may find it useful to create a storyboard. The term *storyboarding* originated in the film industry as a way for directors to illustrate scenes in their films before actually shooting them. Storyboarding can be as simple as sketching what you want your webpage or other document to look like. Figure 11.8 illustrates a storyboard for the homepage of a website.

Acting as a map to your site or a preliminary layout of a document, a storyboard gives you a clearer idea about the structure and navigation of each page of your website or document, and it allows you to plan the text and visuals. Like a draft of a report or proposal, your storyboard will become fuller the more you revise it—adding, moving, and linking content and graphics.

Here are some guidelines for effective storyboarding:

1. **Sketch the site or document pages on a piece of paper.** Label them according to various pages of the site or sections of the document, for example, Homepage, About Our Company, Contact Us, Place an Order.

2. **Plan the layout of each page.** Consider what content each page must have and how that information can be signaled through graphics. For instance, a homepage should include your company's logo, a brief introduction to or history of the company, and information about the goods or services it offers.

3. **Determine basic design elements**. For instance, decide which fonts, backgrounds, and frames will look professional and appealing to customers. Then decide on size and height, and where to place columns, images, sidebars, or highlighted areas.

4. **Build in navigational aids.** Make searching your website quick and effective. Always include hyperlinks to your company's e-mail address so that readers can contact you. Ensure that your website is easy to navigate by drawing arrows on the storyboard between pages that should be linked together, as in Figure 11.8.

Storyboard for a homepage.   **Figure 11.8**

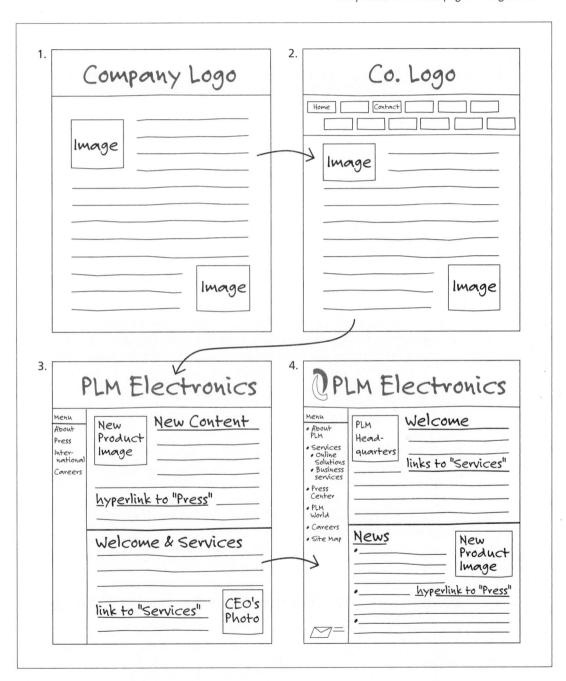

Every page should be linked to the homepage. For brochures or other documents, make sure you insert clear and concise headings and subheadings, and provide contact information.

Tech Note

Using PDFs

A Portable Document Format (PDF) file offers a variety of options to users. Like the Print Preview option in a word-processing program, a PDF allows readers to see a snapshot of a document exactly as the original looks, complete with color, fonts, and visuals. Like a tracked document or a wiki, a PDF is equipped with a function that allows multiple readers to add comments to the document. But the most common use of PDFs is on websites.

No doubt you've frequently run across PDFs while navigating websites. This is because PDFs are ideal for the Web environment. First, they allow documents originally written for print format to be easily posted on the Web without having to make complicated and difficult HTML conversions. Second, they can be locked when posted online so that readers cannot make changes to the document. Having a fixed or set online form is especially helpful for companies or organizations for them to use a consistent and tested format.

But keep in mind that PDFs can be viewed only using the Adobe Reader, which can be downloaded free of charge from the Adobe website at *http://www .adobe.com/products/acrobat/readstep2.html*. In order to create PDF documents, Adobe Acrobat Professional software is required. The process of converting documents into PDF format is fairly simple and offers the writer numerous options.

When writing for and designing websites, then, be cautious about overusing PDFs. Although they are convenient and secure for online posting, remember that your audience reads differently on the Web than they do in reading printed pages. One last and perhaps most important caution: If your website includes too many long PDFs that readers have to scroll through, they will leave and move to another site.

Four Rules of Effective Page Design: A Wrap Up

The following four rules summarize the basic rules for effective document and Web design. Adhere to them and your work, online or in print, will be professional looking.

1. **Keep it simple.** Try not to overuse graphic effects and design styles in an effort to impress your audience; keep it simple so that you do not lose sight of the purpose of the document or website in an over-the-top design.

2. **Be consistent.** Use your layout and terminology consistently throughout your document and website. Repeat your design elements on each page, whether this is the fonts you use, the text alignment, or any colors and borders.
3. **Make it clear.** Make your message clear. While fancy designs may look good on paper, they do not always add to the clarity of your document or website.
4. **Less is more.** Limit the number of items on your page to a maximum of perhaps five, depending on your page size. An item can be a paragraph of text or a picture, as long as it provides a focal point on the page. Too much information on a Web or document page can make the document difficult to digest, and the information itself can be lost in the clutter.

Revision Checklist

Printed Documents

☐ Kept readers' busy schedules and reading time limits in mind when designing a document.
☐ Learned options of desktop publishing program.
☐ Arranged information in the most logical, easy-to-grasp order.
☐ Inserted heads and subheads to organize information for reader.
☐ Included only relevant visuals—clip art, icons, stock photos.
☐ Left adequate, eye-pleasing white space in text.
☐ Provided adequate margins to frame document.
☐ Justified margins, right, left, or center, depending on document and reader's needs.
☐ Maintained pleasing, easy-to-read line length.
☐ Kept line spacing consistent and easy on the readers' eyes.
☐ Chose appropriate typeface for message and document.
☐ Selected serif or sans serif font depending on message and readers' needs.
☐ Did not mix typefaces.
☐ Used effective type size, neither too small (under 10 point) nor too large (over 12 point), for body of text.
☐ Incorporated appropriate visual cues (for example, italics or boldface) for readers.
☐ Made all heads and subheads parallel and grammatically consistent.
☐ Supplied lists, bullets, numbers to divide information.

Websites

☐ Ensured website is easy to locate.
☐ Made sure navigation is clear and logical, not overly complex.
☐ Provided identification for all pages either with headings or text that explains the purpose of each page.
☐ Tested all links to make sure they are functioning, including internal links and links to other websites so customers can easily navigate the site.

Continued

☐ Ensured that the site is informative and content is current and periodically updated.

☐ Used headings, subheadings, and white space to break information into readable chunks.

☐ Made site easy to read for both native and non-native English readers.

☐ Kept the site current.

☐ Provided ways for reader to interact with the site, whether for e-mail subscriptions or to provide feedback.

☐ Used images and icons effectively.

☐ Encouraged visitor interaction by soliciting feedback.

☐ Did not crowd images and text on the same page.

☐ Included memorable icons and other symbols in appropriate places on the page; tested them to determine why and how they contribute to the overall effect of the site.

☐ Made sure the site is ethical.

Exercises

1. Find an example of an effectively designed document, according to the criteria discussed in this chapter. It could be a memo, a brochure, a newsletter, a report, a set of instructions, a section of a textbook, or a website. E-mail or write a short (one-page) memo to your instructor describing the document's design and explaining why it works. Attach a copy of the document or the website to your e-mail or memo.

2. Working with a team of three or four students, bring poorly designed documents to class. As a collaborative venture determine which of the documents the group submits is the hardest to follow, the most unappealing, and the least logically arranged. After selecting that document, collaboratively write a memo to your instructor on what is wrong with the design and what you would do to improve its appearance and organization.

3. Redesign (reformat; add headings, spacing, and visual clues; include relevant clip art; and so on) the document your group selected for Exercise 2 and submit it to your instructor.

4. Find an ineffectively designed document—a form, a set of instructions, a brochure, a section of a manual, a story in a newsletter—and assume that you are a document design consultant. Write a sales letter to the company or agency that prepared and distributed the document, offering to redesign it and any other documents they have. Stress your qualifications and include a sample of your work. You will have to be convincing and diplomatic—precisely

and professionally persuading your readers that they need your services to improve their corporate image, customer relations, and sales or services.

5. Redesign one of the documents on pages 524–525 (on handwashing or detector placement) to make it conform to the guidelines specified in this chapter.

6. Locate two webpages that cover a similar product, service, industry, or other topic. Analyze some of the webpage elements each one uses, comparing each site's strengths and weaknesses. Write a one-page memo to your instructor explaining which is the more effective site.

7. Find a webpage that you feel is ineffective. Using the four keys to effective writing, as well as your knowledge of webpage elements, write a one-page assessment of the site, discussing three or four changes that would make it more effective. Attach a printed hard copy of the webpage to your assessment.

8. The homepage of the Stanley's Accounting Temps website below is not as effective as it could be. Write a one-page memo to your instructor specifying how Stanley's Accounting Temps might better address its online customers. Group your recommendations under the headings of content, design, and navigation. As a supporting document for your memo, design a new homepage for the company by sketching it on a piece of paper or creating it on the computer. (The Stanley's Accounting Temps homepage example was created using word processing software.)

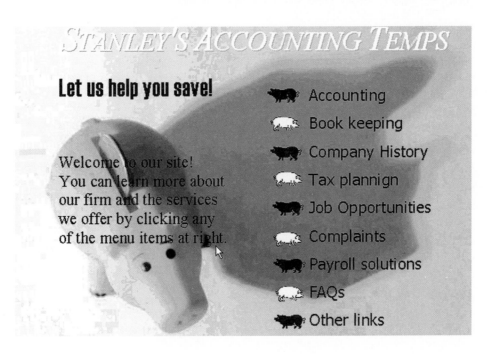

9. As a collaborative exercise, find a business website that promotes a product or service. Your group will work as Web consultants proposing that the company redesign its site. Your presentation should include an evaluation of the existing site—general and specific elements—as well as a proposal for how you would improve the site.

WHY SHOULD YOU WASH YOUR HANDS?
Bacteria and viruses (germs) that cause illnesses are spread when you don't wash your hands.
If you don't wash your hands, you risk acquiring:
The common cold or flu
Gastrointestinal illnesses Shigella or hepatitis A
Respiratory illnesses
Should you wash your hands?
You need to wash your hands several times every day. Some important times to wash your hands are:
BEFORE
Preparing or eating food.
Treating a cut wound.
Tending to someone who is sick.
Inserting or removing contacts.
After
USING THE BATHROOM.
Changing a diaper or helping a child use the bathroom (don't forget the child's hands)
Handling raw meats/poultry/eggs
Touching pets, especially reptiles
Handling garbage
Sneezing or blowing your nose, or helping a child blow his/her nose
Touching any body fluids like blood or mucus
Being in contact with a sick person
Playing outside or with children and their toys

WHEN SHOULD YOU WASH YOUR HANDS?
There is a right way to wash your hands.
Follow these steps and you will help protect yourself and your family from illness.
Like any good habit, proper hand washing must be taught.
Take the time to teach it to your children and make sure they practice.

TTI's in the loop on effective detector placement

Ever sit in bumper-to-bumper traffic and wish they'd widen the roads so people could get through more quickly? Well, that costs a lot of money. Which is why transportation engineers who deal with traffic congestion and the problems it causes look for more cost-effective alternatives to get you where you're going—and faster.

TTI researchers recently completed a TxDOT/FHWA-sponsored study entitled *Effective Detector Placement for Computerized Traffic Management.* The research sought to expand and improve the use of inductance loop detectors (ILDs) to complement traffic signals, signal systems and other advanced traffic management systems. This is a cheaper congestion solution than building or widening a road.

An ILD is an electrical circuit containing a loop of copper wire embedded in the pavement. As a vehicle passes over the wire loop, it takes energy from the loop. If that change is large enough, a detection is recorded. Thus, we are able to collect data on the movement or presence of vehicles on the roadway. Advanced traffic management systems operate best with accurate information on how many vehicles are present and how fast they are traveling.

The primary goal of the recent project was to use loop detectors as an integral part of the congestion-reduction system. Traditional problems with ILDs were addressed—like crosstalk, or interference between two adjacent loops—and innovative new applications for ILDs in advanced traffic systems—like detecting wrong-way HOV-lane movements.

Other applications include using ILDs to move traffic more efficiently at diamond interchanges, at high-volume, high-speed approaches and on the freeway entrance ramps. The long-range contribution of the study is a set of guidelines for using ILDs in the situations listed above. As freeway management systems continue to evolve, the guidelines developed through the nine study reports will provide designers with practical information on the most effective placement of ILDs.

A major finding of the research deals with lead length, or the length of wire necessary to connect the loop to the detector electronic unit. The study showed that the loop can be placed more than 4,000 feet from the point of control—four times the currently accepted distance. This information will give traffic designers much more flexibility when integrating ILDs into their traffic system designs.

The researchers also made some important discoveries about using ILDs to measure speed. They found that the best speed trap is nine meters (two loops interconnected with a timing device and spaced nine meters apart). They also determined that to get reasonably accurate and consistent speeds with an ILD, some things must be the same between a pair of loops: make, type or model of the detector units, sensitivity settings and loop configuration.

The findings from this research facilitate the use of loop detectors in managing traffic. And better management of driver frustration— just as important, even if less measurable than the congestion that causes it—is bound to follow.

Ultimately the three watchwords for this project were optimization, innovation, and implementation. Taking the tried-and-true and finding a better way to use it is, after all, the underlying building block for all engineering endeavors.

To order TTI Research Report 1392-9F, see the back page order form of this issue. For more information on loop detectors, contact Don Woods, 409/845-5792, FAX 409/845-6481 (E-mail: d-woods@tamu.edu).

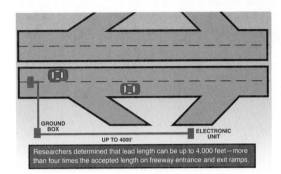

GROUND BOX

ELECTRONIC UNIT

UP TO 4000'

Researchers determined that lead length can be up to 4,000 feet—more than four times the accepted length on freeway entrance and exit ramps.

Source: Texas Transportation Institute's Researcher; article author, Chris Pourteau. Reprinted with permission of the Texas Transportation Institute.

CHAPTER 12

Writing Instructions and Procedures

Visit www .cengage.com/ english/kolin for this chapter's online exercises, ACE quizzes, and Web links.

Clear and accurate instructions are essential to the world of work. Instructions tell—and frequently show—how to do something. They indicate how to perform a procedure (draw blood; change the oil in your car); operate a machine (a pH meter; a digital camera); construct, install, maintain, adjust, or repair a piece of equipment (an incubator; a scanner). Everyone from the consumer to the specialist uses and relies on carefully written and designed instructions.

Instructions and Your Job

As part of your job, you may be asked to write instructions, alone or with a group, for your co-workers as well as for the customers who use your company's services or products. When writing long, complex instructions, you certainly will be part of a team of engineers, programmers, document design experts, marketing specialists, and even attorneys. Your employer stands to gain or lose much from the quality and the accuracy of the instructions you prepare. Note how the employees working at Shay Melka's company have to follow a host of instructions to collect and report business research (see pp. 305–309).

While the purpose of writing instructions is to explain how to perform a task in a step-by-step manner, the purpose of writing procedures is slightly different. Often the two terms are incorrectly used interchangeably. **Procedures** (or policies) describe not a task to be completed, but a set of established rules of conduct to be followed within an organization, such as a business.

This chapter will first show you how to develop, write, illustrate, and design a variety of instructions, and then move into a discussion of writing procedures (pp. 554–557) about job-related duties.

Why Instructions Are Important

Perhaps no other type of occupational writing demands more from the writer than do instructions because so much is at stake—for both you and your reader. The reader has to understand what you write and perform the procedure as well. You

Tech Note

Online Instructions

Online instructions offer added benefits. Most computers have the ability to give you pictures and sounds, but those two features can be limited without additional hardware. With a sound card your computer can provide a much wider range of possibilities, including having someone actually talk you through the various steps of a procedure and alert you to special problem areas. If your computer also has a graphics card, you can speed up the display of images and take advantage of animation that shows the process being done. Such features—visuals and sound—further help instruction writers meet the precise needs of their readers who may need to perform a trial run before attempting a procedure or who may want to double-check a step before going on.

cannot afford to be unclear, inaccurate, or incomplete. Instructions are significant for many reasons, including safety, efficiency, and convenience.

Safety

Carefully written instructions get a job done without damage or injury. Poorly written instructions can be directly responsible for an injury to the person trying to follow them and may result in costly damage claims or even lawsuits. Notice how the product labels in your medicine chest inform users when, how, and why to take a medication safely. Without those instructions, consumers would be endangered by taking too much or too little medicine or by not administering it properly. To make sure your instructions are safe, they must be

- accurate
- consistent
- thorough
- clearly written
- carefully organized
- legally proper

Companies have a legal/ethical obligation to inform customers about what constitutes normal use and what, if any, injuries or malfunctions can result if the product or procedure is misused. Several U.S. agencies notify consumers about products that have been found to be unsafe. The U.S. Consumer Product Safety Commission is a useful source for such information; check out its website at *http://www.cpsc.gov*.

Efficiency

Well-written instructions help a business run smoothly and efficiently. No work would be done if employees did not have clear instructions to follow. For example, without the instructions on flextime (see Figure 12.11 on pp. 556–557) employees would not know when to work. Imagine how inefficient it would be for a business if employees had to stop their work each time they did not have or could not understand a set of instructions. Or, equally alarming, what if employees made a number of serious mistakes because of confusing directions, costing a business lost sales and increased expenses? Giving readers helpful tips to make their work easier will also increase their efficiency. See pages 545–546 for the use of notes.

Convenience

Clear, easy-to-follow instructions make a customer's job easier and less frustrating. In the customer's view, instructions reflect a product's quality and a service's quality and convenience. How many times have you heard complaints about a company because the instructions were hard to follow? Poorly written and illustrated instructions will cost your customers time and you their business. Instructions are also a vital part of "service after the sale." Owners' manuals, for example, help buyers to avoid a product breakdown (and the headache and expense of starting over) and to keep it in good working order.

The Variety of Instructions: A Brief Overview

Instructions vary in length, complexity, and format. Some instructions are one word long: *stop*, *lift*, *rotate*, *print*, *erase*. Others are a few sentences long: "Insert blank disk in external disk drive"; "Close tightly after using"; "Store in an upright position." Short instructions are appropriate for the numerous relatively nontechnical chores performed every day.

Detailed instructions may be as long as a page or a book. When your firm purchases a new mainframe computer or a piece of earth-moving equipment, it will receive an instruction pamphlet or manual containing many steps, cautionary statements, and diagrams.

Instructions can be given in a variety of formats, as Figures 12.1 through 12.4 show. They can be paragraphs (Figure 12.1), employ visuals to illustrate each step (Figure 12.2), or use a numbered list (Figures 12.3 and 12.4). You will have to determine which format is most appropriate for the kinds of instructions you write. For writing that affects policies or regulations, as in Figure 12.11 (on pp. 556–557), you will most often use a hard copy of a memo, which serves as a more permanent and legal notice than would an e-mail.

Assessing and Meeting Your Audience's Needs

Put yourself in the readers' position. In most instances you will not be available for readers to ask you questions when they do not understand something. Consequently, they will have to rely only on your written instructions. Your purpose in

Instructions on how to repair a halyard.   **Figure 12.1**

Easy Temporary Join for Synthetic Ropes

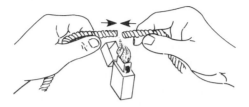

If you are faced with the problem of reeving a new halyard on a flagpole or mast, or through a block or pulley in an inaccessible location, the solution can be easy if both old and new lines are made of nylon or polyester (Dacron, Terylene, etc.). Simply join the ends of the old and new lines temporarily by melting end fibers together in a small flame (a little heat goes a long way). Rotate the two lines slowly as the fibers melt. Withdraw them from the flame before a ball of molten material forms, and if the stuff ignites, blow out the flame at once. Hold the joint together until it is cool and firm.

Directions written in clear, plain language

Source: R. I. Standish, *Parks*.

writing the instructions is to get the readers to perform the same steps you followed and, more important, to obtain the same results you did.

Do not assume that members of your audience have performed the task before or have operated the equipment as many times as you have. (If they had, there would be no need for your instructions.) No writer of instructions ever disappointed readers by making directions too clear or too easy to follow.

Key Questions to Ask About Your Audience

To determine your audience's needs, ask yourself the following questions:

- How and why will my readers use my instructions? (Engineers have different expectations than do office personnel and customers.)
- What language skills do they possess—is English their first (native) language?
- How much do my readers already know about the product or procedure?
- How much background information will I have to supply?
- What steps will most likely cause readers trouble?
- How often will they use my instructions—every day or just as a refresher?
- Where will my audience most likely be following my instructions—in the workplace, outdoors, in a workshop or laboratory equipped with tools, or alone in their homes?
- What resources—such as special equipment or power sources—will my readers need to perform my instructions successfully?

Figure 12.2   Instructions that supply a visual with each step.

Proper Brushing

Proper brushing is essential for cleaning teeth and gums effectively. Use a toothbrush with soft, nylon, round-ended bristles that will not scratch and irritate teeth or damage gums.

1

Place bristles along the gumline at a 45-degree angle. Bristles should contact both the tooth surface and the gumline.

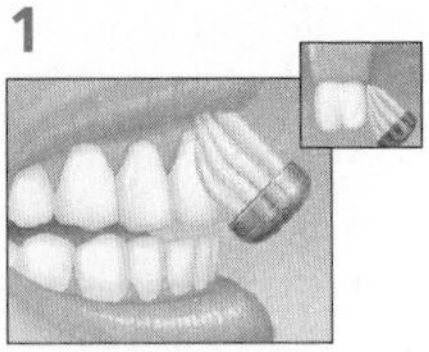

2

Gently brush the outer tooth surfaces of 2–3 teeth using a vibrating back and forth rolling motion. Move brush to the next group of 2–3 teeth and repeat.

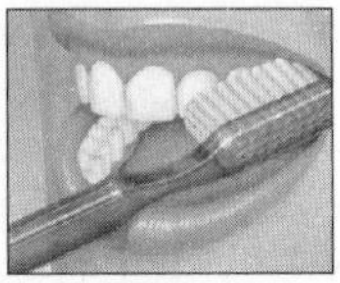

3

Maintain a 45-degree angle with bristles contacting the tooth surface and gumline. Gently brush, using back, forth, and rolling motion along all of the inner tooth surfaces.

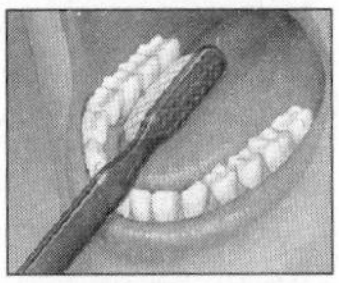

4

Tilt brush vertically behind the front teeth. Make several up and down strokes using the front half of the brush.

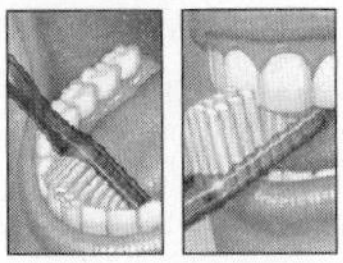

5

Place the brush against the biting surface of the teeth and use a gentle back and forth scrubbing motion. Brush the tongue from back to front to remove odor-producing bacteria.

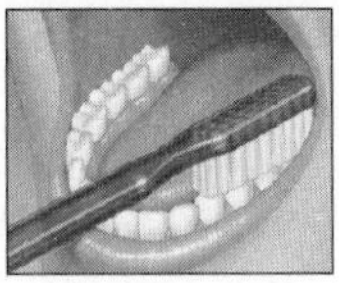

Source: Reprinted by permission of American Dental Hygienists Association. Illustrations adapted and used courtesy of the John O. Butler Company, makers of *GUM* Healthcare products.

Writing Instructions for International Audiences

Keep in mind that the audience for your instructions will often include a worldwide audience of potential customers, many of whom will be non-native speakers of English. Follow these guidelines when writing instructions for this group of readers:

Instructions given in a numbered list describing a sequence of steps.   **Figure 12.3**

Copying Files to a USB Flash from Your PC or Laptop

1. Insert the USB Flash (see illustration at right) into a USB portal of your PC or laptop.

2. Find the folder or file to be copied to the USB Flash and right click on it.
   NOTE: The folder or file will be highlighted and a menu with Open at the top will appear.

3. Within the menu, move your cursor down to the Send To option.
   A list of locations will appear.

4. Choose the USB Flash location. Your folder or file will be automatically copied over.
   CAUTION: DO NOT remove the USB Flash at this point, or you will risk damaging it.

5. Go to My Computer from the computer's Start menu. Double click on the USB Flash to see if the copying was successful.

6. Eject the USB Flash before removing it from the computer. Go to My Computer again, right click on the USB Flash, and select the Eject option from the menu.

7. Remove the USB Flash from the USB portal.

---

1. **Write in plain, simple language.** Use international English (see page 8 and pages 170–187).

2. **Use terms and units of measurement that your readers will understand.** Avoid abbreviations, acronyms, and jargon, and don't assume that international readers will use or understand the units of measurement found in the United States.

3. **Make sure all of your visuals—including clip art—are clear and culturally appropriate.** Photographs and exploded drawings are easier to read than technical diagrams. In some instances, your instructions may be given exclusively through visuals.

4. **Be aware that colors can have different meanings.** Red, yellow, and green, for instance, may not convey the same meaning in other cultures that they do in the United States.

5. **Ask a non-native speaker to review your instructions.** This will help ensure that your instructions are free of difficult-to-understand and inappropriate terms, expressions, or sentences.

6. **Always include contact information.** Providing help either online or over the telephone will enable non-native and native speakers alike to ask questions.

Figure 12.4   Instructions in a numbered list on how to assemble an outdoor grill.

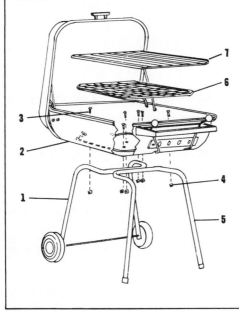

Starts with helpful note and list of tools

Exploded drawing shows relations of parts to each other

Easy-to-follow steps for consumer audience

ASSEMBLY INSTRUCTIONS

The instructions shown below are for the basic grill with tubular legs. If you have a pedestal grill, or a grill with accessories, check the separate instruction sheet for details not shown here.

NOTE: Make sure you locate all the parts before discarding any of the·packaging material.

TOOLS REQUIRED ... A standard straight blade screwdriver is the only tool you need to assemble your new Meco grill. If you have a pedestal grill, you will need a 7-16 wrench or a small adjustable wrench.

1. Before you start, take time to read through this manual. Inside you will find many helpful hints that will help you get the full potential of enjoyment and service from your new Meco grill.

2. Lay out all the parts.

3. Assemble roller leg (1) to bottom rear of bowl (2) with 1¼" long bolts (3) and nuts (4).

4. Assemble fixed leg (5) to bottom front of bowl (2) with 1¼" long bolts (3) and nuts (4).

5. Place fire grate—ash dump (6) in bottom of bowl (2) between adjusting levers.

6. Place cooking grid (7) on top of adjusting levers. Make sure top grid wires run from front to back of grill.

Source: Meco Assembly Instructions and Owners Manual, Metals Engineering Corp., P.O. Box 3005, Greenville, TN 37743. Reprinted by permission.

Two Case Studies on Meeting Your Audience's Needs

The instructions contained in Figures 12.5 and 12.6 are addressed to two different audiences, each with separate needs. The Hercules memo in Figure 12.5 was sent to a technical audience—firefighters and supervisors—who needed instructions on a special process. Cliff Burgess's memo in Figure 12.6, on the other hand, went to all Burton employees, a more diverse, rather than technical, group of readers. His helpful instructions do not require a list of equipment or materials or a description of steps. Instead, his memo consists of an introduction, three bulleted instructions, a conclusion, and a nontechnical visual for his audience.

Instructions alerting a technical audience to special circumstances.   **Figure 12.5**

TO: All Shift Supervisors
               All Firefighters
FROM: Robert Ferguson *R.F.*
SUBJECT: Operating Procedures During Energy Shortages
DATE: October 10, 2006

The following policy has been formulated to help in maintaining required pressures during periods of low wood flow and severe natural gas curtailment.

Explains why instructions are important

All boilers are equipped with lances to burn residue or no. 6 fuel oil as auxiliary fuels. When wood is short, no. 6 oil should be burned in no. 2 and no. 3 boilers at highest possible rate consistent with smoke standards. To do this, take these steps:

Identifies conditions

1. Put no. 6 oil on no. 3 boiler lances.
2. Shut down overfire air.
3. Shut down forced draft.
4. Turn off vibrators.
5. Keep grates covered with ashes or wood until ash cover exists.

Lists directions in numbered steps

These steps will result in an output of 25,000 to 35,000 lbs./hr. steam from no. 3 boiler and will force wood on down to no. 4 boiler.

Alerts readers about what to expect

If necessary, the same procedure can be repeated on no. 2 boiler.

Source: Reprinted by permission of Hercules, Inc.

The Process of Writing Instructions

As we saw in Chapter 2, clear and concise writing evolves when you follow a process. To make sure your instructions are accurate and easy for your audience to perform, plan your steps, do a trial run, write and test your draft, and revise and edit.

Plan Your Steps

Before writing, do some research to understand completely the process you are asking someone else to perform. Make sure you know

- the reason for doing something
- the parts or tools required
- the steps to follow to get the job done
- the results of the job
- the potential risks or dangers

Figure 12.6   An instructional memo listing safety precautions for a general audience.

BURTON WORLDWIDE SYSTEMS
www.burton.com

TO: All Burton Employees
FROM: Cliff Burgess, Environmental Safety   *C.B.*
RE: Video Display Terminal (VDT) Safety Precautions
DATE: September 12, 2008

Introduction emphasizes reasons for instructions

You may experience some possible health risks in using your computer video display terminal (VDT). These risks include sleep disorders, behavioral changes, danger to the reproductive system, and cancer.

Nontechnical explanation offers an analogy

The source of any risks comes from the electromagnetic fields (EMFs) that surround anything that carries an electric current—for example, copiers, circuit breakers, and especially VDTs. Your computer monitor is a major source of EMFs. Magnetic fields can go through walls as easily as light goes through glass.

Although EMFs may affect your health, you can considerably reduce your exposure to these fields by following these three simple steps:

Steps stand out through bullets, bold-face, and spacing

- **Stay at least three feet (an arm's length) away from the front of your VDT.** (The magnetic field is significantly reduced with this amount of distance.)

- **Stay at least four feet away from the sides and back of someone else's VDT.** (The fields are weaker in the front of the VDT but much stronger everywhere else.) Refer to the following sketch, which you may want to post in your office.

Simple visual clarifies instructions

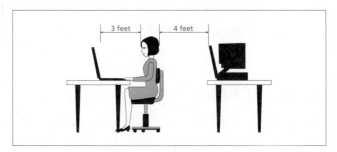

- **Switch your VDT/computer off when you are not using it.** (If the computer has to remain on, be sure to switch off the monitor; screen savers do not affect the exposure to **EMFs**.)

Conclusion reassures readers they are acting safely

Our environmental safety team will continue to monitor and investigate any problems. Observing the guidelines above, however, will help you to take all the necessary precautions in order to minimize your exposure to EMFs.

Lists helpful hyperlink

To view a demonstration of proper safety techniques, go to www.burton.com/vdtsafety/.

1215 Madison • St. Louis, Missouri 63174
314-555-4300 • FAX 314-555-4311

Source: Reprinted by permission.

Tech Note

Using Software to Design Instructions

To make sure your instructions are easy to read both for native as well as non-native speakers of English, take advantage of the following software features when you write a set of instructions:

1. **Use the Print Preview option.** When you view your instructions in Print Preview, you see a snapshot of the overall layout. That way you can make sure the layout of your instructions is clear and easy to follow, that you have included enough space between directions, and that you include necessary visuals in the right places.

2. **Choose a font that is easy to read.** Avoid fancy fonts, particularly flowery or gothic ones that may contain unusual letterings some readers may not recognize.

3. **Include boldface sparingly.** Use it only to emphasize warnings, cautions, and notes (see pages 545–546) so readers will not overlook the crucial messages they contain.

4. **Use numbered and bulleted lists.** Numbering steps helps readers follow the sequence of your instructions and bulleted lists break up text and make your instructions easier to read.

5. **Be careful about what you take from a clip art library.** Use only warning and caution icons that will be immediately recognizable to a broad range of readers. Also, make sure they are the right size—neither so big that they make your instructions look frivolous or so small that readers will not see them clearly at once. Stay a way from cartoon-like warning and caution icons, which readers may not take seriously.

6. **Provide hyperlinks within the instructions when necessary.** Don't ask your readers to do research. If there is anything they should be familiar with before actually using your instructions, provide links to those materials. Hyperlinks are important in the introduction to your instructions and also for contact information.

7. **Use boilerplate materials if available.** Boilerplates are your own company's documents that have already been created and can be reused in whole or in part without acknowledgment or permission. Warranties, guarantees, and specifications are examples of boilerplate materials.

If you are not absolutely sure about the process, ask an expert for a demonstration. Do some background reading and talk to or e-mail colleagues who may have written or followed a similar instruction.

Do a Trial Run

Actually perform the job (assembling, repairing, maintaining, dissecting) yourself or with your writing team. Go through a number of trial runs. Take notes as you go along and be sure to divide the job into simple, distinct steps for readers to follow. Don't give readers too much to do in any one step. Each step should be *complete*, *sequential*, *reliable*, *straightforward*, and *easy* for your audience to identify and perform.

Write and Test Your Draft

Transform your notes into a draft (or drafts) of the instructions you want readers to follow. Then conduct a **usability test** by asking individuals from the intended audience (consumers, technicians) who may never have performed the job to follow your instructions as you have written them. Ask participants to read your instructions aloud and ask questions. Observe where they run into difficulty—or get results different from yours.

Revise and Edit

Based on your observations and user feedback, revise your instructions to avoid:

- missing steps
- too many activities in one step
- steps that are out of order
- unclear or incomplete steps

Edit the final copy of the instructions that you will give to readers. Always consider whether your instructions would be easier to accomplish if you included visuals.

Analyzing the needs and the background of your audience will help you to choose appropriate words and details. A set of instructions accompanying a chemistry set would use different terminology, abbreviations, and level of detail than would a set of instructions a professor gives a class in organic chemistry.

> General Audience: Place 8 drops of vinegar in a test tube with a piece of limestone about the size of a pea.

> Specialized Audience: Place 8 gtts of CH_3COOH in a test tube and add 1 mg of CO_3.

The instructions for the general audience avoid the technical abbreviations and symbols the specialized audience requires. If your readers are puzzled by your directions, you defeat your reasons for writing them.

Using the Right Style

To write instructions that readers can understand and turn into effective action, observe the following guidelines.

1. Use verbs in the present tense and imperative mood. Imperatives are commands that have deleted the pronoun *you*. Note how the instructions in Figures 12.1 through 12.4 contain imperatives—"Move the arrow" instead of "You move the arrow." In instructions, deleting the *you* is not discourteous, as it would be in a business letter or report. The command tells readers, "These steps work, so do them exactly as stated." Choose imperative (action) verbs such as those listed in Table 12.1.

2. Write clear, short sentences in the active voice. Keep sentences short and uncomplicated. Sentences under twenty words (preferably under fifteen) are easy to read. Note that the sentences in Figures 12.1 through 12.4 are, for the most part, under fifteen words.

3. Use precise terms for measurements, distances, and times. Indefinite, vague directions leave users wondering whether they are doing the right thing. Avoid vague words such as *frequently*, *occasionally*, *probably*, and *possibly*. The following vague direction is better expressed through precise revision.

    Vague: Turn the distributor cap a little. (*How much is a little?*)

    Precise: Turn the distributor cap one quarter of a rotation.

TABLE 12.1   Some Helpful Imperative Verbs Used in Instructions

add	determine	include	paste	rotate	tie
adjust	dig	increase	peel	rub	tighten
apply	display	insert	pick up	run	tilt
attach	double-click	inspect	plug	save	trace
blow	download	install	point	scan	transect
call up	drag	lift	pour	scroll	transfer
change	drain	load	press	select	trim
check	drill	log on	prevent	set	turn
choose	drop	loosen	print	send	twist
clean	ease	lower	provide	shift	type
click	eliminate	lubricate	pull	shut off	unplug
clip	enter	measure	push	slide	use
close	exit	mix	raise	slip	ventilate
connect	find	mount	re-boot	spread	verify
contact	flip	move	release	squeeze	wash
copy	flush	navigate	remove	start	weigh
cover	forward	notify	replace	strain	wind
create	gather	oil	reply	switch	wipe
cut	group	open	review	tear	wire
delete	hold	pass	roll	thread	wrap

4. Use connective words as signposts. Connective words specify the exact order in which something is to be done (especially when your instructions are written in paragraphs). Words such as *first*, *then*, *before* help readers stay on course, reinforcing the sequence of the procedures.

5. Number each step when you present your instructions in a list. You also can use bullets. Plenty of white space between steps also distinctly separates them for the reader.

Using Visuals Effectively

Readers welcome visuals in almost any set of instructions. Visuals help readers get a job done more quickly and increase their confidence. A visual can help users

- locate parts, areas, and so on
- identify the size and placement of parts
- show how to assemble parts effectively and easily
- illustrate the right and wrong way to do something
- identify possible sources of danger, injury, or malfunction
- determine whether a problem is serious or minor
- recognize normal and safe limits/ranges

The number and kinds of visuals you include will, of course, depend on the process or equipment you are explaining and your audience's background and needs. Some instructions may require only one or two visuals. The instructions in Figure 12.3 show users how to copy files using a USB Flash. The illustration clearly shows what a USB Flash looks like. In Figure 12.2 each step is accompanied by a visual demonstrating a proper technique of brushing teeth.

Another frequently used visual in instructions is an exploded drawing, like the one in Figure 12.4, which helps consumers see how the various parts of the grill fit together, or the one in Figure 12.7, which labels and shows the relationship of the parts of an industrial extension cord.

Guidelines for Using Visuals in Instructions

Follow these guidelines to use visuals effectively in your instructions:

1. Set visuals off with white space so it is easy to find and read them.
2. Place each visual next to the step it illustrates, not on another page or buried at the bottom of the page.
3. Use a visual that is appropriate for your audience (e.g., a photo showing a person doing a proper stretch for a general audience rather than an elaborate medical illustration of the muscular system).
4. Assign each visual a number (Figure 1, Figure 2) and refer to visuals by figure number in your instructions.
5. Make sure the visual looks like the object the user is trying to assemble, maintain, run, or repair. Don't use a photo of a different model.

Exploded drawing showing how to assemble an industrial extension cord.   Figure 12.7

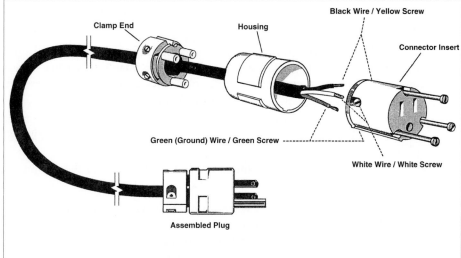

1. Run the end of the cord through the clamp end and then through the center hole of the housing.
2. Pull the cord through until it extends 2 inches beyond the housing.
3. Strip about 1¼ inches of outer insulation from the end of the cord.
4. Twist the exposed ends to prevent stray strands.

Source: Drawing courtesy of Sally Eddy and Georgia-Pacific Company.

6. Always inform readers if a part is missing or reduced in your visual.
7. Where necessary, label or number parts of the visual, as in Figure 12.4.

The Five Parts of Instructions

Except for very short instructions, such as those illustrated in Figures 12.1 through 12.4, or for instructions on policies and regulations (see Figure 12.11, pp. 556–557), a set of instructions generally contains five main parts: (1) an introduction; (2) a list of equipment and materials; (3) the actual steps to perform the process; (4) warnings, cautions, and notes; and (5) a conclusion (when necessary).

Introduction

The function of an introduction is to provide readers with enough *necessary* background information to understand why and how your instructions work. An introduction must make readers feel comfortable and well prepared before they turn to the actual steps.

What to Include in an Introduction

You can do one or all of the following in your introduction. Not every introduction to a set of instructions will contain all four categories of information listed here.

Visit www .cengage.com/ english/kolin for an online exercise, "Rewriting Online Instructions."

Some instructions will require less detail. You will have to judge how much background information to give readers for the specific instructions you write.

Refer to Figure 12.8 (p. 541), which is an introduction to a guide for nurses who have to know how to use an infusion pump.

1. State why the instructions are useful for a specific audience. Many instructions begin with introductions that stress safety, educational, or occupational benefits. Here is an introduction from a set of safety instructions describing protective lockout of equipment.

> The purpose of these instructions is to provide plant electrical technicians with a uniform method of locking out machinery or equipment. This will prevent the possibility of setting moving parts in motion, energizing electrical lines; or opening valves while repair, setup, or cleaning work is in progress.

Note how Figure 12.8 highlights the safety and convenience of the equipment for the nursing staff to meet their patients' needs.

2. Indicate how a particular piece of equipment or process works. An introduction can briefly discuss the "theory of operation" to help readers understand why something works the way your instructions say it should. Such a discussion sometimes describes a scientific law or principle. An introduction to instructions on how to run an autoclave begins by explaining the function of the machine: "These instructions will teach you how to operate an autoclave, which is used to sterilize surgical instruments through the live additive-free stream."

The introduction in Figure 12.8 describes the different modes of the infusion pump and the pump itself.

3. Point out any safety measures or precautions a reader may need to be aware of. By alerting readers early in your instructions, you help them to perform the process much more safely and efficiently. The introduction in Figure 12.8 cautions the nursing staff about an audible alarm signal in the event of a malfunction.

4. Stress any advantages or benefits the reader will gain by performing the instructions. Make the reader feel good about buying or using the product by explaining how it will make a job easier to perform, save the reader time and money, or allow the reader to accomplish a job with fewer mistakes or false starts. Nurses learn from the introduction in Figure 12.8 that the infusion pump is quickly programmed.

Note how the following introduction to a set of instructions on an auto dial/auto answer modem encourages the reader to learn how to operate this system.

Welcome to high-speed telecommunications and congratulations on choosing the Signalman EXPRESSi. You've made an excellent choice. The EXPRESSi is the ideal link between your computer and the ever-expanding world of information utilities, databases, electronic mail, bulletin boards, computer time sharing, and more.

The EXPRESSi can be used in the IBM Personal Computer. And because the EXPRESSi fits inside your PC, it saves valuable desk space and eliminates expensive, bulky cables.[1]

---

[1]Courtesy of Anchor Automation, Inc., Chatsworth, CA.

Introduction to a guide for using an infusion pump.   **Figure 12.8**

1 Overview Orientation

The LifeCare PROVIDER 5500 System is a portable infusion pump, specially designed to deliver analgesic drugs, antibiotics, and chemotherapeutics.

The pump can be programmed in either milligrams or cubic centimeters, and in four different delivery configurations for greater nursing convenience and to tailor precisely the most effective regimen for each patient.

Bolus Mode allows your patient to self-administer analgesia within programmed limits.

This is the traditional PCA delivery, "analgesia-on-demand," based on the patient's need.

Continuous Mode delivers a continuous "background" infusion with no additional PCA doses permitted.

Continuous-plus-Bolus Mode allows the patient to self-administer a Bolus dose in addition to receiving a simultaneous Continuous dose infusion.

Intermittent Mode delivers a specific dose (in cc or mg) at intermittent intervals over 24 hours.

You can also establish the "lockout" interval, the frequency with which a patient may receive a Bolus dose of analgesic drug.

The PROVIDER 5500 System records all settings in memory and can be quickly re-programmed to save nursing time when repeating established protocols or changing fluid reservoirs.

The portable system operates on battery power.

To minimize tampering and discourage theft, there is an optional locking security lockbox that also secures the system to an IV pole.

The audible alarm signals in the event of a malfunction, and the digital readout describes the malfunction.

● Compact and lightweight.

● Delivery rates between 0.1 cc and 250 cc per hour, in 0.1-cc increments.

Display Panel

● Individual display indicators appear only during programming and operation.

● Only on a *selective* basis.

● Tone sounds when activated.

● Runs on BATTERY POWER ONLY.

● Disposable Primary IV set with integral infusion cartridge.

Gives function of equipment

Saves time

Offers security

Describes different modes or options

Explains convenience features

Emphasizes safety features

Source: Reprinted by permission of Abbott Laboratories Hospital Products Division.

List of Equipment and Materials

Clearly, some instructions, as in Figure 12.1, do not need to inform readers about equipment or materials they will need. But when a list is necessary, make it complete and clear. Do not wait until the readers are actually performing one of the steps to tell them that a certain type of drill or a specific kind of chemical is required. They may have to stop what they are doing to find the equipment or material; moreover, the process may fail or present hazards if users do not have the right equipment at the right time. For example, if a Phillips screwdriver is essential to complete one step, specify that type of screwdriver under the heading "Equipment and Materials"; do not list just "screwdriver." See "Tools Required" in Figure 12.4.

Steps for Your Instructions

The heart of your instructions will consist of clearly distinguished steps that readers must follow to achieve the desired results. Figure 12.10 (pp. 547–553) contains a model set of steps on how to set up a printer. Note how each step is precisely keyed to the visual, further helping readers perform the procedure. Refer to Figure 12.10 as you study this section.

Guidelines for Writing Steps

To help your readers understand your steps, observe the following rules:

1. Put the steps in their correct order and number them. If a step is out of order or is missing, the entire set of instructions can be wrong or, worse yet, dangerous. Double-check every step and number each step to indicate its correct place in the sequence of events you are describing.

2. Put only the right amount of information in each step. Make each step short and simple. Giving readers too little information can be as risky as giving them too much. Keep in mind that each step asks readers to perform a single task in the entire process.

3. Group closely related activities into one step. Sometimes closely related actions belong in one step to help the reader coordinate activities and to emphasize their being done at the same time, in the same place, or with the same equipment. See how step 4 accomplishes this in Figure 12.9.

Instructions on how to use a fax machine are clearer when distinct steps are stated separately. The first set of instructions below incorrectly tells users how to transmit a fax by combining steps that must be performed separately. Step 2 asks users to pick up the phone and then dial the number—two separate actions. Step 3 asks users to press the button and return the handset, again two actions that cannot be performed simultaneously.

Incorrect: 1. Load the paper into the outgoing document slot, adjusting the paper guides to the appropriate width.
2. Pick up the telephone handset and listen for a dial tone. When you hear the dial tone, dial the number of the receiving fax machine.
3. When the receiving fax machine answers the ring, press the start button. After the transmission is completed, return the handset to its cradle.

Correct: 1. Load the paper into the outgoing document slot, adjusting the paper guides to the appropriate width.
2. Pick up the telephone handset and listen for a dial tone.

Instructions (with visual) on how to connect a computer monitor.   **Figure 12.9**

Connecting the Monitor to the Computer

1   Make sure the monitor and computer are turned off. (See previous section on safety.)

Steps clearly numbered and keyed to visual below

2   Connect the power cord to the back of the display.

3   Plug the other end of the cable into a grounded outlet.

4   Connect the video cable on the monitor to the 15-pin video graphics connector on the rear panel of the computer, and tighten the fastening screws. (If you have an HP Pavilion computer, this port is marked in orange. For other computers, check your computer manual for the video port location.)

Describes what to look for

Note: *Don't force the cable into the connector; line it up carefully so you don't bend the pins.*

Gives readers a helpful hint

5   Connect the microphone cable to the computer's sound input. On HP computers, this port is yellow. The end of the cable is also yellow.

6   Connect the microphone cable to the monitor. The connector and the cable end are both yellow.

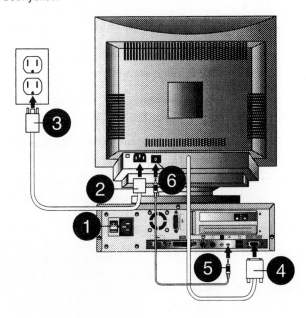

Visual placed on same page as directions to help reader follow steps

   3. When you hear a dial tone, dial the number of the receiving fax machine.

   4. When the receiving fax machine answers the ring, press the start button.

   5. After the transmission is completed, return the handset to its cradle.

Don't divide an action into two steps if it has to be done in one. For example, instructions showing how to light a furnace would not list as two steps actions that must be performed simultaneously to avoid a possible explosion.

Incorrect: 1. Depress the lighting valve.
            2. Hold a match to the pilot light.
Correct: 1. Depress the lighting valve while holding a match to the pilot light.

Similarly, do not separate two steps of a computer command that must be performed simultaneously.

Incorrect: 1. Press the CONTROL key.
            2. Press the ALT key.
Correct: 1. While holding down the CONTROL key, press the ALT key.

4. Give the reader hints on how best to accomplish the procedure. Obviously, you cannot do that for every step, but if there is a chance that the reader might run into difficulties you should provide assistance. Particular techniques on how to operate or service equipment also help readers: "If there is blood on the transducer diaphragm, dip the transducer in a blood solvent, such as hydrogen peroxide, Hemosol, etc." If readers have a choice of materials or procedures in a given step, you might want to list those that would give the best performance: "Several thin coats will give a better finish than one heavy coat."

5. State whether one step directly influences (or jeopardizes) the outcome of another. Because all steps in a set of instructions are interrelated, you could not (and should not have to) tell readers how every step affects another. But stating specific relationships is particularly helpful when dangerous or highly intricate operations are involved. You will save the reader time, and you will stress the need for care. Forewarned is forearmed. Here is an example.

Step 2: Tighten fan belt. Failure to tighten the fan belt will cause it to loosen and come off when the lever is turned in step 5.

Do not wait until step 5 to tell readers that you hope they did a good job in tightening the fan belt in step 2. Information that comes after the fact is not helpful.

6. Where necessary, insert graphics to assist readers in carrying out the step. For example, see the drawings of the printer in Figure 12.10 (pp. 547–553).

Warnings, Cautions, and Notes

At appropriate places in the steps of your instructions you may have to stop the reader to issue a warning, a caution, or a note. Note the examples on the following page as well as those in Figure 12.10, especially in step 4.

Warnings

A warning tells readers that a step, if not prepared for or performed properly, could endanger their safety or even their lives, as here. Note that translations ensure readers' safety.

WARNING: UNPLUG MACHINE BEFORE REMOVING PLATEN GLASS.

ADVERTENCIA: DESENCHUFE LA MAQUINA ANTES DE QUITAR EL VIDRIO.

Spanish translation

Cautions

A caution tells a reader how to avoid a mistake that could damage equipment or cause the process to fail, for instance, "Do not force the plug."

 Caution: **Formatting erases all data on the disk**
小心： 格式化会删除磁盘的所有资料

Chinese translation

Notes

A note does not relate to the safety of the user or the equipment. It provides clarification or a helpful hint on how to do the step most efficiently.

 At 20 degrees F, a battery uses about 68 percent of its power.

Eksi 7 derecede, bir pil enerjisinin yaklasik yuzde 68 ini kullanir.

Turkish translation

Guidelines on Using Warnings, Cautions, and Notes

1. **Do not regard warnings and cautions as optional.** They are vital for legal and safety reasons to protect lives and property. In fact, you and your company can be sued if you fail to notify the users of your product or service of dangerous conditions that will or could result in injury or death.

2. **Put warnings and cautions in the right place.** Place them immediately before the step to which they pertain. If you insert a warning or caution statement too early, readers may forget it by the time they come to the step to which it applies. And if you put the notification too late, you almost certainly expose the reader to danger and the equipment to break down.

3. **Put warnings and cautions in a distinctive format.** Warnings and cautions should be graphically set apart from the rest of the instructions, as the triangle in the box at the top of the page illustrates. There should be no chance that readers will overlook them. Put such statements in capital letters, boldface type, boxes, different colors (red is especially effective for warnings if your readers are native speakers of English, but see p. 481; yellow is often used for caution). Be careful, though, about using colors for non-native speakers of English. Make sure your audience understands an icon or a symbol, such as a skull and

crossbones, an exclamation point inside a triangle, or a traffic stoplight, often used to signal a warning, a hazard, or some other unsafe condition.

4. **Include enough explanation to help readers know what to watch out for and what precautions to take.** Do not just insert the word WARNING or CAUTION. Explain what the dangerous condition is and how to avoid it. Look at the examples of warnings and cautions in Figure 12.10.

5. **Do not include a warning or a caution just to emphasize a point.** Putting too many in your instructions will decrease the dramatic impact they should have on readers. Use them sparingly—only when absolutely necessary—so readers will not be tempted to ignore them.

6. **Use notes only when the procedure calls for them and they will help readers,** as in Figures 12.4 and 12.9.

Tech Note

Using Clip Art

You can use clip art to draw readers' attention to warnings, cautions, and notes. You can find a wide range of clip art with any graphics software package you use. Once you select a piece of clip art, it is easy to paste it into any word processing document. But choose your clip art carefully; use clip art images only when they are functional and unambiguous for your intended audience, including international audiences (see pp. 470–472).

Conclusion

Not every set of instructions requires a conclusion. For short instructions containing a few simple steps, such as those in Figures 12.1 through 12.4, no conclusion is necessary. These instructions usefully end with the last step the reader must perform. For longer, more involved jobs, a conclusion can help readers finish the job with confidence and accuracy.

When they are necessary, conclusions can provide a succinct wrap-up of what the reader has done or end with a single sentence of congratulations, or reassure readers as the conclusion in Cliff Burgess's memo (Figure 12.6) does. A conclusion might also tell readers what to expect once a job is finished, describe the results of a test, or explain how a piece of equipment is supposed to operate.

Model of Full Set of Instructions

Study Figure 12.10, which is a set of instructions on setting up an Epson printer that includes the parts discussed in this chapter: an introduction; a list of materials; numbered steps; and warnings, cautions, and notes. Pay special attention to how the writer coordinates words with visuals to assist readers.

Complete set of instructions with steps, visuals, cautions, notes, warnings.   **Figure 12.10**

Quick Setup

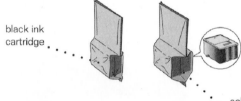

Here's how to set up your new EPSON® printer . . .

1 Unpack the Printer

Remove any packing material from the printer, as described on the Notice Sheet in the box. Save all the packaging so you can use it if you need to transport the printer later. You'll find these items inside:

black ink
cartridge

color ink
cartridge

paper
support

Appropriately brief introduction for this straightforward procedure

Each step begins with imperative verb

List (with visuals) of materials

Exploded drawing assists readers to identify/ assemble parts

Continued

Figure 12.10   (Continued)

Place the printer flat on a stable desk near a grounded outlet. Leave plenty of room in back for the cables and enough room in front for opening the output tray.

Do NOT put the printer:

Bulleted list highlights importance of correct printer placement

- In an area with high temperature or humidity
- In direct sunlight or dusty conditions
- Near sources of heat or electromagnetic interference, such as loudspeakers or cordless telephone base units.

Also, be sure to follow the Safety Instructions in the Introduction of your *User's Guide*.

2 Attach the Paper Support

Insert the paper support in the top slot on the back of the printer.

Groups clearly distinguished steps in the procedure

3 Plug In the Printer

First make sure the power is off. Check the ⏻ power button; it's off when its surface is raised above the printer surface.

Uses appropriate icon to alert reader to possible damage

Caution:
Do not plug the printer into an outlet controlled by a wall switch or timer, or on the same circuit as a large appliance. This may disrupt the power, which can erase memory and damage the power supply.

power

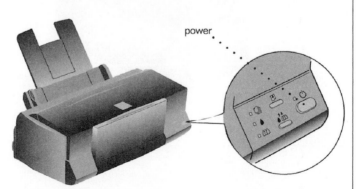

Plug the power cord into a properly grounded outlet.

(Continued) **Figure 12.10**

Install the Ink Cartridges

1. Lower the output tray and raise the printer cover.

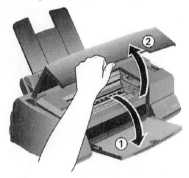

Directional arrows visualize correct movement of parts

2. Press the ⏻ power button to turn on the printer. The ⏻ power light flashes, the ♦ and 🖤 ink out lights come on, and the ink cartridge holders move to the installation position.

3. Pull up the ink cartridge clamps.

Visual clues reinforce right way to perform procedure

Caution:
You must remove the tape seal from the top of the cartridge or you will permanently damage it. Don't remove the tape seal from the bottom or ink will leak.

Warning:
If ink gets on your hands, wash them thoroughly with soap and water. If ink gets in your eyes, flush them immediately with water.

4. Open the ink cartridge packages. Remove the disposable yellow portion of the tape seal on top.

Caution and warning notices inserted in right places with distinctive icons

black ink cartridge . . . . .

color ink cartridge

Continued

Figure 12.10 (Continued)

Callouts make parts easy to find

5. Lower the ink cartridges into their holders with the labels face up and the arrows pointing toward the back of the printer.

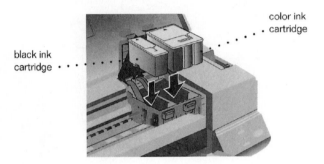

color ink cartridge

black ink cartridge

Generous white space makes steps easy to distinguish and follow

6. Push down the clamps until they lock in place.

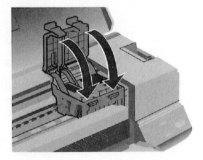

Appropriate icon alerting reader of a possible problem

Caution:
Never turn off the printer when the ⏻ power light is flashing.

7. Press the cleaning button to return the print heads to their home position and charge the ink delivery system. Charging can take up to five minutes, with the ⏻ power light flashing until it's finished.

8. Close the printer cover.

(Continued)   Figure 12.10

5 Load the Paper

1. Slide the left edge guide all the way left and **pull** out the output tray extension.

2. Fan a stack of plain paper and then even the edges.

3. Load the stack with the printable surface face up. Push the paper against the right edge guide.

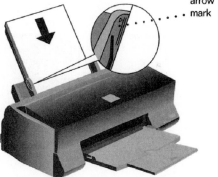

arrow
mark

Includes two visuals, one of them an exploded drawing, helpfully inserted between the steps to which they apply

Offers clear-cut directions on how to load paper

Note:
Don't load paper above the arrow mark inside the left edge guide.

Helpful information to ensure best use of equipment

4. Slide the left edge guide back against the stack of paper.

Continued

Figure 12.10   (Continued)

Check the Printer

1. Turn off the printer.

2. While holding down the load/eject button, turn on the printer. Then release the buttons.

3. A page prints out showing the ROM version and a nozzle check pattern. When it's finished, turn off the printer. If you have any problems with the test, see Chapter 6 in your *User's Guide* for more information.

Provides a useful source for further information on performing step correctly

Inserts a brief introduction that alerts users to follow one of two procedures, depending on the type of computer they have

Connect the Printer to Your Computer

You can connect your EPSON Stylus™ COLOR 600 to either an IBM® compatible PC or an Apple® Macintosh® You'll need a shielded, twisted-pair parallel cable to connect to a PC or an Apple System Peripheral-8 cable to connect to a Macintosh. For a complete list of system requirements, see the Introduction of your *User's Guide*.

Connecting to a PC

1. Turn off the printer and your computer.

2. Connect the cable to the printer's parallel interface; then squeeze the wire clips together until they lock in place. (If your cable has a ground wire, connect it now.)

Chooses informative icon for a note

Note:
The printer is assigned to parallel port LPT1; if you want to use a different port, see your Windows documentation for instructions.

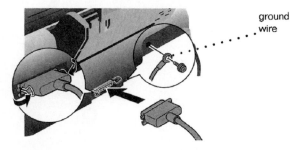

ground wire

Enlarged view of parts helps readers see step

3. Connect the other end of the cable to your computer's parallel port and secure it as necessary.

Connecting to a Macintosh

1. Turn off the printer and your Macintosh.

2. Connect one end of the cable to the serial connector on the back of the printer.

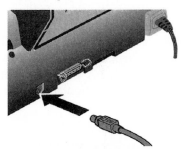

Visual identifies precise location for connecting parts

Note:
If you're using a PowerBook™, connect your printer to the modem port.

3. Connect the other end of the cable to either the modem port ✆ or the printer port 🖶 on your Macintosh.

8 Install the Printer Software

Now you need to install the printer software so you can control printing from your computer.

Functions as both a step—installing software—and a wrap-up

Installing on a PC

You can install the printer software for Windows 95 or Windows 3.1 from the EPSON printer software CD-ROM. If you don't have a CD-ROM drive, you can install the software using the EPSON printer software diskettes.

Installing from the CD-ROM

In addition to the printer driver and utilities, the CD-ROM contains EPSON Answers, a comprehensive online guide that includes:

▶ **How To** for step-by-step printer operating instructions
▶ **Color Guide** with practical color printing information
▶ **Problem Solver** to help you fix printer problems
▶ **Test Print** so you can check your print quality

Follow the instructions inside the CD-ROM case to install the software. To run EPSON Answers, click on its icon in the EPSON program group or folder.

Provides information for further troubleshooting and contact with manufacturer

EPSON ANSWERS
If you have a CD-ROM drive, you can run EPSON Answers, the on-screen guide to your new printer. It puts you on the right track quickly and easily.

Source: Reprinted by permission of Epson America.

▌Writing Procedures for Policies and Regulations

Up to this point, we have concentrated primarily on instructions dealing with how to put things together; how to install, repair, or use equipment; and how to alert readers to mechanical or even personal danger(s). But there is another similar type of writing that deals with guidelines for the world of work: procedures. These concern policies and regulations found in employee handbooks and other internal corporate communications, such as on websites, in memos, or in e-mail messages. (Note, however, that some companies do not disseminate policy via e-mail because it is perceived as less formal than hard copy and is not deletable.) Figure 12.11 shows an example of a company policy on flextime. Also review Figure 1.1, which shows a company procedure about how employees can be reimbursed for job-related writing courses.

Some Examples of Procedures

Procedures deal with a wide range of "how-to" activities within an organization, including the following:

- accessing a company file or database
- preparing for an audit
- applying for family or medical leave
- dressing professionally at work
- routing information
- reserving a company vehicle or facility
- taking advantage of telecommuting options
- filing a work-related grievance
- requesting travel expense reimbursement
- seeking safety in case of a fire, flood, or other emergency
- fulfilling promotion requirements

Policy procedures have a major impact on a company and its workers; they affect schedules, payrolls, acceptable and unacceptable behaviors at work, and a range of protocols governing the way an organization does business internally and externally. Procedures help an organization run smoothly and consistently. Adhering to them, every employee follows the same regulations and standards.

Meeting the Needs of Your Marketplace

As with instructions, you will have to plan carefully for writing procedures. A mistake in business procedures can be as wide ranging and as costly as an error in a set of assembly instructions—perhaps more so since poorly written procedures can land a company and/or its employees in significant legal trouble.

To avoid such difficulties, spell out precisely what is expected of employees—how, when, where, and why they are to perform or adhere to a certain policy. Use the same strategies as for instructions discussed earlier in this chapter. Leave no chance for misunderstanding or ambiguity; be unqualifyingly straightforward and clear-cut. Determine what information employees need to comply with all the regulations you are specifying and supply it.

Many times procedure statements involve a change in the work environment. Help readers by including, whenever necessary, definitions, headings, some prefatory explanations, and an offer to help employees with any questions they may have. Always present a copy of the procedures to management to approve or to revise before sending them to employees.

Figure 12.11 shows a memo from a human resources director notifying employees how they can take advantage of a new flextime arrangement at work, including what they can and cannot do within the framework of flextime. Note how the writer divides her procedures document into an introduction explaining when flextime will go into effect and what choices employees have about it, a section that usefully defines and delineates the concept of flextime, and finally a section containing specific guidelines. These guidelines, while not sequential, function as a series of steps that employees have to follow, and ultimately these steps will affect the entire organization.

The various regulations about what employees cannot do in flextime might be seen as the equivalents of warning and caution statements discussed in "Warnings, Cautions, and Notes" (pp. 545–546). Observe, too, how the writer does not veer off to discuss benefits to the employer or to examine where and how often flextime has been used elsewhere. Finally, this example of procedural writing protects the employer legally by establishing the policies and rules by which an employee's scheduled work time is clearly defined, delineated, and evaluated.

Some Final Advice

Perhaps the most important piece of advice to leave you with is this: Do not take anything for granted when you have to write a set of instructions or procedures. It is wrong and on occasion dangerous to assume that your readers have performed the process before, that they will automatically supply missing or "obvious" information, or that they will easily anticipate your next step. No one ever complained that a set of instructions was too clear or too easy to follow.

Figure 12.11 Instructions on a new scheduling policy.

WebTech, Inc. **www.webtech.com**
4300 Ames Boulevard, Gunderson, CO 81230-0999 (303) 555-9721
FAX (303) 555-9876

TO: All Employees
FROM: Tequina Bowers T.B.
Human Resources

DATE: March 20, 2009
SUBJECT: Opting for a Flextime Schedule

Notifies readers of new policy and states purpose of the memo

Effective sixty days from now, on May 10, 2009, employees will have the opportunity to go to a flextime schedule or to remain on their current 8-hour fixed schedule. This memo explains the flextime option and establishes the guidelines and rules you must follow if you choose this new schedule.

Flextime Defined

Spells out precisely how company defines flextime—uses boldface for most important information

Flextime is based on a certain number of **core hours** and **flexible hours**. Our company will be open twelve hours, from 6:00 a.m. to 6:00 p.m. weekdays, to accommodate both fixed and flextime arrangements. During this 12-hour period, all employees on flextime will be expected to work **eight and one-half consecutive hours,** including a half-hour for lunch. All employees must work a **common core time from 10:00 a.m. to 4:00 p.m.** but flextime employees will be free to choose their own starting and quitting times. For instance, they might elect to arrive at 8:00 a.m. and leave at 4:30 p.m., or they may want to start at 9:30 a.m. and leave at 6:00 p.m.

Flextime Guidelines and Rules

Stresses employee responsibilities and consequences of violating rules

Employees need to understand their individual responsibilities and adjust their schedules accordingly. All employees must adhere strictly to the following regulations and realize that flextime schedule privileges can be revoked for violations.

2.

What Flextime Employees **Must Do**

(1) Be present during core time, arriving and leaving the plant during their flexible work hours.
(2) Observe a minimum unpaid half-hour lunch break each working day.
(3) Cooperate with their supervisors to make sure coverage is provided for their departments from 6:00 a.m. to 6:00 p.m.
(4) Notify supervisors of absences.
(5) Attend monthly corporate meetings even though such meetings may be outside their chosen flextime schedules.
(6) Adhere to dress code during any time they are at work, regardless of their flextimes.
(7) Agree to work on a flextime schedule for a 6-month period.

Carefully outlines what is and is not acceptable according to new policy

Numbered points make policy easier to understand, follow, and refer to in the future

What Flextime Employees **Can't Do**

(1) Be tardy during core time.
(2) Switch, bank, borrow, or trade flextime hours with other employees without the approval of an immediate supervisor.
(3) File for overtime without a supervisor's approval.
(4) Self-schedule a vacation by expanding flextime hours.
(5) Switch back and forth between fixed time and flextime.

Stipulates what new policy will not allow in separate section of memo

How Do You **Sign Up** for Flextime?

If you wish to begin a flextime schedule, first you need to obtain and complete a transfer of hours form from your supervisor. Next, you need to bring that signed form to Human Resources (Admin. 201) to participate officially in this program.

I will be happy to talk to you about this new option and to answer any of your questions. Please call me at ext. 5121, e-mail me at **tbowers@webtech.com**, or come by my office in Admin. 201. Thank you.

Explains steps to begin flextime

Encourages feedback and questions

Revision Checklist

☐ Analyzed my intended audience's background, especially why and how they will use my instructions.

☐ Tested my instructions to make sure they include all necessary steps in their proper sequence.

☐ Made sure all measurements, distances, times, and relationships are precise and correct.

☐ Avoided technical terms if my audience is not a group of specialists in my field.

☐ Used the imperative mood throughout my instructions.

☐ Wrote clear, short sentences.

☐ Chose effective visuals, labeled them, and placed them next to the step(s) to which they apply.

☐ Made my introduction proportionate to the length and complexity of my instructions and suitable for my readers' needs.

☐ Included necessary background, safety, and operational information in the introduction.

☐ Provided a complete list of tools and materials my audience needs to carry out the instructions.

☐ Put instructions in easy-to-follow steps and in the right order.

☐ Used numbers or bullets to label the steps and inserted connective words to indicate order.

☐ Used warnings, cautions, and notes where necessary and in a form that makes them easily seen and read.

☐ Supplied a conclusion that summarizes what readers should have done or reassures them that they have completed the job satisfactorily.

☐ Spelled out clearly responsibilities, benefits, restrictions, and consequences of any procedures for readers.

☐ Defined any terms readers may be unfamiliar with in procedures for policies and regulations.

☐ Gave a copy of procedures to administrators for their approval before distributing to employees.

Exercises

1. Find a set of instructions that does not contain any visuals, but that you think should have some graphic material to make it clearer. Design those visuals yourself and indicate where they should appear in the instructions.

2. From a technical manual in your field or in an owner's manual, locate a set of instructions that you think is poorly written and illustrated. In a memo to your

instructor, explain why the instructions are unclear, confusing, or badly format-ted. Then revise the instructions to make them easier for the reader to carry out. Submit the original instructions with your revision.

3. Write a set of instructions in numbered steps (or in paragraph format) on one of the following relatively simple activities.
   a. tying a shoe
   b. using an ATM
   c. unlocking a door with a key
   d. sending a text file from a cell phone
   e. planting a tree or a shrub
   f. sewing a button on a shirt
   g. removing a stain from clothing
   h. pumping gas into a car
   i. creating a blog
   j. checking a book out of the library
   k. polishing a floor
   l. shifting gears in a car
   m. photocopying a page from a book

4. Write an appropriate introduction and conclusion for the set of instructions you wrote for Exercise 3.

5. Write a set of full instructions on one of the following more complex topics. Identify your audience. Include an appropriate introduction; a list of equip-ment and materials; numbered steps with necessary warnings, cautions, and notes; and an effective conclusion. Also include whatever visuals you think will help your audience.
   a. scanning a document
   b. changing a flat tire
   c. testing chlorine in a swimming pool
   d. shaving a patient for surgery
   e. changing the oil and oil filter in a car
   f. surveying a parcel of land
   g. pruning hedges
   h. jumping a dead car battery
   i. using the Heimlich maneuver to help a choking individual
   j. filleting a fish
   k. creating a logo for a letterhead
   l. taking someone's blood pressure
   m. downloading a homepage from the Web
   n. painting a car
   o. cooking a roast
   p. flossing a patient's teeth after cleaning
   q. creating a computer file

6. The following set of instructions is confusing, vague, and out of order. Rewrite the instructions to make them clear, easy to follow, and correct. Make sure that each step follows the guidelines outlined in this chapter.

Reupholstering a Piece of Furniture

(1) Although it might be difficult to match the worn material with the new material, you might as well try.

(2) If you cannot, remove the old material.

(3) Take out the padding.

(4) Take out all of the tacks before removing the old covering. You might want to save the old covering.

(5) Measure the new material with the old, if you are able to.

(6) Check the frame, springs, webbing, and padding.

(7) Put the new material over the old.

(8) Check to see if it matches.

(9) You must have the same size as before.

(10) Look at the padding inside. If it is lumpy, smooth it out.

(11) You will need to tack all the sides down. Space your tacks a good distance apart.

(12) When you spot wrinkles, remove the tacks.

(13) Caution: in step 11 directly above, do not drive your tacks all the way through. Leave some room.

(14) Work from the center to the edge in step 11 above.

(15) Put the new material over the old furniture.

P.S. Use strong cords whenever there are tacks. Put the cords under the nails so that they hold.

---

7. Write a set of procedures, similar to Figure 12.11, on one of the following policies or regulations at your school or job site:
   a. offering quality customer service over the phone or via the Web
   b. filing a claim for a personal injury on the job
   c. designing an employee's personal space—what is and is not allowed?
   d. using the Internet at work for personal use
   e. ensuring confidentiality at work (to practice professional ethics in the workplace)
   f. enrolling in mandatory courses to maintain a license or certificate
   g. playing music in the workplace
   h. going through an orientation procedure to begin a new job
   i. following an acceptable dress code
   j. allowing tattoos and body piercings in the workplace
   k. registering a domain name for a sponsored group at work or school
   l. receiving reimbursement for carpooling, taking public transportation, or riding a bicycle to work

CHAPTER 13

Writing Winning Proposals

A proposal is a detailed plan of action that a writer submits to a reader or group of readers for approval. The readers are usually in a position of authority—supervisors, managers, department heads, boards of private foundations, military or civic leaders—to endorse or reject the writer's plan. Proposals are among the most important types of job-related writing. Their acceptance can lead to improved working conditions, better use of technology, a more efficient and economical business, additional jobs and business for a company, or a safer environment.

Every proposal you write must exhibit a "can do" attitude, putting the reader and his or her company's needs at the center of your work. As you go through this chapter, keep in mind the slogan of Yates Engineering that has won millions of dollars of business through its reader-centered proposals: "On time . . . within budget . . . to your satisfaction." Time, budget, and your readers' satisfaction and convenience are among the most important ingredients of a winning proposal. Notice how the advertisement in Figure 13.1 appeals to customers' desires to have a variety of security services—from motion detection systems to video recordings to well-trained officers.

Visit www
.cengage.com/
english/kolin for
this chapter's
online exercises,
ACE quizzes, and
Web links.

Writing Successful Proposals

Proposals are written for many purposes and many different audiences; for example:

- to your boss, seeking authorization to hire staff, change a procedure, or to purchase a new piece of equipment for the office (as Marcus Weekley did in Figure 2.5, requesting that his firm upgrade their printers).
- to potential customers, offering a product or a service (such as offering to supply a fire chief with special firefighting gear, or offering an office manager a line of ergonomically designed furniture).
- to a government agency, such as the Department of the Interior, seeking funds to conduct a research project (for instance, requesting money to study the mating and feeding habits of a particular species, or asking for funds to discover ways to detect environmental hazards more quickly).

■ to foundations to raise funds for a nonprofit organization (for example, a small art gallery requesting funding from an arts foundation).

Characteristics of Proposals

Proposals are characterized by three things, which are discussed below:

1. They vary in size and scope.
2. They are persuasive plans.
3. They frequently are collaborative efforts.

Proposals Vary in Size and Scope

Depending on the job, proposals can vary greatly in size and in scope. A proposal to your employer could easily be conveyed in a page or two. To propose doing a small job for a prospective client—for instance, redecorating a waiting room in an accountant's office—a letter with information on costs, materials, and a timetable might suffice. The sales letters in Figures 6.3 and 6.4 (pp. 202 and 204) illustrate short proposals in letter format. But an extremely large and costly job—constructing a ten-story office building, for example—requires a detailed report hundreds of pages long with appendixes on engineering specifications, detailed budgets, and even résumés of all key personnel working on the project.

Figure 13.1   An example of a "can-do" attitude.

A discussion of long, elaborate proposals is beyond the scope of this chapter. But the principles and techniques of audience analysis, organization, and drafting that this chapter does cover apply equally well to any longer project you may have to prepare individually or as part of a team at work or school.

Proposals Are Persuasive Plans

Proposals, whether large or small, must be highly persuasive to succeed. Without your audience's approval, your plan will never go into effect, however accurate and important you think it is. Enthusiasm is not enough to persuade readers; you have to supply hard evidence based on research. Your proposal must convince readers that your plan will help them improve their businesses, make their jobs easier, save them money, enhance their image, improve customer satisfaction, or all of these.

Competition is fierce in the world of work, and a persuasive proposal frequently determines which company receives a contract. Demonstrate to your reader why your plan is better—more efficient, practical, economical—than a competitor's. In a sense, a proposal combines the persuasiveness of a sales letter (see Chapter 6), the documentation of a report (see Chapters 8 and 15), and the binding power of a contract, because if the reader accepts your proposal, he or she will expect you to live up to its terms to the letter.

A proposal is an argument (a plan) you must convince your reader to accept. You cannot write a successful proposal until you

1. fully understand your audience's needs/problems and why solving them is important to your audience's ongoing business
2. formulate a careful, detailed plan to solve these needs/problems
3. prove beyond doubt that you have the logic, time, technology, and personnel to solve the audience's precise problem
4. match your timetable and budget with your reader's

These four goals are not only vital to your persuasive plan but, when incorporated within a proposal that your reader accepts, become part of a legally binding agreement.

Proposals Frequently Are Collaborative Efforts

Like many other types of business and technical writing, proposals often are the product of teamwork and sharing. Even a short in-house proposal is often researched and put together by more than one individual in the company or agency.

Often, individual employees will pull together information from their separate areas (such as graphics and design, finance, marketing, sales, transportation, and even legal) and put it into a proposal that each team member then reads and revises until the team agrees that the document is ready to be released.

■ Types of Proposals

Proposals are classified according to how they originate and where they are sent after they are written. Distinctions are made between *solicited* and *unsolicited* proposals based on how they originate and between *internal* and *external* proposals based on where they are sent. Depending on your audience and your purpose, you may write an internal solicited or unsolicited proposal, or you may write an external solicited or unsolicited proposal. Solicited proposals often involve requests for proposals, or RFPs. Figure 13.2 provides a visual representation of the various types of proposals.

Figure 13.2   Types of proposals.

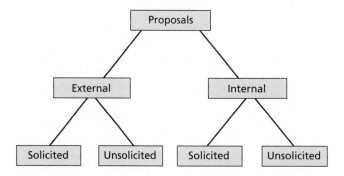

Solicited Proposals and Requests for Proposals

When a company has a particular problem to be solved or a job to be done, it will solicit, or invite, proposals. The company will notify you and other competitors by preparing a **request for proposals (RFP)**, which is a set of instructions that specify the exact type of work to be done along with guidelines on how and when the company wants the work completed.

Some RFPs are long and full of legal requirements and conditions. Others, like the examples in Figures 13.3 and 13.4, are more concise. RFPs are sent to firms with track records in the area the company wants the work done. RFPs are also printed in trade publications and put on the Web (see Tech Note on page 570) to attract the highest number of qualified bidders for the job. The U.S. government publishes RFPs in the *Federal Register* or *Commerce Business Daily* (both are online), while private companies sometimes send their RFPs to *Business Daily*. No two RFPs are alike.

An RFP helps you to know what the customer wants. It is often extremely detailed and even tells you how the company wants the proposal prepared, for example, what information is to be included (on backgrounds, personnel, equipment,

A sample RFP for a smaller project.   Figure 13.3

REQUEST FOR PROPOSALS

Mesa Community College is soliciting proposals to construct and to install fifty individual study carrels in its Holmes Memorial Library. These carrels must be highly serviceable and conform to all specification standards of the American Library Association (ALA). Proposals should include the precise measurements of the carrels to be installed, the specific acoustical and lighting benefits, Internet access, and the types and amount of storage space offered. Work on constructing and installing the carrels must be completed no later than the start of the Fall Semester, August 21, 2009. Proposals should include a schedule of when different phases of work will be completed and an itemized budget for labor, materials, equipment, and necessary tests to ensure high-quality acoustical performance. Contractors should detail their qualifications, including a description of similar recent work and a list of references. Proposals should be submitted in triplicate no later than March 1, 2009, to:

Mrs. Barbara Feldstein-Archer
Director of the Library
Mesa Community College
Mesa, CO 80932-0617
BFeldstein-archer@Mesa.edu

---

budgets), where it needs to appear, and how many copies of the proposal you have to submit.

Your proposal will be evaluated according to how well you fulfill the terms of the RFP. For that reason, follow the directions in the RFP exactly. Note that the solicited proposal in Figure 13.5 (pp. 567–569) directly refers to the terms of the RFP. You should even use the language (specialized terms, specifically stated needs) of the RFP in your proposal to convince readers that you understand their requirements and to get them to accept your plan. If you have any questions, by all means call the agency or company so you do not waste your time or theirs by including irrelevant or unnecessary details in your proposal.

Unsolicited Proposals

With an unsolicited proposal, you—not the reader—make the first move. Unlike a solicited proposal, in which the company to which you are submitting the proposal knows about the problem, your unsolicited proposal has to convince readers that (1) there is a problem and (2) you and your firm are the ones to solve it.

Figure 13.4   A sample of part of an RFP for a larger project.

Fund for Rural America

The Fund for Rural America program supports competitively awarded research, extension, and education grants addressing key issues that contribute to economic diversification and sustainable development in rural areas. The amount available for support of this program is approximately $9,500,000. Approximately 15–20 awards will be made.

Program Description

Preservation of the economic viability of rural communities will be the focus of the Fund for Rural America. The program focuses attention **on rural communities' twin challenges of rural community innovation and demographic change.** Challenges of an aging population, the arrival of new immigrant populations, youth retention, and work force development all are critical issues that affect the rural economy. **Rural communities may propose research, education, and extension/outreach projects that will create an understanding of these demographic forces and develop capacity to turn these challenges into economic promise.**

Rural Community Innovation

This program area seeks research, education, and extension proposals that will **help rural Americans address existing and new problems in innovative ways. The goals of this program area are to generate new knowledge and transfer that knowledge to assist rural communities to diversify their economies, to develop and maintain profitable farms, firms, and businesses, to build community capacity, to aid smart growth, to protect natural resources, and to increase family economic security.**

Projects are encouraged to combine multiple strategies for innovation such as the following:

- community implications of moving to a more bio-based economy (carbon sequestration credits, promoting locally based bio-based industries, linking bio-based materials to a diverse agriculture and community);
- impacts of land use development options (farmland preservation, farming on the urban fringe, rural-urban land use issues);
- policies and institutions that increase profitability among small and minority farmers;
- e-commerce applications for remote rural areas and minority populations, community information networks to support e-commerce and e-communities, and adapting e-commerce strategies for e-community planning and social capital development;
- network capabilities among producers, businesses, entrepreneurs, families, individuals, non-profit groups, community institutions, local government, and state and federal entities;
- nature of new generation of computer-based planning tools (geospatial analysis, information stores, economic and land use blueprints) and ability to apply and tailor them for place-sensitive development.

Doing that is not as difficult as it sounds. See how the writers of the unsolicited proposal in Figure 13.6 on pages 574–577 identify a relevant problem for their readers. If your readers accept your identification of the problem, you have greatly increased the chances of their accepting your plan to solve it. Just remember that you will have to prove that solving the problem carries major benefits for your reader.

A proposal in response to a request from a company.   **Figure 13.5**

Reynolds Interiors • 250 Commerce Avenue S.W. • Portland, OR 97204-2129

January 21, 2009

Mr. Floyd Tompkins, Manager
General Purpose Appliances
Highway 11 South
Portland, OR 97222

Dear Mr. Tompkins:

In response to your request listed on your website for bids
for an appropriate floor covering at your new showroom,
Reynolds Interiors is pleased to submit the following proposal.
We appreciated the opportunity to visit your facility in order to
submit this proposal.

Begins with reference to RFP

After carefully reviewing your specifications for a floor covering
and inspecting your new facility, we believe that **Armstrong
Classic Corlon 900** is the most suitable choice. We are enclosing
a sample of the Corlon 900 so you can see how carefully it is
constructed.

Identifies best solution

Corlon's Advantages
Guaranteed against defects for a full three years, **Corlon** is one
of the most durable floor coverings manufactured by Armstrong.
It is a heavy-duty commercial floor 0.085 inch thick for
protection. Twenty-five percent of the material consists of
interface backing; the other 75 percent is an inlaid wear layer
that offers exceptionally high resistance to everyday traffic.
Traffic tests conducted by the Independent Floor Covering
Institute repeatedly proved the superiority of **Corlon's**
construction and resistance.

Describes product's features that benefit reader

Another important feature of **Corlon** is the size of its rolls. Unlike
other leading brands of similar commercial flooring—Remington
or Treadmaster—**Corlon** comes in 12-foot-wide rather than

Distinguishes product from competitors'

http://www.reynolds.com • 503-555-8733 • Fax: 503-555-1629

Continued

Figure 13.5   (Continued)

2

6-foot-wide rolls. This extra width will significantly reduce the number of seams on your floor, thus increasing its attractiveness and reducing the dangers of the seams splitting.

Installation Procedures

Explains how job is professionally done

The **Classic Corlon** requires that we use the inlaid seaming process, a technical procedure requiring the services of a trained floor mechanic. Herman Goshen, our floor mechanic, has more than fifteen years of experience working with the inlaid seam process. His professional work and keen sense of layout and design have been consistently praised by our customers.

Installation Schedule

Gives realistic timetable

We can install the **Classic Corlon** on your showroom floor during the first week of March, which fits the timetable specified in your request. The material will take three and one-half days to install and will be ready to walk on immediately. We recommend, though, that you not move equipment onto the floor for 24 hours after installation.

Costs

The following costs include the **Classic Corlon** tile, labor, and tax:

Itemizes all costs

750 sq. yards of **Classic Corlon** at $23.50/sq. yd.	$ 17,625.00
Labor (28 hrs @ $18.00/hr.)	$ 504.00
Sealing fluid (10 gals. @ $15.00/gal.)	$ 150.00
Total	$ 18,279.00
Tax (5 percent)	$ 913.95
GRAND TOTAL	**$ 19,192.95**

Our costs are $250.00 under those you specified in your request.

Continued

3

Reynolds's Qualifications

Reynolds Interiors has been in business for more than 28 years. In that time, we have installed many commercial floors in Portland and its suburbs. In the last year, we have served more than 60 customers, including the new multipurpose Tech Mart plant in Portland. We would be happy to furnish you with a list of satisfied customers.

Establishes history of service

Conclusion

Thank you for the opportunity to submit this proposal. We believe you will be pleased with the quality of our work and the appearance and durability of an Armstrong **Corlon** floor. If we can provide you with any further information, please call us or visit us at our website.

Encourages reader to accept

Sincerely yours,

Neelow Singh

Neelow Singh
Sales Manager

Jack Rosen

Jack Rosen
Installation Supervisor

Internal and External Proposals

An internal proposal is written to one or several decision maker(s) in your own organization who have to sign off or approve your plan. As you will see on pages 573–581, an internal proposal can deal with a variety of topics, including changing a policy or procedure, requesting additional personnel, or purchasing or updating equipment or software.

An external proposal, on the other hand, is sent to a decision maker outside your company. It might go to a potential client you have never worked for or to a previous or current client. An external proposal can also be sent to a government funding agency, such as the Department of Agriculture, in its request in Figure 13.4. External proposals tend to be more formal than internal ones.

Guidelines for Writing a Successful Proposal

Visit www
.cengage.com/
english/kolin for
an online exercise,
"Analyzing Two
Sample Online
Proposals."

The following guidelines will help you persuade your audience to approve your plan. Refer to these guidelines and Figures 13.5 and 13.6 both before and while you formulate your plan.

1. Approach writing a proposal as a problem-solving activity. Everything in your proposal should relate to the problem, and the organization of your proposal should reflect your ability to solve problems. Psychologically, make the readers feel confident that you have the knowledge and the experience they must use. Convince your audience you know their needs and will meet them, as Neelow Singh and Jack Rosen do in Figure 13.5.

2. Regard your audience as skeptical readers. Even though you offer a plan that you think will benefit readers, do not be overconfident that they will automatically

Tech Note

Online RFPs

RFPs are not only sent out to firms and published in trade magazines, but they are often posted on numerous websites as well. The U.S. government, through its many departments and agencies, is an important source for RFPs. Fed Biz Opps claims that it offers "a comprehensive listing of government RFP's" (*http://www.fbo.gov*). In addition, the *Federal Register,* published daily by the U.S. government, issues hundreds of legal regulations as well as requests for proposals from various government agencies. You can browse the *Federal Register* online via GPO access—*http://www.gpoaccess.gov/fr.*

Another source of RFPs on the Internet is *Commerce Business Daily* (CBD), which "lists notices of proposed government procurement actions, contract awards, sales of government property and other procurement information. A new edition of the *CBD* is issued every business day, and each edition contains approximately 500–1000 notices." The *CBD* website—*http://cbdnet.gpo.gov*— lists the government's requests for equipment, supplies, and a variety of services from assembling to maintaining equipment to fee-charging dredging.

Here are a few of the agencies you can search in the *CBD* for RFPs:

- Department of Health and Human Services
- United States Department of Energy
- United States Department of Agriculture
- Defense Support Center and Agencies
- National Aeronautics and Space Administration
- Department of Justice and other law enforcement agencies

accept it as the best and only way to proceed. To determine whether your proposal is feasible, readers will question everything you say. If your proposal contains errors or inconsistencies, omits information, or deviates from what they are looking for, your readers will reject it.

3. Research your proposal thoroughly. A winning proposal is *not* based only on a few well-meaning, general suggestions. All your good intentions and enthusiasm will not substitute for the hard facts readers will demand. Spell out your plan or procedure; give the nuts and bolts of how, when, and where the job will be done, as Figure 13.5 does. Concrete examples persuade readers; unsupported generalizations do not. You will have to research your topic by shopping for the best prices, comparing prices and services, verifying schedules, visiting customers, making site visits, and interviewing individuals at your company or your client's company to make sure you are on the right track.

4. Scout out what your competitors are doing. Closely related to guideline 3, assessing your competition is essential to write a winning proposal. Become familiar with your competitors' products or services, have a fair idea about their market costs, and be able to show how your company's work is better. Read their websites and print publications (such as catalogs and marketing brochures) very carefully. Let readers know you have done your homework. See how the writing team for the business report in Chapter 8 (pp. 376–388) studies its competition. Note, too, how the writers in Figure 13.5 prove their product is superior.

Tech Note

Document Design and Your Proposal

As we saw in Chapter 11, the overall design and layout of a document play a major role in the acceptance by an audience. This is especially true of a proposal—a key sales document for you and your company. Keep in mind that your proposal will be competing with many (perhaps 50 or 100) other proposals, and the first impression it makes will be visual. It should be attractive, logically organized, and reader friendly. If it is designed professionally and pleasingly, it will remain in the running. If not, your proposal may be rejected before your audience reads your first sentence.

Here are some guidelines to help you prepare an attractive and carefully designed proposal.

- Make sure you follow the RFP guidelines to the letter—in terms of spacing, title page, number of copies, appendixes, exhibits, and so on.
- Double-check to make sure it looks professional. Use good-quality paper and a sturdy binding.

Continued

Continued

- Organize your proposal into sections that help readers identify and follow the various parts of your proposal—for example, problem, solution, budget, timetable, personnel.
- Use plenty of clearly marked, logically ordered, and consistent headings (or, if necessary, subheadings) to separate sections of your proposal to help readers follow and understand your work easily, quickly, and clearly.
- Insert extra spacing between sections of your proposal so they stand out and show readers your work is organized.
- Use a professional-looking and easy-to-read font and type size. Stay away from script fonts and those with ornate designs. Do not try to cram more information in by resorting to an 8-point type.
- Include easy-to-follow indented lists, each item preceded by a bullet or an asterisk.
- Clearly label and insert all visuals in the most appropriate places in your proposal—see Figure 13.6 (pp. 574–577), which does an especially effective job.
- Put budgets in easy-to-read tables, not buried in a paragraph of prose. Make sure each item in a budget/timetable is identified and highlighted or relegated to a footnote or appendix.
- Keep paragraphs at five to six sentences. Heavy blocks of prose slow readers down and make them think your work is dense and hard to follow. Consider your readers' comfort level.
- Do not fail to include and label any supporting documents or materials that will not be a part of your proposal proper, for example, schedules, surveys, or samples, as in Figure 13.5.

5. Prove that your proposal is workable. The bottom-line question from your reader is "Will this plan work?" Your proposal should contain no statements that say, "Let's see what happens if we do X or Y." Analyze and test each part of your proposal to eliminate any quirks and to revise the proposal appropriately before readers evaluate it. What you propose should be consistent with the organization and capabilities of the company. It would be foolish and risky to recommend, for example, that a small company with fifty employees triple its work force to implement your plan.

6. Be sure your proposal is financially realistic. "Is it worth the money?" is another bottom-line question you can expect from your readers. For example, it would be unrealistic to recommend that your company spend $20,000 to solve a $2,000 problem. Note how Figure 13.5 stresses that the costs are under control. Above all, make readers believe that the benefits are worth the costs.

7. Be ethical. Don't misinterpret your product or service by exaggerating its benefits or neglecting to mention limitations readers need to be aware of (limited warranties, conditions under which a piece of equipment will not work, low energy efficiency, etc.).

8. Package your proposal attractively. Make sure that your proposal is well presented: inviting, attractive, and easy to read. Also be sure that all visuals are clear and appropriately placed. The visual appearance of your proposal can contribute greatly to whether it is accepted. Study the guidelines in the Tech Note on pages 571–572.

Internal Proposals

The primary purpose of an internal proposal, such as the one shown in Figure 13.6, is to offer a realistic and constructive plan to help your company run its business more efficiently and economically.

On your job you may discover a better way of doing something or a more efficient way to correct a problem. You believe that your proposed change will save your employer time, money, or further trouble. Note how Bebe Torelli, Peter Loo, and Faareed Kahn in Figure 13.6 (pp. 574–577) identified and researched a more effective way to lower fuel costs, monitor company trucks, and better meet customer needs at Midwest Maintenance.

Generally speaking, your proposal will be an informal, in-house message, so a brief (usually two- to four-page) memo should be appropriate. You decide to notify your department head, manager, or supervisor, or your employer may ask you for specific suggestions to solve a problem he or she has already identified. Mike Gonzalez's memo in Figure 4.3 (p. 126) responds to such a request from his employer.

Typical Topics for Internal Proposals

An internal proposal can be written about a variety of topics, such as

- purchasing new or more advanced technology to replace obsolete or inefficient computers, transducers, robots, and the like or upgrading equipment
- obtaining document security software and offering training sessions to show employees how to use it
- recruiting new employees, or retraining current ones on a new technique or process
- eliminating a dangerous condition or reducing an environmental risk to prevent accidents—for employees, customers, or the community at large
- improving communication within or between departments of a company or agency to save expenses or to increase jobs
- expanding work space or making it more efficient, private, ergonomically beneficial to employees, or more inviting to customers

As the list shows, internal proposals cover almost every activity or policy that can affect the day-to-day operation of a company or agency.

Figure 13.6   An internal unsolicited proposal.

Midwest Maintenance and Repair
www.midwestmr.com

TO: Allendyne Marks, Director of Operations
FROM: Bebe Torrelli, Peter Loo, and Faareed Kahn
DATE: April 10, 2009
RE: A proposal to lower fuel costs at Midwest
             Maintenance and Repair

PURPOSE

Clearly states why proposal is being sent

We propose a cost-effective solution to a major operating problem at Midwest Maintenance and Repair: rapidly rising fuel costs, combined with increased fuel consumption, have led to problems in meeting our customers' needs. To solve this problem, we recommend that you approve the purchase and installation, within the next month, of a global positioning system (GPS) for each of Midwest's twenty-two service vehicles. Having access to accurate field information about our fleet as a result of the GPS technology will result in substantial fuel savings and a more efficient management of our resources. Using a GPS system is well worth our return on investment and can enhance Midwest's image as a customer-oriented company.

THE PROBLEM

Identifies problem by documenting exact cost increases

In the past three months, Midwest's fuel costs have risen 16 percent, from $7,200.00 per month in January to $8,600.00 per month in March, or an average of $65.00 per truck per month. We have received quotes from four area fuel vendors, but learned that our current supplier, Gasco, although continuing to give us the lowest bid, has understandably raised its prices in response to the current gas crisis. Yet, even more costly, our fuel consumption during this same quarter rose by 18 percent, resulting in an increase of $75.00 per truck each month. While the number of service calls that Midwest made during this quarter compares favorably with the three previous quarters, our technicians are driving more miles to make the same number of service calls.

Divides problem into interrelated parts

Part of the reason for our increased fuel consumption may be traced to the unprecedented expansion of the Henderson metropolitan area into the western parts of the county, forcing our technicians to drive farther to reach new customers in those areas. Moreover, two major construction projects — the new cloverleaf added at the intersections of U.S. 59 and Old Highway 14 and the resurfacing of Collins Blvd. — have

page 2

slowed in-city traffic flow; some former two-way streets have become one-way, numerous detours have been set in place, and traffic light patterns have slowed, resulting in more driving time and fuel consumption. We are experiencing more traffic jams and idling at traffic lights wastes fuel. Still another cause for our increased expenses is the additional overtime hours our technicians are recording because of the expanded service area and these traffic-flow problems. Checking with Chief Dispatcher Kara Trueblood, we learned that our crews are clocking in an average of 50 hours per week in overtime. Yet even with so much overtime, Trueblood documented 31 missed appointments in the past two weeks. Not surprisingly, the number of customer complaints over late or missed appointments has risen.

Overall, then, spiraling energy charges, additional fuel consumption, increased overtime, and the potential loss of business because of customer complaints demand that we find a way to route service vehicles more efficiently and meet customer expectations more promptly. The bottom line is that Midwest is spending 34 percent more to operate this quarter than it did during the same period last year, and we are jeopardizing the customer base it has taken us many years to build.

Emphasizes need to solve problem now

A SOLUTION TO THE PROBLEM

Purchasing and installing a GPS tracking system for each of our twenty-two service vehicles will, we believe, give us a much more effective way to route and monitor our service vehicles. As a result, we can see significant savings in fuel and personnel costs and improve customer service. Specifically, with a GPS system in place, we can:

Shows how a GPS system can solve over-all problem

- Locate the truck closest to a customer using the GPS's "Show Me the Nearest Vehicle" feature, thereby avoiding sending a truck clear across town, reducing customer waiting time, and using less fuel.
- Dispatch trucks using the "Real Time Digital Map" feature (which is updated each month) to the fastest, most efficient routes, considerably reducing excessive travel and downtime locating a customer's address.
- Track any vehicle anytime, anywhere, which will enable us to supply customers who request specific technicians with more accurate arrival times.
- Ensure that our technicians will have access to reliable and accurate routes to specific addresses by using GPS's "Geofencing" feature, thus reducing fuel consumption and overtime.

Relates solutions to individual parts of the problem

Continued

Figure 13.6   (Continued)

page 3

Condenses technical detail into an easy-to-read bulleted list

- Find out precisely how long it takes service vehicles to arrive at particular destinations and use that information for improved customer service to those locations in the future.
- Monitor each technician's start and stop times to determine if company vehicles are being used for side trips, personal business, or excessive idle time during lunch or rest breaks.
- Cut a customer's "estimated waiting time" from the standard "anytime between nine to five inches to just an hour or two, thereby significantly improving customer satisfaction.
- Increase the number of service calls per day by outlining quicker routes and making more efficient use of the technician's driving time.
- Bill customers more accurately by having the technician input, via GPS, the time of the call, the type of service performed, and the parts necessary to complete the job.
- Verify or settle customer complaints about whether service was late, incomplete, or not provided.
- Pass savings to our customers by reducing the "add-on" costs associated with increased operating expenses and the maintenance of over-driven vehicles.

Documents the feasibility of the solution, financially and operationally

Purchasing and installing a GPS unit for each of our service vehicles is both feasible and financially rewarding. The installation time requires less than two hours per vehicle, and we could save even more time by staggering that number. Loading GPS technology onto our current network base would, according to our systems engineer, Pral Sandhi, take no more than one day. All of the GPS vendors we spoke with indicated that installation might even be scheduled for a Saturday or other light service period. Moreover, authorizing the expenditure within the next month will ensure that full GPS service is available before our busy summer season.

Wisely compares vendors' prices and services

COSTS

We found GPSLocator to be the highest quality provider. Not only does their system provide the most frequent and reliable construction updates, but it also comes equipped with an excellent traffic conditions feed. In terms of cost, GPSLocator is also the best option. We asked for and received quotes from the four local GPS providers (Navigator USA, GPS Foundations, and Earthmap, along with GPSLocator), and discovered that, by far, GPSLocator offered the best customized package at a price 20 percent lower than the three other vendors.

(Continued)   Figure 13.6

page 4

The following costs for implementing our proposal are, therefore, based on the GPSLocator quote.

Tracking antenna, transmitter, & connective cables	$359.95
Activation fee	$27.99
Installation fee	$34.95
Automatic software updates fee	$29.99
TOTAL COST PER VEHICLE	$453.00
TOTAL COST	**$453 × 22 vehicles = $9,966.00**

Itemizes costs

These costs will be offset by the following savings:

Overtime savings	50 hours/week @ $18.00/hour = $ 45,000/year
Fuel consumption savings	$75.00/month per vehicle × 22 vehicles × 12 months = $ 19,800/year
ANNUAL SAVINGS	$45,000 + $19,800 = $64,800
TOTAL ANNUAL SAVINGS MINUS INITIAL COSTS	**$64,800 − $9,966 = $54,834.00**

Ends with bottom line cost

When justifying the costs for GPS systems over a five-year period, we found that the one-time charge of $9,966.00 (22 × $453.00) can be reduced to only $1,839.00 per year. That is, we can subtract this one-time cost each year so that after 5 years it is equal to only $1,839.00 per vehicle, further reinforcing the benefits this proposal offers.

Interprets future costs/ benefits

CONCLUSION

Authorizing the purchase and installation of a GPSLocator system for our fleet of twenty-two service vehicles is both feasible and highly cost-effective, given the current state of energy prices and the changing map of the greater Hendersonville metropolitan area. Endorsing this proposal has the potential of saving Midwest more than $55,000 annually by decreasing fuel consumption, reducing overtime, routing/monitoring vehicle and technicians more efficiently, and increasing customer approval.

Reiterates financial benefits and urgency to convince reader to approve plan

We look forward to discussing this proposal with you anytime at your convenience and answering any questions that you may have.

Following Corporate Policy

Writing an internal proposal requires you to be aware of and sensitive to office policies. Meet first with your boss to see if she or he has already identified the problem or has suggestions on how to solve it. Then you might provide your boss with a draft and ask for revisions or feedback.

You cannot assume that your reader(s) will automatically agree with you that there is a problem or that your plan is the only way to tackle it. To be successful, write your internal proposal keeping in mind the needs and likes of your boss and others who may have to sign off on your proposal. Remember that your boss will expect you to be very convincing about both the problem you say exists and the changes you are advocating in the workplace under his or her supervision. Don't step on corporate toes. Similarly, you may need the approval of individuals in other offices, departments, or branches of your organization.

Ethically Resolving and Anticipating Readers' Problems

When you prepare an internal proposal, you need to be aware of the ethical obligations you have and the ways to meet them. Here are some important guidelines:

1. **Consider the implications of your plan company-wide.** The change you propose (transfers, new budgets or technology, new hires) may have sweeping and potentially disruptive implications for another office or division in your company.

2. **Keep in mind what impact your change may have for co-workers from cultural traditions other than your own.** In addition to speaking to your boss, consult your human resources or cultural diversity director.

3. **Find out whether what you are proposing is within your company's budget.** How much can your department, branch, or office spend (e.g., on hardware, software, updates)? Check with your boss, and always monitor the price of your company's stock to make sure any expenditures are likely to be approved.

4. **Realize that your reader may feel threatened by your plan.** After all, you are advocating a change. Be careful. Your plan may sound just right to you, but your boss may regard it as a piece of criticism leveled at him or her, your department, or the company as a whole. Think about the long-term effect your proposed change will have on your boss's duties, responsibilities, or relationships with your co-workers or with his or her own superiors. Don't override your boss or attempt to undermine existing authority.

5. **Take into account that your reader may have "pet projects" or predetermined ways of doing things.** Make every attempt to acknowledge them respectfully. You may even find a way to build on or complement such projects or procedures.

6. **Keep in mind that your boss may have to take your proposal farther up the organizational ladder for commentary and, eventually, approval.** Again, refer to Joycelyn Woolfolk's description in Figure 3.6 of the chain of command at her organization.

7. Never submit an internal proposal that offers an idea you think will work but relies on someone else to supply the specific details on how it will work. For example, do not write an internal proposal that says the marketing, technical support, or human resources department could supply the necessary details for your proposal to work. That unfairly pushes the responsibility onto someone else.

Organization of an Internal Proposal

A short internal proposal follows a relatively straightforward plan of organization, from identifying the problem to solving it. Internal proposals usually contain four parts, as shown in Figure 13.6: **purpose, problem, solution,** and **conclusion**. Refer to the figure (pp. 574–577) as you read the following discussion.

The Purpose

Begin your proposal with a brief statement of why you are writing to your supervisor: "I propose that . . . ." State right away why you think a specific change is necessary now. Then succinctly define the problem and emphasize that your plan, if approved by the reader, will solve that problem.

The Problem

In this section prove that a problem exists. Document its importance for your boss and your company; as a matter of fact, the more you show, with concrete evidence, how the problem affects the boss's work (and area of supervision), the more likely you are to persuade him or her to act.

Here are some guidelines for documenting a problem:

- Avoid vague (and unsupported) generalizations such as, "We're losing money each day with this procedure (piece of equipment)"; "Costs continue to escalate"; "The trouble occurs frequently in a number of places"; "Numerous complaints have come in"; "If something isn't done soon, more problems will result."
- Provide quantifiable details about the problem, such as the amount of money or time a company is actually losing per day, week, or month. Document the financial trouble so that you can show in the next section how your plan offers an efficient and workable solution.
- Indicate how many employees (or work-hours) are involved or how many customers are inconvenienced or endangered by a procedure or condition. The writers in Figure 13.6 researched employee overtime and used this information to document the importance of the problem.
- Verify how widespread a problem is or how frequently it occurs by citing specific occasions. Again, see how Torelli and her collaborative team cite evidence from the interview they conducted with Chief Dispatcher Kara Trueblood.
- Relate the problem to an organization's image, corporate reputation, or influence (where appropriate). Pinpoint exactly how and where the problem lessens your company's effectiveness or hurts its standing in the market. Indicate who is affected and how the problem affects your company's business, as the writers do especially well in the second paragraph of the problem statement in Figure 13.6.

The Solution or Plan

In this section describe the change you propose and want approved. Tie your solution (the change) directly to the problem you have just documented. Each part of your plan should help eliminate the problem or should help increase the productivity, efficiency, or safety you think is possible.

Your reader will again expect to find factual evidence. Be specific. Do not give merely the outline of a plan or say that details can be worked out later. Supply details that answer the following questions: (1) Is the plan workable? and (2) Is it cost-effective?

To get the reader to say "Yes" to both questions, supply the facts you have gathered as a result of your research. For example, if you propose that your firm buy a new piece of equipment, do the necessary homework to locate the most efficient and cost-effective model available, as the proposal writers in Figure 13.6 do about GPS technology.

- Supply the vendors' names, the costs, major conditions of service and training contracts, and warranties.
- Describe how your firm could use the equipment/technology to obtain better or quicker results.
- Document specific tasks the new equipment can perform more efficiently at a lower cost than the equipment now in use.

A proposal to change a procedure must address the following questions:

- How does the new (or revised) procedure work?
- How many employees or customers will be affected by it?
- When will/can it go into operation?
- How much will it cost the employer to change procedures?
- What delays or losses in business might be expected while the company switches from one procedure to another?
- What employees, equipment, technology, or locations are already available to accomplish the change?

As those questions indicate, your reader will be concerned about schedules, working conditions, employees, methods, locations, equipment, and the costs involved in your plan for change. The costs, in fact, will be of utmost importance and you may want to provide a separate "Costs" section in your proposal to demonstrate their importance, as the writers of the proposal in Figure 13.6 do. Make sure you supply a careful and accurate budget. Moreover, make those costs attractive by emphasizing how inexpensive they are compared to the cost of *not* making the change.

It is also wise to raise alternative solutions, before the reader does, and to discuss their disadvantages. Notice how Torelli, Loo, and Kahn do that in Figure 13.6 by stressing that Midwest Maintenance is already buying fuel from the least costly source, but that this alone will not keep costs down.

The Conclusion

Your conclusion should be short—a paragraph or two at the most. Remind readers that (a) the problem is ongoing and serious, (b) the reason for change is justified and beneficial to your organization, and (c) action needs to be taken. Reemphasize the most important benefits as Torelli, Loo, and Kahn do in their proposal in Figure 13.6. Also indicate that you are willing to discuss your plan with the reader, a necessity in arguing for a corporate change at any level.

▌Sales Proposals

A sales proposal is the most common type of external proposal. Its purpose is to sell your company's products or services for a set fee. Whether short or long, a sales proposal is a marketing tool that includes a sales pitch as well as a detailed description of the work you propose to do. Figures 13.5 and 13.7 are both sales proposals.

The Audience and Its Needs

Your audience will usually be one or more executives who have the power to approve or reject a proposal. Unlike readers of an internal proposal, your audience for a sales proposal may be even more skeptical since they may not know you or your work. You can increase your chances of success by trying to anticipate their questions, such as:

- Does the writer's firm understand our problem?
- Can the writer's firm deliver what it promises?
- Can the job be completed on time?
- Is the budget reasonable and realistic?
- Will the job be done exactly as proposed?
- Has the writer demonstrated his or her trustworthiness?

Answer each of these questions by demonstrating how your product or service is tailored to the customer's needs.

Make sure, too, that your proposal has a competitive edge. Readers will compare your plan with those they receive from other proposal writers. Your proposal has to convince readers that the product or service your company offers is more reliable, economical, efficient, and timely than that of another company. Here is where your homework pays off.

The key to success is incorporating the "you attitude" throughout your proposal. Relate your product, ser**vic**e, or personnel to the reader's exact needs as stated in the RFP for a solicited proposal or through your own investigations for an unsolicited proposal. *You cannot submit the same proposal for every job you want to win and expect to be awarded a contract.* Different firms have different needs.

Organizing a Sales Proposal

Most sales proposals include the following elements: introduction, description of the proposed product or service, timetable, costs, qualifications of your company, and conclusion.

Introduction
The introduction to your sales proposal can be a single paragraph in a short sales proposal or several pages in a more complex one. Basically, your introduction should prepare readers for everything that follows in your proposal. The

Figure 13.7   An unsolicited sales proposal.

203-555-6714   203-555-1732   **techsupport@interserv.com**

August 24, 2009

Ms. Alexandra Tyrone-Saunders
Vice President, Operations
Gemini Consultants, Inc.
Hartford, CT 06631-7106

Dear Ms. Tyrone-Saunders:

Begins by thanking Gemini for using Omega's services and suggests ways Omega can further help

Thank you for allowing us to service your Webmax PCs last week. While we were working on the PCs, I saw several ways in which Omega Technologies might improve your network system. Based on our assessment of Gemini's requirements for the most up-to-date automation available, we recommend that you purchase ten Lightstar 686 notebooks, one for each member of your sales staff.
This latest generation of notebooks will provide your staff with a powerful and cost-effective system that will meet Gemini's needs well into the next decade.

PROBLEM AREAS

Clearly identifies and quantifies the problem and why it is important to solve

Each of your ten sales staff spends up to 60 minutes at the end of each business day entering orders he or she has accumulated in the field into the PCs at your office. That amounts to 50 overtime hours per week. This procedure is costly and inefficient, taking your staff away from their primary responsibility of serving your clients.

ADVANTAGES OF THE LIGHTSTAR 686 NOTEBOOK

Provides documented endorsement

The Lightstar offers state-of-the-art laptop technology. It can also convert to a touchscreen "tablet" for fieldwork. According to Felicia Gomez, writing for the Worldtech website (*http://www.worldtech.org*), the Lightstar 686 is "one of the best computer buys of the decade, a powerhouse that will amaze the most particular PC user." The description of Lightstar's features below will show you how its many advantages can help Gemini.

ADVANCED TECHNOLOGY

The Lightstar 686 is equipped with the following standard features:

Gives a detailed list of features and specifications

- 3 GHz Pentium M745 processor
- 1 GB of DDR SDRAM memory
- 80 GB hard drive
- 15-inch LCD panel monitor
- 200 MB graphics card
- 802.11 G wireless networking
- ethernet card
- DVD/CD-RW combo drive

(Continued)   Figure 13.7

2.

Your Lightstar will allow you to fax, e-mail, and connect to the Web. Please visit our website (*http://www.omegatech.com*) for more details about Lightstar's technical specifications.

Cost Effective

Your Lightstar 686 will cost far less than your PCs and offer equal or greater capability. Here are some of the bottom-line savings you can expect:

- costs $500.00 less than one of your PCs did two years ago
- provides software upgrades at no charge
- includes everything integrated into the system—monitor, CPU, software

Assured Security

Lightstar offers a unique protection system to ensure the security of your unit:

- an encryption system requiring a special password to log on and activate your notebook
- a special feature known as "call home" that will alert you to unauthorized use
- virus protection software

Efficient Accessibility and Customer Service

Because the Lightstar is compatible with your office network, your representatives can access all pertinent records at your main office while making sales calls. By having direct access to vital documentation, your staff can more efficiently verify, fulfill, and store information about customer orders. The Lightstar makes the virtual office a reality. And because your sales force would be carrying your office with them, you will enhance their telecommuting capabilities now and into the future.

Power Supply Capabilities

Equipped with a lithium ion battery, the Lightstar 686 operates up to six hours without recharging, enabling your staff to make calls at any location for extended periods of time.

Portability

- weighs only 5.6 pounds—it is ultraportable
- slips easily into a briefcase
- comes with protective, dent-proof case

TRAINING AND SERVICE

Because all software can be preloaded, setup and installation can be brief—ten minutes. Since Lightstar is compatible with Gemini's network, your staff will need only minimal training to know how to use the Lightstar to upload sales orders or download product availability information. To help Gemini with any technical needs, our local tech representative, Ory Mawfic, is available to instruct your staff

Details Lightstar's many advantages, each clearly and persuasively listed under easy-to-follow headings

Shows how Lightstar is compatible with Gemini's current technology and is, therefore, easy to use

Figure 13.7   (Continued)

3.

about the operation and to provide routine maintenance of the Lightstar. If a problem occurs, we offer customers the latest in remote diagnostics.

COSTS

Below is an estimated price list for one Lightstar 686 notebook with upgraded features:

Spells out costs exactly, including any options Gemini might select

- 5 GHz Pentium M 755 processor
- 100 GB hard drive
- Integrated 10/100 NIC (network interface)
- 2 GB of DDR SDRAM memory
- Second lithium ion battery

Uses boldface for bottom-line figures

Total purchase price:	**$2,766.00**

You might also consider adding one or all of these options:

Pentium M processor 770 (4 GHz)	$569.00
External speakers	$48.00
USB 2-button optical mouse	$30.00
Total for options:	**$647.00**

OMEGA TECHNOLOGIES' REPUTATION

Provides a short history of Omega and supplies references

For the past 16 years Omega Technologies has provided state-our-of-the-art computer technology and fast, efficient service after the sale. A list of our references is attached. Please feel free to write or call them.

Closes with reference to past and favorable business to encourage future sales

We appreciate your past confidence in choosing Omega Technologies and look forward to providing further service to Gemini. Please call me if you have any questions about the Lightstar or Omega Technologies. I would be happy to bring a few Lightstar 686's to Gemini for you and your sales staff to use for a few days.

If this proposal is acceptable, please sign and return a copy with this letter.

Sincerely yours,

Marion Copely

Marion Copely

I accept the proposal made by Omega Technologies, Inc.

_____

for Gemini Consultants, Inc.
Encl.

introduction itself may contain the following sections, which sometimes may be combined.

1. **Statement of purpose and subject of proposal.** Tell readers why you are writing and identify the specific subject of your work. If you are responding to an RFP, use specific code numbers or cite application dates, as the proposal in Figure 13.5 does. If your proposal is unsolicited, indicate how you learned of the problem, as Figure 13.7 does. Briefly define the solution you propose. Tell your readers exactly what you propose to do for them. Be clear about what your plan covers and, if there could be any doubt, what it does not.

2. **Background of the problem you propose to solve.** In a solicited proposal like Figure 13.5, this section is usually unnecessary, because the potential client has already identified the problem and wants to know how you would solve it. In that case, just point out how your company would address the problem, mentioning your superiority over your competitors (see the fourth paragraph of Figure 13.5).

In an unsolicited proposal, you need to describe the problem in convincing detail, identifying the specific trouble areas. You may want to focus briefly on the dimensions of the problem—when it was first observed, who/what it most acutely affects, and the specific organizational/community context in which the problem is most troubling. However, if it is an external proposal to a current customer, such as the one in Figure 13.7, it would be unwise to point out previous problems your client may have had with your company's service or products. In Figure 13.7 note how the problems of the staff in the field are described mainly as a way to sell the advantages of the Lightstar 686 notebook.

Description of the Proposed Product or Service

This section is the heart of your proposal. Before spending their money, customers will demand hard, factual evidence of what you claim can and should be done. Here are some points that your proposal should cover.

1. **Carefully show potential customers that your product or service is right for them.** Stress particular benefits of your product or service most relevant to your reader. Blend sales talk with descriptions of hardware, as the sections "Advanced Technology" and "Cost Effective" do in Figure 13.7.

2. **Describe your work in suitable detail.** Specify what the product looks like, what it does, and how consistently and well it will perform in the readers' office, plant, hospital, or agency. You might include a brochure, picture, diagram, or, as the writer of the proposal in Figure 13.5 does, a sample of your product for customers to study. Note how in Figure 13.7 the reader is given an independent website to consult about the Lightstar's capabilities and endorsement.

3. **Stress any special features, maintenance advantages, warranties, or service benefits.** Convince readers that your product is the most up-to-date and efficient one they could select. Highlight features that show the quality, consistency, or security of your work. For a service, emphasize the procedures you use, the terms of the service, even the kinds of tools you use, especially any state-of-the-art equipment.

Timetable

A carefully planned timetable shows readers that you know your job and that you can accomplish it in the right amount of time. Your dates should match any listed in an RFP. Provide specific dates to indicate

- when the work will begin
- how the work will be divided into phases or stages
- when you will be finished
- whether any follow-up visits or services are involved

For proposals offering a service, specify how many times—an hour, a week, a month—customers can expect to receive your help; for example, spraying two times a month for an exterminating service.

Costs

Make your budget accurate, complete, and convincing. Don't underestimate costs in the hope that a low bid will win you the job. You may get the job but lose money doing it, because the customer will rightfully hold you to your unrealistic figures. Accepted by both parties, a proposal is a binding legal agreement. Neither should you inflate prices; competitors will beat you in the bidding.

Give customers more than merely the bottom-line cost. Show exactly what readers are getting for their money so they can determine if everything they need is included. Itemize costs for

- specific services
- equipment/materials
- labor (by the hour or by the job)
- transportation
- travel
- training

If something is not included or is considered optional, say so—additional hours of training, replacement of parts, upgrades, and the like.

If you anticipate a price increase, let the customer know how long current prices will stay in effect. That information may spur them to act favorably now.

Qualifications of Your Company

Emphasize your company's accomplishments and expertise in providing similar products or services. Mention the names of important local firms for whom you have worked that would be able to recommend you. But never misrepresent your qualifications or those of the individuals who work with or for you. Your prospective client may verify if you have in fact worked on similar jobs during the last five or six years.

Conclusion

This is the "call to action" section of your proposal. Encourage your reader to approve your plan by stressing the major benefits of your plan. Offer to answer any questions the reader may have. Some proposals end by asking readers to sign and return a copy of the proposal indicating their acceptance, as the proposal in Figure 13.7

does. And build on any previous goodwill or business, as the last paragraph does in Figure 13.7, to further establish trust with your customer or client.

Proposals for Research Reports

You can also expect to write a proposal to your boss seeking her/his authorization to do research, such as students often do in courses to get a topic approved by an instructor.

The principles guiding internal and sales proposals also apply to research proposals. As with internal and sales proposals, you will be writing to convince the reader—your boss—to approve a major piece of work. Your boss will read your proposal to make sure you write the best possible report to solve a company problem and to increase company profits. In drafting your proposal, make sure of four things:

- that you have chosen a significant topic
- that you will have a sufficiently restricted topic
- that you will investigate important sources of information about that topic
- that you can accomplish your work in the specified time

Your proposal will give your boss an opportunity to spot omissions or inconsistencies and to provide helpful suggestions, as in Figure 13.8 (pp. 588–590).

To prepare an effective proposal for a research project, you must do some preliminary research. You cannot pick any topic that comes to mind or guess about procedures, sources, or conclusions. As other proposal readers do, your boss will want convincing and specific evidence for your choice of topic and your approach to it. Be prepared to cite key facts to show that you are familiar with the topic and that you can write about it confidently and knowledgeably.

Be very clear about the type of research you will do and what resources you intend to use. For example, you need to investigate your topic through the following sources before you write a research proposal for your boss:

- Internet—list relevant websites you have consulted
- Search engines/databases—for a working bibliography
- Books and articles, online and print—but only those that bear directly on your topic
- Interviews—whether in person, over e-mail, instant messaging, blogs, and so on
- Proposed visits to relevant sites—laboratories, salt marshes, health care facilities, plants, offices, and so on

Review pages 310–348 of Chapter 8 on primary and secondary research methods.

Organization of a Proposal for a Research Report

Your proposal for an in-house report, as for a school research project, can be a memo or an e-mail divided into five sections, as illustrated in Figure 13.8: *introduction* (or purpose), *scope of the problem* or topic to be investigated, *methods or procedures, timetable,* and *request for approval.* However, be ready to reorder or expand these sections if your boss wants you to follow a different organizational plan.

Figure 13.8   A proposal to do a research report.

To: Marisol Vega, Associate Director
From: Barbara R. Shoemake
Date: September 17, 2008
Subject: Proposal for a report on the ethical and security issues involved in
            using e-mail

PURPOSE

Concisely states purpose

To help the office managers revise company guidelines, I propose to research and
write a background report on the ethical and security issues involved in using
e-mail in the workplace.

Demonstrates importance of topic with statistics

E-mail is our most frequently used business communication. It has been estimated
that over 750 million e-mail users receive 15 billion messages per day, many for
internal business communications. At West Technology, over 27,000 employees
receive their company publications online through e-mail and listservs. Alexis
Brown claims that "E-mail has eclipsed the reading rate of printed material" (*Public
Relations Journal, http://www.prjournal.com* [Jan. 2008]: 25).

Provides further detail about the scope and significance of the problem to be investigated

Despite its many advantages, e-mail still presents major ethical and legal
challenges. Using e-mail ethically can sometimes be a complicated issue for
employees as well as their employers. E-mail privacy is at the heart of the
controversy. Major legal battles have been waged over who can and should read an
individual's e-mail at work. Similarly, court cases have been fought over companies
using their employees' e-mail as evidence against them.

More and more companies continue to reevaluate their standards for involvement
in employees' e-mail rights, often requiring employees to sign compliance
agreements to allow employers access to e-mail sent at the office. An employee's
ethical use of workplace technology is a crucial part of his/her performance
appraisal. Not surprisingly, a greater number of companies are demanding their
employees attend training sessions regarding e-mail and Internet privileges and
abuses at work. Understanding the ethical considerations and security drawbacks
of e-mail is vital for managers evaluating their departments' internal
communications.

Identifies audience for whom paper is intended

To ensure that our managers base company guidelines on the latest information, I
request permission to do research for a background report.

PROBLEMS TO BE INVESTIGATED

At this preliminary stage of my research, I think my report will need to answer the
following questions:

Highlights questions the paper will explore

(1) How does e-mail differ from conventional communication methods (telephone,
    letters) and more recent technologies (instant messaging, blogs) in terms of
    confidentiality in the workplace?

(2) What right (or responsibility) does an employer have to monitor employees'
    workplace e-mail and when does this become a violation of the employees' right
    to privacy?

(Continued)   Figure 13.8

Shoemake 2

(3) What can be done to establish a more secure e-mail system both to protect confidentiality and to ensure company security?

(4) What types of special training programs—on business communication, netiquette, document security software, virus alerts, and legal issues—are most necessary and effective for e-mail use?

I propose, therefore, to divide the body of my paper according to the four key issues of *Confidentiality*, *Monitoring*, *Security*, and *Training*.

Outlines tentative organization of report

METHODS OF RESEARCH

I will survey recent literature dealing with e-mail and security and then attempt to do some primary research as well by interviewing several local iTech experts. Judging from the vast number of entries found on this general topic through the search engines Google, Lycos, MSN, Yahoo!, and AltaVista, e-mail remains a significant but debated topic. But in order to restrict my topic I narrowed my keyword search to concentrate on e-mail security and ethical issues as they affect managers and company policy handbooks. From a preliminary check of documents available here at our corporate library and through online sources, I believe the following works may be most useful.

Gives detailed lists of primary and secondary sources to be consulted, with rationale

Brown, Alexis. "A Study of E-Mail Versus Printed Material." *Public Relations Journal* (Jan. 2008): 25.

Cruiz-Taylor, Deana. "Is It Corporate or Confidential or Both?: Security and Rights in Sending/Receiving Company E-Mail." *E-Commerce* 19.2 (2007): 14–19.

Fetissov, Nikolai N. "Securing Electronic Mail in a Small Company." *SANS Institute White Paper* 12 (May 2007): 1–15.

Uses proper MLA style for documentation

Hernandez, Pedro, Mary Melka, and G. T. Kaplan. "The Legal Foundations for Monitoring Employee E-Mail." *Ethics and Commerce* 4 (August 2007): 7–12.

Howard Ronald A., and Clinton D. Korver. *Ethics for the Real World: Creating a Personal Code to Guide Decisions in Work and Life.* Cambridge; Harvard Business School, 2008.

"Internet Mail Consortium." 29 Jan. 2008. 4 Feb. 2008 <http://www.imc.org>

Kuri, Jetha. 2007 "New Security Policies for E-Mail Technology." *i-Tech Mag* 30 July– 5 Aug. 2007 <http://www.Itech mag.co.in/views/article/php/360721>

Cites only most current and relevant sources

Ricks, Shameeka. "Training Your Staff to Send Secure and Professional E-Mail." *Office Protocol and Systems* 17 (2007): 75–78.

Shih, Dong, and Hsiu-Sen Chiang. "E-Mail Leaks: How Organizations Can Protect Their E-Mails." *Online Information Review* 28.5 (2008): 356–66.

Continued

Figure 13.8   (Continued)

Singh, Ali. *E-Mail—Legal Issues: A Specially Commissioned Report.* London: Thorogood, 2008.

Spykerman, Mike. "Is E-Mail Monitoring Legal?" *Spydex.com.* 2004. 2 Apr. 2008 <http://www.spydex.com/article-email-monitoring-legality.html>

Sunner, Mark. "The Role of E-Mail Security in Meeting Regulatory Requirements." *Help Net Security* 27 Jan. 2005. 14 Aug. 2008 <http://www.net-security.org/article.php?id=762>

"Weblog: Watch E-Mail Attachments. They Might Transport Sensitive Company Reports." *Corp World* 9 July 2008. <http://www.corpworld.com/weblog>

Zhang, Sherry. *E-Mail: Privacy, Ethics, and the Corporate World.* Boston: Business Publications, 2007.

Identifies need for additional interviews

And I hope to interview two or three human resources and i-Tech managers at other firms in the Springfield area to determine how they have addressed these issues. My first choice is Katarina Kuhn at E-Trade, who has offered employee seminars on e-mail security and ethics in the past year. If she is unavailable, I will turn to managers with impressive credentials in the field, such as Paul Goya at Mercy Hospital and Jen Holka at e-Tech, Inc.

Specifies schedule and how to meet it

SCHEDULE

I hope to complete my research by October 13 and my interviews by October 15. Then I will spend the following two weeks working on a draft, which I will submit by October 22. After receiving your comments on my draft, I will work on revisions and the final copy of my background report and turn it in by November 14. I will submit two progress reports—one when I finish my research and another when I decide on the final organization of my report.

Offers contact information; requests feedback

REQUEST FOR APPROVAL

Thanks for approving my plan. I would appreciate any further suggestions on how you think I might best proceed.

The Introduction

Keep your introduction short—a paragraph, maybe two, pinpointing the subject and purpose of your work.

> I propose to research and write a report about the "hot knife" laser used in treating port wine stains and other birthmarks.

> I intend to investigate the relationship that exists between office design and our employees' need for "psychological space."

Then briefly indicate why the topic or the problem you propose to study is significant. In other words, be prepared to explain why you have chosen that topic and why research on it is relevant or worthwhile for a specific audience or objective. Note how Barbara R. Shoemake in Figure 13.8 states how and why her background report will be useful to office managers.

Supply your boss with a few background details about your topic, for example, the importance of using a laser as opposed to conventional ways of treating birthmarks or why psychological space plays a crucial role in employee productivity and morale. Prove that you have thought carefully about selecting a significant, relevant topic.

The Scope of the Problem or Topic to Be Investigated

The second section, which might be entitled "Problems to Be Investigated" or "Areas to Be Studied," shows how you propose to break the topic into meaningful units. Tell your boss what specific issues, points, or areas you hope to investigate. Doing that, you demonstrate how you will limit your topic.

Some bosses ask researchers to formulate a list of questions their research report intends to answer. The topics included in such questions or in a list of areas or problems to be covered might later become major sections of your report. Make sure the issues or questions do not overlap and that each relates directly to and supports your restricted topic. See how in Figure 13.8, Barbara Shoemake intends to divide her report on e-mail security into four distinct yet related areas.

Methods or Procedures

In the third section of your proposal inform your boss how you expect to find the answers to the questions you raised in the previous section or how you intend to locate information about your list of subtopics. It's not enough to write, "I will gather appropriate information and analyze it." Specify what data you hope to include, where they are located, and how you intend to retrieve them.

In researching information about the problem, and any proposed solutions, you can expect to use both primary and secondary research methods and tools. Review relevant sections of Chapter 8. Certainly, you will gather data from the Web and from literature published in print sources. (In fact, research reports can be based exclusively on literature searches.) The literature can include

- websites/blogs/webfolios
- books
- encyclopedias or other reference materials, such as statistical data found in manuals or almanacs, online and in print
- articles in professional journals in print and online
- newspapers, online and in print
- reviews

Inform your boss what indexes, abstracts, or Internet searches you intend to use (review pp. 340–346) as part of your research. To document your preliminary work, provide your boss with a list of appropriate titles on your topic following the style of documentation used for a Works Cited page (discussed on pp. 358–364).

In addition to these online and print materials, you can also expect to collect information from primary research, including lab experiments, field tests, interviews with experts, surveys, or a combination of any of those sources.

Timetable/Schedule

Indicate when and in what order you expect to complete the different phases of your project. Your boss needs that information to keep track of your progress and to make sure you will complete your work on time. Specify tentative dates for completing your research, draft(s), revisions, and final copy.

Some bosses also want progress reports (see pp. 610–617) at regular intervals. If you are asked to do that, indicate when you will submit the progress reports, as Barbara R. Shoemake does in Figure 13.8.

Request for Approval

End your proposal with a request for approval of your topic and a plan of action. You might also invite suggestions from your boss on how to restrict, research, organize, or write about your topic.

A Final Reminder

This chapter has given you some basic information and specific strategies for writing winning proposals. Keep in mind that a proposal presents a plan to a decision maker for his or her approval. To win that approval, your proposal must be (a) realistic, (b) carefully researched, (c) highly persuasive, and (d) visually appealing and easy to follow. Those essential characteristics apply to internal proposals in memo or e-mail format written to your employer, more formal sales proposals sent to a potential customer, and research proposals submitted to your instructor.

Revision Checklist

- [ ] Established and distinguished roles of collaborative team members involved in the preparation of the proposal.
- [ ] Researched appropriate sources for RFPs and followed their instructions.
- [ ] Identified a realistic problem, one that is restricted and relevant to my topic and my audience's needs.
- [ ] Tried effectively to convince audience that the problem exists and needs to be solved.
- [ ] Used headings, white space, lists, and professional-looking font to make proposal visually attractive and reader friendly.

☐ Incorporated quantifiable details demonstrating the scope and importance of the problem.

☐ Persuasively emphasized benefits of solving the problem according to the proposal; incorporated the "you attitude" throughout.

☐ Investigated and overcame competing alternatives or firms.

☐ Offered a solution that can be realistically implemented—that is, it is both appropriate and feasible for audience.

☐ Wrote clearly so audience can understand how and why my proposal would work.

☐ Researched background of problem.

☐ Used specific figures and concrete details to show how proposal saves time, money.

☐ Double-checked proposal to catch errors, omissions, and inconsistencies.

☐ Avoided exaggerations and underbidding.

☐ Presented information ethically.

☐ Organized proposal with appropriate headings for clarity and ease of reading.

☐ *For internal proposal:* Demonstrated how proposal benefits my company and my supervisor; took into account office politics in describing problem and solution; discussed proposal with co-workers or supervisors who may be affected.

☐ *For sales proposal:* Related my product or service to prospective customer's needs; showed a clear understanding of those needs.

☐ Prepared a comprehensive and realistic budget; accounted for all expenses; itemized costs of products and services in sales proposal.

☐ Provided a timetable with exact dates for implementing proposal.

☐ Cited other successful jobs and satisfied clients to show my company's track record.

☐ Concluded proposal with a summary of main benefits to readers and a call to action.

☐ Proved to my boss that I researched the problem by supplying a list of possible references and sources for a proposal for a research report.

Exercises

1. In two or three paragraphs identify and document a problem (in services, safety, communication, traffic, scheduling) you see in your office or your community. Make sure you give your reader—a civic official (head of a department) or employer (section or department head; manager)—specific evidence that a problem does exist and that it needs to be corrected.

2. Write a short internal proposal, modeled after Figure 13.6, based on the problem you identified in Exercise 1.

3. Write a short internal proposal, similar to that in Figure 13.6, recommending to a company or a college a specific change in procedure, technology, training, safety, personnel, or policy. Make sure your team provides an appropriate audience (college administrator, department manager, or section chief) with specific evidence about the existence of the problem and your solution of it. Possible topics include:
   a. providing more and safer parking/lighting
   b. job sharing for mothers
   c. purchasing new office or laboratory equipment or software
   d. hiring more faculty, student workers, or office help
   e. allowing employees to telecommute
   f. changing the decor/furniture in a student or company lounge
   g. increasing the number of weekend, night, or online classes in your major
   h. adding more health-conscious offerings to the school or company cafeteria menu
   i. altering the programming on a campus radio station
   j. installing a new LCD overhead

4. Write an unsolicited sales proposal, similar to the one in Figure 13.7, on one of the following services or products you intend to sell or on a topic your instructor approves:
   a. providing exterminating service to a store or restaurant
   b. supplying a hospital with rental television sets for patients' rooms
   c. designing websites
   d. obtaining temporary office help or nursing care
   e. supplying landscaping and lawn care work
   f. testing for noise, air, or water pollution in your community or neighborhood
   g. furnishing transportation for students, employees, or members of a community group
   h. offering technical consulting service to save a company money
   i. digging a septic well for a small apartment complex
   j. supplying insurance coverage to a small firm (five to ten employees)
   k. cleaning the parking lot and outside walkways at a shopping center
   l. selling a piece of equipment to a business
   m. making a work area safer
   n. preparing a technology seminar or training program for employees
   o. increasing donations to a community or charitable fund
   p. offering discounted memberships at a fitness center

5. Write a solicited proposal for one of the topics listed in Exercise 4 or for a topic that your instructor approves. You might want to review Figure 13.5. Do this exercise as a collaborative effort.

6. Write an appropriate proposal—internal, solicited sales, or unsolicited sales—based on the information contained in one of the following three articles. Assume that you or your prospective customer's company or community faces a problem similar to one discussed in one of these articles. Use as much of the information in the article as you need and add any details of your own that you think are necessary. This exercise can be done as an individual or collaborative assignment.

(a) Multiuse Campuses: A Plan That Works

Gaylord Community School in Gaylord, Michigan, is a bustling center of activity from the first light of dawn to well after dusk. People of all ages come and go until late into the evening for a multitude of activities that include attending classes and meetings, catching up with friends, getting a flu shot, and seeing a play. That's because in addition to a high school, the campus also includes senior and day-care centers, classrooms for adult education, an auditorium for the performing arts, a community health care site, and even a space that can be booked for weddings and other special occasions.

In Big Lake, Minnesota, elementary, middle, and high school buildings are all situated on one centrally located campus that makes up the entire Big Lake School District. Also included in this innovative layout are a state-of-the-art theater, a community resource center, and a multipurpose athletic arena, all of which are used extensively by the entire community.

Both the Gaylord and Big Lake schools are models of a growing movement toward multiuse community campuses that serve as "anchor[s] in the civic life of our nation," according to U.S. Secretary of Education Richard W. Riley. I recently had the pleasure of visiting Secretary Riley in his office. Also present were AARP President Joe Perkins and National Retired Teachers Association Director Annette Norsman, both of whom are involved in many facets of education and lifelong learning.

We discussed many things, including our concerns about the current increase in the number of students caused by the Baby Boom echo (children of the Boomers) and how that population is going to further stress the already crumbling infrastructure of American schools. We also talked about the need for resources—to employ more teachers, bring technology into the classroom, strengthen educational curriculum and opportunities for all ages—and the pressing need to build and renovate schools. That led to a discussion about the necessity and benefit of involving the whole community in the design and use of new school facilities.

I always thought that it was a shame that the majority of schools are used only a third of the day, three fourths of the year, by only a fifth of the population. Considering that there will be more school construction over the next decade than at any time since the 1950s, it just makes sense to consider the intergenerational and community benefits of multiuse spaces, benefits that include everything from establishing better learning environments to getting more bang for the tax buck.

There are many additional bonuses for multiuse educational complexes: They create an exciting community hub, bring life and culture to a central area, and revitalize and nourish the neighborhood in which they are located.

It's a win-win situation for everyone involved.

(b) Self-Illuminating Exit Signs

The Marine Corps Development and Education Center (MCDEC), Quantico, Virginia, submitted a project recently, to replace incandescent illumination exit signs with self-illuminating exit signs for a cost of $97,238. The first-year savings were anticipated to be about $37,171 with an anticipated payback time of 2.6 years—an excellent prospect. The contractor bid much lower, however, and the actual payback will be about 1.5 years.

What are the benefits of these self-illuminating exit signs? The primary benefit is that virtually all operation and maintenance expense is eliminated for the life of the device, normally from 10 to 12 years. Power failures or other disturbances will not cause them to go out.

In new construction, expensive electrical circuits can be totally eliminated. In retrofits, the release of a dedicated circuit for other use may be of considerable benefit. Initial total cost of installing circuits and conventional devices approximately equals the cost of the self-illuminating signs. Installation labor and expense for the self-illuminating signs is about that of hanging a picture.

The amount of electricity saved varies and depends on whether your existing fixtures are fluorescent (13 to 26 watts) or incandescent (50 to 100 watts). Multiply the number of fixtures $\times$ wattage/fixture $\times$ hours operated/day $\times$ days/year — KWH/year savings. For example, assume:

$$400 \text{ incandescent fixtures}$$
$$\$0.08/\text{KWH} \quad 0.05 \text{ KW/fixture}$$
$$24 \text{ hours/day} \quad 365 \text{ day/year operation}$$
$$400 \times 0.1 \times 365 - 350400 \text{ KWH/year}$$
$$350400 \times 0.08 - \$28,032/\text{year for electricity}$$

Now add in savings achieved from:

- reducing labor to change bulbs
- avoiding bulb material, stocking, and storage costs
- avoiding transportation costs involved in bulb changes
- reusing existing bulbs

The above savings can be significant. For the MCDEC Quantico project, estimates of bulb change interval and savings were 700 hours (29 days) and $13,512/year when all factors were considered.

The cost of a self-illuminating sign depends on whether one or two faces are illuminated primarily and varies between different suppliers. Single-face prices will likely be $100 to $150 while double-face prices may be $250 to $330. The contractor at Quantico found better prices than these ranges indicate. The labor cost should be about $10 per sign.

If you can use an exit-sign system with high dependability, no maintenance, and zero operations cost in your retrofit on new construction projects, try a self-illuminating exit-sign system in your economic analysis today. "Isolite" signs, by Safety Light Corp., are listed as FSC (Fire Safety Code) Group 99, Part IV, Section A, Class 9905 signs and are available through GSA contract. Contact Gerald Harnett, Safety Light Corp., P.O. Box 266, Greenbelt, MD 20070 for more information.

(c) Wheelchair-Lift Switch Covers

In order to ensure year-round access to the Springfield Armory National Historic Site (Massachusetts) museum, Michael C. Trebbe designed the cover for switches on wheelchair lifts. During the extreme New England winters, the switch buttons would freeze, thus making the lift inoperable, which in turn required several hours to thaw. The installation of these covers prevented the freezing of the switch buttons and, therefore, allowed maintenance personnel to attend to matters such as snow removal.

The covers were made of materials found on site, which resulted in the covers being almost cost free. The covers can be quickly built, and they are mounted with the same mounting screws as the switch boxes so as to not destroy any original fabric (in the case of Springfield Armory NHS, brownstone). The materials used included:

- 1/8-in. by 4½-in. by 12-in. piece of rubber mat
- 3½-in. by 6½-in. piece of sheet metal
- three aluminum pop rivets
- primer for the sheet metal
- wheelchair symbol
- white paint for the symbol

The sheet metal is bent to a 90-degree angle at the 5½-inch point. The lowest two holes (see diagram) are drilled to mount the screws of the switch box, which also secure the cover. Triangular cutouts and other holes are drilled for the clearance of the housing screws on the rear of the switch box (see diagram).

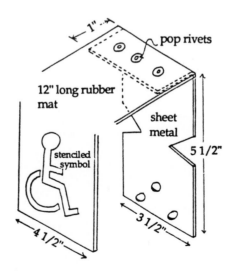

The handicap symbol is stenciled on the front of the piece of rubber mat using white paint.

7. Write a suitable research proposal on which the report on multinational workers in Chapter 15 (pp. 652–667) could have been based.

8. Write a research proposal to your instructor seeking approval for a research-based long report. Do the necessary preliminary research to show that you have selected a suitable topic, narrowed it, and identified the sources of information you have to consult. List at least six relevant articles, two books, and two websites pertinent to your topic.

Writing Effective Short Reports

This chapter shows you how to write short reports, also called informal or semiformal reports. A short report can be defined as an organized presentation of relevant data on any topic—money, travel, time, personnel, equipment, technology, management—that a company or agency tracks in its day-to-day operations. Short reports are practical and to the point. They show that work is being done, and they also show your boss that you are alert, professional, and reliable. Short reports are written to co-workers, employers, vendors, and clients. When they are intended for individuals within your organization, short reports are most often sent as memos. But for clients you will send your reports out as letters.

Businesses cannot function without short written reports. Reports tell whether

- work is being completed
- schedules are being met
- costs have been contained
- sales projections are being met
- unexpected problems have been solved

You may write an occasional report in response to a specific request or question, or you may be required to write a daily or weekly report on routine activities. Short reports are vital to a company's or organization's success. They can make things at work run more efficiently and economically. But when they are poorly done, they can cost your company money and time.

Visit www .cengage.com/ english/kolin for this chapter's online exercises, ACE quizzes, and Web links.

Types of Short Reports

To give you a sense of some of the topics you may be required to write about, here is a list of various types of short reports common in the business world.

appraisal report	design report	incident report
audit report	evaluation report	inventory report
budget report	experiment report	investigative report
construction report	feasibility report	justification report

laboratory report	periodic report	sales report
manager's report	production report	status report
medicine/treatment	progress/activity report	survey report
error report	recommendation report	test report
operations report	research report	trip/travel report

This chapter concentrates on the six most common types of reports you are likely to encounter in your professional work:

1. periodic reports
2. sales reports
3. progress reports
4. trip/travel reports
5. test reports
6. incident reports

The first five reports can be called **routine reports** because they give information about planned, ongoing, or recurring events. The sixth category, **incident reports**, are reports that describe events the writers did not anticipate—accidents, breakdowns, environmental mishaps, delivery delays, or work stoppages. All six, however, can be termed *short* reports because they deal with current happenings rather than long-range forecasts. Short reports focus on the "trees," not the "forest."

Tech Note

Creating Templates for Short Reports

You can use your word processor's template function to create professional-looking reports. Templates, as we saw on pages 493–494, are predesigned formats for page layouts that specify style elements of a document. They allow you to automate designs for periodic, sales, progress, incident, and other types of reports. An existing template will format information to create a professional-looking report. Using templates, you reduce the risk of errors, omissions, and inconsistencies and ensure that you follow your company's style and format.

To create a specific report template, simply set the format options (such as font, margins, and line spacing) for a new document, typing place markers for your text (including headers, footers, and titles), importing custom visuals (such as a company logo), and saving the document as a template. When you need to create a new report using the template, simply open it, insert the content of your report, and save the document under a new file name as a standard word processing file. You can customize the formatting of any or all of the following style elements:

- Headers, footers for a company address, titles, and dates
- List formatting, such as bullet-point style and numbered lists

- Line spacing and text justification
- Font style, size, and color
- Standard graphics, for example, a flowchart or line divisions
- Margin size, paragraph indentation, and columns
- Standard visual elements such as tables, which can also be filled with the data appropriate to your report
- Automatic table of contents based on the titles used in your report

But keep the following precautions in mind when using templates:

- Be sure to follow your company's style when creating templates, particularly those using a company logo.
- Double-check everything in the template to make sure it is appropriate for the type of report you're writing. Differences between reports will require visual adjustments.
- When creating a template with place markers to indicate the position of certain textual elements, insert them in boldface for emphasis. Using brackets (for example, **[type title here]** or **[body of text here]**) will highlight and thus emphasize these place holders.

Guidelines for Writing Short Reports

Although there are many kinds of short reports, they all are written for readers who need factual information so they can get a job accomplished. The following guidelines will help you write any short report successfully.

Do the Necessary Research

An effective short report needs the same careful planning that goes into other on-the-job writing. Your research may be as simple as instant messaging, e-mailing, or leaving a voice mail for a colleague or inspecting a piece of equipment. Or you may have to test or inspect a product or service or assess the relative merits of a group of competing products or services. Some frequent types of research you can expect to do on the job include:

- checking data in reference manuals or code books
- searching archives or databases for recent discussions of a problem or procedure
- reading background information in professional and trade journals
- reviewing a client's file
- testing equipment
- performing an experiment or procedure

- conferring with colleagues, managers, vendors, or clients
- visiting a site and taking field notes
- attending a conference

Follow these guidelines to do effective research for your short reports:

1. Take careful notes, either by hand, on your laptop, or on your PDA.
2. Record all necessary background information that helps readers understand why the report is being written.
3. Collect relevant factual data—dates, places, costs, technology, personnel, conditions.
4. Interview or consult key individuals whose comments and assessments are important to the success of your report. Don't leave essential viewpoints or people out of the loop.

You may want to review relevant sections of Chapter 8 to develop your research strategies further.

Anticipate How an Audience Will Use Your Report

Consider how much your audience knows about your project and what types of information they need most. If your reader requests the report, he or she may be knowledgeable about your topic and approach. But you can't always assume readers will have such information or backgrounds. Employers, the largest audience for your reports, may not always know (or be interested in) the technical details of your work. Instead, they want bottom-line information—costs, personnel, organizational structure, problems, or delays. While co-workers may be familiar with your project, colleagues in other departments, consumers, or individuals outside your company (such as site inspectors or clients) may not. Accordingly, those readers may require more background information, definitions, and examples. Regardless of your audience, your report should answer these five essential questions: who, what, where, why, and how.

All audiences will also expect you to do the following in your report writing:

- **Begin with the most important information**. Do not keep your audience in suspense. Give readers what they want as quickly as possible Your first sentence should focus on why the report was written and why it is important. See how Figures 14.6 and 14.7 later in this chapter supply effective opening paragraphs.
- **Give only essential information**. Don't pad your report with unnecessary details or trivia that will slow down or annoy readers. Focus on how your evidence supports the plan or project. Concentration the big picture.
- **Document information** you give readers about schedules, costs, personnel, sales, materials, the environment, and so forth.

Tech Note

Using the Web to Write Short Reports

Many government agencies provide the statistical raw data that go into various types of short reports—employment figures, population data, environmental statistics, and so on. Downloading and incorporating information from these and other relevant sites give readers the necessary documentation to accept a conclusion or recommendation.

Other businesses also post information that may be relevant to your report for your company. Large corporations such as General Motors or IBM help you see the progress of stocks and mutual funds, or find out other industry-wide information. Finally, ConsumerReports.org (*http://www.consumerreports.org*) evaluates different brands or products so that you don't have to do testing yourself.

Here are some sites that publish appropriate information for short reports:

BizStats (*http://www.bizstats.com*)
Bureau of Labor Statistics (*http://www.bls.gov*)
Bureau of Transportation Statistics (*http://www.bts.gov*)
Federal Reserve (*http://www.federalreserve.gov*)
National Center for Educational Statistics (*http://www.nces.ed.gov*)
National Center for Health Statistics (*http://www.cdc.gov/nass*)
New York Stock Exchange (*http://www.nyse.com*)
Statistics.com (*http://www.statistics.com*)
UNESCO Institute for Statistics (*http://www.uis.unesco.org*)
United Nations Statistics Division (*http://www.un.org/unsd*)
U.S. Census Bureau (*http://www.census.gov*)

Be Objective and Ethical

Your readers will expect you to report the facts objectively and impartially—costs, sales, eyewitness accounts, observations, statistics, test measurements, and descriptions. Your reports should be truthful, accurate, and complete. Here are some guidelines.

- Avoid guesswork. If you don't know or have not yet found out, say so and indicate how you intend to find out.
- Do not substitute impressions or unsupported personal opinions for careful research.
- Provide a straightforward and honest account; don't exaggerate or minimize. Using biased, skewed, or incomplete data is unethical.
- Double-check all facts, figures, and specifications. A mistake here could be costly or potentially dangerous.

Your employer may also ask you to keep your report confidential and share the results with only key members of your organization. You may also want to review the section on ethics in business writing (pp. 27–35).

Organize Carefully

Organizing a short report effectively means that you include the right amount of information in the most appropriate places for your audience. Many times a simple chronological or sequential organization will be acceptable for your readers. Readers will expect your report to contain information on its purpose, findings, recommendations, and a conclusion, as described in the following sections.

Purpose

Always begin by telling readers why you are writing (your goal) and by alerting them to what you will discuss. Provide all necessary background information. For example, give your readers a summary of key events and details at the beginning to help them follow the remainder of the report quickly. When you establish the scope (or limits) of your report, you help readers zero in on specific times, places, procedures, or problems.

Findings

This should be the longest part of your report and contain the data you have collected—facts about prices, personnel, equipment, events, locations, incidents, the results of a survey, the status of a project, or tests. Gather the data from your research; personal observations; interviews; and/or conversations with co-workers, employers, or clients. Choose only those details that are most important to your readers.

Conclusion

Your conclusion tells readers what you think the data mean. A conclusion can summarize what has happened; review what actions were taken; or explain the outcome or results of a test, a visit, or program.

Recommendations

A recommendation informs readers what specific actions you think your company or client should take—market a new product, hire more staff, institute safety measures, select among alternative plans or procedures, prioritize options, and so on. Recommendations must be *relevant, creditable,* and *reliable.* If they are not, your report could be responsible for sending your company down the wrong path. Recommendations need to be based on the data you collected, the resources (budget) and schedule of your department and company, and the conclusions you have reached. They need to show how all the pieces fit together.

Write Clearly and Concisely

Writing clearly and concisely is essential in all business reports. Busy readers, who may need to look at several other reports the day yours comes in, have little time to waste. They will appreciate your making their job easier and faster. If a short report is longer than two or three pages, it may be too long. Ask your boss or co-workers who have done short reports for your company about appropriate length, style, and documentation. Look at previous, similar reports to get a sense of your company's style.

Here are a few guidelines to help you write clearly and concisely.

1. **Use an informative title/subject line that gets to the point right away.** "Software Options" is not as clear as "Which Software Options Are Most Economical and Why."

2. **Write in plain English.** Make every word count, avoid jargon, and keep your writing simple and straightforward. Prune business clichés such as "at the end of the day" or "to put a fine point on it."

3. **For global readers, make sure you use international English.** Keep your sentences short and write in the active voice. Avoid sentence fragments. Do not use U.S. idioms, slang, or abbreviations that could be misunderstood. See pages 173–180.

4. **Adopt a professional yet personal tone.** Avoid being overly formal, or too casual—strike a balance between these two extremes. Don't be arrogant by writing in a tone that suggests you alone have the final authority—you can be assertive without being arrogant.

5. **Do not include unnecessary background information.** You may be tempted to demonstrate that you have done your research thoroughly, but your manager and other readers do not need or want all of the technical details, lengthy project histories, or a technical appendix. Such information will overburden a busy decision maker. Make sure, though, that you document all of the sources you relied on to write the report.

Use an Appropriate Format and Visuals

The way your report looks will influence how your readers will respond to it and to you. Here are some useful guidelines (you may also want to review Chapters 10 and 11 on visuals and document design):

1. **Make your report look professional, readable, and easy to follow.** Don't flood your report with color. See the Tech Note on page 600. Avoid using flashy color or fancy fonts that are hard to read. Also, don't try to squeeze too much text onto the page. Always leave comfortable margins. Design your report to assure your reader that it is well thought out and self-explanatory.

2. **Help readers locate and digest information quickly.** Use headings, subheadings, bullets, and numbered lists to show how you have organized your report clearly. Doing this, you break large portions of text into easy-to-read parts. Your headings and subheadings give readers the big picture at a glance. Many reports in this chapter demonstrate how headings and bulleted/numbered lists assist readers, for instance, see Figures 14.2, 14.3, and 14.10.

3. **Be consistent in your design and format.** Use the same font throughout the text of your report and a consistent typeface for headings and subheadings.

4. **Include only the most essential visuals.** Use visuals only if they make the reader's job easier, reinforcing or summarizing key data quickly, as the table in Figure 14.2 and the map in Figure 14.8 do. Keep visuals simple and relevant— a picture or drawing to illustrate a point. Do not get carried away with inserting numerous visuals for a short report.

5. **Place visuals in the most appropriate place in your report.** Design or import visuals so that they are easy to read, and place them close to the text that they help explain.

Case Study: A Poor and an Effective Short Report

Figure 14.1 is an example of a short report that does not follow the guidelines for writing short reports above. But Figure 14.2, a revised version of Figure 14.1, does. Read through both reports, keeping in mind the differences below:

- **Research.** Although Figure 14.1 includes department statistics, it does not explain or provide a recommendation based on this data. Figure 14.2, however, provides explanations, supplies more detail, and gives recommendations. In addition, this figure looks at other sources rather than just presenting the reader with "uncrunched" numbers.
- **Audience analysis.** Instead of providing the most important information first, analyzing the statistics, or offering recommendations, the report in Figure 14.1 simply throws facts at the audience with only a vague consideration of why they are important. Figure 14.2, however, consistently takes the audience's needs into account by focusing on how Captain Martin will use the statistical information about the crime rate during the second quarter of the year. To help Captain Martin, Figure 14.2 presents the most important information first, then explains and analyzes the numbers for the audience, and provides realistic and direct recommendations.
- **Objectivity/ethics.** In boasting, "lets the facts speak for themselves," Figure 14.1 does not give Captain Martin the full picture. In addition, since details have not been checked against any other sources, the report is incomplete and possibly inaccurate. Figure 14.2, however, eliminates the guesswork and, more ethically, offers solutions, careful analysis, and a variety of relevant sources.
- **Organization.** Figure 14.1 makes no attempt to organize the report in a reader-friendly manner. The purpose statement is vague, the findings are thrown at the audience in three dense and disorganized paragraphs, and the conclusion does not summarize the facts. Figure 14.2, on the other hand, supplies a clear purpose statement, organizes and summarizes the facts concisely, and helpfully classifies recommendations.
- **Writing style and tone.** Figure 14.1 is just an accumulation of numbers, making it hard for readers to access or understand their importance. The tone is smug, arrogant, and self-congratulatory. But Figure 14.2 is easy to follow and to understand. It uses helpful connective words and phrases ("compared to last quarter," "however," "overall," "yet") and includes important contexts ("were less than last quarter"). This report is both professional and reader-friendly at the same time.
- **Format and visuals.** Figure 14.1 does not include headers, bullets, visuals, or consistent fonts and paragraph indentation. Figure 14.2 instead supplies clear heads, breaks the text beneath into easy-to-digest subheads, places numerical data in visual form (Table 1), and uses bulleted lists to help Captain Martin better understand the significance and scope of the report.

An example of a poorly written, poorly organized, and poorly formatted short report.   **Figure 14.1**

To Serve and To Protect

Greenfield Police Department

Emergency 555-1000   **Administration** 555-1001   **Traffic** 555-1002

TO: Capt. Alice Martin
FROM: Sergeants Daniel Huxley, Jennifer Chavez,
                    and Ivor Paz
SUBJECT: Crimes
DATE: July 12, 2008

This report will let you know what happened this quarter as opposed to what happened last quarter as far as crimes are concerned in Greenfield. This **report is based on statistics** the department has given us over the quarter.

Here we'll let the facts speak for themselves. From Jan.–Mar. we saw 126 robberies while from Apr.–June we had 106. Home burglaries for this period: 43; last period: 36. 33 cars were stolen in the period before this one; now we have 40. Interestingly enough, **last year at this time we had only 27** thefts. Four of them involved heirlooms.

Homicides were 9 this time versus 8 last quarter; assault and battery charges were 92 this time, 77 last time. Carrying a concealed weapon 11 (10 last quarter). We had 47 arrests (55 last quarter) for charges of possession of **a controlled substance**. Rape charges were 8, **1 less than last quarter.** 319 citations this time for moving violations: speeding 158/98, and failing to observe the signals 165/102 last quarter. DUIs this quarter—only 45, or 23 fewer than last quarter.

Misdemeanors this time: disturbing peace 53; vagrancy/ public drunkenness 8; violating leash laws 32; violating city codes 39, including **dumping trash**. Last quarter the figures were 48, 59, 21, 43.

We believe this report is **complete and up-to-date**. We further hope that this report has given you all the **facts** you will need.

Poor format—some paragraphs indented, some not; boldfacing not helpful

Vague subject line

Introduction doesn't tell reader anything about overall picture

Throws facts out without any sense of reader's needs

Irrelevant data

No analysis or guided commentary— just undigested numbers

Hard-to-follow comparisons and contrasts

Conclusion provides no summary or recommendation

Figure 14.2   A well-prepared report, revised from Figure 14.1.

To Serve and To Protect

Greenfield Police Department

Emergency 555-1000   **Administration** 555-1001   **Traffic** 555-1002

TO: Captain Alice Martin
FROM: Sergeants Daniel Huxley, Jennifer Chavez, and Ivor Paz
SUBJECT: Crime rate for the second quarter of 2008
DATE: July 12, 2008

From April 1 to June 30, 852 crimes were committed in Greenfield, representing a 5 percent increase over the 815 crimes recorded during the previous quarter.

TYPES OF CRIME

The following report, based on the table below, discusses the specific types of crimes, organized into four categories: **robberies and theft, felonies, traffic,** and **misdemeanors**.

Table 1. Comparison of the 1st and 2nd Quarter Crime Rates in Greenfield

CATEGORY	1st Quarter	2nd Quarter
ROBBERIES and THEFT		
Commercial	63	75
Domestic	36	43
Auto	33	40
FELONIES		
Homicides	16	13
Assault and battery	77	92
Carrying a concealed weapon	10	11
Poss. of a controlled substance	55	47
Rape	9	8
TRAFFIC		
Speeding	165	158
Failure to observe signals	102	98
DUI	78	45
MISDEMEANORS		
Disturbing the peace	48	53
Vagrancy	40	48
Public drunkenness	19	40
Leash law violations	21	32
Dumping trash	35	37
Other	8	12

Robberies and Theft

The greatest increase in crime was in robberies, 20 percent more than last quarter. Downtown merchants reported 75 burglaries, exceeding $985,000. The biggest theft occurred on May 21 at Weisenfarth's Jewelers when three armed robbers stole more than $97,000 in merchandise. (Suspects were

Page 2

apprehended two days later.) Home burglaries accounted for 43 crimes, though the thefts were not confined to any one residential area. We also had 40 car thefts reported and investigated.

Felonies
Homicides decreased slightly from last quarter—from 16 to 13. Charges for battery, however, increased—15 more than we had last quarter. Arrests for carrying a concealed weapon were nearly identical this quarter to last quarter's total. But the 47 arrests for possession of a controlled substance were appreciably down from the first quarter. Arrests for rape for this quarter also were less than last quarter's. Three of those rapes happened within one week (May 6–12) and have been attributed to the same suspect, now in custody.

Provides essential background and statistical information and comparative analyses

Traffic
Traffic violations for this period were lower than last quarter's figures, yet this quarter's citations for moving violations (335) represent a 5 percent increase over last quarter's (322). Most of the citations were issued for speeding (158) or for failing to observe signals (98). Officers issued 45 citations to motorists for DUIs, an impressive decrease over the 78 DUIs issued last quarter. The new state penalty of withholding a driver's license for six months of anyone convicted of driving while under the influence appears to be an effective deterrent.

Easy-to-read sentences

Misdemeanors
The largest number of arrests in this category were for disturbing the peace—53. Compared to last quarter, this is an increase of 10 percent. There were 88 charges for vagrancy and public drunkenness, an increase from the 59 charges made last quarter. We issued 32 citations for violations of leash laws, which represents a sizable increase over last quarter's 21 citations. Thirty-seven citations were issued for dumping trash at the Mason Reservoir.

Includes only data reader needs

CONCLUSION
Overall, while the crime rate has decreased in traffic (especially DUIs) and possession of controlled substances this quarter, we have seen a marked increase in arrests for robberies and battery.

Summarizes findings of report

RECOMMENDATIONS
To help deter robberies in the downtown area, we recommend the following:

- increasing surveillance units in the area
- offering merchants our workshop on safety and security precautions, as we did during the first quarter

Historically, battery arrests have risen during the second quarter. Our recommendations to counter this trend include:

- continuing to work closely with Neighborhood Watch Groups
- providing more foot and bicycle patrols in the neighborhoods with the highest incidence of battery complaints

Offers specific actions/changes department should make based on conclusion of report in bulleted lists

Periodic Reports

Periodic reports, as their name signifies, provide readers with information at regularly scheduled intervals—daily, weekly, bimonthly (twice a month), monthly, quarterly. They help a company or agency keep track of the quantity and quality of the services it provides and the amount and types of work done by employees. Information in periodic reports helps managers make schedules, order materials, assign personnel, budget funds, and, generally speaking, determine corporate needs.

Figure 14.2 provides essential police information clearly and concisely. As we saw, it summarizes, organizes, and interprets the data collected over a three-month period from individual activity logs. Because of this report, Captain Alice Martin will be better able to plan future protection for the community and to recommend changes in police services.

Sales Reports

Sales reports provide businesses with a necessary and ongoing record of accounts, online and mail purchases, losses, and profits over a specified period of time. They help businesses assess past performance and plan for the future. In doing that, they fulfill two functions: **financial** and **managerial**. As a financial record, sales reports list costs per unit, discounts or special reductions, and subtotals and totals. Like a spreadsheet, sales reports show gains and losses. They may also provide statistics for comparing two quarters' sales.

Sales reports are also a managerial tool because they help businesses make both short- and long-range plans. The restaurant manager's sales report illustrated in Figure 14.3 guides the owners in planning which popular entrées to highlight and which unpopular ones to change or delete. Note how the recommendations follow logically from the figures Sam Jelinek gives to Gina Smeltzer and Alfonso Zapatta, the owners of The Grill.

Progress Reports

Progress reports inform readers about the status of ongoing projects. They let readers know how much and what type of work has been done by a particular date, by whom, how well, and how close the work is to being completed. A progress report emphasizes whether you are

- keeping within your schedule
- staying within your budget
- using the proper equipment
- making the right assignments
- identifying an unexpected problem or possible glitch
- providing adjustments in schedules, personnel, and so on
- completing the job efficiently, correctly, and according to codes

Almost any kind of ongoing work can be described in a progress report—construction of an apartment complex, creation of a website, documentation of a patient's rehabilitation, and so on. These reports are often prepared at key phases, or milestones, in a project.

A sales report to a manager.   **Figure 14.3**

Dayton, OH 43210 • (813) 555-4000 • (813) 555-4100 fax ⎍ www.Thegrill.com

TO: Gina Smeltzer   DATE: June 28, 2009
        Alfonso Zapatta, Owners
FROM: Sam Jelinek S.J.   SUBJECT: Analysis of entrée sales,
        Manager   June 10–16 and June 17–23

As we agreed at our monthly meeting on June 4, here is my analysis of entrée sales for two weeks to assist us in our menu planning. Below is a record of entrée sales for the weeks of June 10–16 and June 17–23 that I have compiled and put in a table for easier comparisons.

Begins with purpose and scope of report reader needs

	Portion size	June 10–16		June 17–23		2 weeks combined	
		Amount	**Percentage**	**Amount**	**Percentage**	**Amount**	**Percentage**
Cornish Hen	6 oz.	238	17	307	17	545	17
Stuffed Young Turkey	8 oz.	112	8	182	10	294	12
Broiled Salmon Steak	8 oz.	154	11	217	12	371	13
Brook Trout	12 oz.	182	13	252	14	434	9
Prime Rib	10 oz.	168	12	198	11	366	11
Lobster Tails	2–4 oz.	147	10	161	9	308	10
Delmonico Steak	10 oz.	56	4	70	4	126	4
Moroccan Chicken	6 oz.	343	25	413	23	756	24
		1,400	**100**	**1,800**	**100**	**3,200**	**100**

Organizes findings of the report in helpful table

Boldfaces totals

Recommendations
Based on the figures in the table above, I recommend that we do the following:

1. Order at least 100 more pounds of prime rib each two-week period to be eligible for further quantity discounts from the Northern Meat Company.
2. Delete the Delmonico steak entrée because of its low acceptance.
3. Introduce a new chicken or fish entrée to take the place of the Delmonico steak; I would suggest grilled lemon chicken to accommodate those patrons interested in tasty, low-fat, lower-cholesterol entrées.

Offers precise and relevant recommendations

Please give me your reactions within the next week. It shouldn't take more than a few days to implement these changes.

Requests authorization to implement recommendations

Audience and Length for a Progress Report

A progress report is intended for individuals who are not generally working alongside you but who need a record of your activities to coordinate them with other individuals' efforts and to learn about problems or changes in plans. For example, since supervisors (or non-native speakers of English who manage overseas offices) may not be in the field or branch office, they will rely on your progress report for much of their information. Customers, such as a contractor's clients, expect reports on how carefully their money is being spent, if schedules are being met, and whether there is a risk of going over the budget.

The length of the progress report will depend on your audience and on the complexity of the project. A short e-mail about organizing a time management workshop, such as that in Figure 14.4, might be all that is necessary. A report to a

Figure 14.4   A one-time progress report sent as an e-mail.

| Reply | Save | Forward | Print | Delete |

To: <ksands@multiplex.org>
From: <pjavon@multiplex.org>
Date: Oct. 12, 2008 1:23:31 PM EDT
Subject: Preparations for Time Management Workshop

Refers to reader's needs

As you requested last week, I e-mailed the managers of all departments in both our Trenton and Frankfurt, Germany, offices on Monday, October 9, to remind them of the time management workshop we will be offering on November 9 by teleconference.

Summarizes work done by citing specific dates, places, names, and means of collaboration

I have confirmed the date and the operation of the technical links and relays with Carmen Suarez in Technical Services and have also e-mailed Jürgen Weiss in Frankfurt to make sure things are in place there.

I have reserved the corporate conference center for November 9 and have ordered DTP copies of all the packets we will need. The packets going to Germany will be Jet-Expressed, overnight delivery, on November 5 so they will be in Frankfurt two days before the teleconference.

Assures reader that project is on schedule

By tomorrow, I will prepare a list of all those employees scheduled to participate in the workshop and send it to you.

Plans are going according to schedule.

Philip Javon, Projects Coordinator
Trenton Branch
3800 S. Morris
Hershey, PA 12346
VOICE: 657-555-1719
<Pjavon@multiplex.org>

boss about the progress an employee is making on a research report easily could be handled in a memo, such as Barbara Shoemake's progress report in Figure 14.5 or her research report described in the proposal in Chapter 13 (pp. 588–590). Similarly, Dale Brandt's assessment of the progress his construction company is making in renovating Dr. Burke's office is given in a two-page letter in Figure 14.6.

A progress report regarding a research report.   **Figure 14.5**

TO: Marisol Vega
FROM: Barbara R. Shoemake B.R.S.
DATE: Oct. 10, 2008
SUBJECT: First Progress Report on Research Report

This is the first of two progress reports that you asked me to submit about my background research report on the ethical and security issues of using e-mail.

Clearly states purpose of the report

From March 8 until April 6, I gathered information from library holdings, the Internet, and an interview. Of the fourteen references listed in my proposal, I found only ten. Articles by Cruiz-Taylor ("Corporate or Confidential") and Hernandez et al. ("Legal Foundations for Monitoring Employee E-Mail") are not available at the library or on the Internet. Two of my Internet sites—"Internet Mail Consortium" and Seetha Kury's "New Security Policies" —are under construction. But *i-TechMag* will be up and running in the next few days. In the meantime, I'll try to replace "Internet Mail Consortium."

Provides update by citing individual references and actions taken

On September 23, I had an extended interview (1½ hours) with Katrina Kuhn, a human resources manager. She gave me some seminar handouts as well as a copy of a report on e-mail protocols that she wrote for the Society of Midwest Business Communicators—materials I hope to incorporate in my report.

Details results of primary research

Because of an extended trip to Denver, Paul Goya could not meet with me. At his suggestion, I am trying to schedule an interview with Robert Sims, the Documents Manager at Mid-Atlantic Power Company. Sims has given several seminars on e-mail security. Even if Mr. Goya cannot meet with me, Ms. Kuhn gave me enough information about a business manager's view of systems. However, not currently having the articles and websites listed above may slow, but not stop, my work.

Gives detailed descriptions of steps in the research process

Starting tomorrow, I will begin my report and can submit a draft by Oct. 21. You will receive my second progress report by Oct. 15.

Concludes with next steps

Figure 14.6   The second of three progress reports from a contractor to a customer.

Brandt Construction Company

Halsted at Roosevelt, Chicago, Illinois 60608-0999 • 312-555-3700 • Fax: 312-555-1731
http://www.brandt.com

April 28, 2009

Dr. Pamela Burke
1439 Grand Avenue
Mount Prospect, IL 60045-1003

Dear Dr. Burke:

Begins with key information: project status

Here is my second progress report about the renovation work being done at your new clinic at Hacienda and Donohue. Work proceeded satisfactorily in April according to the plans you had approved in March.

Review of Work Completed in March

Recaps for background

As I informed you in my first progress report on March 31, we tore down the walls, pulled the old wiring, and removed existing plumbing lines. All the gutting work was finished in March.

Work Completed During April

Summarizes current accomplishments, including attention to greening the building

By April 7, we had laid the new pipes and connected them to the main septic line. We also installed the two commodes, the four standard sinks, and the utility basin. The heating and air-conditioning ducts were installed by April 13. From April 16–20, we erected soundproof walls in the four examination rooms, the reception area, your office, and the laboratory. Throughout we used environmentally safe materials. We had no problems reducing the size of the reception area by five feet to make the first examination room larger, as you had requested.

Problems with the Electrical System

Identifies problems, how they were solved, and costs involved

We had difficulty with the electrical work, however. The number of outlets and the generator for the laboratory equipment required extra-duty power lines that had to be approved by both Con Edison and Cook County inspectors. The approval slowed us down by three days. Also, the vendor, Midtown Electric, failed to deliver the recessed lighting fixtures by April 27 as promised. Those fixtures and the generator are now being installed. Moreover, the cost of those fixtures will increase the material budget by **$5,288.00**. The cost for labor is as we had projected—**$89,450**.

Work Remaining

Specifies what work remains and when it will be completed

The finishing work is scheduled for May. By May 11, the floors in the examination rooms, laboratory, washrooms, and hallways should be tiled and the reception area

page 2

and your office carpeted. By May 13, the reception area and your office should be paneled and the rest of the walls painted. If everything stays on schedule, touch-up work is planned for May 14–18. You should be able to move into your new clinic by May 21.

Projects suc-cessful comple-tion of work

You will receive a third and final progress report by May 14. Thank you again for your business and the confidence you have placed in our company.

Promises to keep reader informed

Sincerely yours,

Dale Brandt

Tech Note

RSS Feeds

Really simple syndication (RSS) feeds help readers keep up with constantly changing information on the Internet. Available through a subscription service, RSS feeds track updates from a variety of sources including blogs, podcasts, news sites, weather reports, financial information, stock quotations, sales trends, product information, medical websites, or even new books in your local library. RSS feeds are distributed in "streams" or "channels," and are collected by an RSS "aggregator," a software program that allows users to identify, gather, organize, and read updated information on a regular basis in areas most relevant to their job.

Most feeds generally consist of a headline, a summary/abstract, and a link to the original website, as in the feed below:

```
<item>
<title>Nutrition Labels: How Precise Are They?</title>
<description>The FDA allows food to have 20 percent more diet-damaging
nutrients than the labels state.</description>
<link>http://abcnews.go.com</link>
</item>
```

RSS feeds may also contain photographs, videos, sound, or other media content. Often, too, the user's RSS aggregator will offer several helpful features and tools for sharing e-mail, blog posts, and the like and indicate that the user has already seen or read a particular post.

Continued

Continued

RSS feeds benefit report writers in several ways by:

- listing the latest headlines in an area of interest
- providing summaries and update time stamps so readers do not have to visit large numbers of websites each day to see if anything new has been added
- helping users quickly determine if they need to visit the updated websites themselves
- offering custom searches saved as RSS feeds
- organizing feeds for convenient browsing

RSS feeds are especially helpful to writers who have to generate short reports based on assimilating data from a wide variety of sources. For instance, when an advertising accounts manager has to create a weekly report about automotive industry data to determine what advertisements should be shown in a particular market, she can use an RSS aggregator to learn about trends and the most recent industry news from a variety of the most relevant sources. An aggregator helps the manager group feeds by subject, date, or by another area relevant to her report, making it a relatively simple task to quickly find, scan, and extract the data she needs for her report.

Steven Turner

Frequency of Progress Reports

Progress reports can be written daily, weekly, monthly, quarterly, or annually. Your specific job and your employer's needs will dictate how often you have to keep others informed of your progress. A single progress report is sufficient for Philip Javon's purpose in Figure 14.4. Barbara Shoemake was asked to submit two progress reports, the first of which is found in Figure 14.5. Contractor Brandt determined that three reports, spaced four to six weeks apart, would be necessary to keep Dr. Burke posted; Figure 14.6 is the second of those reports.

Parts of a Progress Report

Progress reports should contain information on (1) the work you have done, (2) the work you are currently doing, and (3) the work you will do.

How to Begin a Progress Report
In a brief introduction

- indicate why you are writing the report
- provide any necessary project titles or codes and specify dates
- help readers recall the job you are doing for them

If you are writing an initial progress report, supply background information in the opening. Philip Javon's first sentence in Figure 14.4, for example, quickly establishes his purpose by reminding Kathy Sands of their discussion last week.

Similarly, Barbara Shoemake in the first paragraph in Figure 14.5 states the purpose and scope of her work.

If you are submitting a subsequent progress report, inform your reader about where your previous report left off and where the current one begins. Make sure you clearly specify the period covered by each report. Note how Dale Brandt's first paragraph in Figure 14.6 calls attention to the continuity of his work.

Body

The body of the report should provide significant details about costs, materials, personnel, and times for the major stages of the project.

- Emphasize completed tasks, not false starts. If you report that the carpentry work or painting is finished, readers do not need an explanation of paint viscosity or geometrical patterns.
- Omit routine or well-known details ("I had to use the library when I wanted to read the back issues of *Safety News* that were not on the Web").
- Describe in the body of your report any snags you encountered that may affect the work in progress. See Dale Brandt's section on electrical problems in Figure 14.6. It is better for the reader to know about trouble early in the project, so appropriate changes or corrections can be made.

Conclusion

The conclusion should give a timetable for the completion of duties or submission of the next progress report. Give the date by which you expect work to be completed. Be realistic; do not promise to have a job done in less time than you know it will take. Readers will not expect miracles, only informed estimates. Even so, any conclusion must be tentative. Note that the good news Dale Brandt gives Dr. Burke about moving into her new clinic is qualified by the words "If everything stays on schedule." He is also well aware of the "you attitude" by thanking Dr. Burke again for her business.

◼ Trip/Travel Reports

Reporting on the trips you take is an important professional responsibility. In documenting what you did and saw, trip reports keep readers informed about your efforts and how they affect ongoing or future business. Trip reports also should be written after you attend a convention or sales meeting or call on customers.

Questions Trip/Travel Reports Answer

Specifically, a trip report should answer the following questions for your readers:

- Where did you go?
- When did you go?
- Why did you go?
- Whom did you see?
- What did they tell you?
- What did you do about it?

For a business trip you are also likely to have to inform readers how much the trip cost and to supply them with receipts for all your expenses.

Common Types of Trip/Travel Reports

Trip reports can cover a wide range of activities and are called by different names to characterize those activities. Most likely, you will encounter the following three types of trip/travel reports:

1. **Field trip reports.** These reports are written after a visit to a laboratory, military installation, office complex, hospital, detention center, or other facility to show what you have learned about the operation of those places. You will be expected to describe how an institution is organized, the technical procedures and/or equipment it uses, pertinent ecological conditions, or the ratio of one group to another. The emphasis in these reports is on the educational value of the trip, as Mark Tourneur's report in Figure 14.7 (pp. 619–620) demonstrates.

2. **Site inspection reports.** These reports inform managers about conditions at a branch office or plant, a customer's business, or on the advisability of relocating an office or other facility. After visiting the site, you will determine whether it meets your employer's (or customer's) needs.

Site inspection reports tell how equipment or production procedures are working or provide information about the physical plant, the environment (air, soil, water, vegetation), technology services, or financial operations.

Figure 14.8 (pp. 621–622), which begins with a recommendation, is a report written to a district manager interested in acquiring a new site for a fast-food restaurant.

3. **Home health or social work visits.** Nurses, social workers, and probation officers report daily on their visits to patients and clients. Their reports describe clients' lifestyles, assess needs, and make recommendations. Figure 14.9 (pp. 623–624) is a report from a social worker to a county family services agency.

How to Gather Information for a Trip/Travel Report

Regardless of the kind of trip/travel report you have to write, your assignment will be easier and your report better organized if you follow these suggestions:

1. Before you leave on the field trip, site inspection, or visit, be sure you are prepared as follows:
   a. Obtain all necessary names; street, e-mail, and website addresses; and relevant telephone, cell, and fax numbers.
   b. Check files for correspondence, case studies, previous trip/travel reports, or terms of contracts or agreements.
   c. Download work orders, instructions, or other documents pertinent to your visit, for example, websites and ads.
   d. Bring a laptop, PDA, or notebook with you.
   e. Locate a map of the area and get the directions you'll need beforehand (both maps and directions can be obtained easily on the Internet).

A field trip report.   Figure 14.7

TO: Katherine Holmes, RN, MSN
            Director, RN Program
FROM: Mark Tourneur M.T.
            RN Student
DATE: November 19, 2008
SUBJECT: Field Trip to Water Valley Extended Care Center

On Monday, November 10, I visited the Water Valley Extended Care
Center, 1400 Medford Boulevard, in preparation for my internship in
an extended care facility next semester.

Explains purpose of trip

Philosophy and Organization

Before my tour started, the director, Sue LaFrance, explained the
holistic philosophy of health care at Water Valley and emphasized the
diverse kinds of nursing practiced there. She stressed that the agency
is not restricted to geriatric clients but admits anyone requiring
extended care. She pointed out that Water Valley is a medium-sized
facility (150 beds) and contains three wings: (1) the Infirmary, (2) the
General Nursing Unit, and (3) the Ambulatory Unit.

Describes the facility

Primary Client Services

My tour began with the Infirmary, staffed by one RN and two LPNs,
where I observed a number of life-support systems in operation:

- IVs
- oxygen setups
- feeding tubes
- cardiac monitors

Summarizes what he observed, whom he met, and the various units of the facility

Then I was shown the General Nursing Unit (40 beds), staffed by
three LPNs and four aides. Clients can have private or semiprivate
rooms; bathrooms have wide doors and lowered sinks for patients
using wheelchairs or walkers. The Ambulatory Unit serves 90 clients
who can provide their own daily care.

Continued

Figure 14.7   (Continued)

*Usefully pro-
vides subheads
to organize his
report*

Additional Client Services
Dietetics
Before lunch in the main dining room, I was introduced to Jack Isoke, the dietitian, who explained the different menus he coordinates. The most common are low-sodium and ADA (American Diabetic Association) restricted-calorie. Staff members eat with the clients, reinforcing the holistic focus of the agency.

Pharmacy
After lunch, Kendra Tishner, the pharmacist, discussed the agency's procedures for ordering and delivering medications. She also described the client teaching she does and the in-service workshops she conducts.

*Documents
conferences
with staff*

Physical/Spiritual Therapy
I then observed clients in both recreational and physical therapy. Water Valley's full-time physical therapist, Tracy Cook, works with stroke and arthritic clients and helps those with broken bones regain the use of their limbs. In addition to a weight room, Water Valley has a small sauna that most of the clients use at least twice a week.

*Cites various
locations in the
facility*

The clients' spiritual needs are not neglected, either. A small chapel is located just south of the Ambulatory Unit.

Benefits for My Internship
From my visit to Water Valley, I learned a great deal about the health care delivery system at an extended care facility. I was especially pleased to have been given so much information on emergency procedures, medication orders, and physical therapy programs. My forthcoming internship should be even more productive, since I now have firsthand knowledge about these various services.

*Concludes with
the importance
of the trip*

-2-

A site inspection report using a map.   **Figure 14.8**

VAIL's
Chicken House

TO: Pretha Bandi   *B.A.*   DATE: June 30, 2008
FROM: Beth Armando   SUBJECT: New Site for Vail's #7
           Development Department

Recommendation
The best location for the new Vail's Chicken House is the vacant Dairy World restaurant at the northeast corner of Smith and Fairfax avenues—1701 Fairfax. I inspected this property on June 19 and 20 and also talked to Kim Shao, the broker at Crescent Realty representing the Dairy World Company.

The Location
Please refer to the map below. Located at the intersection of the two busiest streets on the southeast side, the property will allow us to take advantage of the traffic flow to attract customers. Being only one block west of the Cloverleaf Mall should also help business.

Noah's
Zanwood Av.
Dairy World   Cloverleaf Mall
McGonagles   South Kildare   Smith Av.   Fairfax Av.   Pacific St.   Grant Av.
Geary St.

Another benefit is that customers will have easy access to our location. They can enter or exit Dairy World from either Smith or Fairfax. Left turns on Smith are prohibited from 7 a.m. to 9 a.m., but since most of our business is done after 11 a.m., the restriction poses few problems.

Denver, CO 87123   (303) 555-7200   http://www.vails.com

Begins with most important detail

Gives contact information

Provides necessary background details

Includes map and traffic flow information essential for reader's purpose

Continued

Figure 14.8   (Continued)

Area Competition

Assesses the location, in light of the competition

Only two other fast-food establishments are in a one-mile vicinity. McGonagles, 1534 South Kildare, specializes in hamburgers; Noah's, 703 Zanwood, serves primarily seafood entrées. Their offerings will not directly compete with ours. The closest fast-food restaurant serving chicken is Johnson's, 1.8 miles away.

Parking Facilities

Gives only the most essential facts audience needs on parking, seating capacity, and alterations

The parking lot has space for 45 cars, and the area at the south end of the property (38 feet × 37 feet) can accommodate 14–15 cars. The driveways and parking lot were paved with asphalt last March and appear to be in excellent condition. We will also be able to make use of the drive-up window on the north side of the building.

The Building

Waiting is clear and concise

The building has 3,993 square feet of heated and cooled space. The air-conditioning and heating units were installed within the last fifteen months and seem to be in good working order; nine more months of transferable warranty remain on these units.

Paragraphs are easy to follow

The only major changes we must make are in the kitchen. To prepare items on the Vail's menu, we need to add three more exhaust fans (there is only one now) and expand the grill and cooking areas. The kitchen also has three relatively new sinks and offers ample storage space in the sixteen cabinets.

Does not overwhelm with petty details

The restaurant has a seating capacity of up to 54 persons; 10 booths are covered with red vinyl and are comfortably padded. A color-coordinated serving counter could seat 8 to 10 patrons. The floor does not need to be retiled, but the walls will have to be painted to match Vail's color decor.

A social worker's visit report.   *Figure 14.9*

**GREEN COUNTY
FAMILY SERVICES**

Randall, VA 21032
703-555-4000
greenfam@msn.net

TO: Margaret S. Walker, Director
             Green County Family Services
FROM: Jeff Bowman, Social Worker
SUBJECT: Visit to Mr. Lee Scanlon
DATE: October 10, 2008

PURPOSE OF VISIT
At the request of the Green County Home Health Office, I visited Mr. Lee
Scanlon at his home at 113 West Diversy Drive on Monday, October 6. Mr.
Scanlon and his three children (ages six, eight, and eleven) live in a two-bedroom
apartment above a garage. Last week, Mr. Scanlon was discharged from
Lutheran General Hospital after leg surgery, and he has asked for financial
assistance.

DESCRIPTION OF VISIT
Mr. Scanlon is a widower with no means of support except unemployment
compensation of $1,084 a month. He lost his job at Beaumont Industries when
the company went out of business four weeks ago and wants to go back to
work, but Dr. Marilyn Canning-Smith advised against it for six to seven
weeks. His oldest child is diabetic and the six-year-old daughter must have a
tonsillectomy. Mr. Scanlon also told me the problems he is having with his
refrigerator; it "is off more than it is on," he said.

Mrs. Alice Gordon, the owner of the garage, informed me that Mr. Scanlon
had paid last month's rent but not this month's. She also stressed how much
the Scanlons need a new refrigerator and that she had often let them use hers
to store their food.

*Starts with
why, where,
when, and
with whom the
visit was made*

*Lists informa-
tion obtained
from client,
doctor, and
owner*

Continued

Figure 14.9   (Continued)

Itemizes infor-mation to make it easy to follow and compare

Margaret S. Walker
October 10, 2008
Page 2

Here is a breakdown of Mr. Scanlon's monthly bills:

Expenses	Income
$ 550 rent	$1,084 unemployment compensation
150 utilities	
450 food	
120 drugs	
80 transportation	
$ 1,350	

ACTION TAKEN

To assist Mr. Scanlon, I have done the following:

1. Set up an appointment (10/19/08) for him to apply for food stamps.
2. Conferred with Blanche Derringo regarding Medicaid assistance.
3. Requested that the State Employment Commission help him to find a job as soon as he is well enough to work. My contact person is Wesley Sahara; his e-mail address is **wsahara@empcom.gov.**
4. Visited Robert Hong at the office of the Council of Churches to obtain food and money for utilities until federal assistance is available; he also will try to find the Scanlons another refrigerator.
5. Telephoned Sharon Muñoz at the Green County Health Department (555-1400) to have Mr. Scanlon's diabetic daughter receive insulin and syringes gratis.

Documents research done to assist client rather than giving reader recommenda-tions to follow

f. If necessary, get permission to inspect the site and represent your company/agency.

g. Bring a record of appointment times and locations, as well as the job titles of the people whom you expect to meet.

h. Bring a camcorder, tape recorder, camera, or calculator, if necessary, to record important data.

2. During your trip, keep the following tips in mind:

a. Actually visit the location. Do not rely on virtual tours or another person's impressions.

b. Meet with all contact people and relevant participants.

c. Collect important documents that may not have been available before you left—case studies, blueprints, surveys.

d. Pay attention to important matters that you could not observe beforehand—business protocols, technical equipment or resources, and the like.

3. When you return, keep the following hints in mind as you compile your report:
   a. Write your report promptly. If you put it off, you may forget important items.
   b. When a trip takes you to two or more widely separated places, note in your report when you arrived at each place and how long you stayed.
   c. Exclude irrelevant details, such as whether the trip was enjoyable, what you ate, or how delighted you were to meet people.
   d. Check to make sure you have listed names and calculated figures correctly.

Test Reports

Much physical research (the discovery and documentation of facts) is communicated through short reports variously called **experiment, investigation, laboratory,** or **operations reports**. They all record the results of tests, whether the tests were conducted in a forest, computer center, laboratory, shopping mall, or soybean field. You may be asked to test an existing or new product or procedure or verify certain physical or environmental conditions for a class or employer.

Objectivity and accuracy are essential ingredients in a test report. Readers want to know about your empirical research (the facts), not about your feelings. Record your observations without bias or guesswork in a laboratory journal or log book and always document the results with precise measurements using the standard symbols and abbreviations of your profession.

Questions Your Report Needs to Answer

Readers will expect your test report to supply the following information:

- why you performed the test—an explanation of the reasons, your goals, and who may have authorized you to perform the test
- how you performed the test—under what circumstances or controls you conducted the test; what procedures and equipment you used
- what the outcomes were—your conclusions
- what implications or recommendations follow from your test—what you learned, discovered, confirmed, or even disproved or rejected

When you sign the final copy of your report, you certify that things happened exactly when, how, and why you say they did.

Case Study: Two Sample Test Reports

Figure 14.10 is a relatively simple and short test report in memo format regarding sanitary conditions at a hospital psychiatric unit. The report follows a direct and useful pattern of organization:

- statement of purpose—*why?*
- findings—*what happened?*
- recommendations—*what next?*

Submitted by an infection control officer, the report does not provide elaborate details about the particular laboratory procedures used to determine whether bacteria were

Figure 14.10   A test report with recommendations.

States why tests were performed and how

Gives results of tests conducted in different locations

Provides detailed directions based on results

Charleston Central
HOSPITAL

TO: James Dill, Supervisor   FROM: Janeen Cufaude   *J.C.*
       Housekeeping   Infection Control Officer

DATE: December 10, 2007   SUBJECT: Routine sanitation inspection

As part of the monthly check of the psychiatric unit (11A), the following areas were swabbed and tested for bacterial growth. The results of the lab tests of these samples are as follows:

AREA	FINDINGS
1. cabinet in patients' kitchen	1. positive for 2 colonies of strep germs
2. rug in eating area	2. positive for food particles and yeasts and molds
3. baseboard in dayroom	3. positive for particles of dust
4. medicine counter in nurses' station	4. negative for bacteria—no growth after 48 hours
5. corridor by south elevator	5. positive for 4 colonies of staph germs isolated

ACTIONS TO BE TAKEN AT ONCE
1. Clean the kitchen cabinet with K-504 liquid daily, 3:1 dilution.
2. Shampoo rug areas bimonthly with heavy-duty shampoo and clean visibly soiled areas with Guard-Pruf as often as needed.
3. Wipe all baseboards weekly with K-12 spray cleanser.
4. Mop heavily traveled corridors and access areas with K-504 cleanser daily, 1:1 dilution.

Charleston, WV 25324-0114   ●   (304) 555-1800   ●   www.cch.org

---

present; nor does it describe the pathogenic (disease-causing) properties of the bacteria. Such descriptions are unnecessary for the audience (the housekeeping department) to do its job.

A more complex example of a short test report is found in Figure 14.11 (pp. 627–629), which studies the effects of four light periods on the growth of

A short test report published in a scientific journal.   **Figure 14.11**

Paulownia Seedlings Respond to Increased Daylength

M. J. Immel, E. M. Tackett, and S. B. Carpenter

Abstract

Paulownia seedlings grown under four photoperiods were evaluated after a growing period of 97 days. Height growth and total dry weight production were both significantly increased in the 16- and 24-hour photoperiods.

Begins with informative abstract

Introduction

Paulownia (*Paulownia tomentosa* [Thunb.] Steud.), a native of China, is a little known species in the United States. Recently, however, there has been increased interest in this species for surface mine reclamation (*1*).* Paulownia seems to be especially well adapted to harsh micro-climates of surface mines; it grows very rapidly and appears to be drought-resistant. In Kentucky and surrounding states, paulownia wood is actively sought by Japanese buyers and has brought prices comparable to black walnut (*2*).

This increased interest in paulownia has resulted in several attempts to direct seed it on surface mines, but little success has been achieved. The high light requirements and the extremely small size of paulownia seed (approximately 6,000 per gram) may be the limiting factors. Planting paulownia seedlings is preferred; but, because of their succulent nature, seedlings are usually produced and outplanted as container stock rather than bareroot seedlings. Daylength is an important factor in the production of vigorous container plants (*5*).

Our study compares the effects that four photoperiods—8, 12, 16, and 24 hours—had on the early growth of container-grown paulownia seedlings over a period of 97 days.

Gives background, purpose, and scope of study

Materials and Methods

Seeds used in this study were stratified in a 1:1 mixture of peat moss and sand at 4°C for 2 years. Following cold storage, seeds were placed on a 1:1 potting soil–sand mix and mulched with cheesecloth. They were then placed under continuous light until germination occurred. Germination percentages were high, indicating paulownia seeds can survive long periods of storage with little loss of viability (*3*).

Thirty days after germination, 3- to 4-centimeter seedlings were transplanted into 8-quart plastic pots filled with an equal mixture of potting soil, sand, and peat moss.

Describes steps taken: procedures, conditions, and equipment used

---

* To save space, the References section has been omitted.

Continued

Figure 14.11   (Continued)

Seventy-five seedlings were randomly assigned to each of the four treatments. Treatments were for 4 photoperiods—8, 12, 16, and 24 hours—and were replicated three times in 12 light chambers. Each chamber was 1.2- by 1.2-meters with an artificial light source 71 centimeters above the chamber floor.

Uses technical terms and symbols audience expects

The light source consisted of eight fluorescent lights: four 40-watt plant growth lamps alternated with four 40-watt cool white lamps. Light intensity averaged 550 foot-candles (1340μ einsteins/m²/s) at the top of each pot and the temperature averaged 23°C (±2°C).

Seedlings were watered and fertilized after transplanting with a 6-gram 14-4-6 agriform container tablet. Beginning 1 month after transplanting, two seedlings were randomly selected and harvested from each chamber for a total of 24 trees. Height, root collar diameter, length of longest root, and oven-dry weight (at 65°C) were determined for each seedling. Harvests continued every week for 5 additional weeks.

Results and Discussion

Results indicate that early growth of paulownia is influenced by photoperiod, as shown in Table 1.

Includes a visual to summarize results and then explains what happened

TABLE 1. Height Diameter, Root Length, Total Dry Weight, and R/S Ratio for Paulownia Seedlings Grown Under Four Photoperiods After 97 Days.

Photo period (hrs.)	Height (cm)	Diameter (cm)	Root length (cm)	Total dry weight (gm)	R/S ratio
8	13.1b	0.48	16.0	1.65c	0.18
12	17.8b	0.67	34.7	7.27b	0.32
16	27.3a	0.93	31.1	15.92a	0.39
24	29.2a	0.90	43.9	18.56a	0.33

Expanding the photoperiod from 8 to either 16 or 24 hours increased height growth by 100 percent. Height growth in the 12-hour treatment also increased, but did not differ significantly from the 8-hour treatment. Heights under photoperiods of 8, 12, 16, and 24 hours were 13.1, 17.8, 27.3, and 29.2 centimeters, respectively.

Cites related studies

Previous studies have also shown that photoperiod affects the growth of paulownia seedlings (*4, 6*). Sanderson (*6*), for example, found that paulownia seedlings grown under continuous light averaged 27.2 centimeters in height after 101 days compared with 29.2 centimeters for our 24-hour seedlings. Other corresponding photoperiods were equally comparable. Downs and Borthwick (*4*) also concluded that height growth of paulownia was affected by extending the photoperiod.

(Continued)   Figure 14.11

Page 3

The greatest treatment differences were shown in total dry weight production. Refer again to Table 1. The mean weight of 1.65 grams for seedlings in the 8-hour treatment was significantly less than that of any of the other photoperiods. The 16- and 24-hour treatments did not differ significantly. In fact, they more than doubled the average weight for seedlings in the 12-hour treatment.

Root-to-shoot ratio (R/S) indicates the relative proportion of growth allocated to roots versus shoots for the seedlings in each photoperiod. In this study, shoots were developing at nearly three times the rate of the roots for seedlings in the 12-, 16-, and 24-hour photoperiods.

Provides accurate measurements in a clear, objective tone

The 0.18 R/S ratio for seedlings in the 8-hour treatment was much lower, indicating that relative growth of the shoot is approximately five times that of the root. The shorter photoperiod therefore decreased root development relative to shoot development as well as significantly reduced total dry weight production.

Although root collar diameter and root length did not significantly differ under the different photoperiods after 97 days, there was a trend for greater diameter and root growth with longer photoperiods.

Conclusions

Results indicate that the growth of paulownia seedlings is affected by changes in the photoperiod. Increasing the photoperiod significantly increased height growth and total dry matter production. The distribution of dry matter (R/S ratio) was altered by increasing the photoperiod; the ratio was larger in the longer photoperiods. In contrast to earlier studies (4), we found paulownia seedlings subjected to extended photoperiods were still growing after 97 days.

Interprets the significance of the results

paulownia seedlings (a flowering tree cultivated in China). The report, published in a scientific journal, is addressed to specialists in forestry. Such a test report follows a different, more detailed pattern of organization than the report in Figure 14.10 and includes an informative abstract, an introduction, a materials and methods section, a results and discussion section, and a list of references cited in the study.

To meet the needs of an expert audience, the writers of the report in Figure 14.11 had to include much more information than did Janeen Cufaude, the infection control officer who wrote the report in Figure 14.10, about the way the test was conducted and the types of scientific data the audience needs. The researchers did not have to define technical terms for their audience, and they could confidently use scientific symbols and formulas as well.

Incident Reports

The short reports discussed thus far in this chapter have dealt with routine work. They have described events that were anticipated or supervised. But every business or agency runs into unexpected trouble that delays routine work or results in personal injury. These circumstances need to be documented in an incident report. The audience for an incident (or accident) report can be within your organization or outside it. Employers use incident reports to make changes so that the problem does not occur again or so that a job can be done more effectively and safely. On some occasions government inspectors, insurance agents, and attorneys must be informed about those events that have interfered with or threatened normal, safe operations.

When to Submit an Incident Report

An incident report is submitted when there is, for example,

- an accident—fire, automobile, physical injury
- a law enforcement offense
- an environmental danger, including a computer virus
- a machine breakdown
- a delivery delay
- a cost overrun
- a production slowdown

Figure 14.12 is an incident report about a train derailment submitted by the engineer on duty. This report is in memo format, but some companies or agencies require you to fill out a special form. Because it can contain legally sensitive information needed in hard copy, an incident report should not be sent as an e-mail.

Parts of an Incident Report

Include the following information in your incident report. Note how Figure 14.12 includes precise and accurate information for each of these parts.

1. **Identification details.** Indicate who and what was involved and gather all relevant data—names, contact information, model/serial numbers, and so on. Record titles, department, and employment identification numbers. Indicate if you or your fellow employees were working alone. For customers or victims, record home addresses, phone numbers, and places of employment. Insurance companies will also require policy numbers.

2. **Type of incident.** Briefly identify the incident—personal injury, fire, burglary, equipment failure. Identify any part(s) of the body precisely. "Eye injury" is not enough; "injury to the right eye, causing bleeding" is better. "Dislocated right shoulder" or "punctured left forearm" is descriptive and exact. A report on damaged equipment should list model numbers.

3. **Time and location of the incident.** Include precise date (not "Thursday") and time (a.m. or p.m.).

4. **Description of what happened.** This section is the longest part of the report. Let readers know exactly what happened and why, how it occurred, and what led up to the incident.

An incident report in memo format. **Figure 14.12**

THE GREAT HARVESTER RAILROAD
Des Moines, IA 50306-4005
http://www.ghrr.com

TO: Angela O'Brien, District Manager
James Day, Safety Instructor
FROM: Nick Roane, Engineer *N.R.*
DATE: May 10, 2009
SUBJECT: Derailment of Train 28 on May 10, 2009

DESCRIPTION OF INCIDENT

At 7:20 a.m. on May 10, 2009, I was driving Engine 457 traveling north at a speed of fifty-two miles an hour on the single main-line track four miles east of Ridgeville, Illinois. Weather conditions and visibility were excellent. Suddenly the last two grain cars, 3022 and 3053, jumped the track. The train automatically went into emergency braking and stopped immediately. There were no injuries to the crew. But the train did not stop before both grain cars turned at a 45° angle. After checking the cars, I found that half the contents of their loads had spilled. The train was not carrying any hazardous chemical shipments.

I notified Supervisor Bill Purvis at 7:40 a.m., and within forty-five minutes he and a section crew arrived at the scene with rerailing equipment. The section crew removed the two grain cars from the track, put in new ties, and made the main-line track passable by 9:25 a.m. At 9:45 a.m. a vacuum car arrived with Engine 372 from Hazlehurst, Illinois, and its crew proceeded with the cleanup operation. By 10:25 a.m. all the spilled grain was loaded onto the cars brought by the Hazlehurst train. Bill Purvis notified Barnwell Granary that their shipment would be at least three hours late.

CAUSES OF INCIDENT

Supervisor Purvis and I checked the stretch of train track where the cars derailed and found it to be heavily worn. We believe that a fisher joint slipped when the grain cars hit it, and the track broke. You can see the location of the cracked fisher joint in the drawing below.

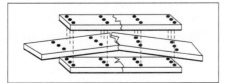

Gives precise time, location

Describes what happened

Explains what was done

Determines likely cause

Supplies easy-to-read exploded visual

Continued

Figure 14.12   (Continued)

*Offers recom-
mendations
to solve or pre-
vent problem
from recurring*

Page 2

RECOMMENDATIONS

We made the following recommendations to the switch yard in Hazlehurst to be carried out immediately.

1. Check the section of track for ten miles on either side of Ridgeville for any signs of defective fisher joints.
2. Repair any defective joints at once.
3. Instruct all engineers to slow down to five to ten mph over this section of the track until the rail check is completed.

5. **What was done after the incident.** Describe the action you took to correct conditions, how things got back to normal, and what was done to treat the injured, make the environment safer, speed a delivery, repair damaged equipment.

6. **What caused the incident.** Make sure your explanation is consistent with your description of what happened. Pinpoint the trouble. In Figure 14.12, for example, the defective fisher joint is listed under the heading "Causes of Incident."

7. **Recommendations.** Recommendations about preventing the problem from recurring may involve repairing any broken parts (as in Figure 14.12), calling a special safety meeting, asking for further training, adapting existing equipment, doing emergency planning, or modifying schedules. Make sure all of your recommendations are consistent with what actually happened.

Protecting Yourself Legally

An incident report can be used as legal evidence, and it then becomes part of a permanent legal record that can be used by law enforcement and attorneys in court to establish negligence and liability on your and your company's part. It can also be used by an employer to determine employee responsibility. An incident report frequently concerns the two topics over which powerful legal battles are waged—health and property.

You have to be very careful about collecting and recording details. Make sure your report is not sketchy or incomplete. To avoid these errors, you may have to interview employees or bystanders; travel to the incident site; check manuals, code books, or other guides; consult safety experts; or research records/archives.

To ensure that what you write is legally proper, follow these guidelines.

1. **Submit your report promptly and sign or initial it.** Any delay might be seen as a cover-up. Send your report to the appropriate parties immediately after you have gathered the necessary information and had it reviewed by your supervisor.

2. **Be accurate, objective, and complete.** Recount clearly what happened in the order it took place. Never omit or distort facts; the information may surface later, and you could be accused of a cover-up. Do not just write "I do not know" for an answer. If you are not sure, state why. Also be careful that there are no discrepancies in your report.

3. **Give facts, not opinions.** Provide a factual account of what actually happened, not a biased interpretation of events or one based on speculation or hearsay. Vague words such as "I guess," "I wonder," "apparently," "perhaps," or "possibly" weaken your objectivity. Stick to details you witnessed or that were seen by eyewitnesses. Identify witnesses or victims by giving names, addresses, places of employment, and so on. Keep in mind that stating what someone else saw is regarded as hearsay and therefore is not admissible in a court of law. State only what *you* saw or heard. When you describe what happened, avoid drawing uncalled-for conclusions. Consider the following statements of opinion and fact:

   Opinion: The patient seemed confused and caught himself in his IV tubing.
      Fact: The patient caught himself in his IV tubing.

   Opinion: The equipment was defective.
      Fact: The bolt was loose.

   Be careful, too, about blaming someone. Statements such as "Baxter was incompetent" or "The company knew of the problem but did nothing about it" are libelous remarks.

4. **Do not exceed your professional responsibilities.** Answer only those questions you are qualified to answer. Do not presume to speak as a detective, an inspector, a physician, a supervisor, or a judge. Do not represent yourself as an attorney or a claims adjuster in writing the report. And don't take sides.

Short Reports: Some Final Thoughts

To prepare successful short reports, keep in mind the rules of short report writing discussed in this chapter. Always take into account your readers' needs and expectations at every stage of your writing, document carefully what you write about, take accurate and complete notes, write objectively and ethically, present complicated data clearly and concisely, provide background and contexts where necessary, and include specific recommendations, where called for, based upon the facts. Remembering these basic rules will earn you praise and possibly promotions at work.

Visit www
.cengage.com/
english/kolin for
an online exercise,
"Creating Report
Templates Using
Your Word
Processor."

Revision Checklist

☐ Had a clear sense of how my readers will use my report.
☐ Did appropriate research to give my audience enough information to help them make careful decisions.
☐ Provided significant information about costs, materials, personnel, and times so readers will know that my work consists of facts, not impressions.
☐ Double-checked all data—names, costs, figures, dates, places, and equipment models and numbers.
☐ Verified necessary statistics and trends so that my periodic and sales reports are thorough and accurate.
☐ Made sure all my comments and recommendations were ethical.
☐ Followed all agency or organization guidelines.
☐ Adhered to all legal requirements.
☐ Eliminated unnecessary details or those too technical for my audience.
☐ Kept report concise and to the point.
☐ Used headings whenever feasible to organize and categorize information.
☐ Supplied relevant visuals to help readers understand my message and crunch any numbers.
☐ Employed underlining, boldface, or italics to set headings apart or to emphasize key ideas.
☐ Began report with statement of purpose that clearly described the scope and significance of my work.
☐ Incorporated tables and other pertinent visuals to display data whenever appropriate.
☐ Explained clearly what the data mean.
☐ Determined that recommendations logically follow from the data and that recommendations are relevant and realistic.

Exercises

1. Bring to class an example of a periodic report from your present or previous job or from any community, religious, or social organization to which you belong. In an accompanying memo to your instructor, indicate who the audience is and why such a report is necessary, stressing how it is organized, what kinds of factual data it contains, what visuals were used, and how it might be improved in content, organization, style, and design.

2. Assume that you are a manager of a large apartment complex (200 units). Write a periodic report based on the following information—26 units are vacant, 38

soon will be vacant, and 27 soon will be leased (by June 1). Also add a section of recommendations to your supervisor (the head of the real-estate management company for which you work) on how vacant apartments might be leased more quickly and perhaps at increased rents. Consider such important information as decorating, advertising, and installing a new security system.

3. Assume you work for a household appliance store. Prepare a sales report based on the information contained in the following table. Include a recommendation section for your manager.

Product	Number Sold	
	October	November
Kitchen Appliances		
Refrigerators	72	103
Dishwashers	27	14
Freezers	10	36
Electric Ranges	26	26
Gas Ranges	10	3
Microwave Ovens	31	46
Laundry Appliances		
Washers	50	75
Dryers	24	36
Air Treatment		
Room Air Conditioners	41	69
Dehumidifiers	7	2

4. Write a progress report on the wins, losses, and ties of your favorite sports team for the past season. Address the report to the director of publicity for the team and stress how the director might use those facts for future publicity. As part of your report, indicate what might be an effective lead for a press release about the team's efforts.

5. Submit a progress report to your writing teacher on what you have learned in his or her course so far this term, which writing skills you want to develop in greater detail, and how you propose doing so. Mention specific memos, e-mail, letters, instructions, reports, or proposals you have written or will soon write.

6. Compose a site inspection report on any part of the college campus or plant, office, or store in which you work that might need remodeling, expansion, rewiring for computer use, or new or additional air-conditioning or heating work.

7. You and your collaborative team have been asked to write a short preliminary inspection report on the condition of a historic building for your state historical society. Inspecting the building, the home of a famous late-nineteenth-century

governor, you discover the problems listed below. Include all these details in your report. Also supply recommendations for your readers—a director of the state historical society, a state architect, and four representatives of the subcommittee on finance from your state legislature. Design two appropriate visuals to include in your report.

- The eight front columns are all in need of repair; two of them may have to be replaced.
- The area below each bottom window casement needs to be excavated for waterproofing.
- The slate tile on the roof has deteriorated and needs immediate replacement.
- The front stairs show signs of mortar leaching and require attention at once.
- Sections of gutter on the northwest and northeast sides of the house must be changed; other gutters are in fair shape.
- Wood shutters need to be repainted; four of the twelve may even need to be replaced.
- All trees around the house need pruning; an old elm in the backyard shows signs of decay.
- The siding is in desperate need of preparation and painting.
- The brick near the front entrance is dirty and moss-covered.

8. Write a report to an instructor in your major about a field trip you have taken recently—to a museum, laboratory, health care agency, correctional facility, radio or television station, agricultural station, or office. Indicate why you took the trip, name the individuals you met on the trip, and stress what you learned and how that information will help you in course work or on your job.

9. Submit a test report on the purpose, procedures, results, conclusions, and recommendations of an experiment you conducted on one of the following subjects:

   a. soil
   b. Internet access
   c. water
   d. automobiles
   e. textiles/clothing
   f. animals
   g. recreational facilities
   h. computer hardware or software
   i. forests
   j. food
   k. air quality
   l. transportation
   m. blood
   n. noise levels

10. Write an incident report about a problem you encountered in your work or at home in the last year. Document the problem and provide a solution. Use the memo format in Figure 14.12.

11. Write an incident report about one of the following problems. Assume that it has happened to you. Supply relevant details and visuals in your report. Identify the audience for whom you are writing and the agency you are representing or trying to reach.

   a. After hydroplaning, your company car hits a tree and has a damaged front fender.

b. You have been the victim of an electrical shock because an electrical tool was not grounded.

c. You twist your back lifting a bulky package in the office or plant.

d. A virus has infected your company's intranet and it will have to be shut down for twelve hours to debug it.

e. The crane you are operating breaks down and you lose a half-day's work.

f. The vendor shipped the wrong replacement part for your computer, and you cannot complete a job without buying a more expensive software package.

g. An electrical storm knocked out your computer; you lost 1,000 mailing addresses and will have to hire additional help to complete a mandatory mailing by the end of the week.

h. A scammer has stolen some sensitive files (documents) about a new product or service your company had hoped to launch next month.

i. An irate customer threatens one of your sales clerks, but there was no physical violence.

12. Choose one of the following descriptions of an incident and write a report based on it. The descriptions contain unnecessary details, vague words, insufficient information, unclear cause-and-effect relationships, or a combination of those errors. In writing your report, correct the errors by adding or deleting whatever information you believe is necessary. You may also want to rearrange the order in which information is listed. Use a memo format, like that in Figure 14.12, to write the report.

a. After sliding across the slippery road late at night, my car ran into another vehicle, one of those fancy imported cars. The driver of that car must have been asleep at the wheel. The paint and glass chips were all over. I was driving back from our regional meeting and wanted to report to the home office the next day. The accident will slow me down.

b. Whoever packed the glass mugs did not know what he or she was doing. The string was not the right type, nor was it tied correctly. The carton was too flimsy as well. It could have been better packed to hold all those mugs. Moreover, since the bus had to travel across some pretty rough country, the package would have broken anyhow. The best way to ship these kinds of goods is in specially marked and packed boxes. The value of the box contents was listed at $575.

CHAPTER 15

Writing Careful Long Reports

Visit www
.cengage.com/
english/kolin for
this chapter's on-
line exercises,
ACE quizzes, and
Web links.

This chapter introduces you to long reports—how they differ from short reports, how they are written, and how they are organized. It is appropriate to discuss long reports in one of the last chapters of *Successful Writing at Work*. A major project in the world of work, the long report gives you an opportunity to use and combine many of the writing skills and research strategies you have already learned. In business, a long report is the culmination of many weeks or months of hard work on an important company project.

The following skills will be most helpful to you as you prepare to study long reports; appropriate page numbers appear for those topics that have already been discussed.

- assessing and meeting your audience's multiple needs (pp. 8–20)
- gathering and summarizing information, especially from print and online sources, and conducting interviews (pp. 310–348; 401–407)
- generating, drafting, revising, and editing your ideas (pp. 43–70)
- reporting the results of your research accurately and concisely (pp. 56–58; 625–629)
- creating and introducing visuals and designing documents (pp. 437–482)
- using an appropriate method of documentation (pp. 353–374)
- preparing an informative abstract (pp. 417–420)

Having improved those skills, you should be ready to write a successful long report. The writer of the model long report in Figure 15.3 (pp. 652–667) uses all of the above skills in her work for RPM Technologies.

How a Long Report Differs from a Short Report

Both long and short reports are invaluable tools in the world of work. Basic differences exist, though, between the two types of reports. A short report is not a watered-down version of a long report; nor is a long report simply an expanded version of a short one. The two types of reports differ in scope, research, format, timetable, audience, and collaborative effort.

The following section explains some of the key differences between these two reports. By understanding these differences, you will be better able to follow the rest of Chapter 15 as it covers the process of writing a long report and the organization and parts of such a report. Study the long report at the end of this chapter to study how these two types of documents differ.

Scope

A long report is a major study that provides an in-depth view of a key problem or idea. It might be eight to twenty pages long or even much longer, depending on the scope of the subject. The implications of a long report are wide-ranging for a business or industry—relocating a plant, adding a new network, changing a programming operation, or adapting the workplace for multinational employees, as in Figure 15.3.

Long reports examine a problem in detail, while short reports cover just one part of the problem. Unlike a short report, a long report may discuss not just one or two current events, but rather a continuing history of a problem or idea (and the background information necessary to understand it in perspective). For example, the short test report on paulownia in Figure 14.11 (pp. 627–629) would be used with many other test reports in a long report for a group of environmentalists or a government agency on the value of planting those trees to prevent soil erosion.

The titles of some typical long reports further suggest their extensive (and in some cases exhaustive) coverage:

- *A Master Plan for the Recreation Needs of Dover Plains, New York*
- *The Transportation Problems in Kingford, Oregon, and the Use of Monorails*
- *Promoting More Effective E-Commerce and E-Tailing at TechWorld*
- *The Use of Virtual Reality Attractions in Theme Parks in Jersey City, New Jersey*
- *Public Policy Implications of Expanding Health Care Delivery Systems in Tate County*
- *Internet Medicine in Providing Health Care in Rural Areas: Ways to Serve Southern Montana*

Research

A long, comprehensive report requires much more extensive research than a short report does. Information can be gathered over time from primary and secondary research—Internet searches, surveys, books, articles, statistical sources, laboratory experiments, site visits and tests, interviews, and the writer's own observations. For a long report, you will have to do a great deal of research and possibly interviewing to track down the relevant background information and to discover what experts have said about the subject and what they propose should be done. Note how much research went into the business report in Chapter 8 and the long report in Figure 15.3.

Information gathered for many short reports can also help you prepare a long report. In fact, as the example of paulownia in Figure 14.11 again shows, a long report can use the experimental data from a short report to arrive at a conclusion. Also for a long report, writers often supply one or more progress reports (one type of short report).

Finally, preparing a proposal can lead to writing a long report. You might suggest a change to an employer, who then would ask you to write a long report containing the research necessary to implement that change.

Tech Note

Using Government-Sponsored Research

The U.S. government conducts or sponsors a great deal of research that you might find relevant for your reports. Government research is at the heart of science, technology, health care, and so on. Below is a short list of some major government and other relevant websites.

- *http://www.epa.gov* will lead you to press releases, test guidelines, and information about grants, contracts, and job opportunities at the Environmental Protection Agency.
- *http://www.osha.gov* is the website for the Occupational Safety and Health Administration, where you'll find information about OSHA standards, news releases and fact sheets, publications, technical information, and safety links.
- *http://www.astd.org* is the homepage of the American Society for Training and Development, which oversees research and maintains databases of information about employee training in business, industry, education, and government.
- *http://www.sba.gov* leads to the Small Business Administration's webpage, where you'll find guides on beginning a small business, the opportunity to include your business in the national register, and contact information for local SBA offices.
- *http://www.bls.gov* is the site of the Bureau of Labor Statistics. This organization compiles and maintains financial information such as charts on U.S. dollar inflation, growth or decline of the number of businesses, and wage increases.
- *http://www.gpoaccess.gov* is a service of the U.S. Government Printing Office that provides free electronic access to a variety of federal documents, including legislative, judicial, and executive resources.
- *http://www.archives.gov* is the address for National Archives and Records Administration, an independent federal agency that oversees the management of all federal records.

Format

A long report is too detailed and complex to be adequately organized in a memo or letter format. The product of thorough research and analysis, the long report gives readers detailed discussions and interpretations of large quantities of data. To present the information in a logical and orderly fashion, the long report contains more parts, sections, headings, subheadings, documentation, and supplements (appendixes) than would ever be included in a short report. Look at the Table of Contents for Figure 15.3 (p. 652) to see at a glance the various divisions and parts of a long report.

Timetable

The two types of reports differ in the time it takes to prepare them. Writers of these two reports work under different expectations from their readers and under different deadlines. A long report explores with extensive documentation a subject involving personnel, locations, costs, safety, or equipment. Many times a long report is required by law, for example, investigating the feasibility of a project that will affect the ecosystem. A short report is often written as a matter of routine duty, with the writer sometimes given little or no advance notice. The long report, however, may take weeks or even months to write. Below is a timetable for a typical long report.

Do research and conduct interviews	Outline or Draft	Conferences for Revisions and Further Research	Revise and Prepare Figures	Proof-read and Polish	Submit Long Report
4 weeks	3 weeks	1 week	3 weeks	1 week	Due Date

Audience

The audience for a long report is generally broader—and includes individuals higher up in an organization's hierarchy—than that for a short report. Your short report may be read by co-workers, a first-level supervisor, and possibly that person's immediate boss, but a long report is always intended for people in the top levels of management—presidents, vice presidents, superintendents, directors—who make executive, financial, and organizational decisions. These individuals are responsible for long-range planning, seeing the big picture, so to speak.

Collaborative Effort

Unlike many short reports, the long report in the world of business may not always be the work of one employee. Rather, it may be a collaborative effort, the product of a committee or group whose work is reviewed by a main editor to make sure that the final copy is consistently and accurately written. Individuals in many departments within a company—computer programming, document design, engineering, graphics, legal affairs, public relations, safety—may cooperate in planning, researching, drafting, revising, and editing a long report.

The group should estimate a realistic timeframe necessary to complete the various stages of their work—when drafts are due and when editing must be concluded, for example. A project schedule based on that estimate should then be prepared. **But remember: Projects always take longer than initially planned.** Prepare for a possible delay at any one stage. The group may have to submit written progress reports (see pp. 610–617) to its members, as well as to management.

To be successful a collaborative writing team should also observe the guidelines and procedures for collaborative writing (review Chapter 3 on pp. 76–115).

The Process of Writing a Long Report

As we just saw, writing a long report requires much time and effort. Since your work will be spread over many weeks, you need to see your report not as a series of static or isolated tasks but as an evolving project. Before you embark on that project, review the information on the writing process in Chapter 2 (pp. 44–69). You may also want to study the flow chart in Figure 10.14, which illustrates the different stages in writing a research paper/report. The following guidelines will also help you to plan and write your long report.

1. Identify a significant topic. You'll have to do preliminary research—reading, online searching, compiling a questionnaire or survey, conferring with and interviewing experts—to get an overview of key ideas, individuals involved, and the implications for your company and/or community. Note the kinds of research Terri Smith Ruckel and her group did for her long report in Figure 15.3. As she did, expect to search a variety of print and Internet sources, to read and evaluate them, and to incorporate them in your work.

2. Expect to confer regularly with your supervisor(s) and/or team members. In these meetings, be prepared to ask pertinent and researched questions to pin down exactly what your boss wants and how your writing team can accomplish this goal. Focus your questions on the company's use of your report, how the company wants you to express certain ideas, and the amount of information it needs. Your supervisor may want you to submit an outline before you draft the report and may expect several more drafts for his or her and other executives' approval before you write the final version.

3. Revise your work often. Be prepared to work on several outlines and drafts. Your revisions may sometimes be extensive, depending on what your boss or collaborative team recommends. You may have to consult new sources and arrive at a new interpretation of those sources. As you narrow your purpose and scope, you may find yourself deleting information or modifying its place in your work. This shifting around as well as adding and deleting information will help you to arrive at a carefully organized report. At the later stages, you will be revising and editing your words, sentences, and paragraphs.

4. Keep the order flexible at first. Even as you work on your drafts and revisions, remember that a long report is not written in the order in which the parts will finally be assembled. You cannot write in "final" order—abstract to recommendations. Instead, expect to write in "loose" order to reflect the process in which you gathered information and organized it for the final copy of the report. Usually, the body is written first, the introduction later so the authors can make sure they have not left anything out. The abstract, which appears very early in the report, is always written after all the facts have been recorded and the recommendations made or the conclusions drawn. The title page and the table of contents are always prepared last.

5. Prepare both a day-to-day calendar and a checklist. Keep both posted where you do your work—above your desk or computer, or use your computer's built-in calendar program, if available—so you can track your progress. Make sure your collaborative team is following the same calendar and using the same checklist. The calendar should mark **milestones**, that is, dates by which each stage of your work must be completed. Match the dates on your calendar with the dates your employer may have given you to submit an outline, progress report(s), and the final copy. Your checklist should list the major parts of your report. As you complete each section, check it off. Before assembling the final copy of your report, use the checklist to make sure you have not omitted something.

Parts of a Long Report

A long report may include some or all of the following twelve parts, which form three categories: *front matter* (letter of transmittal, title page, table of contents, list of illustrations, abstract); *report text* (introduction, body, conclusion, recommendations); and *back matter* (glossary, references cited, appendixes).

Front Matter

As the name implies, the front matter of a long report consists of everything that precedes the actual text of the report. Such elements introduce, explain, and summarize to help the reader locate various parts of the report. Use lowercase roman numerals for front matter page numbers, not Arabic numbers.

Letter of Transmittal

This three- or four-paragraph (usually only one-page) letter states the purpose, scope, and major recommendation of the report. It highlights the main points of the report that your readers would be most interested in. Always be willing to answer any questions the reader has about your report. Figure 15.1 shows a sample letter of transmittal for a business report.

Figure 15.1   A letter of transmittal for a long report.

α
ALPHA CONSULTANTS

■ 1400 Ridge ■ Evanston, California 97214-1005 ■
■ 805-555-9200 ■ FAX 805-555-0221 ■
www.alpha.com

August 3, 2009

Dr. K. G. Lowry, President
Coastal College
San Diego, CA 93219-2619

Dear Dr. Lowry:

Indicates why report was done

We are happy to offer you the enclosed report, **A Study to Determine New Directions in Women's Athletics at Coastal College**, which you commissioned us to prepare. The report contains our recommendations about strengthening existing sports programs and creating new ones at Coastal College.

Gives two key recommendations

Our recommendation is that Coastal should engage in more active recruitment to establish a more competitive women's baseball team, to offer additional athletic activities in women's track and field by August 2010, and to create a new interdisciplinary program between the Athletic Department and the Women's Studies Program.

Encourages response and questions

We hope that you find our report useful in meeting students' needs at Coastal College. If you have any questions or if you would like to discuss any of our recommendations, please call us.

Sincerely yours,

Barbara Gilchrist
Barbara Gilchrist

Lee T. Sidell
Lee T. Sidell

Encl. Report

Title Page

Since MLA and APA have different formats for title pages, find out what your boss prefers. Basically, though, your title page should contain the following:

- the full title of your report; tell readers what your topic is and how you have restricted it in time, space, or method. Your title determines if and how you have done your work. Avoid titles that are vague, too short, or too long.

Vague Title:	A Report on the Internet: Some Findings
Too Short:	The Internet
Too Long:	A Report on the Internet: A Study of Dot-Com Companies, Their History, Appeal, Scope, Liabilities, and Their Relationship to Ongoing Work Dealing with Consumer Preferences and Protection Within the Last Five Years in the Midwest

- the name of the company or agency preparing the report
- the name(s) of the report writer(s)
- the date of the report
- any agency, order, or grant number
- the name of the firm for which the report was prepared

Make sure that your title page looks professional. Center your title and graphically subordinate any subtitles. Do not use abbreviations (e.g., *bldgs.* for *buildings, gov't* for *government, bus.* for *business*) or acronyms (e.g., *AMS* for *Association of Marketing Students; PTAs* for *Physical Therapist Assistants*).

Table of Contents

The table of contents lists the major headings and subheadings of your report and tells readers on which pages they can be found. Essentially, a table of contents shows how you organized your report. In Figure 15.2, for example, the reader can see how the report "A Study to Determine New Directions in Women's Athletics at Coastal College" is divided into four chief parts (Introduction, Discussion, Conclusion, and Recommendations). A table of contents emerges from many outlines and drafts. The items on those outlines frequently expand, shrink, and move around until you decide on the formal divisions and subdivisions of your report.

Tech Note

Automatically Formatting the Long Report

You can save a considerable amount of formatting time when writing your long report by using your word processing program's automatic formatting features. Automatic formatting can be applied either as you keyboard or after you've completed your report, by turning on the automatic formatting

Continued

Continued

options and choosing which types of formats you'd like to apply to the report. You can

- automatically format the most basic parts of your paper
- apply bulleted or numbered lists
- automatically format the heading structure of your report by using your word processing program's tagging function
- create a table of contents via your tagged headers
- produce an index by tagging individual words

It will take only a short time to familiarize yourself with your word processing program's autoformatting features, and by doing so you can reduce the cumbersome process of formatting each item individually. You should always carefully proofread your report to make sure that the autoformatting has been correctly applied.

Include front matter components in your table of contents, but never list the contents page itself, the letter of transmittal, or the title page in your table of contents. Never have just one subheading under a heading. You cannot divide a single topic by one.

Incorrect: EXPANDING THE SPORTS PROGRAM
   Basketball
  BUILDING A NEW ARENA
   The West Side Location

Correct: EXPANDING THE SPORTS PROGRAM
   Basketball
   Track and Field
  BUILDING A NEW ARENA
   The West Side Location
   Costs

List of Illustrations

This list of all the visuals indicates where they can be found in your report.

Abstract

As discussed in Chapter 9, an abstract presents a brief overview of the problem and conclusions; it summarizes the report. An informative abstract is far more helpful to readers of a report than is a descriptive one, which gives no conclusions or results.

Not every member of your audience will read your entire report, but almost everyone will read the abstract. For example, the president of the corporation or the director of an agency may use the abstract as the basis for approving the report and passing it on for distribution. Thus, the abstract may be the most important part of your report. See how the abstract Terri Smith Ruckel and her staff prepared for her readers on page 655 succinctly gives conclusions, or findings.

A table of contents for a long report.   Figure 15.2

CONTENTS

LIST OF ILLUSTRATIONS	**iii**
ABSTRACT	**iv**
INTRODUCTION	**1**
Background	1
Problem	1
Purpose of Report	4
Scope of Report	5
DISCUSSION	**6**
Women's Athletics at Area Colleges	7
Baystown Community College	7
California State University of Arts and Sciences	9
Central Community College	10
Coastal's Sports Facilities	12
North Campus	12
South Campus	14
Coastal's Coaching Staff	16
First-Year Teams	16
Varsity Teams	18
Performance of Coastal's Women Athletes	19
Coastal's Recruitment Strategies	20
Women's Athletics and the Curriculum	22
Interdisciplinary Women's Studies Program	23
Physical Fitness Program	25
CONCLUSION	**28**
Views of the Students	29
Views of the Coaches	31
RECOMMENDATIONS	**32**
REFERENCES	**34**
APPENDIXES	**37**
A. Sample Questionnaire for Coaches	38
B. Sample Questionnaire for Athletes	39
C. Profile of Women Athletes' Achievements	42
D. Cost Estimates for Physical Plant Additions	47

Uses Roman numerals for front matter

Records key sections and subsections of report

Indents subsections with subheads

Boldfaces major sections

List references on separate pages in report

Gives titles to documents in appendixes

Abstracts may be placed at various points in long reports—on the title page, on a separate page, or as the first page of the report text.

Text of the Report

The text of a long report consists of an introduction, the body, conclusions, and sometimes recommendations.

Introduction

The introduction may constitute as much as 10 or 15 percent of your report, but usually it is not any longer. If it were, the introduction would be disproportionate to the rest of your work, especially the body section. The introduction is essential because it tells readers why your report was written and thus helps them to understand and interpret everything that follows. See how Terri Smith Ruckel and her staff in the introduction to Figure 15.3 emphasize the importance of their reseach for their employer, RPM Technologies, on pages 656–658. Do not put your findings, conclusions, or recommendations in your introduction.

Do not regard the introduction as one undivided block of information. It includes the following related parts, which should be labeled with subheadings. Keep in mind, though, that your employer may ask you to list these parts in a different order.

1. Background. To understand why your topic is significant and hence worthy of study, readers need to know about its history. This history may include information on such topics as who was originally involved, when, and where; how someone was affected by the issue; what opinions have been expressed on the issue; what the implications of your study are. Note how the long report in Figure 15.3 provides useful background information on when, where, how, and why multinational employees here entered the U.S. work force.

2. Problem. Identify the problem or issue that led you to write the report. Your problem needs to be significant to warrant a long report on it. Because the problem or topic you investigated will determine everything you write about in the report, your statement of it must be clear and precise. That statement may be restricted to a few sentences. Here is a problem statement from a report on how construction designs have not taken into account the requirements of disabled Americans.

> The construction industry has not satisfactorily met the needs for accessible workplaces and homes for all age and physical ability groups. The industry has relied on expensive and specialized plans to modify existing structures rather than creating universally designed spaces that are accessible to everyone.

3. Purpose statement. The purpose statement, crucial to the success of the report, tells readers why you wrote the report and what you hope to accomplish or prove. It expresses the goal of all your research. In explaining why you gathered information about a particular problem or topic, indicate how such information might be useful to a specific audience, company, or group. Like the problem statement, the purpose statement does not have to be long or complex. A sentence or two will suffice. You might begin simply by saying, "The purpose of this report is . . .".

4. Scope. This section informs readers about the specific limits—number and type of issues, time, money, locations, personnel, and so forth—you have placed on your investigation. You inform readers about what they will find in your report or what they won't through your statement about the scope of your work. The long report in Figure 15.3 concentrates on adapting the U.S. workplace to meet the communication and cultural needs of a work force of multinational employees, not trends in the international employment market—two completely different topics.

The Body

Also called the *discussion*, this section is the longest, possibly making up as much as 70 percent of your report. Everything in this and all the other sections of your report grows out of your purpose and how you have limited your scope. The body contains statistical information, details about the environment, and physical descriptions, as well as the various interpretations and comments of the authorities whose work you consulted or individuals whom you have interviewed as part of your research. The body can also identify and describe the range of options you surveyed and earmark the most appropriate, as the business report in Chapter 8 does in listing various print and online publications in which to advertise the new apartment complex.

The body of your report should:

- be carefully organized to reveal a coherent and well-defined plan
- separate material into meaningful parts to identify the major issues as well as subissues in your report
- clearly relate the parts to each other
- use headings to help your reader identify major sections more quickly

Your organization should reflect the different headings (and even subheadings) included in your report. Use them throughout your report to make it easy to follow. Organizational headings will also enable someone skimming the report to find specific information quickly. The headings, of course, will be included in the table of contents. (Note how Figure 15.3 is carefully organized into sections.)

In addition to headings, use transitions to reveal the organization of the body of your report. At the beginning of each major section of the body, tell readers what they will find in that section and why. Summary sentences at the end of a section will tell readers where they have been and prepare them for any subsequent discussions. The report in Figure 15.3 does an effective job of providing internal summaries.

Conclusion

The conclusion should tie everything together for readers by presenting the findings of your report. Findings, of course, will vary depending on the type of research you do. For a research report based on a study of sources located through various reference searches, the conclusion should summarize the main viewpoints of the authorities whose works you have cited. For a report done for a business, you must spell out the implications for your readers in terms of costs, personnel, products, location, and so forth.

Regardless of the type of research you do, your conclusions should do the following:

- be based on the information and documentation in the body of the report
- corroborate the evidence/information you gave in the body of your report
- grow out of the work you describe in the body of the report
- keep to the areas that your report covers, and not stray into areas it did not

In essence, to write an effective conclusion, you will have to summarize carefully a great deal of information accurately and concisely. Notice how the following conclusion of a long report on the Japanese tuna market clearly summarizes the market opportunities explained in the report.

Conclusion

The U.S. tuna industry has great potential to expand its role in the Japanese market. This market, currently 400,000 tons a year and growing rapidly, is already being supplied by imports that account for 35 percent of all sales. Our report indicates that not only will this market expand but its share of imports will continue to grow. The trend is alarming to Japanese tuna industry leaders because this important market, close to a billion dollars a year, is increasingly subject to the influence of foreign imports. Decreasing catches by Japan's own tuna fleet as well as an increased preference for tuna by affluent Japanese consumers have contributed significantly to this trend.

Recommendations

The most important part of the report, after the abstract, is the recommendations section, which tells readers

- what should be done about the findings recorded in the conclusion
- how you want them to solve the problem your report has focused on
- what equipment to purchase, when to expand a market, how to expand a website, or whom to recruit and retain, as in Figure 15.3

The report on the Japanese tuna market mentioned above uses a numbered list to make its recommendations.

Recommendations

Based on our analysis of the Japanese tuna market, we recommend five marketing strategies for the U.S. tuna industry:

1. Farm greater supplies of bluefin tuna to export.

2. Market our own value-added products.

3. Sell fresh tuna directly to the Tokyo Central Wholesale Market.

4. Sell wholesale to other Japanese markets.

5. Advertise and supply to Japanese supermarket chains.

Back Matter

Included in the back matter of the report are all the supporting data that, if included in the text of the report, would bog the reader down in details and cloud the main points the report makes.

Glossary

The glossary is an alphabetical list of the specialized vocabulary used in the report and the definitions. A glossary might be unnecessary if your report does not use a highly technical vocabulary or if *all* members of your audience are familiar with the specialized terms you do use.

Citations List

Any sources cited in your report—websites, books, articles, television programs, interviews, reviews, audiovisuals—are usually listed in this section (see Chapter 8 on preparing a Works Cited or Reference list). Also, ask your employer how he or she wants information to be documented. Sometimes, in the world of work, employers prefer all information to be documented in footnotes or cited parenthetically in the text. Note that the long report in Figure 15.3 uses the American Psychological Association (APA) system of documentation. Although APA no longer recommends a table of contents or list of illustrations, these pages are supplied here as models for students who are asked to use them.

Appendix

An appendix contains supporting materials for the report—tables and charts too long to include in the discussion, sample questionnaires, budgets and cost estimates, correspondence about the preparation of the report, case histories, transcripts of telephone conversations. Group like items in an appendix, as the examples under "Appendixes" in Figure 15.2 show. (Note that the plural of *Appendix* is *Appendixes,* not *Appendices.*)

▌A Model Long Report

The following long report in Figure 15.3 was written by a senior training specialist, Terri Smith Ruckel, and her staff for the vice president of human resources, who commissioned it. The main task facing Ruckel and her staff was to demonstrate what RPM Technologies should do to meet the needs of multinational workers and thus promote diversity in the workplace. She gathered relevant data from both primary and secondary research, including journal articles, websites, personal interviews, government documents, and even in-house publications from newsletters and company records, all on the RPM intranet.

Figure 15.3 contains all the parts of a long report discussed in this chapter except a glossary and an appendix. Intended for a general reader interested in learning more about the problems multinational workers face, the report does not contain the technical terms and data that would require a glossary and an appendix. Note how the cover letter introduces the report and its significance for RPM Technologies and the abstract succinctly identifies the main points of the report.

Figure 15.3   A long report.

RPM Technologies

4500 Florissant Drive   St. Louis, MO 63174

314.555.2121   www.rpmtech.com

May 5, 2008

Jesse Butler
Vice President, Human Resources
RPM Technologies

Dear Vice President Butler:

Begins with major recommendation of report

With this letter I am enclosing the report my staff and I prepared on effective ways to recruit and retain a multinational work force for RPM Technologies, which you requested six weeks ago. The report argues for the necessity of adapting the RPM workplace to meet the needs of multinational employees, including promoting cultural sensitivity and making business communications more understandable.

Presents findings of report

Multinational workers undoubtedly are playing a major role in U.S. business and at RPM. With their technical skills and homeland contacts, they can help RPM Technologies successfully compete in the global marketplace.

Alerts reader to major issue

Businesses like RPM need to recruit qualified multinational workers aggressively and then provide equal opportunities in the workplace. But we must also be sensitive to cultural diversity and communication demands. By including cross-cultural training—for native and non-native English-speaking employees alike—businesses like RPM promote cultural sensitivity. In-house language programs and plain-English or translated versions of key corporate documents can further improve the workplace environment for our multinational employees.

Offers to answer questions

I hope you will find this report useful and relevant in your efforts to recruit additional multinational employees for RPM Technologies. If you have any questions or want to discuss any of our recommendations or research, please call me at extension 5406 or e-mail me at the address below. I look forward to receiving your suggestions.

Sincerely yours,

Terri Smith Ruckel

Terri Smith Ruckel
Senior Training Specialist
truckel@rpmtech.com

Enclosure

Adapting the RPM Workplace for Multinational Employees in the New Millennium

Title page is carefully formatted and uses boldface

Terri Smith Ruckel

**Senior Training Specialist
RPM Technologies**

Identifies writer and job title

Ruckel presents report from entire staff— writing for another's signature

Prepared for

**Jesse Butler
Vice President, Human Resources**

RPM executive who assigned the report

May 5, 2008

Date submitted

2

Table of Contents

LIST OF ILLUSTRATIONS 3

ABSTRACT 4

INTRODUCTION 5
    Background 5
        Global Immigration 5
        Today's High-Tech Immigrants 5
    Problem 7
    Purpose 7
    Scope 7

DISCUSSION 7
    Providing Equal Workplace Opportunities for Multinational Employees 7
    Aggressive Recruitment of IT Professionals from Diverse Cultures 7
        Commitment to Ethnic Representation 9
    Promoting and Incorporating Cultural Awareness Within the Company 9
        Cross-Cultural Training 9
        Promotion of Cultural Sensitivity 10
    Making Business Communication More Understandable for
    Multinational Employees 11
        Translation of Written Communications 11
        Exchange of Language Learning: Some Options 13
        Language Training Must Be Reciprocal 13

CONCLUSION 13

RECOMMENDATIONS 14

REFERENCES 15

Major divisions of report in all capital letters

Subheadings indicated by different typeface, indentations, and italics

Page numbers included for major sections of report

3

List of Illustrations

Figure 1: Major Ethnic Groups Immigrating to New York City: 2007 6

Figure 2: Immigration to Silicon Valley in 2007 6

Figure 3: Union Bank of California:
Growth in Percentage of Multinational Employees .8

Figure 4: A Multicultural Calendar 12

Identifies each figure by title and page number

4

Abstract

This report investigates how U.S. businesses such as RPM must gain a competitive advantage in today's global marketplace by recruiting and retaining a multinational work force. This current wave of immigrants is in great demand for their technical skills and economic ties to their homeland. Yet many companies like ours still operate by policies designed for native speakers of English. Instead, we need to adapt RPM's company policies and workplace environment to meet the cultural, religious, social, and communication needs of these multinational workers. To do this, we need to promote cultural sensitivity training, both for multinationals and employees who are native speakers of English. Additionally, as other U.S. firms have, RPM should adapt vacation schedules and daycare facilities for its multicultural work force. Equally important, RPM should ensure, either through translations or plain-English versions, that all documents can be easily understood by multinational workers. RPM might also offer non-native speakers of English in-house language instruction while providing foreign language training for employees who are native speakers of English.

Concise, informative abstract states purpose of report and why it is important

Uses helpful transitional words

5

Introduction

Background

The U.S. work force is undergoing a remarkable revolution. The U. S. Bureau of Labor Statistics predicts that by 2012 the labor force in the United State will comprise 162 million workers who must fill 167 million jobs (2007). According to the U.S. Chamber of Commerce, a shortage of skilled workers is sure to increase "even more heavily in the future when many of the baby boomers begin to retire" (2007).

The most dramatic effect of filling this labor shortage will be in hiring greater numbers of highly skilled multinational employees, including those joining RPM. Currently, "one of every five computer specialists [and] one of every six persons in engineering or science occupations . . . is foreign born" (Kaushal & Fix, 2007). This new wave of immigrants will make up 37 percent of the labor force by 2015 and continue to soar afterward. By 2025 the number of international residents in the United States will rise from 26 million to 42 million, according to the U.S. Chamber of Commerce (2007). The Congressional Budget office predicts that "over the next decade net immigration will average between 500,000 and 1.5 million people annually" (2007). As Alexa Quincy aptly puts it, "the United States is becoming the most multiculturally diverse country in the global economy" (2008).

The following commissioned report explores the implications that this new multinational work force will have on RPM and what we must do to accommodate these workers.

Global Immigration

Unlike earlier generations, today's immigrants actively maintain ties with their native countries. These new immigrants travel back and forth so regularly they have become global citizens, exercising an enormous influence on a business like RPM. Figure 1 reveals the major ethnic groups immigrating to New York city in 2007. Demographers Crane and Boaz claim, "Immigration [will] give America an economic edge in the global economy . . . most notably in the Silicon Valley and other high-tech centers. They provide business contacts with other markets, enhancing [a company's] ability to trade and invest profitably abroad" (2007).

Today's High-Tech Immigrants

Undeniably, many immigrants today often possess advanced levels of technical expertise. A report by the Kaiser Foundation found that California's Silicon Valley has significantly benefited from the immigrants who have arrived with much needed technical training. Asian, Indian, Pakistani, and Middle Eastern scientists and engineers, who have relocated from a number of countries, now hold more than 40 percent of the region's technical positions ("Immigrants Attain," 2007). Figure 2 indicates the countries of origin for Silicon Valley's immigrants in 2007 and records each nationality's percentage.

6

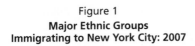

Numbers and titles figure

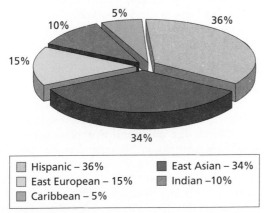

Figure 1
**Major Ethnic Groups
Immigrating to New York City: 2007**

5%

10%

15%

36%

34%

	Hispanic – 36%		East Asian – 34%
	East European – 15%		Indian –10%
	Caribbean – 5%		

Source: Brown, 2007.

Pie chart reveals differences in immigrant workers

Cites source for visual

Figure 2
Immigration to Silicon Valley in 2007

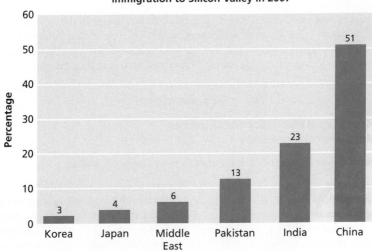

Source: Asian American Business Journal.

Describes new workers and documents their significance to businesses

Relevc
in ap;
plac

7

Information technology (IT) employers are enthusiastically searching for skilled international workers. The annual immigration limit, set at 65,000 by the controversial new H-1B guest worker visa, has been met on the very first day of a company's fiscal year by many IT firms (Frauenheim, 2006). Murali Krishna Devarakonda, president of the Immigrants Support Network in Silicon Valley, states why: "We contribute significantly to the research and development of a company's technology" (personal communication, April 3, 2008).

E-mail not included in APA Reference list

Problem

RPM, like other e-companies, is experiencing a critical talent shortage of IT professionals, making the recruitment of a multinational work force a vital concern of U.S. businesses. Meeting the cultural and communication demands of these workers poses serious challenges for RPM. The traditional workplace has to be transformed to respect the ways multinational employees communicate about business. Native English-speaking employees will also have to be better prepared to understand and appreciate their international co-workers. Unfortunately, many corporate policies and programs at RPM, and at other U.S. companies, have been created for native-born, English-speaking employees (Martin, 2005; Reynolds, 2008). Rather than rewarding multinational workers, such policies unintentionally punish them.

Identifies a major problem and why it exists

Two separate, corroborating sources

Purpose

The purpose of this report is to show that because of the increasing numbers of multicultural employees in the workplace, RPM must adapt its business environment for this essential and diverse work force.

Concisely states why the report was ritten

Scope

This report explores cultural diversity in the U.S. workplace in the new millennium and suggests ways for RPM to compete successfully in the global village by providing equal employment opportunities for multinational workers, fostering cross-cultural literacy, and improving training in intercultural communication.

der ill

Discussion

iding Equal Workplace Opportunities ltinational Employees

e Recruitment of IT Professionals from Diverse Cultures

t visual ropriate in text

al work force is essential if RPM wants to compete successfully in a cultur-
bal market. But firms such as RPM must be prepared to adapt or modify
nd procedures to attract these multinational employees, beginning with
ruitment and retention policies. Routine visits to U.S. campuses by
or "specialized international recruiters" can help to identify and to
ob candidates (Hamilton, 2008).

8

RPM should consider visiting universities abroad with distinguished technical programs to attract talented multinational employees. We should encourage students and recent graduates from these universities to apply for a 1-J visa to learn more about RPM through an internship program here. As Catherine Bolgar reports, "Boeing went to Russia for specialist software engineers it couldn't find in the U.S" (personal interview, June 18, 2007). These searches should be combined with our websites and executive blogs to emphasize RPM's commitment to globalization. Working more cooperatively with the Immigration and Naturalization Service will unquestionably help us to retain the most qualified, professional staff.

Emphasizes recruiting multinational workers and suggests how to do so

Capitalizing on a diverse work force, RPM can more effectively increase its multicultural customer base worldwide. Logically, customers buy from the people they can relate to culturally. RPM can take a lead from Union Bank of California, a business that effectively serves a diverse West Coast population, especially its Asian and Hispanic customers. The bank has a successful recruitment history of hiring employees with language skills in Japanese, Vietnamese, Korean, and Spanish. In fact, Union Bank is ranked fourth overall as an employer of minorities ("*Federation* magazine," 2008). Figure 3 charts the increase in the percentage of multinational employees from these language groups hired by Union Bank.

Identifies specific benefits for RPM

Another successful business, Darden Restaurants, Inc., selected Richard Rivera, a Hispanic, to serve as president of Red Lobster, the nation's largest full-service seafood chain. Overseeing 680 restaurants nationwide, Rivera was the "most powerful minority in the restaurant industry" (R. Jackson, personal interview, January 15, 2007). Under

Interview supports the need for recruitment

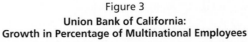

Figure 3
**Union Bank of California:
Growth in Percentage of Multinational Employees**

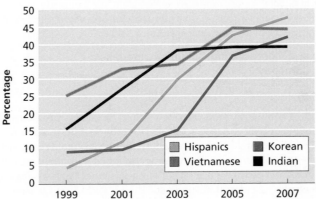

Tracks key information in a clear and concise graph

Source: Hamilton, 2008.

9

Rivera's leadership, Red Lobster established an impressive history of hiring more international employees—totaling more than 35 percent of its work force—than it had in previous years. Rivera believes that management cannot properly respond to customers from different ethnic backgrounds if the majority of its employees is limited to native speakers of English. Closer to RPM in St. Louis, Whitney Abernathy—manager of Netshop, Inc.—found that contracts from Indonesia increased by 17 percent after she hired Jakarta native Safja Jacoef (personal communication, April 2, 2007).

Personal interview not included in APA References list

Commitment to Ethnic Representation

Many companies have strong mission statements on diversity and multinational employees in the workplace. The most progressive of these companies—Toyota, American Airlines, IBM, and Wal-Mart—promote multinationals as mentors and interpreters. Kodak, which is committed to all-inclusive workplace, has eight employee network groups, including the Hispanic Organization for Leadership and Advocacy, or HOLA, which "helps all Hispanic employees reach their full potential while developing a rewarding career" ("Careers," 2008). Such a proactive program, which we would do well to incorporate at RPM, recognizes the Leadership abilities of multinational employees. Moreover, "glass ceilings," which in the past have prevented women and ethnic employees from moving up the corporate ladder, are being shattered. The recruitment and promotion of non-native speakers of English is vital for corporate success. Tesfaye Aklilu, vice president at United Technologies, astutely observes:

Stresses other business precedents that encourage recruiting these employees

Indent quotation of forty or more words

> In a global business environment, diversity is a business imperative. Diversity of cultures, ideas, perspectives, and values is the norm of today's business environment. The exchange of ideas from different cultural perspectives gives a business additional, valuable information. Every employee can see his/her position from a global vantage point. ("Diversity at UTC," 2008)

RPM would do well to follow the lead of one of our chief vendors, Ablex Plastics, which created the position of Vice President for Diversity last year and for which Ablex won an award from the International Business Foundation for hiring more Asian American women managers (Ablex, 2008).

Second major section

Promoting and Incorporating Cultural Awareness Within the Company

Cross-Cultural Training

Cites business incentive to adapt as competitors did

Many of RPM's competitors are creating cultural awareness programs for international employees as well as native speakers. Employees find it easier to work with someone whose values and beliefs they understand, while employers benefit from effective on-the-job collaboration. Such a program could have prevented the problem RPM confronted last quarter when one of our non-native English-speaking employees was offended by a cultural misunderstanding ("RPM first Quarter," 2008). We may want to model our cultural sharing programs after those at American Express, which has a work force representing 40 nations, or those at Extel Communications

References internal document from company intranet

10

with its large percentage of Hispanic and Vietnamese employees. United Parcel Service (UPS) profitably pairs a native English-speaking employee with someone from another cultural group to improve on-the-job problem solving and communication. Jamie Allen, a UPS employee since 1995, found her work with Lekha Nfara-Kahn to be one of the most rewarding experiences of her job (Johnson, 2008).

Although they need to encourage cultural sensitivity training, U.S. firms like RPM also need to be cautious about severing international workers' cultural ties—a delicate balance. When management actively promotes bonds among employees from similar cultures, workers are less fearful about losing their identity and becoming "token" employees. The law firm of Latour and Lleras recommends that new multinationals join a "buddy system" to assist their transition to American culture while retaining their cultural identity (U.S. Visa News, 2007).

Globe Citizens Bank has long mobilized culturally similar groups by asking workers of shared cultural heritages to network with each other (Gordon & Rao, 2008). Employees of Turkish backgrounds from Globe's main New York office go to lunch twice a month with Turkish-born employees from the Newark branches. Globe hosts these luncheons and in return receives a bimonthly evaluation of the bank's Turkish and Middle Eastern policies (Hamilton, 2008).

Cultural education must go both ways, though. Both sides have things to learn about doing business in a global environment. The U.S. business culture has conventions, too, and few international employees would want to ignore them, but they need to know what those conventions are (Johnson, 2008). A frequent problem with U.S. corporations such as RPM is that we assume everyone knows how we do things and how we think—it never occurs to us to explain ourselves. Problems often stem from easy-to-correct misunderstandings about business etiquette. For example, native speakers of English are typically comfortable within a space of 1.5 to 2 feet for general personal interactions in business. But workers from Taiwan or Japan, who prefer a greater conversational distance, feel uncomfortable if their desks are less than a few feet away from another employee's workspace (Johnson, 2008; "Taiwanese Business Culture," 2007).

Promotion of Cultural Sensitivity

Corporate efforts to validate different cultures might also include the recognition of an ethnic group's holidays or memorable historical events. RPM has just begun to do this by hosting cultural events, including Cinco de Mayo and Chinese New Year celebrations. Techsure, Inc., an Illinois software firm, allows Muslim employees to alter their work schedules during Ramadan, Islam's holy month of fasting (M. Saradayan, personal communication, February 28, 2008). Many companies honor National Hispanic Heritage Month in September, which coincides with the independence celebrations of five Latin American countries (U.S. Department of Equal Employment Opportunity [USEEO], 2007). GRT Systems sends New Year's greetings at Waisak (the Buddhist Day of Enlightenment) to its Chinese employees and to Indian workers at Rama Dipawli. Chemeka Taylor, GRT's operations manager, wisely points out: "We send native-born employees

Comment included in a source

Transition to new subdivision—networking of employees with similar cultural backgrounds

Relevant source on topic of immigration

Offers two examples RPM could follow

Another clear transitional sentence

Identifies key RPM problem

Gives cultural example

Gives precise ways RPM can incorporate cultural sensitivity into the workplace

Includes appropriate calendar of ethnic holidays

11

Christmas cards; why shouldn't we honor our international work force, too?" (personal communication, April 10, 2008). Figure 4 (on page 8) provides a multicultural calendar that RPM needs to follow in developing cultural sensitivity policies.

Successful U.S. firms have been sensitive to the needs of their English-speaking employees for years. Flexible scheduling, telecommuting options, daycare, and preventive health programs have become part of corporate benefit plans to take care of employees. Many of these options and benefits have already been in place at RPM. But an international work force presents additional cultural opportunities for management to respond to with sensitivity. For example, company cafeterias might easily accommodate the dietary restrictions of vegetarian workers or those who abstain from certain foods, such as dairy products or meat. At Globtech, soybean and fish entrees are always available (Reynolds, 2008). Adding ethnic items at RPM would express our cultural awareness and respect for multinational employees.

Daycare remains a key issue in hiring and retaining skilled employees, whether they are native or non-native speakers of English. RPM's facilities at our high-tech park in San Luis Obispo have brought us much positive publicity over the past eight years ("RPM Daycare Facilities," 2007). But by modifying child care that reflects our workers' culturally diverse needs, RPM can give a multinational work force greater peace of mind and better enable them to do their jobs. A pacesetter in this field is DEJ Computers, which insists that at least two or three of its daycare workers must be fluent in Korean or Hindi (Parker, 2007). Another culturally sensitive employer, Techsure assists its Hispanic work force by hiring bilingual daycare workers and by serving foods the children customarily eat at home (Gordon & Rao, 2008).

Making Business Communication More Understandable for Multinational Employees

Translation of Written Communications

All employees must be able to understand business communications affecting them. Among the essential documents causing trouble for multicultural readers are company handbooks, insurance and health care obligations, policy changes, and OSHA and EPA safety regulations (Hamilton, 2008). To ensure maximum understanding of these documents by a multinational work force, RPM needs to provide a translation or at least a plain-English version of them. To do this, RPM could solicit the help of employees who are fluent in the non-native English speakers' languages as well as contract with professional translators to prepare appropriate documents.

Workplace signs in particular, especially safety messages, must consider the language needs of international workers. In the best interest of corporate safety, RPM needs to have these signs translated into the languages represented by multinationals in the workplace and/or to post signs that use global symbols. Unquestionably, we need to avoid signs that workers might find hard or even impossible to decipher. For example, a large **P** for "parking" or an **H** for "hospital" might be unfamiliar to non-native speakers of English (Parker, 2007).

12

Figure 4
A Multicultural Calendar

December 2008

	1	2	3	4	5	6
						Feast of St. Nicholas (some European countries)
7 Advent begins (Christian)	**8** Bodhi Day (Rohatsu-Buddhist)	**9** Eid al-Adha (Islam) Gita Jayanti (Hindu)	**10**	**11**	**12** Datta Jayanti (Hindu); Feast Day—Our Lady of Guadalupe (Catholic)	**13**
14	**15**	**16** Posada begins (Hispanic) (ends Dec. 24)	**17**	**18**	**19**	**20**
21	**22** Ekadashi (Hindu); Hanukkah begins (Jewish) (ends Dec. 29)	**23**	**24** Nocha Buena (Hispanic)	**25** Christmas (Christian)	**26** Kwanzaa begins (Interfaith) (ends Jan. 1)	**27**
28	**29** Muharram (Islam)	**30**	**31** New Year's Eve (Western)			

Some Information About December's Holidays

Bodhi Day: Buddhists celebrate Prince Gautama's vow to attain enlightenment.

Christmas: Christians celebrate the birth of Jesus Christ.

Datta Jayanti: Celebration of the birth of Lord Datta.

Eid al-Adha: Feast of Sacrifice, commemorating the prophet Abraham's sacrifice of his son Ishmael to Allah.

Ekadashi: Twice monthly purifying fast.

Feast of Our Lady of Guadalupe: (Mexico) Catholic Christian holiday honoring the appearance of the Virgin Mary near Mexico City in 1531.

Feast of Saint Nicholas: Christmas celebration in Austria, Belgium, France, Germany, the Netherlands, Czech Republic, Poland, Russia, and Switzerland.

Gitta Jayanti: Celebration of the anniversary of the Bhagavad Gita, the Hindu scripture.

Hanukkah: Eight-day celebration commemorating the rededication of the Temple of Jerusalem.

Kwanzaa: African American and Pan-African celebration of family, community, and culture. Seven life virtues are presented.

Muharram: Islamic New Year.

Nocha Buena: The final day of the Posada celebration.

Posada: Nine-day Latin American Christian celebration, symbolizing the journey to find an inn.

Source: Johnson (2008).

Major visual with a great deal of detail merits a full page to make it readable

Pays attention to major world holidays

Provides descriptions of cultural significance of dates

Lists source

13

Exchange of Language Learning: Some Options

Outlines specific benefits for RPM

Providing English language instruction for employees also has obvious advantages for the RPM workplace in the new millennium. Should RPM elect not to provide in-house language training, the company should consider reimbursing employees for necessary courses, books, tapes, and software. Moreover, we might make available an audio library of specific vocabulary and phrases commonly used on the job as well as multilingual dictionaries with relevant technical terms. Hiromi Naguchi, an e-marketing analyst at PowerUsers Networking, praised the English language training she received on her job. Although she had ten years of English in the Osaka (Japan) schools, Naguchi developed her listening and speaking skills on the job, benefiting her employers, co-workers, and customers. As a result of her diligence, Naguchi was recognized as her company's "Most Valuable Employee for 2006" for her interpersonal skills (H. Naguchi, personal communication, April 3, 2007).

Points to apt example

A recent international survey of executive recruiters showed that being bilingual is critical to success in business ("Developing Foreign Language," 2008). According to this survey, 85 percent of European recruiters, 88 percent of Asian recruiters, and 95 percent of those in Latin America stressed the importance of speaking at least two languages in today's global workplace. As Jennifer Torres, CEO of Pacifica Corp., emphasized: "Although English remains the dominant language of international business, multilingual executives clearly have the competitive advantage. This advantage will only increase with the continued globalization of commerce" (cited in "Developing Foreign Language," 2008). Stockard Plastics, one of RPM's major vendors, boasts that its employees know more than seventy languages, everything from Albanian to Yoruba.

Cites business survey to confirm the necessity of change at RPM

Identifies quote within a source

Language Training Must Be Reciprocal

But language training has to be reciprocal—for native as well as non-native speakers—if communication is to succeed in the global marketplace. Sadly, most Americans do not know a foreign language and "fewer than 1 in 10 students at U.S. colleges major in [a] foreign language" (Cutshall, 2005). Unfortunately, this is the case at RPM. However, many of the international workers RPM needs to recruit are bi- or even trilingual. In India or South Africa, for example, the average worker speaks two or more languages every day to conduct business.

Argues that reader must consider both sides

Since RPM needs to recruit such workers, we need to learn more about the cultures and languages of these global employees. RPM management should consider contracting with one of the companies specializing in language instruction for business people found at *http://www.selfgrowth.com/foreignlanguage.html*. We also need to network with international employee groups to solicit their help and advice.

Conclusion

Following the body, conclusion concisely summarizes the highlights of the report without repeating the documentation

To compete in the global marketplace, RPM must emphasize cultural diversity in its corporate mission and throughout the workplace. Through its policies and programs, RPM must aggressively recruit and retain an increasing number of technologically ed-

14

ucated and experienced multinational workers who are in great demand today and will be even more so over the next ten to twenty years. These workers can help RPM increase our international customer base and advance the state of our technology. But we need to ensure that the workplace is sensitive to their cultural, religious, and communication needs. Providing equal opportunities, diversity training and networking, and easy-to-understand business documents will keep RPM globally competitive in recruiting and retaining these essential employees. We can expand only by making a commitment to learn about and respect other cultures and their ways of doing business.

Recommendations

By implementing the following recommendations, based on the conclusions reached in this report, RPM Technologies can succeed in hiring and promoting the IT multinational professionals our company needs for future success in the millennial economy.

1. Recruit multinational workers more effectively through our website, international hiring specialists, and visits to college and university campuses here and abroad while working more closely with the Immigration and Naturalization Service (INS) to retain multinationals.
2. Establish a mentoring program to identify and foster leadership abilities in multinational employees, resulting in retaining and promoting these workers.
3. Promote cultural sensitivity and networking groups comprising both multinationals and native English-speaking employees.
4. Foster a group's cultural ties by actively supporting such work-related organizations as HOLA.
5. Develop educational materials for employees who are native speakers of English about the cultural traditions of their multinational co-workers.
6. Reassess and adapt RPM's daycare facilities to meet the needs of children of multinational employees.
7. Supply relevant translations and plain-English versions of company handbooks, manuals, new regulations, insurance policies, safety codes, and other HR documents.
8. Develop foreign language training programs to enhance communication and collaboration as well as to equip future RPM managers for overseas assignments.

Numbered list format is easy to read for busy executives

Provides specific, relevant recommendations, based on conclusions, to solve the problem at RPM

Uses strong persuasive verbs to introduce such recommendation

15

REFERENCES

Ablex appoints its first vice president for diversity. (2008, February). Retrieved from http://www.ablexinter.org

Bolgar, C. (2007, June 18). Corporations need a global mindset to succeed in today's multipolar business world. *Wall Street Journal Online.* Retrieved from http://www.accenture.com/global/highperformancebusiness_business /multipolarbusinessworld.htm

Brown, P. (2007, December 15). History of U.S. immigration. Retrieved from http://immigration.ucn.edu

Careers at Kodak. (2008, February). Retrieved from www.kodak.com/US /employment/corp/career/whyemploymentworks.jhtmc

Congressional Budget Office. (2006, June). Projections of net immigration to the U.S. Retrieved February 20, 2008, from http://www.cbo.gov/ftpdocs / 72xx/doc7249/0606/Immigration.pdf

Crane, E. H., & Boaz, D. (Eds.). (2007). *Cato Handbook on Policy* (6th ed.). Washington, DC: Cato.

Cutshall, S. (2005, January). Why we need "the year of languages." *Educational Leadership, 62*(4), 20–23.

Developing foreign language skills is good business. (2008, February 3). *Business World* . Retrieved from http://www.businessworld.ca/article.cfm/newsID /7109.cfm

Diversity at UTC. (2008). *United Technologies Corporation.* Retrieved from http://www.utc.com/careers/diversity/index4.htm

Federation magazine ranks Union Bank one of the best companies for minorities. (2008, June 23). *Union Bank of California.* Retrieved from http://www.uboc.com/about /main/0,3250,2485_11256_502261585,00.html

Frauenheim, E. (2006, October 1). H-1B visa limit for 2007 already reached. *C-Net News.* Retrieved from http://news.com.com/H-1B+visa+limit+for+2005+ already+reached/2100–1022_3–5392917.html?tag=nl

Gordon, T., & Rao, P. (2008, April). Challenges ahead for American companies. *National Economics Review, 11*, 38–42, 56.

Hamilton, B. E. (2008, April 15). Diversity is the answer for today's work force. *The New Business Journal, 10*(38), 35–37.

Immigrants attain the American dream in Silicon Valley. (2006, December 19). *The Taipei Times*, p. 12.

Labels list of sources actually cited

All entries listed alphabetically by author's last name, or (if no author) by first word of title

Double spaces between entries

Date of publication included for every entry after author's name or title if author's name not given

Only first word and proper nouns capitalized in title

16

Johnson, V. M. (2008). Growing multinational diversity in business sparks changes. *Business Across the Nation, 23*(7), 43–48.

Kaushal & Fix. (2007). Global engineers' online marketplace. Retrieved from http://www.omkts.com

Martin, P. (2007). Migration news. Retrieved from http://www.migration.ucdavis.edu

Parker, M. (2007, May). Immigration facts. Retrieved from http://immigration.org

Quincy, A. (2008, March 3). Multiculturalism makes for a good business. *Workforce, Inc. 14*(2): 5–8.

Reynolds, P. (2008, March 3). Serving up culture. Retrieved from http://www.culture.org

RPM daycare facilities rated high. (2007, January 15). *RPM News*, 31–32.

RPM first quarter activity report. (2008). *RPM Internal Reports*. Retrieved from RPM intranet.

Taiwanese business culture. (2007). *Executive Planet*. Retrieved from http://www.executiveplanet.com/business-culture-in/132438266669.html

U.S. Bureau of Labor Statistics. (2007). Tomorrow's jobs. Retrieved from http://www.bls.gov/oco/oco2003.htm

U.S. Chamber of Commerce. (2007, June). Chamber, labor leaders renew call for immigration reform. *USChamber.com*. Retrieved from http://www.uschamber.com/issues/index/immigration/uscc_migrationpolicy_statement/htm

U.S. Department of Equal Employment Opportunity. (2007). *Hispanic employment program report for 2006* (EEO Publication No. B56.2245). Washington, DC: U.S. Government Printing Office.

U.S. Visa News. (2007). Latour and Lleras immigration attorneys. Retrieved from http://www.usvisanews.com/relbuddy.shtml

Page provided for print sources

Second and subsequent lines indented 5 spaces

Works without author alphabetized by first word of title

Full Web addresses are given for verification and to make source easy to find

Final Words of Advice About Long Reports

Perhaps no piece of writing you do on the job carries more weight than the long report. You can simplify your job and increase your chances for success by following these guidelines for scheduling, researching, and collaborating:

1. Plan and work early—do not postpone work until a deadline draws near.
2. Do a thorough search among Internet, print, and other resources.
3. Divide your workload into meaningful units—reassure yourself that you do not have to write the report or even an entire section of a report in a day or two.
4. Set up mini-deadlines for each phase of your work and then meet them.
5. Confer often and carefully with others in your group office.

Revision Checklist

☐ Concentrated on a major problem—one with significant implications for my major, neighborhood, city, or employer.

☐ Identified, justified, and described the significance of the main problem.

☐ Did sufficient research—in the library, on the Internet, through interviewing, from personal observation and/or testing.

☐ Became familiar with key terms and major trends in the field.

☐ Anticipated how various managers and other decision makers will use and profit from my report for their long-range planning.

☐ Made sure I understand what employer/teacher/reader is looking for.

☐ Adhered to all specified schedules.

☐ Followed company's/instructor's schedule for completing various stages of long report.

☐ Divided and labeled the parts of long report to make it easy for readers to follow and to show a careful plan of organization.

☐ Supplied an informative abstract that leaves no doubt in readers' minds about what report deals with and why.

☐ Designed attractive title page that contains all the basic information—title, date, for whom the report is written, my name.

☐ Gave readers all the necessary introductory information about background, problem, purpose of report, and scope.

☐ Included in body of report the weight of all my research—the facts, statistics, interview comments, and descriptions—that my readers need.

☐ Included subheadings to reflect the major divisions into which I have organized the research that forms the nucleus of the text.

☐ Wrapped up report in succinct conclusion. Told readers what the findings of my research are and accurately interpreted all data.

☐ Supplied a recommendations section (if required) that tells readers concretely how they can respond to the problem using the data. Offered recommendations that are realistic and practical and related directly to the research and topic.

☐ Included in the final copy of report all the parts listed in table of contents.

☐ Supplied a one-page letter of transmittal or cover letter informing readers why the report was written and describing its scope and findings.

Exercises

1. Send an e-mail to your instructor on how one of the short reports in Chapter 14 could be useful to someone who has to write a long report.

2. Using the information contained in Figures 15.1 and 15.2, draft an introduction for the report "A Study to Determine New Directions in Women's Athletics at Coastal College." Add any details you think will be relevant.

3. What kinds of research did Terri Smith Ruckel and her staff do to write the long report in Figure 15.3? As part of your answer, include the titles of any specific reference works you think the writer may have consulted.

4. Study Figure 15.3 and answer the following questions based on it:
   a. Why can the abstract be termed informative rather than descriptive?
   b. How have Ruckel and her staff successfully limited the scope of the report?
   c. Where have Ruckel and her staff used internal summaries especially well?
   d. Where and how have the writers adapted their technical information for their audience (a general reader)?
   e. What visual devices do the writers use to separate parts of the report?
   f. How do the writers introduce, summarize, and draw conclusions from the expert opinions in order to substantiate the main points?
   g. How have the writers documented information?
   h. What functions does the conclusion serve for readers?
   i. How do the recommendations follow from the material presented in the report? How are they both distinct and interrelated?

5. Come to class prepared to discuss a major community problem suitable for a long report, e.g., traffic, crime, air and water pollution, housing, transportation. Then write a letter to an appropriate agency or business, requesting a study of the problem and a report.

6. Write a report outline for one of the problems you decided on in Exercise 5. Use major headings. Include a cover letter with your outline.

7. Have your instructor look at and approve the outline you prepared for Exercise 6. Then write a long report based on the outline, either individually or as part of a collaborative writing team.

CHAPTER 16

Making Successful Presentations at Work

Visit www
.cengage.com/
english/kolin for
this chapter's
online exercises,
ACE quizzes, and
Web links.

Almost every job requires employees to have and to use carefully developed speaking skills. In fact, to get hired, you have to be a persuasive speaker at your job interview. And to advance up the corporate ladder, you will have to continue to be a confident, well-prepared, and persuasive speaker. The goal of this chapter is to help you be a more successful speaker.

Types of Presentations

On the job you will have numerous presentation responsibilities that will vary in the amount of preparation they require, the time they last, and the audience and the occasion for which they are intended. Here are some frequent types of presentations you can expect to make as part of your job:

- sales appeals to prospective customers
- evaluations of products or policies
- progress reports to your boss and clients
- reports to superiors about your job accomplishments
- justifications of your position or even your department
- appeals and/or explanations before elected officials
- presentations at professional conferences
- explanation of a procedure, decision, or plan before a community/civic group

Whatever type of presentation you are asked to deliver, this chapter gives you practical advice on how to become a better, more assured communicator in both informal briefings, including on the telephone, and formal presentations.

Informal Briefings

If you have ever given a book report or explained laboratory results in front of a class, you have given an informal briefing. Such semiformal reports are a routine part of many jobs. Here are some of the typical informal briefings you may be asked to deliver at work:

- a status report on your current project
- an update or end-of-shift report, like those nurses and police officers give
- an explanation of a policy to co-workers
- a report on a conference you attended
- a demonstration of new procedure or piece of equipment/software
- a follow-up session on equipment or procedures
- a summary of a meeting you attended

Such presentations are usually short (one to seven minutes, perhaps), and you won't always be given advance notice. When the boss tells you to "say a few words about the new website" (or the new programming procedure), you will not be expected to give a lengthy formal speech.

Guidelines for Preparing Informal Briefings

Follow these guidelines when you have to make an informal briefing:

- Make your comments brief and to the point.
- Keyboard a few bulleted items you plan to cover.
- Highlight key phrases and terms you need to stress.
- Include in your notes only the major points you want to mention.
- Arrange your points in chronological order or from cause to effect.

Figure 16.1 is an informal outline with key facts used by an employee who is introducing Diana J. Rizzo, a visiting speaker, to a monthly meeting of safety directors.

Using Telephones and Cell Phones Effectively

When you use the telephone, you represent your company. *How* you say something can be as crucial as *what* you say. Your voice and your attitude symbolize your employer's service or product for the person on the other end of the line.

Telephone Etiquette

Here are some basic rules of telephone etiquette:

1. Answer calls on the second or, at the latest, the third ring.
2. Never tie up a company phone with personal business. One large bank put it this way: "To a customer a busy signal is almost the same as a locked door."
3. Avoid shouting, whispering, mumbling. Keep a smile in your voice. Never chew gum or finish your lunch while on the phone.
4. Answer courteously. Identify yourself, your title, your department. "Hello, this is John Loo, assistant programmer at Northwest Megalink." Ask, "How may I help you?" Responses like "thank you," "please," and "you're welcome" epitomize customer service.
5. Eliminate disturbing background noises—radios, televisions, and printers.

Figure 16.1   Some notes for an informal briefing to introduce an engineer to a group of safety directors.

- Diana J. Rizzo, Chief Engineer of the Rhode Island State Highway Department for twelve years

- Experience as both a civil engineer and safety expert

- Consultant to Secretary Habib, Department of Transportation

- Member of the National Safety Council and author of "Field Test Procedures in Highway Safety Construction"

- Instrumental in revising state safety commission's website

- Received "Award for Excellence" from the Northeastern Association of Traffic Engineers in May 2008

6. Never put someone on hold unless you ask his or her permission. It is rude to say "Hold on. I'll look it up," and then drop the phone on your desk.
7. Prepare for a call by having copies of any correspondence, reports, price lists in front of you. Pull up relevant documents on your computer so they are in front of you as you speak.
8. Check your voice mail every day, just as you would your e-mail. In not doing so, you may miss making a sale or attending a conference.
9. Never give intricate instructions over the phone. E-mail, fax, or snail mail such documents and use a telephone call to follow up and see if the individual has any questions.
10. If you record a message to greet callers when you are not available, make it friendly and helpful, not gruff and rude—"I'm not in. Please leave your message at the tone."

Cell Phone Courtesy

When you use a cell phone, follow all the rules of telephone etiquette just discussed in addition to heeding these guidelines:

1. Do not use your personal cell phone (or beeper) at work.
2. Silence the ringer on any cell phone during a conference; sales meeting; business breakfast, lunch, or dinner. Use the cell phone vibrator option to notify you of a call.

3. When you use a cell phone for business-related calls, select a professional-sounding ring tone. "The Charge of the Light Brigade" or a melody from a cartoon is inappropriate.

4. Turn cell phones off when you enter a hospital (where they can interfere with life-support equipment), or attend a religious service, legal hearing (courtroom), or professional conference.

5. Never allow a ringing cell phone to interrupt a meeting with a customer. It makes you look like you are not interested in the customer's needs.

6. It is unethical to charge personal long-distance calls to your company cell phone number without authorization from your boss.

7. Always make sure your cell phone is charged and working. It's embarrassing to lose contact with your boss or a client in the middle of a conversation.

Formal Presentations

Whereas an informal briefing is likely to be short, generally conversational, and intended for a limited number of people, a formal presentation is much longer, far less conversational, and may be intended for a wider audience. It involves much more preparation.

Avoid speaking "off the cuff." The professional speechmaker may be comfortable with this approach, but for most of us, the worst way to make a presentation is to speak without any preparation whatsoever. You only fool yourself if you think you will have all the necessary details and explanations in the back of your head. Once you start talking, generally everything does not fall into place smoothly. Without preparation, you are likely to confuse important points or forget them entirely. Mark Twain's advice is apt here: "It takes three weeks to prepare a good impromptu speech."

Expect to spend several days preparing your presentation. You cannot just dash it off. You will need time to

- research the subject
- interview key resource individuals
- prepare, time, and sequence visuals
- coordinate your talk with presentations by co-workers or your boss
- rehearse your presentation

Many of us are uncomfortable in front of an audience because we feel frightened or embarrassed. Much of that anxiety can be eased if you know what to expect. The two areas you should investigate thoroughly before you begin to prepare your presentation are (1) who will be in your audience and (2) why they are there.

Analyzing Your Audience

The more you learn about your audience, the better prepared you will be to give them what they need. Just as you do for your written work, for your oral presentation you will have to do some research about the audience, emphasizing the "you attitude" and establishing your own credibility.

Consider Your Audience as a Group of Listeners, Not Readers

While audience analysis pertains both to readers of your work and to listeners of your presentation, there are several fundamental differences between these two groups. Unlike a reader of your report, the audience for your presentation

- is a captive audience
- may have only one chance to get your message
- has less time to digest what you say
- has a shorter attention span
- can't always go back to review what you said or jump ahead to get a preview
- is more easily distracted—by interruptions, chairs being moved, people coughing, and so on
- cannot absorb as many of the technical details as you would include in a written report

Take all of these differences into account as you plan your presentation and assess who constitutes your audience.

Guidelines for Analyzing Your Audience

Here are five key questions to ask when analyzing your audience.

1. **How much do they know about your topic?**

   - consumers with little or no technical knowledge
   - technical individuals who understand terms, jargon, and background
   - business managers looking only for the bottom line

2. **What unites them as a group?**

   - members of the same profession
   - customers using the same products
   - employees of the company you work for

3. **What is their interest level/stake in your topic?**

   - highly motivated
   - neutral—waiting to be informed, entertained, or persuaded
   - uninterested in your topic—only there because attendance is mandatory
   - uncooperative and antagonistic, likely to challenge you

4. **What do you want them to do after hearing your presentation?**

   - buy a product or service
   - adopt a plan
   - change a schedule
   - learn more about your topic
   - sign a petition
   - accept a new policy (e.g., safety procedures)

5. **What questions are they likely to raise?**

   - about money
   - about personnel
   - about locations
   - about schedules/time tables

Special Considerations for a Multinational Audience

Given the international makeup of audiences at many business presentations, you may have to address a group of listeners (international co-workers or customers) whose native language is not English or even make a presentation before individuals in a country other than your own. Consider your audience's particular cultural taboos and protocols. Do they accept your looking at them directly, or do they see eye contact as an invasion of their privacy? Will they expect you to stand in one place, or will they be comfortable if you move about the room while you speak?

As you prepare a talk before a multinational audience, keep the following points in mind:

1. Brush up on your audience's culture, especially accepted ways they communicate with one another (see pp. 170–187).
2. Find out what constitutes an appropriate length for a talk before your audience. (German listeners might be accustomed to hearing someone read a thirty- to forty-page paper, while members from another culture would regard that practice as improper.)
3. Be especially careful about introducing humor—avoid anything that is based on nationality, dialect, religion, or race.
4. Think twice about injecting anything autobiographical into your speech. Some cultures regard such intimacy as highly inappropriate.
5. Steer clear of politics; you risk losing your audience's confidence.
6. Choose visuals with universally understood icons.

The Parts of Formal Presentations

As you read this section, refer to Marilyn Claire Ford's PowerPoint presentation in Figure 16.2. Note how effectively she used the PowerPoint format to convince a potential client, GTP Systems, to purchase a service contract provided by World Tech, her employer. Her talk consists of seven slides that contain relevant images and concise text.

The Introduction

The most important part of a presentation is your introduction, which should capture the audience's attention by answering these questions: (1) Who are you? (2) What are your qualifications? (3) What specific topic are you speaking about? and (4) How is the topic relevant to us?

Your first and most immediate goal is to establish rapport with your audience, win their confidence, and elicit their cooperation. Since your audience is probably at their most attentive during the first few minutes of your presentation, they will

Figure 16.2   A sample PowerPoint presentation.

Introduces speaker, provides reason for presentation

Type size and font are clear

> # Switching to Videoconferencing: A Wise Choice for GTP
>
> Marilyn Claire Ford
> **World Tech**
> Desktop Videoconferencing
> November 14, 2008

Chooses short title for each slide

Succinctly lists benefits in short, easy-to-read bulleted points

> # What We Offer
>
> - User-friendly desktop videoconferencing
> - Cutting-edge communication technology
> - Flexible low-cost networking
> - Single network for data, voice, and video

First of four slides that make up the body of the presentation

Develops first key sales feature

Relevant visual does not mask text

> # Easy to Use
>
> With a simple telephone call you can
> - Arrange a meeting with colleagues around the globe
> - Use your computer to access World Tech's conferencing system
> - See, hear, and talk with all participants

(Continued)   Figure 16.2

Cost Effective

- Dramatically reduces costs for travel
- Upgrades current computer system for less than the price of an existing computer
- Cuts data-processing expenses by 60%

Second point gives only essential facts

Does not overwhelm listeners with numbers

Appropriate icon emphasizing "cutting" costs

Improves Staff Efficiency

- Brings people together at the right time
- Enhances communication when employees see and hear one another
- Whiteboard technology aids collaboration

Third sales feature appropriately mentions specific technology

Leaves generous margins

Icon shows staff interacting

Advantages in the Global Marketplace

- Global sales increase
- Clear recording of all meetings
- Safety risks lessened from decreased employee travel

Addresses security as last, most emphatic point in list of advantages

Continued

Figure 16.2   (Continued)

Conclusion

Issues a call to action; makes contact easy through website, e-mail, and telephone

> # Conclusion
>
> - Recap of technology benefits
> - Thank you for listening
> - Please sign up this week—it is easy
> - Just log on to **www.WorldTech.com**
> - Any questions?
>
> **mcf@WorldTech.com**
> **1-800-271-5555**

pay close attention to everything about you and what you say. Seize the moment and build momentum.

An effective introduction should be proportional to the length of your presentation. A ten-minute speech requires no more than a sixty-second introduction; a twenty-minute speech needs no more than a two- or three-minute introduction. Note how Marilyn Claire Ford in slides 1 and 2 introduces herself, her company, and its benefits for GTP.

How to Begin
You can begin by introducing yourself, emphasizing your professional qualifications and interests. (A self-introduction is unnecessary if someone else has introduced you or if you know everyone in the room.)

Give Listeners a Road Map
Give listeners a "road map" at the beginning of your presentation so that they will know where you are, where you are going, and what they have to look forward to or to recall. Indicate what your topic is and how you have organized what you have to say about it.

> My presentation today on fraud identification software will last about 20 minutes and is divided into three parts. First, I will outline briefly recent software changes. Second, I will give a detailed review of how those changes directly affect our company. Third, I will show how our company can profitably implement those changes. At the end of my presentation, there will be time for your questions and comments.

The most informative presentations are the easiest to follow. Restrict your topic to ensure that you will be able to organize it carefully and sensibly—for example, a tasty diet under 1,000 calories a day or a course in learning Java or another software package.

Capture the Audience's Attention

Use any of the following strategies to get your audience to "bite the hook":

- Ask a question. "Did you know that every thirty minutes a foreign-owned business opens in China?" or "Do you know what's in your bottled water besides water?"
- Start with a quotation. Winston Churchill said, "We get things to make a living but we give things to have a life." (Consult *Bartlett's Familiar Quotations* online at *http://www.bartleby.com.*)
- Give an interesting statistic. "In 2009, two million heart attack victims will live to tell about it." (Go to the *World Almanac and Book of Facts* online at *www.worldalmanac.com* to find something relevant to your presentation topic.)
- Relate an anecdote or story. Be sure it is relevant and in good taste; make your audience feel at ease and friendly toward you by establishing a bond with them.

Be careful about using humor in a business talk. It could backfire—the audience may not get the point or may even be offended by it.

The Body

The body is the longest part of your presentation, just as it is in a long report. Make it persuasive and relevant to your audience by (1) explaining a process, (2) describing a condition, (3) solving a problem, (4) arguing a case, or (5) doing all of these. See how the body of Marilyn Claire Ford's presentation in Figure 16.2 is organized around the customer benefits of GTP's switching to desktop videoconferencing. In slides 3–6 she outlines how easy, economical, and efficient such technology is to use in a global marketplace.

To get the right perspective, recall your own experiences as a member of an audience. How often did you feel bored or angry because a speaker tried to overload you with details or could not stick to the point?

Ways to Organize the Body

Here are a few helpful ways you can present and organize information in the body of your presentation. In writing a report, design your document to help readers visually, supplying headings, underscoring, bullets, necessary white space, and headers and footers. In a speech, switch from those purely visual devices to aural ones, such as the following:

1. **Give signals (directions) to show where you are going or where you have been.** Enumerate your points: *first, second, third.* Emphasize cause-and-effect relationships with *subsequently, therefore, furthermore.* When you tell a story, follow a chronological sequence and fill your speech with signposts: *before, following, next, then.*

2. **Comment on your own material.** Tell the audience if some point is especially significant, memorable, or relevant. "This next fact is the most important thing I'll say today."

3. **Provide internal summaries.** Spending a few seconds to recap what you have just covered will reassure your audience and you as well.

   We have already discussed the difficulties in establishing a menu repertory, or the list of items that the food service manager wants to appear on the menu. Now we will turn to ways of determining which items should appear on a menu and why.

4. **Anticipate any objections or qualifications your audience is likely to have.** Address the issues with relevant facts about costs, personnel, and/or equipment in your presentation.

The Conclusion

Plan your conclusion as carefully as you do your introduction. Stopping with a screeching halt is as bad as trailing off in a fading monotone. An effective conclusion should leave the audience feeling that you and they have come full circle and accomplished what you promised. Let readers know you are near the end of your presentation.

What to Put in a Conclusion

A conclusion should contain something memorable. Never introduce a new subject or simply repeat your introduction. A conclusion can contain the following:

- a fresh restatement of your three or four main points
- a call to action, just as in a sales letter—to buy, to note, to agree, to volunteer
- a final emphasis on a key statistic (for example, "The installation of the stainless steel heating tanks has, as we have seen, saved our firm 32 percent in utility costs, since we no longer have to run the heating system all day.")

End your presentation, as Marilyn Claire Ford does in slide 7, with a concise summary of the main points and urge listeners to invest company money in your product or service—for example, World Tech Desktop Videoconferencing system.

Mean It When You Say, "Finally"

When you tell your audience you are concluding, make sure you mean it. Saying, "In conclusion," and then talking for another ten minutes frustrates listeners and makes them less receptive to your message.

Visit www .cengage.com/ english/kolin for two online exercises, "Creating Presentations Using Presentation Software" and "Tailoring Presentation Software Presentations for International Audiences."

Presentation Software

As Marilyn Claire Ford's presentation in Figure 16.2 demonstrates, business talks very frequently rely on PowerPoint, Corel Presentation, and other software. Using these graphics packages is a crucial skill your employer will expect you to have. These software packages enable you to create an electronic slide show with concise text and carefully chosen visuals that can be created on any computer. They allow you to plan, write, and add visuals to your slides all at once.

Presentation Software Capabilities

With PowerPoint, Corel, and other presentation software, you can

- format and edit text
- import visuals, photos, digital art, clip art
- incorporate a variety of shapes and symbols—arrows, cylinders, pyramids, flow charts, pictograms
- offer animation, sound bites, movie/video clips
- design and insert logos and letterheads
- develop overhead and onscreen transparencies
- reuse and revise your presentation anytime

Presentation software will also help you as a speaker. For instance, you can keyboard your notes so that they scroll at the bottom of your computer screen. These notes are not projected for the audience, but they are available if you need to glance at them. You also can print parts of your presentation or your entire program in color or gray scale (black and white) to reinforce your presentation with professional-looking handouts.

Editing with presentation software allows you to customize any text or visual for any audience. You can copy, move, alter, and delete text, graphics, or sound bites. For example, you could edit the number of categories (bars, lines of a graph, items in a table) of an imported visual for a consumer audience that does not need much detail, or you could change the levels and positions of an organizational chart to suit your agency's needs. You can also change the color or texture of a background, saving you time and effort in recreating a visual.

Organize the Presentation

Map your presentation before you actually create your slides. One way is to prepare an outline (see pp. 48–49 in Chapter 2). First, identify the large topics you want to cover, as Marilyn Claire Ford did in presenting information to save GTP time and money, increase staff efficiency, and help the company compete in a global economy. Then you can organize your topics logically and persuasively. Divide your presentation into major topics that best accomplish your objective—whether to inform, to persuade, or to document. Include only those supporting details that relate directly to your topic and to your audience's needs. These key points should help you determine the number of slides and visuals you use. But don't overwhelm audiences with too much information. Note that Marilyn Claire Ford needs only seven slides in Figure 16.2. Also choose your visuals carefully. Resist the temptation to dazzle your audience with electronic special effects. Your goal is not to create a glitzy show but to represent your company professionally.

Test the Technology

Be prepared technologically as well as organizationally. Find and eliminate any bugs at the rehearsal stage. Call in advance to confirm the room and any equipment you may need. Save and preview your presentation in the format in which you will be giving it. Bring your own computer to the meeting and set up the projector and your laptop together in advance. Make sure your software is compatible with the equipment you will use. Be sure any websites you use are (1) valid and (2) actually connected to the Internet, not broken links.

Prepare Transparencies and Handouts

It's always wise to have backup transparencies or handouts with you in case your computer malfunctions—before or during your presentation. Transparencies of your presentation will enable you to continue to project your slides while you speak. Handouts of your slides (make sure you keep a copy for yourself) will allow you to continue with your presentation, and they will also help your audience follow your presentation and take notes about it while you speak. Make sure your handouts match your presentation exactly by providing your audience with a print-out of each slide.

Guidelines on Using Presentation Software Effectively

Here are some tips to ensure that the design, organization, and delivery of your presentation go smoothly. Refer to Figure 16.2 as you study them.

Readability

- Make sure each slide is easy to read—clear and uncluttered.
- Use a type size that is crisp and easy to see, even from a distance. For a small presentation on your laptop, use 24-point type or larger. Increase your type size for headings, as in Figure 16.2.
- Keep your type style and size consistent. Don't switch from one font to another.
- Avoid ornate or script type, and do not put everything in boldface or in all capital letters. See how Marilyn Claire Ford boldfaces only her headings in Figure 16.2.

Text

- Keep text short and simple. Use easy-to-recall names, words, and phrases. Your audience will not have the time to read long, complex messages.
- Use bulleted lists instead of unbroken paragraphs. But put no more than five bulleted lines on a slide and limit each line to seven or eight words. Don't squeeze words on a line. Include no more than 40 words per slide.
- Double-space between bulleted items and leave generous margins on all sides.
- Title each slide—using a question, a statement, or key name or phrase.

Sequencing Slides

- Keep your slides in the correct order in which you need to show them.
- Don't digress from your sequence or jump ahead.
- Limit your presentation to twelve slides, maximum.
- Don't show a slide if you aren't going to discuss it.
- Keep the same transition (cover left or straight right) from slide to slide to avoid visual confusion.

- Spend about two to three minutes per slide, but don't read each slide verbatim. Summarize main ideas or concisely expand them while looking at your audience, not the slide.
- Adjust your presentation to fit your time limit. Always time your slides so your audience can easily read them. Never continue to show a slide after you have moved on to a new topic.
- Avoid showing a visual until you are ready to comment on it.
- Determine if you will take questions between slides or only after your presentation has been completed. If you invite audience participation and interaction, build in extra time between your slides.
- Don't show blank screens to transition from one point to another.

Background/Color

- Find a pleasant contrasting background to make your text easy to read. Avoid extremely light or dark backgrounds that make your slides hard to read. Stay away from stark backgrounds, such as using cold white images on a black screen.
- Use the same background for each slide.
- Avoid shadowing your text for "decorative" visual effect.
- Use color sparingly and make sure it is professionally appropriate. Don't turn each slide into a sizzling neon sign.

Graphics

- Keep graphics clear, simple, and positioned appropriately on the slide.
- Be sure visuals do not cover or shadow text.
- Show only those visuals that support your main points. Not every slide requires a visual. For example, slides 1 and 7 in Figure 16.2 do not use visuals.
- Include no more than one graphic per slide; otherwise, your text will be more difficult to read.
- Include easy-to-follow graphs and charts instead of complicated tables, elaborate flow charts, or busy diagrams.
- Use clip art sparingly. Remember: Less is more. Clip art must be functional, not distracting. Each piece of clip art in Figure 16.2 is functional.
- Incorporate animation, sound effects, or movie/video clips only when they are persuasive, relevant, and undeniably professional.
- Don't bother with borders; they do not make a slide clearer.

Quality Check

- Be sure your spelling, grammar, names, dates, costs, and sources are correct.
- Double-check all math, tables, equations, and percentages.

Noncomputerized Presentations

You can expect to incorporate visuals in a variety of other media besides Power-Point presentations. There will be situations in which you may have to use a conventional chalkboard, a flip chart (where large pieces of white paper are anchored on an easel and flipped over like pages in a notebook or tablet), a slide projector, or an overhead projector to make a presentation instead of a computer. You will almost surely be asked to prepare handouts that include text, visuals, or a combination to accompany your presentation, and to distribute them either before or during your talk.

Regardless of the medium you use, make sure your visuals are

- easy to see
- easy to understand
- self-explanatory
- relevant
- accurate

Getting the Most from Your Noncomputerized Visuals

The following practical suggestions will help you get the most from your visuals when time and space may prohibit using computer setups.

1. **Do not set up your visuals before you begin speaking.** The audience will be wondering how you are going to use them and so will not give you their full attention. When you are finished with a visual, put it away so that your audience will not be distracted by it or tempted to study it instead of listening to you.

2. **Firmly anchor any maps or illustrations.** Having a map roll up or a picture fall off an easel during a presentation is embarrassing.

3. **Never obstruct the audience's view by standing in front of your visuals.** Use a pointer or a laser pointer (a pen-sized tool that projects a bright red spot up to 150 feet) to direct the audience's attention to your visual.

4. **Avoid crowding too many images onto one visual.** Use more than one flip chart page, slide, or transparency instead.

5. **Do not put a lot of writing on a visual.** Elaborate labels or wordy descriptions defeat your reason for using the visual. Your audience will spend more time trying to decipher the writing than attempting to understand the visual itself. If any writing must appear on one of your visuals, enlarge it so your audience can read it quickly and easily.

6. **Be especially cautious with a slide projector.** Check beforehand to make sure all your slides are in order and are right side up. Most important, make sure the projector is in good working order. Practice changing from one transparency to another. And test your tape recorder if you are using one as part of your presentation.

Rehearsing Your Presentation

Don't skip rehearsing your presentation thinking it will save you time. Rehearsing will actually help you become more familiar with your topic and overall message, building your confidence. Rehearsing will also help you to acquire more natural speech rhythms—pitch, pauses, and pacing. Here are some strategies to use as you rehearse your speech.

- Know your topic and the various parts of your talk.
- If possible, practice in the room where you will make your presentation.
- Speak in front of a full-length mirror for at least one rehearsal to see how an audience might view you.
- Talk into a tape recorder to determine whether you sound friendly or frantic, poised or pressured. You can also catch and correct yourself if you are speaking too quickly or too slowly. A rate of about 120 to 140 words a minute is easy for an audience to follow.
- Time yourself so that you will not exceed your allotted time or fall far short of your audience's expectations.
- Practice with the presentation software, visuals, equipment, or projector that you intend to use in your speech for valuable hands-on experience.
- Monitor the type of gestures (neither too many nor too few) you can use for clarity and emphasis in your talk.
- Videotape your final rehearsal and show it to a colleague or instructor for feedback.

Delivering Your Presentation

Poor delivery can ruin a good presentation. You will be evaluated by your style of presentation just as you are in your written work. When you speak before an audience, you will be evaluated on the image you project: how you look, how you talk, and how you move (your body language). Do you mumble into your notes, never looking at the audience? Do you clutch the lectern as if to keep it in place? Do you shift nervously from one foot to the other? All those actions betray your nervousness and detract from your presentation.

First impressions are crucial. Research shows that people decide what they think of you in the first two or three minutes of your presentation. The way you dress is important but so is your body language. In fact, 75 percent of your audience's impressions are influenced by your body language. The nonverbal signals you send affect how your audience will regard your leadership abilities, your sales performance, even your sincerity. Pay attention to gestures, hand movements, how you stand, and so on. No matter how many hours you have worked to get your message ready, if your nonverbal presentation is misleading or inappropriate, the impact of what you say will be lost.

The following suggestions on how to deliver a presentation will help you to be a well-prepared, poised speaker.

Settling Your Nerves Before You Speak

Being nervous before your presentation is normal—a faster heartbeat, sweaty palms, shaking. But don't let your nerves stand in your way of delivering a highly successful talk.

Here are some ways you can calm yourself before you deliver your presentation and get some healthy doses of confidence.

- Give yourself plenty of time to get there. The more you have to rush, the more anxious you will be.
- Don't bring anything with you that is likely to spill, such as coffee or a soft drink.
- Avoid caffeine for a few hours before your talk if it makes you jittery.
- Take some deep breaths and then hold your breath while you count to ten. Then exhale. This will slow your heart rate and lower your blood pressure.
- Remind yourself that you have spent hours preparing. Your hard work will pull you through.
- Try to chat with one or two members of the audience ahead of time and relax. See your audience as friends—people who can help your career.

To get helpful experience in making presentations, consider joining Toastmasters International, an educational organization that has transformed many frightened speakers into accomplished speech makers. Members of Toastmasters meet at least once a month to listen to and evaluate one another's speeches. From feedback and strategic training, you will grow to be a more confident public speaker. Check out their website at *http://www.toastmasters.org*.

Finally, you could try a panic control method advocated by Don Greene in his book *Fight Your Fear and Win* (New York: Broadway Books, 2002): "Picture the person you would most like to emulate, someone whose confidence or style you admire," and then combine his or her grace, speaking voice, and persuasion with your own qualifications for a successful, less-stressed presentation.

Tech Note

Whiteboards

A whiteboard (or smartboard) looks like a chalkboard but is really an interactive digital screen with computer capabilities. Instead of using chalk, though, you write with a digital device on the whiteboard screen, which is connected to a computer through a USB or other wireless link. You can use your finger like a mouse and the board like a keyboard, which allows you to search, scan, show, and edit any document or graphic. With just a touch of your hand or with a digital device, you can move elements around, such as numbers, graphs, and words, and enlarge or shrink elements. For instance, you can alter the shape/content of a visual such as a pie chart, table, or graph to reflect statistical changes or election outcomes, as newscasters did during U.S. presidential primaries in 2008. You can also insert, edit, update, and store content.

Whiteboards have a variety of uses in the world of work. They are invaluable when you have to make a presentation. Usually mounted on a wall or panel, a whiteboard can function as a projector screen on which you can scan, show, edit, share, and store information from your computer. You can also enlarge an image or figure and just by touching it, you can insert call outs, or pull up a website or company publication from an intranet to illustrate a point or answer a question.

This interactive technology can also improve collaborative writing by making brainstorming, researching, or planning much easier; for instance, you can send each member of your group the information that has been developed and displayed on a whiteboard. Further, whiteboards have been used extensively in training sessions in which employees have to see and learn about a great deal of information in a short time. Using a whiteboard, you can give new employees or visitors an impressive virtual tour of your company and its various divisions and departments.

Guidelines for Making Your Presentation

Everyone is nervous before a talk. Accept that fact and even allow a few seconds for "panic time." Then put your nervous energy to work for you. Chances are, your audience will have no idea how anxious you are; they cannot see the butterflies in your stomach. Again, see your audience as friends, not enemies. Remember to do the following:

1. **Establish eye contact with your listeners.** Look at as many people in your audience as possible to establish a relationship with them. Never bury your head in notes or keep your eyes fixed on a computer screen or keyboard. You will only signal your lack of interest in the audience or your fear of public speaking. Some timid speakers think that if they look only at some fixed place or object in the back of the room, the audience will regard this as eye contact. But that kind of cover-up does not work. Even during a short PowerPoint presentation, try to establish rapport with each person in the room.

2. **Adjust to audience feedback.** Watch your listeners' reactions and respond appropriately—nodding to agree, pausing a moment, paraphrasing to clarify a confusing point. Know your material well so that if someone asks you a question or wants you to return to a point, you are not fumbling through your notes or trying feverishly to locate the right screen.

3. **Use a friendly, confident tone.** Speak in a natural, pleasant voice, but avoid verbal tics ("you know," "I mean") and fillers ("um," "ah," "er") repeated several times each minute. Such nervous habits will make your audience nervous and your speech less effective. Use pauses instead.

4. **Vary the rate of your delivery.** Use your natural speaking voice. Vary your rate and inflection to help you emphasize key points and make transitions. Talk slowly enough for your audience to understand you, yet quickly enough so that you don't sound as if you are belaboring or emphasizing each word.

5. Adjust your volume appropriately. Talking in a monotone, never raising or lowering your voice, will lull your audience to sleep or at least inattention. Talk loudly enough for everyone to hear, but be careful if you are using a microphone. Your voice will be amplified, so if you speak loudly, you will boom rather than project. Every word with a *b, p,* or *d* will sound like an explosive in your listener's ear. Watch out for the other extreme—speaking so softly that only the first two rows can hear you.

6. Watch your posture. Don't shift from one foot to another. But do not slouch or look wooden either. If you stand motionless, looking as if rigor mortis has set in, your speech will be judged cold and lifeless, no matter how lively your words are. Be natural yet dynamic; smile, nod your head. Refer to an object on the screen by touching it, or use a pointer to emphasize something on an overhead projector.

7. Use appropriate body language. Be natural and consistent. Do not startle an audience by suddenly pounding on the lectern or desk for emphasis. Avoid gestures that will distract or alienate your audience. For example, don't fold your arms as you talk, a gesture that signals you are unreceptive (closed) to your audience's reactions. Also, avoid nervous habits that can divert the audience's attention: clicking a ballpoint pen, scratching your head, rubbing your nose, twirling your hair, pushing up your glasses, fumbling with your notes, tapping your foot, or drumming your fingers on a desk. Nor do you have to remain still or step with robotlike movements. The remote control for a PowerPoint presentation allows you to casually walk around the room as you click and change screens.

8. Dress professionally. Do not wear clothes or clanking jewelry (such as a necklace or bracelet charms) that call attention to themselves. Follow your company's dress code. Unless it specifies otherwise (e.g., casual Fridays), wear clothes that are the business norm. Women should wear a businesslike dress or suit; men should choose a dark suit or sports coat, white or blue shirt, and a tasteful tie.

When You Have Finished

Don't just sit down before your computer, walk back to your place on the platform or in the audience, or, worse yet, march out of the room. Thank your listeners for their attention and stay at the lectern or at your laptop for audience applause or questions, as Marilyn Claire Ford did in Figure 16.2. If appropriate, give the person who introduced you a chance to thank you while you are still in front of the group.

If a question-and-answer session is to follow your speech, anticipate questions your audience is likely to ask. But it's a good idea to give your audience a time limit. For example, you might say, "I'll be happy to answer your questions now before we break for lunch in ten minutes." By setting limits, you reduce the chances of a lengthy debate with members of the audience, and you also can politely leave after your time elapses.

Evaluating Presentations

A large portion of this chapter has given you information on how to construct and deliver a formal presentation. As a way of reviewing that advice, study Figure 16.3—an evaluation form similar to those used by instructors in communications classes. Note that the form gives equal emphasis to the speaker's performance or delivery and to the organization, content, and sequence of the presentation.

An evaluation form for a presentation.   Figure 16.3

Name of Speaker: ―――――――   Date of Presentation: ――――――

Title of Presentation: ――――――   Length of Presentation: ――――――

PART I: THE SPEAKER'S DELIVERY
(Circle the appropriate number using the scale below.)

1. Appearance	1 unprofessional	2	3	4	5 well groomed
2. Eye contact	1 poor	2	3	4	5 effective
3. Voice	1 monotonous	2	3	4	5 varied
4. Diction	1 slurred	2	3	4	5 clear
5. Posture	1 poor	2	3	4	5 natural
6. Gestures	1 distracting	2	3	4	5 appropriate
7. Self-confidence	1 nervous	2	3	4	5 poised
8. Audience interaction	1 minimal	2	3	4	5 engaging

PART II: THE PRESENTATION ITSELF
(Circle the appropriate number: 1 = poor; 5 = superior.)

1. Made sure topic was relevant to audience	1	2	3	4	5
2. Began with clear statement of purpose	1	2	3	4	5
3. Followed logical organization	1	2	3	4	5
4. Gave audience cues for transitions between sections	1	2	3	4	5

Continued

Figure 16.3   (Continued)

5. Matched technical level of audience	1	2	3	4	5
6. Provided convincing supporting evidence	1	2	3	4	5
7. Did not digress	1	2	3	4	5
8. Concluded with a summary of main points	1	2	3	4	5
9. Stayed within time limits	1	2	3	4	5
10. Allowed time for questions	1	2	3	4	5

PART III: USE OF VISUALS (SLIDES OR OTHER GRAPHICS)
(Circle the appropriate number: 1 = poor; 5 = superior.)

1. Used appropriate number of visuals	1	2	3	4	5
2. Ensured visuals were relevant	1	2	3	4	5
3. Carefully timed the sequence of visuals	1	2	3	4	5
4. Made sure audience could see visuals clearly	1	2	3	4	5
5. Selected visuals that clarified or simplified a point	1	2	3	4	5
6. Referred to visuals and indicated why they were important	1	2	3	4	5
7. Acknowledged the source of visuals	1	2	3	4	5

Revision Checklist

☐ Anticipated audience's background, interest, or even potential resistance, and questions about message of both informal and formal presentations.

☐ Used effective telephone/cell phone etiquette.

☐ Prepared outline and identified and corrected any weak or redundant areas.

☐ Prepared introduction to provide "road map" of presentation and to arouse audience interest.

☐ Started with interesting and relevant statistics, a question, an anecdote, or similar "hook" to capture audience attention.

☐ Limited body of presentation to main points.

☐ Sequenced main points logically and made connections among them.

☐ Used supporting examples and illustrations appropriate to audience.

☐ Made sure conclusion contains summary of main points of my presentation and/or specific call for action.

☐ Designed visuals that are clear, easy to read, and relevant for audience.

☐ Experimented successfully with presentation software and applications before including it in a presentation.

☐ Used appropriate number of slides.

☐ Showed slides in right, carefully timed sequence.

☐ Prepared transparencies and handout in case of computer trouble.

☐ Rehearsed presentation thoroughly to become familiar with its organization and visuals.

☐ Monitored volume, tone, and rate to vary delivery and emphasize major points.

☐ Rehearsed gestures to make them relevant and nonintrusive.

☐ Timed presentation, complete with visuals, to run close to allotted time.

Exercises

1. Prepare a three- to five-minute presentation explaining how a piece of equipment that you use on your job works. If the equipment is small enough, bring it with you to class. If it is too large, prepare an appropriate visual or two for use in your talk.

2. You have just been asked to talk about the students at your school or the employees where you work. Narrow the topic and submit an outline to your instructor, showing how you have limited the topic and gathered and organized evidence. Use two or three appropriate visuals (tables, photographs, maps, charts, icons, or even videos) to give a PowerPoint presentation. Follow the format of the presentation in Figure 16.2.

3. Prepare a ten-minute presentation on a controversial topic that you would present before a civic group—the PTA, the local chapter of an organization, a post of the Veterans of Foreign Wars, a synagogue, a mosque, or a church club.

4. Using the information contained in the report on marketing research in Chapter 8 (pp. 375–387) or the long report on multinational workers in Chapter 15 (pp. 652–667), prepare a short presentation (five to seven minutes) for your class.

5. Using the evaluation form in Figure 16.3, evaluate a speaker—a speech class student, a local politician, or a co-worker delivering a report at work. Specify the time, place, and occasion of the speech. Pay special attention to any visuals the speaker uses.

6. Deliver a formal presentation (fifteen to twenty minutes) on the various uses and advantages of the Internet for individuals in your career field. Restrict your topic and divide it into four key issues or parts, as in Figure 16.2. Use at least three visuals with your talk. Submit an outline to your instructor.

A Writer's Brief Guide to Paragraphs, Sentences, and Words

To write successfully, you must know how to create effective paragraphs, write and punctuate clear sentences, and use words correctly. This guide succinctly explains some of the basic elements of clear and accurate writing.

▌Paragraphs

Writing a Well-Developed Paragraph

A paragraph is the basic building block for any piece of writing. It is (1) a group of related sentences (2) arranged in a logical order (3) supplying readers with detailed, appropriate information (4) on a single important topic.

A paragraph expresses one central idea, with each sentence contributing to the overall meaning of that idea. The paragraph does that by means of a *topic sentence*, which states the central idea, and *supporting information*, which explains the topic sentence.

Supply a Topic Sentence

The topic sentence is the most important sentence in your paragraph. Carefully worded and restricted, it helps you to generate and control your information. An effective topic sentence also helps readers grasp your main idea quickly. As you draft your paragraphs, pay close attention to the following three guidelines.

   1. Make sure you provide a topic sentence. In their rush to supply readers with facts, some writers forget or neglect to include a topic sentence. The following paragraph, with no topic sentence, shows how fragmented such writing can be.

> No topic sentence: Sensors found on each machine detect wind speed and direction and other important details such as ice loading and potential metal fatigue. The information is fed into a small computer (microprocessor) in the nacelle (or engine housing). The microprocessor automatically keeps the blades turned into the wind, starts and stops the machine, and changes the pitch of the tips of the blades to increase power under varying wind conditions. Should any part of the wind turbine suffer damage or malfunction, the microprocessor will immediately shut the machine down.

Only when a suitable topic sentence is added—"The MOD-2 wind turbine is programmed to run completely by computer"—can readers understand what the technical details have in common.

Tech Note

Some Helpful Grammar and Writing Resources

In addition to the guidelines in this section of *Successful Writing at Work*, refer to the following websites to brush up on grammar, punctuation, and usage:

- Purdue Online Writing Lab (OWL): *http://owl.english.purdue.edu*
  A well-known online writing resource, this site covers a range of writing topics, from mechanics to English as a second language (ESL).
- Paradigm Online Writing Assistant: *http://www.powa.org*
  Provides information on editing documents.
- University of Calgary—The Basic Elements of English: *http://www.ucalgary.ca/UofC/eduweb/grammar*
  This interactive guide covers the essentials—parts of speech, sentence elements, punctuation, and word use.
- University of Oregon—The Tongue Untied: *http://grammar.uoregon.edu/toc.html*
  This online handbook is subtitled "A Guide to Grammar, Punctuation, and Style."

2. Put your topic sentence first. Place your topic sentence at the beginning—not the middle or end—of your paragraph because the first sentence occupies an emphatic position. Burying the key idea in the middle or near the end of the paragraph makes it harder for readers to comprehend your purpose or act on your information.

3. Be sure your topic sentence is focused. If restricted, a topic sentence discusses only one central idea. A broad or unrestricted topic sentence leads to a shaky, incomplete paragraph for two reasons.

- The paragraph will not contain enough information to support the topic sentence.
- A broad topic sentence will not summarize or forecast specific information in the paragraph.

The following example of a carefully constructed paragraph contains a clear topic sentence in an appropriate position (highlighted in color) and adequate supporting details.

Fat is an important part of everyone's diet. It is nutritionally present in the basic food groups we eat—meat and poultry, dairy products, and oils—to aid growth or development. The fats and fatty acids present in those foods ensure proper metabolism, thus helping to turn what we eat into the energy we need. Those same fats and fatty acids also act as carriers for important vitamins like A, D, E, and K. Another important role of fat is that it keeps us from feeling hungry by delaying digestion. Fat also enhances the flavor of the food we eat, making it more enjoyable.

Three Characteristics of an Effective Paragraph

Effective paragraphs have **unity**, **coherence**, and **completeness**.

Unity

A unified paragraph sticks to one topic without wandering. Every sentence, every detail, **supports**, **explains**, or **proves** the central idea. A unified paragraph includes only relevant information and excludes unnecessary or irrelevant comments.

Coherence

In a coherent paragraph all sentences flow smoothly and logically to and from each other like the links of a chain. Use the following three techniques to achieve coherence.

1. Use transitional words and phrases. Some useful connective, transitional words, along with the relationships they express, are listed in Table A.1.

> Paragraph with connective words: Advertising a product on the radio has many advantages over using television. *For one thing*, radio rates are much cheaper. *For example*, a one-time 60-second spot on television can cost $750. *For that money*, advertisers can purchase nine 30-second spots on the radio. *Equally attractive* are the low production costs for radio advertising. *In contrast*, television advertising often includes extra costs for models and voice-overs. *Another* advantage radio offers advertisers is immediate scheduling. *Often* the ad appears during the same week a contract is signed. *On the other hand*, television stations are *frequently* booked up months in advance, so it may be a long time *before* an ad appears. *Furthermore*, radio gives advertisers a greater opportunity to reach potential buyers. *After all*, radio follows listeners everywhere—in their homes, at work, and in their cars. *Although* television is very popular, it cannot do that.

2. Use pronouns and demonstrative adjectives. Words like *he, she, him, her, they*, and so on contribute to paragraph coherence and increase the flow of sentences.

> Paragraph with pronouns: Traffic studies are an important tool for store owners looking for a new location. These studies are relatively inexpensive and highly accurate. They can tell owners how much traffic passes by a particular location at a particular time and why. Moreover, they can help owners to determine what particular characteristics the individuals have in common. Because of their helpfulness, these studies can save owners time and money and possibly prevent financial ruin.

TABLE A.1   Transitional, or Connective, Words and Phrases

Addition	again	besides	moreover
	additionally	first, second, third	next
	along with	furthermore	together with
	also	in addition	too
	and	many	what's more
	as well as		
Cause/effect	accordingly	consequently	on account of
	and so	due to	since
	as a result	hence	therefore
	because of	if	thus
Comparison/ contrast	but	in contrast	on the other hand
	conversely	in the same way	similarly
	equally	likewise	still
	however	on the contrary	yet
Conclusion	all in all	in brief	on the whole
	altogether	in conclusion	to conclude
	as we saw	in short	to put into perspective
	at last	in summary	to summarize
	finally	lastly	to wrap up
Condition	although	granted that	provided that
	depending	if	to be sure
	even though	of course	unless
Emphasis	above all	for emphasis	of course
	after all	indeed	surely
	again	in fact	to repeat
	as a matter of fact	in other words	to stress
	as I said	obviously	unquestionably
Illustration	for example	in other words	that is,
	for instance	in particular	to demonstrate
	in effect	specifically	to illustrate
Place	across from	below	over
	adjacent to	beyond	there
	alongside of	here	under
	at this point	in front of	where
	behind	next to	wherever
Time	afterward	formerly	previously
	at length	hereafter	soon
	at the same time	later	simultaneously
	at times	meanwhile	subsequently
	beforehand	next	then
	currently	now	until
	during	once	when
	earlier	presently	while

3. **Use parallel (coordinated) grammatical structures.** Parallelism means using the same *kind of* word, phrase, clause, or sentence to express related concepts.

> Orientation sessions accomplish four useful goals for trainees. First, they introduce trainees to key personnel in accounting, data processing, maintenance, and security. Second, they give trainees experience logging into the database system, selecting appropriate menus, editing core documents, and getting off the system. Third, they explain to trainees the company policies affecting the way supplies are ordered, used, and stored. Fourth, they help trainees understand their responsibilities in such sensitive areas as computer security and use.

Parallelism is at work on a number of levels in the paragraph above, among them

- The four sentences about the four goals start in the same way grammatically ("... they introduce/give/explain/help ...") to help readers categorize the information.
- Within individual sentences, the repetition of *present participles* (log*ging*, selec*ting*, edi*ting*, get*ting*) and of *past participles* (order*ed*, us*ed*, stor*ed*) helps the writer to coordinate information.
- Transitional words—*first, second, third, fourth*—provide a clear-cut sequence.

Completeness

A complete paragraph provides readers with sufficient information to **clarify, analyze, support, defend,** or **prove** the central idea expressed in the topic sentence. The reader feels satisfied that the writer has given necessary details.

> Skimpy paragraph: Farmers are turning their crops and farm wastes into cost-effective fuels. Much grown on the farm is being converted to energy. This energy can have many uses and save farmers a lot of money in operating expenses.

> Fully developed paragraph: Farm crops and wastes are being turned into fuels to save farmers' operating costs. Alcohol can be distilled from grain, sugar beets, and corn. Converted to ethanol (90 percent gasoline, 10 percent ethanol), this fuel runs such farm equipment as irrigation pumps, feed grinders, and tractors. Similarly, through a biomass digestion system, farmers can produce methane from animal or crop wastes as a natural gas for heating and cooking. Finally, cellulose pellets, derived from plant materials, become solid fuel that can save farmers money in heating barns.

Sentences

Constructing and Punctuating Sentences

The way you construct and punctuate your sentences can determine whether you succeed or fail in the world of work. Your sentences reveal a lot about you. They tell readers how clearly or how poorly you can convey a message. And any message is only as effective and as thoughtful as the sentences of which it is made.

What Makes a Sentence

A sentence is a complete thought, expressed by a subject and a verb that can make sense standing alone.

subject	verb	
Websites	sell	products.

The Difference Between Phrases and Clauses

The first step toward success in writing sentences is learning to recognize the difference between phrases and clauses. A **phrase** is a group of words that does not contain a subject and a verb; phrases cannot make sense standing alone. Phrases cannot be sentences.

in the park	No subject: Who is in the park?
	No verb: What was done in the park?
for every patient in intensive care	No subject: Who did something for every patient?
	No verb: What was done for the patients?

A **clause** does contain a subject and a verb, but *not every clause is a sentence.* Only **independent** (or **main**) **clauses** can stand alone as sentences. Here is an example of an independent clause that is a complete sentence.

subject	verb	object
The president	closed	the college.

A **dependent** (or **subordinate**) **clause** also contains a subject and a verb, but it does not make complete sense and cannot stand alone. Why? A dependent clause contains a subordinating conjunction—*after, although, as, because, before, even though, if, since, unless, when, where, whereas, while*—at the beginning of the clause. Such conjunctions subordinate the clause in which they appear and make the clause dependent for meaning and completion on an independent clause.

After	
Before	
Because	} the president closed the college
Even though	
Unless	

"After the president closed the college" is not a complete thought but a dependent clause that leaves us in suspense. It needs to be completed with an independent clause telling us what happened "after."

dependent clause	independent clause		
	subject	verb	phrase
After the president closed the college,	we	played	in the snow.

Avoiding Sentence Fragments

An incomplete sentence is called a **fragment**. Fragments can be phrases or dependent clauses. They either lack a verb or a subject or have broken away from an independent clause. A fragment is isolated: It needs an overhaul to supply missing parts to turn it into an independent clause or to glue it back to an independent clause to have it make sense.

To avoid writing fragments, follow these rules. *Note that incorrect examples are preceded by a minus sign, correct revisions by a plus sign.*

1. Do not use a subordinate clause as a sentence. Even though it contains a subject and a verb, a subordinate clause standing alone is still a fragment. To avoid this kind of sentence fragment, simply join the two clauses (the independent clause and the dependent clause containing a subordinating conjunction) with a comma—*not* a period or semicolon.

- Unless we agreed to the plan. (What would happen?)
- Unless we agreed to the plan; the project manager would discontinue the operation. (A semicolon cannot set off the subordinate clause.)
+ Unless we agreed to the plan, the project manager would discontinue the operation.
- Because safety precautions were taken. (What happened?)
+ Because safety precautions were taken, ten construction workers escaped injury.

Sometimes subordinate clauses appear at the end of a sentence. They may be introduced by a subordinate conjunction, an adverb, or a relative pronoun (*that, which, who*). Do not separate these clauses from the preceding independent clause with a period, thus turning them into fragments.

- An all-volunteer fire department posed some problem<u>s. E</u>specially for residents in the western part of town.
+ An all-volunteer fire department posed some problem<u>s,</u> especially for residents in the western part of town. (The word *especially* qualifies <u>posed</u>, referred to in the independent clause.)

2. Every sentence must have a subject telling the reader who does the action.

- Being extra careful not to spill the solution. (Who?)
+ The technician was being extra careful not to spill the solution.

3. Every sentence must have a complete verb. Watch especially for verbs ending in *-ing*. They need another verb (some form of *to be*) to make them complete.

- The machine running in the computer department. (Did what?)

You can change that fragment into a sentence by supplying the correct form of the verb.

+ The machine *is running* in the computer department.
+ The machine *runs* in the computer department.

Or you can revise the entire sentence, adding a new thought.

+ The machine running in the computer department processes all new accounts.

4. Do not detach prepositional phrases (beginning with *at, by, for, from, in, to, with,* and so forth) **from independent clauses.** Such phrases are not complete thoughts and cannot stand alone. Correct the error by leaving the phrases attached to the sentence to which they belong.

– By three o'clock the next day. (What was to happen?)
+ The supervisor wanted our reports by three o'clock the next day.

Avoiding Comma Splices

Fragments occur when you use only bits and pieces of complete sentences. Another common error that some writers commit involves just the reverse kind of action. They weakly and wrongly join two complete sentences (independent clauses) with a comma as if those two sentences were really only one sentence. Such an error is called a **comma splice.** Here is an example.

– Gasoline prices have risen by 15 percent in the last month, we will drive the car less often.

Two independent clauses (complete sentences) exist:

+ Gasoline prices have risen by 15 percent in the last month.
+ We will drive the car less often.

A comma alone lacks the power to separate independent clauses.

As the preceding example shows, many pronouns—*I, he, she, it, we, they*—are used as the subjects of independent clauses. A comma splice will result if you place a comma instead of a semicolon between two independent clauses where the second clause opens with a pronoun.

– Maria approved the plan, she liked its cost-effective approach.
+ Maria approved the plan; she liked its cost-effective approach.

However, relative pronouns (*who, whom, which, that*) are preceded by a comma, not a period or a semicolon, when they introduce subordinate clauses.

– She approved the plan. Which had the cost-effective approach.
+ She approved the plan, which had the cost-effective approach.

Four Ways to Correct Comma Splices

1. Remove the comma separating two independent clauses and replace it with a period. Then capitalize the first letter of the first word of the new sentence.

+ Gasoline prices have risen by 15 percent in the last month. We will drive the car less often.

2. Insert a coordinating conjunction (*and, but, or, nor, so, for, yet*) **after the comma.** Together, the conjunction and the comma properly separate the two independent clauses.

+ Gasoline prices have risen by 15 percent in the last month, and so we will drive the car less often.

3. Rewrite the sentence (if it makes sense to do so). Turn the first independent clause into a dependent clause by adding a subordinate conjunction; then insert a comma and add the second independent clause.

+ <u>Because</u> gasoline prices have risen by 10 percent in the last mont<u>h,</u> we will drive the car less often.

4. Delete the comma and insert a semicolon.

+ Gasoline prices have risen by 15 percent in the last mont<u>h;</u> we will drive the car less often.

Of the four ways to correct the comma splice, sentences 3 and 4 are equally suitable, but sentence 3 reads more smoothly and so is the better choice.

The semicolon is an effective and forceful punctuation mark when two independent clauses are closely related, that is, when they announce contrasting or parallel views, as the two following examples reveal.

+ The union favored the new legislation; the company opposed it. (contrasting views)
+ Night classes help the college and the community; students can take more credit hours to advance their careers. (parallel views)

How *Not* to Correct Comma Splices

Some writers mistakenly try to correct comma splices by inserting a conjunctive adverb (*also, consequently, furthermore, however, moreover, nevertheless, then, therefore*) after the comma.

– Gasoline prices have risen by 15 percent in the last mont<u>h,</u> consequently we will drive the car less often.

Because the conjunctive adverb (*consequently*) is not as powerful as the coordinating conjunction (*and, but, for*), the error is not eliminated. If you use a conjunctive adverb—*consequently, however, nevertheless*—you still must insert a semicolon or a period before it, as the following examples show.

+ Gasoline prices have risen by 15 percent in the last mont<u>h;</u> consequently, we will drive the car less often.
+ Gasoline prices have risen by 15 percent in the last mont<u>h. C</u>onsequently, we will drive the car less often.

Avoiding Run-On Sentences

A **run-on sentence** is the opposite of a sentence fragment. The fragment gives the reader too little information, the run-on too much. A run-on sentence forces readers to digest two or more grammatically complete sentences without the proper punctuation to separate them.

Run-on: The Internet has become a primary source of information and students and other researchers are right to call it a virtual library this library is not like the collections of books and magazines that are carefully shelved always waiting for students to check and recheck them too often a website disappears or changes considerably and without a backup file or a hard copy the researcher has no document to quote from and no exact citation to prove that he or she consulted an authentic source.

Revised: The Internet has become a primary source of information. Students and other researchers are right to call it a virtual library, although this library is not like the collections of books and magazines that are carefully shelved, waiting for students to check and recheck them out. But too often a website disappears, is under construction, or changes considerably. Without a backup file or a hard copy of the site, the researcher has no document to quote from and no exact citation to prove that he or she consulted an authentic source.

As the revision above shows, you can repair a run-on sentence by (1) dividing it into separate, correctly punctuated sentences and (2) by adding coordinating conjunctions (*and, but, yet, so, or, nor*) between clauses.

Making Subjects and Verbs Agree in Your Sentences

A subject and a verb must agree in number. A singular subject takes a singular verb, whereas a plural subject requires a plural verb.

Singular Subject	Plural Subjects
the engineer calculates	engineers calculate
a report analyzes	reports analyze
a policy changes	policies change

You can avoid subject-verb agreement errors by following several simple rules.

1. **Disregard any words that come between the subject and its verb.**

   Faulty: The customer who ordered three parts want them shipped this afternoon.
   Correct: The <u>customer</u> who ordered three parts <u>wants</u> them shipped this afternoon.

2. **A compound subject** (two parts connected by *and*) **takes a plural verb.**

   Faulty: The engineering department and the safety committee prefers to develop new guidelines.
   Correct: The engineering department and the safety committee prefer to develop new guidelines.

3. **When a compound subject contains** *neither . . . nor* **or** *either . . . or,* **the verb agrees with the subject closest to it.**

   Faulty: Either the residents or the manager are going to file the complaint.
   Correct: Either the residents or the manager is going to file the complaint.
   Correct: Either the manager or the residents are going to file the complaint.

4. **Use a singular verb after collective nouns** (like *committee, crew, department, group, organization, staff, team*) **when the group functions as a single unit.**

   Correct: The crew was available to repair the machine.
   Correct: The committee asks that all recommendations be submitted by Friday.

   but

   Correct: The staff were unable to agree on the best model.
   (The staff acted as individuals, not a unit, so a plural verb is required.)

5. Use a singular verb with indefinite pronouns (such as *anyone, anybody, each, everyone, everything, no one, somebody, something*).

> Each of the programmers has completed the seminar.
> Somebody usually volunteers for that duty.

Similarly, when *all, most, more,* or *part* is the subject, it requires a singular verb.

> Most of the money is allocated.
> Part of the equipment was salvageable.

6. Words like *scissors* and *pants* are plural when they are the true subject.

> Faulty: A pair of trousers were available in his size. (*Pair* is the singular subject.)
> Correct: The trousers were on sale.

7. Some foreign plurals (*curricula, data, media, phenomena, strata, syllabi*) **always take a plural verb.**

> The data conclusively prove my point.
> The media are usually the first to point out a politician's weak points.

8. Use a singular verb with fractions.

> Three-fourths of her research proposal was finished.

Writing Sentences That Say What You Mean

Your sentences should say exactly what you mean, without double talk, misplaced humor, or nonsense. Sentences are composed of words and word groups that influence each other.

Writing Logical Sentences
Sentences should not contradict themselves or make outlandish claims. The following examples contain errors in logic; note how easily the suggested revisions solve the problem.

> Illogical: Steel roll-away shutters make it possible for the sun to be shaded in the summer and to have it shine in the winter. (The sun is far too large to shade; the writer meant that a room or a house, much smaller than the sun, could be shaded with the shutters.)
> Revision: Steel roll-away shutters make it possible for owners to shade their living rooms in the summer and to admit sunshine during the winter.

Using Contextually Appropriate Words
Sentences should use the combination of words most appropriate for the subject.

> Inappropriate: The members of the Nuclear Regulatory Commission saw fear radiated on the faces of the residents. (The word *radiated* is obviously ill advised in this context; use a neutral term.)
> Revision: The members of the Nuclear Regulatory Commission saw fear reflected on the faces of the residents.

Writing Sentences with Well-Placed Modifiers

A **modifier** is a word, phrase, or clause that describes, limits, or qualifies the meaning of another word or word group. A modifier can consist of one word (a *green* car), a prepositional phrase (the man *in the telephone booth*), a relative clause (the woman *who won the marathon*), or an *-ing* or *-ed* phrase (*walking three miles a day*, the student was in good shape; *seated in the first row*, we saw everything on stage).

A **dangling modifier** is one that cannot logically modify any word in the sentence.

   – When answering the question, his calculator fell off the table.

One way to correct the error is to insert the right subject after the *-ing* phrase.

   + When answering the question, he knocked his calculator off the table.

You can also turn the phrase into a subordinate clause.

   + When he answered the question, his calculator fell off the table.
   + His calculator fell off the table as he answered the question.

A **misplaced modifier** illogically modifies the wrong word or words in the sentence. The result is often comical.

   – Hiding in the corner, growling and snarling, our guide spotted the frightened cub. (Is our guide growling and snarling in the corner?)
   – All travel requests must be submitted by employees in red ink. (Are the employees covered in red ink?)

The problem with both of those examples is word order. The modifiers are misplaced because they are attached to the wrong words in the sentence. Correct the error by moving the modifier to where it belongs.

   + Hiding in the corner, growling and snarling, the frightened cub was spotted by our guide.
   + All travel requests by employees must be submitted in red ink.

Misplacing a relative clause (introduced by relative pronouns like *who, whom, that, which*) can also lead to problems with modification.

   – The salesperson recorded the merchandise for the customer that the store had discounted. (The merchandise was discounted, not the customer.)
   – The salesperson recorded the merchandise that the store had discounted for the customer. (The salesperson recorded for the customer; the store did not discount for the customer.)
   + The salesperson recorded for the customer the merchandise that the store had discounted.

Always place the relative clause immediately after the word it modifies.

Correct Use of Pronoun References in Sentences

Sentences will be vague if they contain a faulty use of pronouns. When you use a pronoun whose **antecedent** (the person, place, or object the pronoun refers to) is unclear, you risk confusing your reader.

   Unclear: After the plants are clean, we separate the stems from the roots and place them in the sun to dry. (Is it the stems or the roots that lie in the sun?)

Revision: After the plants are clean, we separate the stems from the roots and place the stems in the sun to dry.

Unclear: The park ranger was pleased to see the workers planting new trees and installing new benches. This will attract more tourists. (The trees or the benches or both?)

Revision: The park ranger was pleased to see the workers planting new trees and installing new benches, because the new trees and benches will attract more tourists.

Words

Spelling Words Correctly

Your written work will be judged in part on how well you spell. A misspelled word may seem like a small matter, but on an employment application, e-mail, incident report, letter, or short or long report it stands out to your discredit. A spelling mistake will look careless or, even worse, uneducated to a client or a supervisor. Readers will inevitably question your other skills if your spelling is incorrect.

The Benefits and Pitfalls of Spell Checkers

Computer **spell checkers** can be handy for flagging potential problem words. But beware! Spell checkers recognize only those words that have been listed in them. A proper name or infrequently used word may be flagged as an error even though the word is spelled correctly. Moreover, a spell checker will not differentiate between such homonyms as *too* and *two* or *there* and *their*. A spell checker identifies only misspelled words, not misused words. In short, do not rely exclusively on spell checkers to solve all your spelling and word-choice problems.

Consulting a Dictionary

Always have a dictionary handy. Two online dictionaries to consult are *Merriam-Webster On-Line* at *http://www.m-w.com* and *The American Heritage Dictionary of the English Language* at *http://www.bartleby.com/61.*

Using Apostrophes Correctly

Apostrophes cause some writers special problems. Basically, apostrophes are used for four reasons: (1) contractions, (2) possessives, (3) plurals, and (4) abbreviations. The guidelines below will help you to sort out those uses.

1. In a **contraction**, the apostrophe takes the place of the missing letter or letters: *I've = I have; doesn't = does not; he's = he is; it's = it is.* (*Its* is a possessive pronoun (the dog and *its* bone), not a contraction. There is no such form as *its'*.)

2. To form a **possessive**, follow these rules.

a. If a singular or plural noun does not end in an *-s*, add *'s* to show possession.

Mary's locker	the woman's jacket
children's books	the women's jackets
the staff's dedication	the company's policy

   b. If a singular noun ends in *-s*, add *'s* to show possession.

   the class's project   the boss's schedule

   c. If a plural noun ends in *-s*, add just the *'* to indicate possession.

   employees' benefits   computers' speed
   lawyers' fees   stores' rates

   d. If a proper name ends in *-s*, add *'s* to form the possessive.

   Jones's account   Keats's poetry
   the Williams's house   James's contract

   e. If it is a compound noun, add an *'* or *'s* to the end of the word.

   brother-in-law's business   Ms. Allison Jones-Wyatt's order

   f. To indicate shared possession, add just *'s* to the last name.

   Warner and Kline's Computer Shop   Juan and Anne's major

   g. To indicate separate possession, add *'s* to each name.

   Juan's and Tia's transcripts   Shakespeare's and Byron's poetry

3. To form the plural of numbers and capital letters used as nouns, including abbreviations without periods, just add *s*. To avoid misreading some capital letters, however, you may need to add the apostrophe.

   during the 1980s   all perfect 10s
   their SATs   several local YMCAs
   the 3 Rs   straight A's

4. For abbreviations with periods and for lowercase letters used as nouns, form the plural by adding *'s*.

   his *p*'s and *q*'s   Ph.D.'s

Using Hyphens Properly

Use a hyphen (- as opposed to a dash —) for

- **compound words**
  four-part lecture   heavy-duty machine   hand-held PA

- **most words beginning with *self***
  self-starting   self-defense   self-regulating   self-governing

- **fractions used as adjectives**
  at the three-quarter level   two-thirds majority   three-dimensional drawing

Using Ellipses

Sometimes a sentence or passage is particularly useful, but you may not want to quote it fully. You may want to delete some words that are not really necessary for your purpose. These omissions are indicated by using an *ellipsis* (three spaced dots within the sentence to indicate where words have been omitted). Here is an example.

Full Quotation: "Diet and nutrition, which researchers have studied extensively, significantly affect oral health."

Quotation with Ellipsis: "Diet and nutrition . . . significantly affect oral health."

Using Numerals Versus Words

Write out numbers as words rather than numerals

- **to begin a sentence**

Nineteen ninety-nine was the first year of our recruitment drive.

- **to list the first number when two numbers are used together**

The company needed eleven 9-foot slabs.

But use numerals, not words,

- **with abbreviations, percentages, symbols, units of measurement, dates**

17%	11:30 a.m.	70 ml
Dec. 3, 2003	$250.00	50 K

- **for page references**

pp. 56–59

- **for large numbers**

3,000,000	23,750	1,714

Use both numerals and words when you want to be as precise as possible in a contract or a proposal.

We agreed to pay the vendor an extra twenty-five dollars ($25.00) per hour to finish the job by the 18th of May.

For information on conventions of writing and using numbers for an international audience, see Chapter 5.

Matching the Right Word with the Right Meaning

The words in the following list frequently are mistaken for one another. Some are true homonyms; others are just similar in spelling, pronunciation, or usage. The part of speech is given after each word. Make sure you use the right word in the right context.

accept (v) to receive, to acknowledge: *We accept your proposal.*
except (prep) excluding, but: *Everyone attended the meeting except Neelou.*

advice (n) a recommendation: *I should have taken Xi's advice.*
advise (v) to counsel: *Our lawyers advised us not to sign the contract.*

affect (v) to change, to influence: *Does the detour on Route 22 affect your travel plans?*
effect (n) a result: *What was the effect of the new procedure?*
effect (v) to bring about: *We will try to effect a change in company policy.*

all ready (adj) two-word phrase *all* + *ready*; to be finished; to be prepared: *We are all ready for the inspector's visit.*
already (adv) previously, before a given time: *Our webmaster had already constructed the sites.*

ascent (n) move upward: *We watched the space shuttle's ascent.*
assent (n) agreement: *She won the teacher's assent.*
assent (v) to agree: *The committee asked the company to assent to the new terms.*

attain (v) to achieve, to reach: *We attained our sales goal this month.*
obtain (v) to get, to receive: *You can obtain a job application on their website.*

cite (v) to document: *Please cite several examples to support your claim.*
site (n) place, location: *They want to build a parking lot on the site of the old theater.*
sight (n) vision: *His sight improved with bifocals.*

coarse (a) rough: *The sandpaper felt coarse.*
course (n) subject of study: *Sharonda took a course in calculus this fall.*

complement (v) to add to, enhance: *Her graphs and charts complemented my proposal.*
compliment (v) to praise: *The customer complimented us on our courteous staff.*

continually (adv) frequently and regularly: *This answering machine continually disconnects the caller in the middle of the message.*
continuously (adv) constantly: *The air conditioning is on continuously during the summer.*

council (n) government body: *The council voted to increase salaries for all city employees.*
counsel (n) advice: *She gave the trainee pertinent counsel.*

discreet (adj) showing respect, being tactful: *The manager was discreet in answering the complaint letter.*
discrete (adj) separate, distinct: *Put those figures into discrete categories for processing.*

dual (adj) double: *A clock-radio serves a dual purpose.*
duel (n) a fight, a battle: *The argument almost turned into a duel.*

eminent (adj) prominent, highly esteemed: *Dr. Felicia Rollins is the most eminent neurologist in our community.*
imminent (adj) about to happen: *A hostile takeover of that company is imminent.*

fair (n) convention, exhibition: *The technological fair featured a DVD-CD home theater with five satellite speakers.*
fair (adj) honest: *Their price was fair.*
fare (n) cost for a trip: *She was able to get a discount on a round-trip fare.*
fare (n) food: *They ate East Asian fare.*

foreword (n) preface, introduction to a book: *The foreword outlined the author's goals and objectives in her business report.*
forward (adv) toward a time or place; in advance: *We moved the time of the visit forward on the calendar so we could meet the overseas manager.*
forward (v) to send ahead: *We forwarded her e-mail to her new server.*

imply (v) to suggest: *The supervisor implied that the mechanics had taken too long for their lunch break.*
infer (v) to draw a conclusion: *We can infer from these sales figures that the new advertising campaign is working.*

it's (pronoun + verb) contraction of *it* and *is*: *Do you think it's too early to tell?*
its (adj) possessive form of *it*: *That old printer is on its last legs.*

knew (v) (past tense of *know*): *She knew the new regulations.*
new (a) never used before: *The subwoofer was new.*

lay/laid/laid (v) to put down: *Lay aside that project for now. He laid aside the project. He had already laid aside the project twice before.*
lie/lay/lain (v) to recline: *I think I'll lie down for a while. He lay there for only two minutes before the firefighter rescued him. She has lain out in the sun too often.*

lose (v) to misplace, to fail to win: *Be careful not to lose my calculator. I hope I don't lose my seat on the planning board.*
loose (adj) not tight: *The printer ribbon was too loose.*

miner (n) individual who works in a mine: *His uncle was a miner in West Virginia.*
minor (n) someone under legal age: *The law forbids the sale of tobacco to minors.*

pare (v) to cut back: *Sandoval pared the skin from the apple.*
pair (n) a couple: *They offered a pair of resolutions.*
pear (n) a fruit: *Alphonso ate a pear with lunch.*

passed (v) went by (past tense of *pass*): *He passed me in the hall without recognizing me.*
past (n) time gone by: *We've never used their services in the past.*

personal (adj) private: *The manager closes the door when she discusses personal matters with one of her staff.*
personnel (n) staff of employees: *All personnel must participate in the 401(k) retirement program.*

perspective (n) view: *From the customer's perspective, we are an honest and courteous company.*
prospective (adj) expected, likely to happen or become: *E-mail the prospective budget to district managers.*

plain (adj) simple, not fancy: *He ate plain food.*
plane (n) airplane: *The plane for Dallas leaves in an hour.*
plane (v) to make smooth: *The carpenter planed the wood.*

precede (v) to go before: *A slide show will precede the open discussion.*
proceed (v) to carry on, to go ahead: *Proceed as if we had never received that letter.*

principal (adj) main, chief: *Sales of new software constitute their principal source of revenue.*
principal (n) the head of a school: *She was a high school principal before she entered the business world.*
principal (n) money owed: *The principal on that loan totaled $32,800.*
principle (n) a policy, a belief: *Sales reps should operate on the principle that the customer is always right.*

quiet (adj) silent, not loud: *He liked to spend a quiet afternoon surfing the Net.*
quite (adv) to a degree: *The officer was quite encouraged by the recruit's performance.*

stationary (adj) not moving: *Miguel rides a stationary bicycle for an hour every morning.*
stationery (n) writing supplies, such as paper and envelopes: *Please stop off at the stationery store and buy some more address labels.*

than (conj) as opposed to (used in comparisons): *He is a faster keyboarder than his predecessor.*
then (adv) at that time: *First she called the vendor; then she summarized their conversation in an e-mail to her boss.*

their (adj) possessive form of *they*: *All the lab technicians took their vacations during June and July.*
there (adv) in that place: *Please put the printer in there.*
they're (pronoun + verb) contraction of *they* and *are*: *They're our two best customer service representatives.*

who's (pronoun + verb) contraction of *who* and *is*: *Who's up next for a promotion?*
whose (adj) possessive form of *who*: *Whose idea was that in the first place?*

you're (pronoun + verb) contraction of *you* and *are*: *You're going to like their decision.*
your (adj) possessive form of *you*: *They agree with your ideas.*

◼ Proofreading Marks

⌢o	Correct a typo.	Correct a typo.
r⌢/m⌢/⌢o	Correct more than one typo.	Correct more than one typo.
t	Insert a letter.	Insert a letter.
or words	Insert a word.	Insert a word or words.
๙	Make a deletion.	Make a deletion.
๛	Delete and close up space.	Delete and close up space.
◡	Close up extra space.	Close up extra space.
#	Insert proper spacing.	Insert proper spacing.
#/◡	Close up and insert space.	Close up and insert space.
eq #	Regularize proper spacing.	Regularize proper spacing.
tr	Transpose letters indicated.	Transpose letters indicated.
tr	Transpose as words indicated.	Transpose words as indicated.
tr	Reorder shown as words several.	Reorder several words as shown.
⸦	⸦ Move text to left.	Move text to left.
⸧	⸧ Move text to right.	Move text to right.
¶	Indent for paragraph.	Indent for paragraph.
no ¶	⸦ No paragraph indent.	No paragraph indent.
//	// Align type vertically.	Align type vertically.
run in	Run back turnover lines.	Run back turnover lines.
	Break line when it runs far too long.	Break line when it runs far too long.
⊙	Insert period here.	Insert period here.
⌃	Commas, commas everywhere.	Commas, commas everywhere.
⌄	Its in need of an apostrophe.	It's in need of an apostrophe.
⌄/⌄	Add quotation marks, he begged.	"Add quotation marks," he begged.
;	Add a semicolon don't hesitate.	Add a semicolon; don't hesitate.
:	She advised "You need a colon."	She advised: "You need a colon."
?	How about a question mark.	How about a question mark?
(/)	Add parentheses as they say.	Add parentheses (as they say).
lc	Sometimes you want Lowercase.	Sometimes you want lowercase.
caps	Sometimes you want upperCASE.	Sometimes you want UPPERCASE.
ital	Add italics instantly.	Add italics *instantly*.
bf	Add boldface if necessary.	Add **boldface** if necessary.
wf	Fix a wrong font letter.	Fix a wrong font letter.
sp	Spell out all 3 terms.	Spell out all three terms.
⌃	Change x to a subscript.	Change x to a subscript.
⌄	Change y to a superscript.	Change y to a superscript.
stet	Let stand as is.	Let stand as is. (To retract a change already marked.)

INDEX

abbreviations
apostrophes in, 706
in e-mail, 135
in instant messages, 142
in instructions, 536
international readers and, 175, 531
numerals in, 707
in surveys, 323
on title pages, 645
abstracts, 417–419
in databases, 332
descriptive, 420, 421
informative, 419
in long reports, 643, 646–647, 655
vs. summaries, 417–419
in test reports, 627, 629
academic advisors, 242
Academic Info, 340
academic libraries, 327–329
acceptance letters, 286
accuracy
of measurements, 23–24, 35
of research sources, 300, 302, 347
in résumés, 252
revising for, 56
acronyms, 175, 531, 645
action-oriented writing, 20
active verbs, 61–62, 201
for instructions, 537
in résumés, 252
active voice, 537
address books, 136, 168
addresses
inside letters, 157–159
in résumés, 254
adjectives, 60, 695, 706
adjustment letters, 207, 218–226
"no" letters, 222–226
"yes" letters, 219–222
Adobe Acrobat Professional, 520
Adobe Dreamweaver, 509
Adobe InDesign, 492
Adobe Reader, 520
adverbs, 699, 701
agendas, meeting, 108–109

age-related discrimination, 69
almanacs, 335
Alta Vista, 341
American Fact Finder, 336
American Heritage Dictionary of the English Language, The, 335
American Institute of Graphic Arts (AIGA), 479
American Psychological Association style. *See* APA style
American Society for Training and Development, 640
anecdote, opening with, 679
antecedents, 704–705
APA style, 365–374
basic features of, 355
formats for specific sources, 368–374
general guidelines for, 368, 369
in-text citations, 365–368
for long reports, 651
Reference list, 355, 651, 666–667
See also documenting sources
apologies, in customer relations letters, 222
apostrophes, 705–706
appendixes, 651
Apple Pages, 492
application letters, 273–280
body of, 276–280
closing paragraph, 280
compared with résumés, 273
education and experience in, 276–280
guidelines for, 273–274
opening paragraph, 274–276
"you attitude" in, 274
area graphs, 454
articles in periodicals. *See* journal articles; magazine articles
Ask.com, 341, 346
asterisks
as graphic device, 22
in keyword searches, 345
atlases, 335

audience
anticipating objections of, 680
attitude toward writer, 14
for blogs, 143, 146–147
"can do" attitude for, 561, 562
choosing visuals for, 439
complimenting, 205
customers' rights, 208
document design and, 495
English comprehension of, 13
evaluating relevance to, 414
for executive summaries, 408–412
eye contact with, 687
format expectations of, 14
getting attention of, 203–205, 274, 438, 675–678
identifying, 10–15
for incident reports, 630
for instructions, 528–532, 539–540
interest level of, 674
for internal proposals, 578–579
for letters, 165–168, 187, 195–196, 233
for long reports, 641
for memos, 124–127
for news releases, 427
persuading, 24–27
for presentations, 673–675
prior knowledge of, 13, 674
for progress reports, 611–613
for proposals, 561–562, 563, 570–571
questions to ask about, 12–14, 674–675
readers vs. listeners, 674
reasons for reading, 13–14
for research, 299, 301
revising for, 56
for sales letters, 201
for sales proposals, 581
for short reports, 602, 606
using reader psychology, 201, 207–208
in website design, 510
writer's expectations of, 14, 674

writing for multiple, 13, 14–15, 93, 127
"you attitude" for, 164–170
See also international readers/correspondence
authors, documenting
in APA style, 365–367, 368–374
corporate authors, 357, 361, 366, 370
in edited collections, 360–361, 370
in-text citations, 356–357, 365–367
in MLA style, 356–357, 359, 360–364
multiple authors, 356–357, 360, 365–366, 368
unknown, 366–367
authors, evaluating credibility of, 302, 346

background information
in instructions, 529, 539
in long reports, 648, 656–658
in research proposals, 591
in sales proposals, 585
in short reports, 602, 604, 605
in summaries, 401
back matter, in long reports, 651
bad news messages
in adjustment letters, 222–226
buffer openings for, 223
contents of, 207
indirect approach for, 209, 223
in news releases, 422–423
refusal-of-credit letters, 226–228
Bagin, Don, 3
bar charts, 456–458
ethical use of, 476
examples of, 457–460
uses for, 440
Becker, Kim, 5
biased language. *See* stereotypical language
bibliographic information
in databases, 332
list of sources, 355
in note taking, 349
See also documenting sources; Reference list (APA); Works Cited list (MLA)
BizStats, 339
Blackwell Encyclopedia of Management, 334
blogs, 142–148
characteristics of, 121–122
documenting, 148, 363, 372

ethics for, 143
external, 143, 145–148
internal, 142–145
revision checklist for, 150
subject line for, 144
Bluebook: A Uniform Style of Citation, The (Harvard Law Review), 354
boastful language, 201
body
of letter, 159, 162, 163, 164, 185
of long report, 649, 658–664
of memo, 129
of new release, 426–427
of presentation, 679–680
of progress report, 617
typefaces for, 502, 507
body language
in job interviews, 284
in presentations, 685, 688
boilerplate material, 535
boldface
in document design, 22, 503, 508
for warnings and cautions, 535, 545
bookmarks, for note taking, 348, 349
book reviews, 415–417, 418
books
bibliographic notes for, 349
documenting, 360–361, 368–370
online versions, 360, 368
Boolean connectors, 344
boxes, in document design, 505
brackets, 352
brainstorming, 45–47, 48, 80, 306
brochures, documenting, 363, 372
buffer openings, 223
bulleted lists, 22, 53
autoformatting for, 646
in instructions, 535
in presentations, 682
in research reports, 381
in website design, 517
bullet résumés, 261
bullying, 29, 33
bureaucratic language, 169–170, 171
Bureau of Labor Statistics, 339, 640
Business.com, 340
business etiquette, 281
business letters. *See* letters
business periodical databases, 331, 333–334
business travel sites, 173

calendars. *See* scheduling; timetables
callouts, 470

call to action, 206–207, 680
campus placement office, 244
capitalization
in complimentary close, 159
in e-mail, 134
as graphic device, 22, 508
for warnings and cautions, 545
captions, 444–445, 505
career counselors, 242, 244
career objective, in résumés, 254–256
career planning, 242
career search. *See* employment process
caution statements, 36, 544–546
clip art for, 535, 546
international icons for, 479
CBE Manual for Authors, Editors, and Publishers, The (Council of Biology Editors), 354
cell phone courtesy, 672–673
chalkboards, 683, 686–687
charts, 455–463
bar, 456–458, 459–460
circle, 455–456
computer-generated, 452
ethical use of, 476, 477, 478
flow, 458–460, 462
graphs vs., 455, 457
organizational, 458, 461
chat rooms, library, 332
chronological résumé, 261
chunking, 491, 517
circle charts, 440, 455–456
clarity
in document design, 521
in short reports, 604–605
in stating purpose, 15
clauses
dependent, 698
independent, 698
relative, 704
wordy, 62, 63
clip art, 470–472, 473
in desktop publishing, 495, 505
guidelines for using, 441, 471–472
for international readers, 480
for warnings and cautions, 535, 546
closings. *See* complimentary closes; conclusions
clustering, 45, 46, 48
coherence, in paragraphs, 695–697
collaborative writing, 76–107
advantages of, 78–80
being team player for, 77, 84–85
career success and, 76–78
case study of, 91–97, 98–99

collaborative writing (*continued*)
communication in, 32
cooperative model, 88
cultural diversity in, 87
document tracking systems for, 103,
104
e-mail for, 100–103
ethics and, 29
of evaluative summaries, 415,
416–417
feedback in, 78, 85–86, 100
functional model, 88–91
groupware for, 97–107, 109
guidelines for setting up, 83–84
integrated model, 91–97, 105–106
for long reports, 641–642
managing group conflict, 85–87
meetings and, 107–114
for proposals, 563
psychological benefits of, 78
revision checklist for, 114–115
sequential model, 88
timetables for, 84, 87
whiteboards for, 687
wikis for, 103–105
writing process for, 80–83, 91–97
See also computer-supported collaboration; meetings
collection letters, 207, 228–232
color
cultural symbolism of, 182, 481, 531
in document design, 506
in presentation slides, 683
for warnings and cautions, 545
Columbia Encyclopedia, 334
columns
in document design, 501
in tables, 448, 449, 450, 451
comma splices, 700–701
Commerce Business Daily, 564, 570
common knowledge, 354
company directories, 336–338
company image
letters and, 153
promoting, 24–25
routine messages and, 122, 131
company logos. *See* logos
company names
in international correspondence, 178
in letters, 157, 158
company qualifications, in sales proposals, 586
competition
assessing, for proposals, 571, 581
cultural views of, 185–187

complaint letters, 207, 212–218
guidelines for, 215–218
resources for complaints, 215
things to avoid in, 212
tone for, 212–215
complimentary closes, 135–136, 158,
159–160, 178
compound words, 706
Computer Ethics Institute, 29, 30
computer skills section, in résumés, 260
computer-supported collaboration,
97–107
advantages of, 99–100
avoiding pitfalls of, 106–107
in integrated model, 105–106
scheduling meetings, 109
in sequential model, 106
types of, 100–105
virtual meetings, 113
See also collaborative writing
computer technology
automatic formatting, 60, 645–646
avoiding global search and replace,
60
backing up and saving files, 5, 50,
55, 137
bookmarks, 348, 349
clip art, 471, 495
for collaborative writing, 97–107
for designing instructions, 535
desktop publishing programs,
492–495, 505–506
digital photography, 465–466
documenting software, 354
document tracking systems, 55, 99,
103, 104
for draft writing, 50
for editing, 59–60
e-mail safety and security, 136–137
ethics and, 29, 30
for graphs and charts, 452
icons, 494–495
intranets, 326, 327
knowledge about, 5
letter wizards, 162
for mailing letters, 168
"mind-mapping" software, 48
for note taking, 114, 348, 349, 350
for planning documents, 48
Portable Document Format (PDF)
files, 520
presentation software, 676–678,
680–683
printing/printers, 161, 273–274,
495–496

print preview option, 56, 161, 520,
535
for revision, 55–56
RSS feeds, 615–616
scanners, 444
short report templates, 600–601
spell checkers, 59–60, 705
for summaries, 401
for virtual meetings, 113
virus protection, 5, 136
for visuals, 21, 444, 446, 452,
494–495, 505–506
watermarks, 168
for website design, 509
whiteboards, 686–687
See also computer-supported
collaboration; e-mail; instant
messages; Internet; Internet
resources; website design
conciseness
in blog writing, 144
in document design, 521
editing for, 62–65
in e-mail, 136
in minutes of meetings, 110
in news releases, 427
in résumés, 251, 252
in short reports, 604–605
in website design, 512
conclusions
in application letters, 280
in instructions, 546
in letters, 162, 163, 164, 187
in long reports, 649–650, 665
in memos, 129
in presentations, 680, 688
in progress reports, 617
in proposals, 580–581, 586–587
in research reports, 386
in short reports, 604
in summaries, 399, 400, 412
concrete language, 21, 201
confidentiality
in collaborative writing, 87, 107
e-mail and, 124, 130, 133, 134
faxes and, 131
honoring, at work, 32
of letters, 154
conflict, managing group, 85–87
congratulations letters, 207
conjunctions. *See* coordinating
conjunctions; subordinating
conjunctions
conjunctive adverbs, 701
connective words, 538, 695, 696

consistency, in document design, 521, 605

Consumer Product Safety Commission, 527

ConsumerReports.org, 603

contact information
in instructions, 531, 535
in résumés, 254
for website design, 517–518

contemporary language, in letters, 169–170

contractions, 175, 705

cooperative collaborative model, 88

coordinating conjunctions, 700, 701, 702

copy notations (*cc:*), 160–161

copyright laws, 29, 445, 467–469

Corel Presentation, 680, 681

Corel Word Perfect, 162, 493

Corporate Information directory, 281, 337

corporate libraries, 325–327

costs
document design and, 496
mentioning in sales letters, 206
in proposals, 572, 578, 580, 586
in sales reports, 610

country-specific search engines, 345

courtesy
in customer relations letters, 208, 219
in e-mails, 136
in international correspondence, 183
in letters, 169
in meetings, 109–110
telephone etiquette, 671–673

courtesy titles, 159

cover letter
for electronic résumés, 272
for faxes, 130
for long reports, 643, 644, 652
for research report, 375

credentials, in résumés, 256–259

credibility, of websites, 346

critiques, 412. *See also* evaluative summaries

cross-hatched bar charts, 457–458

cultural diversity
business presentations and, 675
in collaborative writing, 87
within companies, 8, 578
company image and, 25
global marketplace and, 6–8
guides to, 182
resources for world cultures, 173

respecting cultural traditions, 181
stereotypical language and, 68
See also international readers/correspondence

currency
in international correspondence, 176
of Internet sources, 346, 347
of research sources, 300, 302
in résumés, 252

curriculum vitae, 251. *See also* résumés

customer relations letters, 197, 207–232
adjustment letters, 218–226
bad news messages, 207, 209–211
collection letters, 228–232
complaint letters, 212–218
customers' rights and, 208
diplomacy and reader psychology, 207–208
direct and indirect approaches, 208–209
follow-up letters, 212
good news messages, 207, 209
planning, 208
refusal-of-credit letters, 226–228, 229–230

customer surveys, 307–308

cutaway drawings, 441, 469, 472

Cyborlink.com, 173

dangling modifiers, 704

databases
periodical, 330–334
for research, 327
for résumés, 245–246, 267, 268
See also periodical databases

date line
in international correspondence, 176
in letters, 157, 158
in memos, 128

dates
numerals for, 707
in source documentation, 359, 365, 369

Deardorff's Glossary of International Economics, 335

decision making
in collaborative writing, 78
ethical, 31–33

delimiters, 344–345

delivery, of presentations, 685–688

demonstrative adjectives, 695

dependent clauses, 698, 699–700

descriptive abstracts, 417–418, 420, 421, 646

design. *See* document design; website design

desktop publishing programs, 492–495, 505–506

Dewey decimal system, 326

dictionaries, 335, 705

digital photography, 465–466

diplomacy, in customer relations letters, 207–208, 219, 223–226

directional signs, cultural differences in, 481–482

directions. *See* instructions

directness
in American communication, 183
for good news messages, 208–209

direct observation, 310, 312

directories, company, 336–338

disabled persons, stereotypical language and, 69

discriminatory language. *See* stereotypical language

distribution
document design and, 496
of memos, 124, 128

documentation
importance of, 29, 32
letters as, 153, 154
in short reports, 602
See also documenting sources

document design, 491–508
basic features of, 496
characteristics of effective, 492, 520–521
columns, 501
common mistakes in, 507–508
desktop publishing programs, 492–495, 505–506
example of, 497–500
heads and subheads, 504–505, 507
justification, 503–504
for letters, 161–162, 187–188, 274
line length, 501, 507
margins, 501, 507
page layout, 496–501
planning stage for, 495–496
print preview option for, 535
for proposals, 571–572, 573
for résumés, 251–252, 267
revision checklist for, 521
for short reports, 600–601, 605–606
storyboards for, 518–520
templates for, 493–494, 600–601
typography, 493, 501–505
using color, 506
visual organization for, 491–492

document design (*continued*)
visuals, 443, 445–446, 494–495, 505–506, 507
white space, 496, 507
See also formats; templates; website design
documenting sources, 353–374
APA style, 355, 365–374
for collaborative writing, 84
documentation styles, 354–355
ethics and, 353–354
on external blogs, 148
importance of, 353
in-text citations, 355, 356–358, 365–368
in long reports, 651
material to document, 353–354
material to not document, 354
MLA style, 355–364
in note taking, 349
plagiarism and, 35
Reference list (APA), 355, 368–374, 666–667
in research process, 298, 303
in research proposals, 591–592
revision checklist for, 390
sample report using MLA style, 374–387
in summaries, 403
in test reports, 629
for visuals, 354, 364, 374, 445, 450, 451, 463
in website design, 518
Works Cited list (MLA), 355, 358–364, 387–388
See also APA style; authors, documenting; MLA style
document tracking systems, 55, 99, 103, 104, 107
domain names
obtaining, 515
searching by, 346
dossier preparation, 246–248
drafting, 50–54
for collaborative writing, 81–83, 93–97
computer technology for, 50
for instructions, 536
for research proposals, 587
for summaries, 402, 403
drawing programs, 494–495. *See also* graphic design programs
drawings, 469–470
ethical use of, 476
guidelines for using, 441, 470

in instructions, 532, 538, 539
See also figures
dress
for job interviews, 284
for presentations, 685, 688

EBSCOhost, 333
editing
in collaborative writing, 81–83
computer technology for, 59–60
proofreading marks, 711
sentence construction, 60–62
for stereotypical language, 65–69
for wordiness, 62–65
in writing process, 59–69
See also revision
education
in application letters, 276–280
discussing in job interviews, 282
in résumés, 253–254, 256–258
electronic sources. *See* Internet resources
e-libraries, 329–330
ellipses, 35, 351, 482, 706–707
e-mail, 131–140
address books, 136
alternative addresses for, 5
business vs. personal, 131
characteristics of, 121–122
for collaborative writing, 97, 100–103
confidentiality and, 124, 130, 133, 134
documenting, 367
ethics and, 134
format and style for, 134–136
forwarding, 134
functions of, 131
guidelines for, 134–137
informal tone for, 131
vs. instant messages, 140–141
as legal records, 132–134
length of, 134
memos as, 124, 128, 130
passwords, 137
presenting research in, 300
for progress reports, 612
responding to, 135
revision checklist for, 149
rules of netiquette for, 135
safety and security issues, 136–137
sending attachments, 137, 141
sending résumés via, 267, 272
for sharing meeting notes, 114
signature blocks, 135
when not to use, 140

emoticons, 135
emotions
avoiding, in writing, 21
in complaint letters, 212
employment agencies, 244–245, 246
employment process, 241–289
accepting/declining job offers, 286, 287–288
analyzing strengths and goals, 243
application letters, 273–280
career planning, 242
dossier preparation, 246–248
ethics for, 29, 34
follow-up letters, 285, 286
job interviews, 281–288
letters of recommendation, 247–248
personal websites for job search, 262
portfolios/webfolios, 246, 248–251
resources for finding jobs, 243–246
résumé preparation, 251–273
revision checklist for, 289
salary discussions, 283
steps employers follow in, 241
steps for getting hired, 242
See also résumés
Encarta, 334, 335
enclosure line, 160
Encyclopedia Britannica, 334
Encyclopedia of Banking and Finance, The, 335
encyclopedias, 334–335
envelopes
addressing, 168
document design and, 496
equipment and materials list, 541–542
Erdlen, John D., 243
ethics for workplace, 27–36
for blogs, 143
bullying, 29, 33
bywords signifying, 27
computer-related, 29, 30
documenting sources, 353–354
e-mail and, 134
employers' insistence of, 30–31
ethical decision making, 31–33
ethical dilemmas, 33–34
ethical job requirements, 29
international readers and, 29–30
monitoring of employees for, 30–31
for research, 300–301, 311
for sales letters, 203
for short report writing, 603–604, 606

for visuals, 443, 472–478
  in website design, 518
  writing ethically, 34–36
evaluation, of research sources, 302, 346–348
evaluative summaries, 412–417
  book reviews, 415–417, 418
  evaluating content, 414
  evaluating style, 414
  guidelines for, 413–414
EverNote, 350
executive summaries, 408–412, 413
experience
  in application letters, 276–280
  in résumés, 253–254, 256, 258–259
experiment reports, 625
exploded drawings, 441, 470, 473, 532, 538, 539
external proposals, 569
eye contact
  in job interviews, 284
  in presentations, 675, 687

facts, 21. *See also* objectivity
faxes
  characteristics of, 121–122
  cover sheets for, 130
  guidelines for, 130–131
  revision checklist for, 149
  vs. scanners, 444
Fed Biz Opps, 570
Federal Register, 564, 570
Fedstats, 339
feedback, 36
  in collaborative writing, 78, 85–86, 88, 100
  customer, 307–308
  in presentations, 687
  for website design, 517–518
field trip reports, 618, 619–620
figures, 448, 451–472
  charts, 455–463
  clip art, 470–472, 473
  drawings, 469–470
  graphs, 451–455
  maps, 463–464
  numbering, 445, 538
  photographs, 464–469
  pictographs, 461–463
  See also visuals
financial information, in sales reports, 610. *See also* costs
findings, in short reports, 604
flaming, 134, 212

flip charts, 683–684
flow charts, 440, 458–460, 462
focus groups, 308–309, 311–313, 317–318
follow-up letters, 212, 213–214, 285, 286
fonts. *See* typography
footnotes, for tables, 449
foreign words, plural, 703
formal presentations. *See* presentations
formal tone, 18–19, 153. *See also* tone
formats
  audience expectations about, 14
  for collaborative writing, 84
  computer technology for, 60, 645–646
  for e-mail, 100, 134–135
  for instructions, 528
  for letters, 154–157, 161–162, 183, 187–188
  for long reports, 641
  for memos, 122, 127–128
  for online résumés, 267
  for portfolios/webfolios, 250–251
  for research documents, 299–300
  for short reports, 600–601, 605–606
  See also document design; templates
fractions, 703, 706
fragments, sentence, 699–700
"from" line, in memos, 128
front matter, for long reports, 643–647, 652–655
full block letter format, 155
functional collaborative model, 88–91
functional résumé, 261–266

gambling metaphors, 175
GeoHive, 339
gestures
  cultural differences in, 481
  in presentations, 685
gift offers, in sales letters, 205
global marketplace
  collaborative writing for, 78–80
  formal letter writing in, 170–187
  writing for, 6–8
  See also international readers/ correspondence
glossary, in long reports, 651
goals. *See* purpose
GoBinder, 350
good news messages
  in adjustment letters, 219–222
  contents of, 207
  directness in, 209

goodwill
  in customer relations letters, 215, 219
  in special request letters, 197
  See also "you attitude"
Google, 341, 343, 344, 345
Google Directory, 340
Google Earth, 335
Google Notebook, 350
Google U.S. Government Search, 336
government documents
  documenting, 357, 363, 366, 372
  requests for proposals, 564
  for research, 336
Government Printing Office, 640
government websites
  employment sites, 244–245
  for requests for proposals, 564, 570
  for research, 336
  research-related, 640
  for statistical data, 603
  on world cultures, 173
grade point average (GPA), in résumés, 258
grammar and usage, 59–60, 693–710
graphic design programs
  desktop publishing, 494–495, 505–506
  presentation software, 680–683
  See also visuals
graphic devices
  for clarifying information, 21–23
  in e-mail, 135
  lists, 505
  misusing, 508
  in online résumés, 267
  for warnings and cautions, 545–546
  in website design, 510
graphics. *See* visuals
graphics expert, in collaborative writing, 91
graphs, 451–455
  area, 454
  charts vs., 455, 457
  computer-generated, 452
  ethical use of, 474–476
  functions of, 440, 451
  guidelines for, 454–455
  parts of, 452
  simple and multiple, 452–454
  tables vs., 452
gray literature, 339
Greene, Don, 686
green philosophy, 32–33

group dynamics, 85–87. *See also* collaborative writing
groupware, 97–107. *See also* computer-supported collaboration
guarantees, 29, 206

handbooks, 338
hand gestures. *See* gestures
handouts, for presentations, 681, 682
headers, in memos, 128
headings and subheadings
    in document design, 504–505, 507, 646
    as graphic device, 21, 53
    in long reports, 649
    in proposals, 572
    running heads, 520
    in table of contents, 645–646
    typefaces for, 502
    in website design, 517
headline, in news releases, 423
home health visit reports, 618
homepages
    designing, 509, 512–518
    library, 328, 330
    summaries on, 398
    See also website design
homonyms, 705, 707
honesty
    in résumés, 251
    in sales letters, 203
    in workplace, 27
honors and awards, in résumés, 260
hook, in news releases, 426
Hoover's, 281, 337, 338
HTML (hypertext markup language), 22, 516
human resource departments, 245
humor, in presentations, 675, 679
hyperlinks
    in instructions, 535
    in website design, 510, 512, 513, 518–520
hyphens, 706

icons
    in desktop publishing, 494–495
    as graphic device, 22
    for international readers, 479, 480
    public service, 495
    for warnings and cautions, 535
    in website design, 517
identity theft, 136–137, 262, 268, 307–308

idiomatic expressions, 174
illustrations. *See* visuals
imperative verbs, 537
impressions, avoiding, 21, 603
IMs. *See* instant messages
incident reports, 310, 629–632
    as legal records, 632–633
    parts of, 630–632
    when to submit, 630
independent clauses, 698, 700–701
indexes, 646
indirectness
    in Asian communication, 183
    for bad news messages, 208–209, 223
informal briefings, 670–673
informal reports. *See* short reports
informal tone, 18, 19–20
    for blogs, 143
    in instant messages, 142
    for routine messages, 121, 123, 127, 131
    See also tone
Information Please, 335
informative abstracts, 417–419, 646
InfoTrac, 333
in-house writing, 25–27, 573–581
inquiry letters, 197, 198
instant messages (IMs)
    for business use, 140–142
    characteristics of, 121–122
    documenting, 367
    vs. e-mail, 140–141
    library chat rooms, 332
    revision checklist for, 149–150
instructions, 526–554
    audience considerations, 528–532, 539–540
    conclusions for, 546
    convenience of, 528
    efficiency of, 528
    equipment and materials list, 541–542
    ethics and, 527
    formats for, 528
    imperative verbs for, 537
    importance of, 526–528
    for international readers, 530–531
    introduction, 539–541
    lists in, 531, 532, 535, 538
    model of full set of, 546–553
    online, 527
    planning, 533–536
    in policies and regulations, 554–557
    precise statements in, 24

revising and editing, 536
revision checklist for, 557–558
safety issues, 527, 534, 540
software for designing, 535
steps section, 542–544
style guidelines for, 537–538
testing, 536
types of, 528
visuals in, 530, 538–539, 542, 543, 544
warnings, cautions, notes, 544–546
writing process for, 533–536
integrated collaborative model, 91–97, 105–106
internal proposals, 573–581
    chain of command and, 578
    for changing procedures, 580
    conclusions, 580–581
    costs in, 578, 580
    following corporate policy for, 578
    organization of, 579–581
    purpose of, 573, 579
    reader considerations, 578–579
    revision checklist for, 593
    solutions and plans, 580
    typical topics for, 573
    See also proposals
international English, 8, 173–175
International Organization for Standardization (ISO), 479
international readers/correspondence, 170–187
    abbreviations, acronyms, contractions usage, 175
    avoiding ambiguity, 174
    avoiding idioms, 174
    avoiding sports and gambling metaphors, 175
    avoiding stereotypes, 68, 181–182
    closings and signatures in, 178
    collaborative writing for, 78–80
    color and symbols usage, 182, 480, 481, 531, 545–546
    culturally sensitive questions for, 170–173
    currency references, 176
    date line in, 176
    descriptions of place and space, 175
    e-mail, 136
    English comprehension of, 13
    ethical writing for, 29–30, 36
    examples of, 179–180, 183–187
    format for, 183
    global marketplace and, 6–8
    guidelines for communicating, 173–178

guides to cultural diversity, 182
instructions, 530–531
Internet resources for, 173, 345
letters, 154, 170–187, 188
masculine pronouns in, 67
number conventions for, 177
punctuation symbols for, 482
refusal-of-credit letters, 228,
    229–230
resources for learning about, 173
respecting cultural traditions, 181
respecting race and nationality,
    181–182
revising for, 178–181
salutations in, 178
sentence construction for, 61,
    175–178
short reports, 605
style and tone for, 178
technical terms and, 174
temperature and season references
    in, 175, 177
time references in, 177
units of measure in, 23, 175,
    176–177
view of competition and, 185–187
visuals for, 11, 182, 439, 479–482,
    531
warnings and caution statements,
    544–546
website design for, 517
See also cultural diversity
Internet
    blogs, 142–148
    for collaborative writing, 103–105
    instant messages, 140–142
    online résumés, 245–246, 266–273
    personal websites for job search,
        262
    visual thinking and, 437
    See also computer technology;
        e-mail; Internet resources;
        website design; websites
Internet Public Library (IPL), 329,
    330
Internet resources
    almanacs and atlases, 335
    APA documentation style for, 367,
        368, 369, 370–371, 372
    book reviews, 415
    business information, 603
    company directories, 336–338
    dictionaries, 335, 705
    e-libraries, 329–330
    encyclopedias, 334

evaluating, 346–348
government documents, 336
government research sites, 640
for grammar and writing, 694
gray literature, 339
for international readers, 173, 345
intranets, 326, 327
job resources, 242, 244
MLA documentation style for, 357,
    359, 360, 361, 362, 364
online instructions, 527
periodical databases, 330–334
for registering complaints, 215
requests for proposals (RFP), 570
as research tool, 340–348
revision checklist for, 389
RSS feeds, 615–616
search engines, 341
short report information, 603
statistical resources, 339, 603
subject directories, 340
for visuals, 345, 446, 471
See also databases; search engines
    and strategies
interpolations, 352
interviews, job, 281–288
    dos and don'ts for, 284
    follow-up letters, 285, 286
    preparing for, 281
    questions interviewers can't ask,
        283–284
    questions to ask, 283
    questions to expect, 282–283
    requesting, 280
    salary discussions, 283
interviews, research, 299, 311–317
    conducting, 316
    documenting, 363, 367, 372–373
    drafting questions for, 315–316
    following up after, 316–317
    preparing for, 313
    setting up, 313
    for short reports, 602
    transcript excerpt of, 314–315
in-text citations, 355, 651
    APA style, 365–368
    MLA style, 356–358
    for quotations, 358, 367–368
    sample research paper with,
        377–387
intranets, 326, 327
introductions
    for application letters, 274–276
    for instructions, 539–541
    for international readers, 183

for letters, 162, 163, 164, 183
for long reports, 648, 656–658
for memos, 129
for news releases, 426
for presentations, 675–679
for progress reports, 616–617
for research proposals, 590–591
to research report, 377
for sales proposals, 581–585
for summaries, 400
for test reports, 627, 629
investigation reports, 625
italics, in document design, 22, 503,
    508

JapanZone, 173
jargon
    in blogs, 147
    dictionaries for, 335
    in e-mail, 135
    international readers and, 531
    in news releases, 427
    in summaries, 401
    in surveys, 323
job fairs, 242
job hunting. *See* employment
    process
job interviews. *See* interviews, job
journal articles
    abstracts of, 420, 421
    APA documentation style for,
        370–371
    bibliographic notes for, 349
    databases for, 330–334
    MLA documentation style for,
        361–362
    online versions, 361, 362, 370–371
journals
    job resources in, 245
    research from, 309, 327
justification, 503–504

keys/legends
    for graphs, 454
    for international readers, 482
    for maps, 464
keyword searches
    in databases, 332
    online résumés and, 268, 269
    for search engines, 341–345
    using multiple words, 343
    website design and, 516
    See also search engines and
        strategies
knowledge-oriented writing, 20

labels, in drawings, 470, 539
laboratory reports, 625
laboratory tests, 311
language
  concrete, 21, 201
  contemporary, 169–170
  courteous, 169
  for international readers, 173–175
  pompous and bureaucratic,
    169–170, 171
  sexist, 65–68
  style and tone, 10, 18–20
  technical, 174
  See also stereotypical language;
    word choice; words
leadership, in collaborative writing,
  84, 88
leads, in news releases, 426
lectures/speeches, documenting, 364,
  373
left-justified text, 503
legalese, 169
legal issues
  for e-mail, 132–134
  ethics and, 29
  in external blogs, 148
  incident reports and, 632–633
  letters as records, 153
  in procedural writing, 555
legends. *See* keys/legends
letterhead stationery, 154, 158, 162,
  506
letter of transmittal. *See* cover letter;
  transmittal letter
letters, 153–188
  audience considerations, 165–168,
    187, 195–196
  bad news messages, 207, 209–211,
    222–226
  body of, 159
  compared with memos, 128
  comparison of cultural approaches
    to, 183–187
  complimentary close, 159–160
  contemporary language for,
    169–170, 171–172
  continuing pages, 157
  copy notations, 160–161
  cost of writing, 3–4
  courtesy in, 169
  date line, 157
  enclosure line, 160
  format and design for, 154–157,
    161–162, 183, 187–188, 274
  full block, 154, 155

good news messages, 207, 209,
  219–222
guidelines for writing, 233
importance of, 153–154
inside address, 157–159
for international readers, 170–187
key roles in writing, 196
as legal records, 153
mailing labels and envelopes, 168
making good impression with,
  164–170
modified block, 154, 156
organization of, 162–164, 183–187,
  188, 195, 233
parts of, 157–161
printing, 161, 168, 273–274
revision checklist for, 233–234
salutations, 159
signature, 160
tone for, 153, 164–170, 188, 195, 233
watermarks, 168
"you attitude" for, 164–170
See also international readers/corre-
  spondence; letter types
letters of recommendation, 247–248,
  260–261
letter types, 196–234
  accepting/declining job offers, 286,
    287–288
  adjustment letters, 218–226
  application letters, 273–280
  collection letters, 228–232
  complaint letters, 212–218
  customer relations letters, 207–232
  follow-up letters, 212, 213–214, 285,
    286
  inquiry letters, 197, 198
  refusal-of-credit letters, 226–228,
    229–230
  sales letters, 201–207
  special request letters, 197–201
letter wizards, 162
LexisNexis Academic, 333
Librarians' Index to the Internet, 340
Library of Congress, 332, 336
library resources, 325–330
  chat rooms, 332
  corporate libraries, 325–327
  e-libraries, 329–330
  homepages, 328, 330
  librarians, 327, 332
  online catalogs, 328–329
  periodical databases, 330–334
  public and academic libraries,
    327–329

reference materials, 334–339
search strategies for, 329
line graphs, 440, 452–454. *See also*
  graphs
line length, 501, 507
line spacing, 507
links. *See* hyperlinks
listening, in meetings, 85, 109
list of illustrations, 646, 655
lists
  in document design, 22, 505, 646
  in instructions, 531, 532, 535, 538
  in presentations, 682
  in proposals, 572
logos, 437, 506, 601
long reports, 638–668
  abstracts in, 419, 643, 646–647, 655
  audience for, 641
  automatic formatting for, 645–646
  back matter, 651
  body of, 649, 658–664
  collaborative writing of, 641–642
  conclusion, 649–650, 665
  documenting sources, 651, 666–667
  example of, 651–666
  executive summaries for, 408
  format for, 641
  front matter, 643–647, 652–655
  guidelines for, 667–668
  Internet resources for, 640
  introduction, 648, 656–658
  organization of, 643–651
  presenting research in, 300
  as proposals, 562–563
  recommendations, 650, 665
  research for, 639–640, 642
  revision checklist for, 668
  scope of, 639, 648, 658
  short reports vs., 638–642
  skills needed for, 638
  summarizing, 399
  timetables for, 641, 643
  transmittal letter for, 643, 644, 652
  visuals in, 646, 655
  writing process for, 642–643

magazine articles, documenting, 362,
  371
mailing labels, 168
mail merge, 168
mail surveys, 319
main clauses, 698
main idea
  in summaries, 401–402
  in topic sentences, 693

managerial functions, of sales reports, 610

manipulation, in writing, 34, 35–36

manuals, 338

MapQuest, 335

maps, 441, 463–464, 465
documenting, 364, 374
resources for, 335, 345

margins, 161, 501, 507

measurements
accuracy of, 23–24, 35
in instructions, 537
for international readers, 23, 175, 176–177, 531
metric system, 176
numerals for, 707

media releases, 420–427

meetings, 107–114
agendas for, 108–109
introductory, 83
managing group conflict, 85–87
note taking, 113–114
observing courtesy, 109–110
planning, 107–108
scheduling, 84, 109
virtual, 99, 113
writing minutes for, 110–112
See also collaborative writing

memos, 122–130
audience for, 124–127
characteristics of, 121–122
compared with letters, 128
e-mail vs. hard copy, 124, 128
format for, 122, 127–128
functions of, 122
headers in, 128
for incident reports, 630, 631–632
for internal proposals, 573
organization for, 129
presenting research in, 300
for progress reports, 613
protocols for, 122–124
research proposal as, 587
revision checklist for, 149
routing protocols, 124, 128, 129–130
for short reports, 599
style and tone for, 123, 124, 127

mentors, 242

messages, formulating, 10, 16–17

metasearch engines, 341

metric system, 176

Microsoft Excel, 452

Microsoft Expression Web, 509

Microsoft OneNote, 350

Microsoft Publishing, 492

Microsoft Word, 99, 103, 162, 452

milestones
in collaborative writing, 84
in long reports, 643

"mind-mapping" software, 48

minutes of meetings, 110–112

misplaced modifiers, 704

misquotations, 34, 35

misrepresentation, in writing, 34, 35–36, 203, 476

MLA Handbook for Writers of Research Papers, 354–355

MLA style, 355–364
basic features of, 355
formats for specific sources, 360–364
general guidelines for, 358–359
in-text citations, 356–358
sample report using, 374–387
Works Cited list, 355, 387–388, 651

Modern Language Association style. *See* MLA style

modified block letter format, 154, 156

modifiers, 62, 704

Monthly Business Reviews, 306–307

Moody's Manuals, 338

multiband graphs, 454

multinational corporations
ethics in workplace and, 29–30
international correspondence and, 6–8
See also international readers/correspondence

multiple-bar charts, 457, 459, 460

multiple-line graphs, 452–454

MWBE directory, 337

names and titles
in application letters, 274
courtesy titles, 159
international readers, 178, 181
in letters, 159
in memos, 128
on title page, 645
using first names, 128, 136, 178

National Archives and Records Administration, 640

National Geographic Map Machine, 335

navigation, in website design, 510, 513, 515, 518–520

negative letters, 197

netiquette rules, 135

networking
career, 243–244
for research, 299, 306, 307

neutral letters, 197

neutral words, 66–67

NewsBank, 333

newspapers
documenting sources from, 362, 371
job listings in, 244

news releases, 398, 399, 420–427
bad news messages, 422–423
organization for, 423–427
style and tone for, 427
topics for, 420–422

New York Times Almanac, The, 335

New York Times Glossary of Financial and Business Terms, The, 335

"no" adjustment letters, 222–226

Noble, David F., 3

non-native speakers of English
refusal-of-credit letters, 228, 229–230
speaking to, 675
writing for, 6–8, 13, 170–187
See also cultural diversity; international readers/correspondence

notes, in instructions, 544–546

note taking
computer technology for, 114, 348, 349, 350
guidelines for effective, 348
importance of, 348
for informal briefings, 671, 672
for meetings, 113–114
paraphrases in, 352
quotations in, 350–352
for research, 298, 311, 318, 348–352
revision checklist for, 389
for short reports, 602
what to record, 349–350

nouns
collective, 702
for electronic résumés, 268
possessive forms of, 705–706

numbered lists, 531, 532, 535, 538, 646

numbers
accuracy of, 23–24, 35
as graphic device, 22
in international correspondence, 177
plural form of, 706
subject-verb agreement with, 703
for visuals, 445, 449, 538
writing out vs. numerals, 707

objectivity, 21
evaluating, 414
in incident reports, 633
in research, 311

objectivity (*continued*)
  in short reports, 603, 606
  in test reports, 625
Occupational Outlook Handbook,
  242, 283, 338
Occupational Safety and Health Administration, 640
omissions, ellipses for, 351, 706–707
online catalogs, 328–329
online instructions, 527
online job services, 244
online resources. *See* Internet resources
online résumés, 266–273
  database services for, 245–246, 267, 268
  e-mail address for, 272
  guidelines for formatting, 267
  keywords in, 268, 269, 516
  personal websites for, 262
  privacy and security issues, 268–272
  search-engine ready, 268
  sending, 267, 272
  testing and proofing, 272
  See also résumés
online surveys, 319, 320–321
operations reports, 625
opinions
  in short reports, 603, 633
  in summaries, 400, 412
  words indicating, 402
optical character recognition (OCR) software, 444
organization
  for executive summaries, 412
  for internal proposals, 579–581
  for letters, 162–164, 183–187, 188, 195, 233
  for long reports, 643–651
  for memos, 129
  for news releases, 423–427
  for presentations, 675–680
  for research proposals, 587–592
  for résumés, 252, 261–265
  revising, 56
  for sales proposals, 581–587
  for short reports, 604, 606
  for summaries, 412
organizational charts, 440, 458, 461
originality, of research sources, 300
outlining
  for collaborative writing, 80
  computer technology for, 48
  for customer relations letters, 208
  meeting agendas, 108

in planning stage, 48–49
for presentations, 681
owners' manuals, 528

page layout, 496–501. *See also* document design
page layout software, 492–495
page numbers
  in APA documentation, 369
  for in-text citations, 356–358
  in MLA documentation, 356–358, 359
  numerals for, 707
paragraphs, 693–697
  chunking of, 491
  coherence in, 695–697
  completeness in, 697
  in e-mail, 135
  in letters, 154, 162, 164, 276
  parallelism in, 697
  in proposals, 572
  topic sentence of, 402, 693–695
  unity in, 695
  in website design, 517
parallel construction, 504, 697
paraphrasing, in note taking, 349, 352
parenthetical documentation. *See* in-text citations
passive voice, 67, 178
patchworking, 354
PDF files. *See* Portable Document Format (PDF) files
periodical databases, 329, 330–334
  business and specialty, 331, 333–334
  frequently used, 333
  information found in, 332
periodicals. *See* journal articles; journals
periodic reports, 600, 610
permissions, 148
  for job references, 248
  in photography, 467–469
  in primary research, 311
  for site inspections, 624
personal communications, documenting, 367
persuasive writing
  in customer relations letters, 207–208
  for in-house communication, 25–27
  for promoting company image, 24–25
  in proposals, 563
  in sales letters, 201
  visuals and, 439

"phishing," 136–137
photographs, 464–469, 470
  digital, 465–466
  ethical use of, 474
  guidelines for taking, 467
  for international readers, 480
  scanning, 444
  stock, 495
  uses for, 21, 441, 464
phrases, 698
  prepositional, 63–64
  transitional, 695, 696
  wordy, 62, 63
physical references, in sexist language, 68
pictographs, 441, 461–463
pie charts, 455–456
  ethical use of, 476, 478
  uses for, 440
plagiarism, 35, 518
  guidelines for avoiding, 353–354
  in research, 301, 303
planning
  brainstorming, 45–47
  clustering, 45, 46
  for collaborative writing, 80
  computer technology for, 48
  for customer relations letters, 208
  for document design, 495–496
  instructions, 533–536
  outlining, 48–49
  for research, 309–310, 311, 340
  in writing process, 45–49
Plunkett's Business Almanac, 335
plurals
  apostrophes with, 705–706
  subject-verb agreement with, 702, 703
points, typographic, 502
policies and regulations
  following company, 29
  instructions in, 528
  news releases for, 422
  writing procedures for, 554–557
pompous language, 169–170, 171
Portable Document Format (PDF) files, 520
portfolios/webfolios, 246, 248–251
  business, 251
  documents for, 248–250
  formats for, 250–251
  in interviews, 281
positive letters, 197
possessives, 705–706

PowerPoint, 21, 108, 300, 442, 676–678, 680, 681
Power Think sessions, 306
practical writing, 20
prepositional phrases
  avoiding unnecessary, 63–64
  as fragments, 700
  as modifiers, 704
prepositions, in Internet searches, 343
presentations, 670–691
  analyzing audience, 673–675
  audience questions, 688
  body of, 679–680
  capturing audience's attention, 675–678
  conclusion, 680, 688
  delivering, 685–688
  evaluating, 688–690
  formal, 673–690
  handouts and transparencies, 681, 682, 684
  informal briefings, 670–673
  introduction, 675–679
  to multinational audiences, 675
  noncomputerized, 683–684
  offering "road map" in, 678
  organizational strategies, 679–680
  parts of, 675–680
  preparing for, 673
  presentation software for, 676–678, 680–683
  rehearsing, 684–685
  revision checklist for, 691
  settling nerves before, 685–686
  types of, 670
  visuals in, 680–684
  whiteboards for, 686–687
presentation software, 680–683
  background and color, 683
  capabilities of, 681
  checking quality, 683
  example of, 676–678
  organizing material, 681
  presenting text, 682
  readability of, 682
  sequencing slides, 682–683
  testing, 681–682
  using graphics, 683
press releases, 420–427
primary research, 298, 303–325
  case study of, 305–309
  direct observation, 310
  guidelines for, 311
  interviews and focus groups, 308–309, 311–318

planning in, 309–310
site visits, 310
surveys, 307–308, 318–325
tests, 311
See also research
printing/printers
  document design and, 495–496
  letters, 161, 273–274
  mailing labels and envelopes, 168
print preview option, 56, 161, 520, 535
privacy protection
  for e-mail, 136–137
  for online résumés, 268–272
problem presentation
  in long reports, 648, 658
  in proposals, 579, 585
  in research proposals, 591
problem solving, in proposals, 570
procedures, 526
  for meeting marketplace needs, 554–555
  for policies and regulations, 554–557
  proposals for changing, 580
  revision checklist for, 558
products/services
  comparing in sales letters, 205
  describing in complaint letters, 215–218
  describing in sales proposals, 585
  highlighting appeal of, 205–206
  inquiries about, 197
  misrepresentation of, 36, 203
  news releases for, 422, 423
professional journals. See journals
professional organizations, for job searches, 245
programming flow chart, 460, 462
progress reports, 600, 610–617
  audience for, 611–613
  frequency of, 616
  functions of, 610
  length of, 612–613
  parts of, 616–617
pronouns
  coherence and, 695
  comma splices and, 700
  indefinite, 703
  references, 704–705
  relative, 699, 700
  sexist language and, 67
proofreading, 272, 711
proposals, 561–593
  assessing competition, 571, 581

audience for, 561–562, 563, 570–571, 578–579, 581, 587
characteristics of, 562–563
collaborative writing of, 563
conclusions, 580–581, 586–587
costs in, 572, 578, 580, 586
document design for, 571–572, 573
ethics and, 573
executive summaries for, 408
external, 569
guidelines for writing, 570–573
internal, 569, 573–581
organization of, 579–587
persuasive writing in, 563
problem solving in, 570, 579, 585
purpose of, 561–562, 579, 585
reader-centered, 561
requests for proposals, 564–565, 570
research for, 571, 587
research paper, 587–592
revision checklist for, 592–593
sales, 581–587
size and scope of, 562–563
solicited, 564–565, 567–569
solutions and plans, 580
types of, 564–569
unsolicited, 565–566, 574–577
visuals in, 572
See also internal proposals; sales proposals
ProQuest, 333
Publication Manual of the American Psychological Association, 354–355
public libraries, 327–329
public speaking. See presentations; speaking skills
punctuation, 59
  apostrophes, 705–706
  for comma splices, 700–701
  ellipses, 706–707
  hyphens, 706
  for international readers, 482
  for run-on sentences, 701–702
  for sentence fragments, 699
purpose
  document design and, 495
  establishing, 10, 15–16
  evaluating, 414
  for letters, 233
  of meetings, 108
  for proposals, 561–562
  in research, 299, 301, 303
  revising for, 56
  for short reports, 604

purpose (*continued*)
  for summaries, 399, 400, 412
  of visuals, 438–439
purpose statement
  getting to the point in, 15
  in long reports, 648, 658
  for proposals, 579, 585

QuarkXPress, 492
questionnaires. *See* surveys
questions
  for analyzing audience, 12–14,
    674–675
  following presentations, 688
  in job interviews, 282–283
  leading or biased, 323
  opening with, 205, 679
  in research, 302–303, 315–316
  for surveys, 322–323
quotation marks, in keyword searches,
  345
quotations
  block, 358, 367–368, 385
  documenting, 353, 367
  ellipses in, 351
  interpolations in, 352
  in-text citations for, 358, 367–368
  in news releases, 427
  in note taking, 348, 349, 350–352
  opening with, 679
  selective misquoting of, 35

race and nationality. *See* cultural
    diversity; international
    readers/correspondence
radio programs, documenting, 364
reader psychology, 201, 207–208. *See
    also* audience
recommendations
  based on research, 303
  in incident reports, 632
  in long reports, 650, 665
  in short reports, 604
  in summaries, 399, 400, 412
redundant expressions, 63
Reference list (APA), 355, 365, 651
  example of, 666–667
  general guidelines for, 368, 369
  sample entries for, 368–374
  See also APA style
reference materials, 334–339
  documenting, 362, 371–372
references
  letters of recommendation, 247–248
  in résumés, 260–261, 272

refusal letters, job, 286, 288
refusal-of-credit letters, 226–228,
    229–230
rehearsing presentations, 684–685
relative clauses, 704
religious imagery, 480
repetitious words, 63
reports. *See* long reports; short reports
request for action, in sales letters,
    206–207
request letters. *See* special request
    letters
requests for proposals (RFP), 564–565,
    566
  online, 570
  responding to, 567–569, 571
research, 297–390
  asking questions, 302–303
  audience and purpose for, 299, 301,
    303
  basic process for, 301–303
  case study of workplace, 305–309
  characteristics of effective, 300–301
  for collaborative writing, 80, 88
  documenting sources, 303, 353–374
  ethics and, 301, 311
  evaluating, 414
  evaluating sources for, 302, 346–348
  fundamental areas of, 298
  government websites for, 640
  gray literature for, 339
  information retrieval in, 302
  Internet searches, 340–348
  library resources, 325–330
  for long reports, 639–640, 642
  note taking in, 348–352
  periodical databases, 330–334
  planning for, 309–310, 311, 340
  primary, 303–325. *See also* primary
    research
  for proposals, 571, 587, 591
  reference materials, 334–339
  revision checklist for, 388–390
  scenarios requiring, 298
  school vs. workplace, 299–300
  secondary, 303–309, 325–348
  for short reports, 601–602, 606
  skills needed for, 297
  types of, 44–45, 303–309
  in writing process, 44–45
  See also Internet resources; library
    resources; search engines and
    strategies
research paper proposals, 587–592
  audience for, 587

  organization of, 587–592
  preliminary research for, 587
  requests for approval, 592
  timetables for, 592
  topics and methodology, 591–592
research reports
  formats for, 299–300
  sample report (MLA style), 374–387
  test reports, 625–629
responsibilities
  in collaborative writing, 84
  ethics and, 31–32
  in incident reports, 633
  precise statements of, 24
résumés
  action verbs in, 252
  analyzing skills and strengths, 253
  career objective, 254–256
  chronological, 261
  compared with application letters,
    273
  contact information, 254
  credentials, 256–259
  database services for, 245–246, 267,
    268
  education and experience, 253–254,
    256–259
  e-mailing, 267, 272
  essential characteristics for, 251–253
  ethics and, 29, 251
  functional/skills, 261–266
  honors and awards, 260
  length of, 251, 252
  listing computer skills, 260
  material to exclude from, 254
  online, 245–246, 266–273
  organization of, 252
  parts of, 254–261
  references, 260–261
  related skills and achievements, 260
  revision checklist for, 289
  writing effective, 253–254
  See also employment process; online
    résumés
revision, 55–58
  allowing time for, 55
  case study of, 56–58
  in collaborative writing, 81–83
  computer technology for, 55–56
  of content, 56
  for instructions, 536
  key questions for, 56
  for long reports, 642–643
  of organization, 56
  for summaries, 402–403

of tone, 56
of visuals, 442–443
See also editing
routine reports, 600. *See also* short
  reports
RSS (really simple syndication) feeds,
  615–616
rules
  in document design, 506
  in tables, 448, 451
running heads, 520
run-on sentences, 701–702

safety, instructions and, 527, 534, 540,
  544–546
salary discussions, 254, 283
sales letters, 201–207
  comparison of cultural approaches
    to, 183–187
  describing product's application,
    206
  ethics and, 203
  follow-up letters, 212
  four A's of, 203–207
  getting reader's attention, 203–205
  highlighting product's appeal,
    205–206
  mentioning costs in, 206
  as positive letters, 197
  preliminary guidelines for, 201–203
  request for action in, 206–207
  as short proposals, 562
sales proposals, 581–587
  audience for, 581
  company qualifications, 586
  conclusion, 586–587
  introduction, 581–585
  product/service descriptions, 585
  revision checklist for, 593
  timetable and costs, 586
  See also proposals
sales reports, 600, 610, 611
salutations
  in e-mail, 135–136
  in international correspondence, 178
  in letters, 158, 159
  punctuation for, 159
  sexist, 68, 159
scanners, 444
scheduling
  meetings, 84, 108, 109
  for research, 303
  See also timetables
scope
  for long reports, 639, 648, 658

of message, 16–17
for proposals, 562–563
search engines and strategies
  advanced search options, 345
  designing websites for, 515, 516
  for finding visuals, 471
  ignoring prepositions, 343
  keyword searches, 341–345
  for library catalogs, 329
  list of Internet search shortcuts,
    345–346
  list of search engines, 341, 342
  narrowing subject, 342–343
  for online résumés, 268
  reading summaries, 398
  refining search, 343–344
  revision checklist for, 389
  search engines, 341
  searching vs. surfing, 340
  for specific domains, 346
  subject directories, 340
  using Boolean connectors, 344
  using delimiters, 344–345
  using links, 329
  using multiple words, 343
  using relevant synonyms, 344
  See also Internet resources;
    keyword searches
seasonal differences, international
  correspondence and, 175, 177
secondary research, 298, 303–309,
  325–348. *See also* research
segmented bar charts, 457–458, 459,
  460
sentences
  avoiding complex and lengthy,
    60–61, 175
  avoiding short, choppy, 61
  beginning with "I," 276
  combining, 61, 62
  comma splices, 700–701
  components of, 698
  confusing noun modifiers in, 62
  constructing and punctuating,
    697–705
  contextually appropriate words in,
    703
  editing, 59, 60–62
  fragments, 699–700
  in instructions, 537
  in international correspondence,
    175–178
  logical, 703
  modifier placement in, 704
  opening, importance of, 203–205

pronoun references in, 704–705
  run-on, 701–702
  subject-verb agreement in, 702–703
  topic sentences, 693–695
  using active verbs in, 61–62
  wordiness in, 62–65
sequential collaborative model, 88, 106
services. *See* products/services
sexist language, 65–68
  masculine pronouns, 67
  physical references, 68
  in salutations, 68, 159
  using neutrals words, 66–67
  words ending in -ess or -ette, 67
sexual orientation, stereotypical language and, 69
shadows/shading, as type style, 503
short reports, 599–634
  audience for, 602, 606, 641
  case study of, 606–609
  clarity and conciseness in, 604–605
  data findings, 604
  format for, 599, 600–601, 605–606
  guidelines for writing, 601–606
  incident reports, 629–632
  Internet sources for, 603
  long reports vs., 638–642
  objectivity and ethics for, 603–604,
    606
  organization of, 604, 606
  periodic reports, 610
  progress reports, 610–617
  recommendations, 604
  research for, 601–602, 606, 640
  revision checklist for, 634
  RSS feeds for, 615–616
  sales reports, 610, 611
  scope of, 639
  statistical data for, 603, 616
  style and tone for, 606
  templates for, 600–601
  test reports, 625–629
  timetables for, 641
  trip/travel reports, 617–625
  types of, 599–600
  visuals in, 605–606
signature blocks, in e-mail, 135
signatures
  in international correspondence, 178
  in letters, 160
signpost words, 538
site inspection reports, 618, 621–622
site visits, 299, 310, 312
skills résumé, 261–265
slashes, 482

slide presentations, 680–683
slide projectors, 684
slugs, in news releases, 423
Small Business Administration, 640
smartboards, 686–687
smartphones, 140, 148
social work visit reports, 618, 623–624
software. *See* computer technology
solicited proposals, 564–565, 567–569
solutions section, in proposals, 580
sources
    evaluating, 302, 346–348
    list of, 355
    for research, 299, 301
    See also documenting sources; In-
        ternet resources; library
        resources; primary research;
        research; secondary research
spacing
    as graphic device, 22
    in letters, 161
    line, 507
    in proposals, 572
    See also white space
spam, 137
speaking skills
    for formal presentations, 687–688
    informal briefings, 670–673
    for multinational audiences, 675
    rehearsing, 684–685
    telephone communication, 671–673
    See also presentations
special request letters, 197–201
spelling, 59–60, 705
sports metaphors, 175
spreadsheet graphics, 448, 449, 452
*Standard and Poor's Corporate De-
    scriptions*, 338
statistical data
    documenting, 354
    evaluating, 302
    in graphs, 451
    opening with, 679
    resources for, 338–339, 603
    for short reports, 603, 616
stereotypical language
    age-related, 69
    disabled persons, 69
    ethics and, 29
    race and nationality, 68, 181–182
    sexism, 65–68
    sexual orientation and, 69
    visuals and, 443, 480
    on websites, 348, 518
storyboards, 518–520

stubs, in tables, 448, 449, 451
style, 10, 18–20
    for collaborative writing, 84
    for e-mail, 135–136
    evaluating, 414
    for instructions, 537–538
    for international readers, 178
    for letters, 188, 233
    for memos, 121, 127
    for news releases, 427
    for short reports, 606
    technical and nontechnical, 18–20
style manuals, 354–355
subheadings. *See* headings and sub-
    headings
subject, of sentence, 698
    agreement with verb, 702–703
    compound, 702
    missing, in fragments, 699
subject directories, 327, 340
subject line
    for blogs, 144
    for e-mail, 134
    for memos, 128
subject searches, 329, 342–345
subject-specific search engines, 341
subject-verb-object (s-v-o) sentence
    pattern, 61, 178
subordinate clauses. *See* dependent
    clauses
subordinating conjunctions, 698, 699
summaries, 398–428
    abstracts, 417–419
    book reviews, 415–417, 418
    case study of, 403–411
    computer technology for, 401
    contents of, 399–401
    evaluative, 412–417
    executive, 408–412
    identifying main points for,
        401–402, 403
    length of, 400, 403
    news releases, 398, 420–427
    in presentations, 680
    revision checklist for, 427–428
    steps for preparing, 401–403
    visuals as, 439, 448
    on websites, 398
    what to omit from, 400–401
supporting information, 693
suppressed zero graph, 455
surveys, 318–325
    choosing recipients, 323–324
    compiling and analyzing results,
        324–325

creating questions for, 322–323
    customer, 307–308
    determining medium for, 319
    documenting, 363, 373
    increasing participation in, 324
Sweet, Donald H., 243
symbols
    cultural differences in, 182, 480
    icons, 494–495
    internationally recognized, 479
    numerals in, 707
synonyms, in keyword searches, 344

table of contents
    abstracts from, 419
    in long reports, 643, 645–646, 647,
        654
tables, 382, 448–451
    charts vs., 457
    computer-generated, 452
    graphs vs., 452
    guidelines for, 449–450
    numbering, 445, 449
    parts of, 448, 451
    uses for, 21, 440
teamwork. *See* collaborative writing
technical writing
    glossary for, 651
    in instructions, 536
    for international readers, 174
    scope and details for, 16–17
    style and tone for, 18–19
technology. *See* computer technology
teleconferencing, 113
telephone communication, 671–673
telephone surveys, 319
television programs, documenting,
    364
temperature, in international corre-
    spondence, 177
templates
    in desktop publishing, 493–494
    for graphs and charts, 452
    letter wizards, 162
    for short reports, 600–601
testimonials, in sales letters, 206
testing
    instructions, 536
    online résumés, 272
    presentation software, 681–682
test reports, 600, 625–629
tests, research, 311, 312
thank-you letters, 207, 212
time, in international correspondence,
    177

timetables
    for collaborative writing, 84, 87, 642
    for long reports, 641, 643
    for progress reports, 617
    for research proposals, 592
    for sales proposals, 586
title page, 376, 643, 645, 653
titles
    in APA documentation, 369
    for long reports, 639
    in MLA documentation, 359
    for presentation slides, 682
    for short reports, 605
    for visuals, 444–445, 449, 451
Toastmasters International, 686
"to" line, in memos, 128
tone, 10, 18–20
    for blogs and IMs, 142, 143
    for collection letters, 228
    for complaint letters, 212–215
    for customer relations letters,
        208–209, 219
    formal and informal, 18–20
    for international readers, 178
    for letters, 153, 164–170, 188, 195,
        233
    for news releases, 427
    for "no" letters, 222–223
    for presentations, 687
    revising, 56, 59
    for routine messages, 121, 124, 127,
        131
    for short reports, 605, 606
    "you attitude," 164–170
topics
    audience knowledge of, 674
    for internal proposals, 573
    for long reports, 642
    for news releases, 420–422
    in research proposals, 591
    search strategies by, 329
topic sentences, 402, 693–695, 697
trade journals. See journals
transitions
    in body of reports, 649
    words and phrases for, 401–402,
        695, 696
transmittal letter, 643, 644, 652. See
    also cover letter
travel search sites, 173
trip/travel reports, 310, 600, 617–625
    common types of, 618
    gathering information for, 618–625
    questions answered by, 617–618
typography, 501–505

for captions and visuals, 445
in desktop publishing, 493
heads and subheads, 504–505
for instructions, 535
justification, 503–504
for letters, 161
line length and, 501
line spacing, 507
mixing typefaces, 507
for online résumés, 267
for presentation software, 682
for proposals, 572
typefaces, 501–502
type size, 502, 507
type styles, 503
for warnings and cautions, 545
in website design, 517
See also graphic devices

underlining, in document design, 503,
    508
unity, in paragraphs, 695
unsolicited proposals, 565–566,
    574–577, 582–584
URLs (uniform resource locators),
    359, 369
usability tests, 536
USAJOBS, 245, 267

vague language, 219, 343, 537
verb phrases, avoiding, 61–62
verbs
    active, 61–62, 201, 252, 537
    agreement with subjects, 702–703
    for instructions, 537
    in sentence structure, 698, 699–700
videoconferencing, 78, 113
virtual meetings, 113, 317
visuals, 437–483
    audience considerations, 439
    charts, 455–463
    choosing effective, 439–442
    clip art, 470–472
    computer technology for, 444, 452,
        494–495
    in conjunction with text, 439–442,
        445–446
    in document design, 443, 445–446,
        494–495, 505–506, 507
    documenting, 354, 364, 374, 445,
        450, 451, 463
    in drafting stage, 51
    drawings, 469–470, 471, 472, 473
    ethical use of, 36, 443, 472–478
    evaluating, 347, 414, 442

graphic design programs, 494–495,
    505–506
graphs, 451–455
identifying, 444–445, 505
importance of, 21–23, 437
ineffective use of, 443
in instructions, 530, 538–539, 542,
    543, 544
for international readers, 182, 439,
    479–482, 531
Internet resources for, 345, 446, 471
introducing, 447
list of illustrations for, 646, 655
maps, 463–464, 465
numbering, 445, 447, 449, 538
overuse of, 507
photographs, 464–469
pictographs, 461–463
in presentations, 680–684
in proposals, 572
purpose of, 438–439
revising and editing, 442–443
revision checklist for, 483
scanning, 444
in short reports, 605–606
for simplifying concepts, 438
size of, 445, 446
tables, 448–451
types of, 439, 440–441, 448–472
for warnings and cautions, 546
for website design, 509, 517
visual thinking, 437–438
vocabulary. See language; word choice;
    words

warning statements, 36, 544–546
    clip art for, 535, 546
    international icons for, 479
warranties, 29, 206, 585
Washington Post Business Glossary,
    The, 335
watermarks, 168
Web conferencing, 113
webfolios. See portfolios/webfolios
Webopedia, 335
website design, 491, 508–520
    basic organization, 508–509
    characteristics of effective, 492,
        520–521
    converting print documents,
        510–511
    document design vs., 509–510
    documenting sources, 518
    ease of finding and downloading,
        515

website design (*continued*)
  effective writing for, 509, 517
  ethics and, 518
  homepages, 509, 512–518
  informativeness in, 516–517
  for job searches, 262
  keeping updated, 517
  navigation in, 510, 513, 515, 518–520
  obtaining domain name, 515
  revision checklist for, 521–522
  software for, 509
  soliciting feedback, 517–518
  storyboards for, 518–520
  using PDF files, 520
  visual organization for, 491–492
  visuals for, 509, 517
  for website reading style, 510
  writing for search engines, 516
  See also document design
websites
  About Me/About Us page, 347
  documenting, 357
  evaluating, 346–348
  summaries on, 398
  See also Internet; Internet resources
whiteboards, 686–687
white space
  around headings, 504, 505
  around visuals, 446
  in document design, 496, 507
  in letters, 161
  in website design, 517
  See also spacing
Wikipedia, 334
wikis, for collaborative writing, 99, 103–105, 107
"wild-card characters," in Internet searches, 344–345
word choice
  avoiding ambiguity, 174
  contextually appropriate, 703
  for customer relations letters, 208

editing for, 59
ethical bywords, 27
evaluating, 414
for international readers, 173–175
style and tone, 18–20
for "you attitude," 169–170
See also language
wordiness, editing for, 62–65
word processing programs. *See* computer technology
words
  commonly confused, 707–710
  compound, 706
  connective, 538, 695, 696
  hyphenated, 706
  neutral, 66–67
  signaling main ideas, 401–402
  transitional, 401–402, 695, 696
Works Cited list (MLA), 355, 356, 592, 651
  example of, 387–388
  general guidelines for, 358–359
  sample entries for, 360–364
  See also MLA style
WorldCat, 328
World Factbook, 173
World Wide Web. *See* Internet; Internet resources; website design
writing, job-related
  accurate measurements in, 23–24
  career success and, 3–5, 36–37, 196
  characteristics of, 20–26
  establishing purpose of, 15–16
  ethics and, 27–36
  formulating scope and details, 16–17
  identifying audience for, 10–15
  importance of, 3–6
  keys to effective, 8–20, 36–37, 43
  objectivity in, 21
  offering recommendations, 26
  persuasion in, 24–27
  practical approach to, 20
  revision checklist for, 37

stating responsibilities precisely in, 24
  style and tone of, 18–20
  using visuals in, 21–23
Writing for the NASW Press: Information for Authors, 354
writing process, 44–69
  for collaborative writing, 80–83, 91–97
  defined, 43
  drafting, 50–54
  editing, 59–69
  final version, 53–54
  for instructions, 533–536
  Internet resources for, 694
  for long reports, 642–643
  misconceptions about, 44
  paragraphs, 693–697
  planning, 45–49
  for proposals, 570–573
  researching, 44–45
  for résumés, 253–254
  revising, 55–58
  revision checklist for, 70
  sentences, 697–705
  for short reports, 601–606
  spelling skills, 705
  for summaries, 401–403
  for website design, 509

Yahoo!, 341, 345
Yahoo! Search, 340
Yahoo! Travel, 173
"yes" adjustment letters, 219–222
"you attitude," 164–170
  for application letters, 274
  for customer relations letters, 208
  example of, 167
  guidelines for achieving, 165–167
  for sales proposals, 581

Zacks, 281, 337